# ATLAS OF BUTTERFLIES
# AND DIURNAL MOTHS
# IN THE MONSOON TROPICS
# OF NORTHERN AUSTRALIA

# ATLAS OF BUTTERFLIES AND DIURNAL MOTHS IN THE MONSOON TROPICS OF NORTHERN AUSTRALIA

M.F. BRABY, D.C. FRANKLIN,
D.E. BISA, M.R. WILLIAMS,
A.A.E. WILLIAMS, C.L. BISHOP
AND R.A.M. COPPEN

Australian
National
University

PRESS

**ANU PRESS**

Published by ANU Press
The Australian National University
Acton ACT 2601, Australia
Email: anupress@anu.edu.au

Available to download for free at press.anu.edu.au

ISBN (print): 9781760462321
ISBN (online): 9781760462338

WorldCat (print): 1076493355
WorldCat (online): 1076493279

DOI: 10.22459/ABDM.12.2018

Cover design and layout by ANU Press. Cover photographs by M.F. Braby.

# Contents

# Contributors

**Associate Professor Michael Braby** is an Honorary Associate Professor in the Division of Ecology and Evolution at The Australian National University and a Visiting Scientist at the Australian National Insect Collection. He is recognised internationally for his research on the biodiversity of butterflies—particularly their taxonomy, systematics, biogeography, conservation biology and ecology—and has published four books, 120 scientific research papers and numerous popular articles. His research aims to better understand and document the composition, biogeographic patterns and evolutionary history of butterflies and the underlying processes shaping their assembly on the Australian continent, as well as management actions needed for their conservation. Michael has lived and worked in the Australian Monsoon Tropics for 15 years, having completed a PhD at James Cook University at Townsville (1989–93) before working for the Northern Territory Government in Darwin (2005–14), where he designed and managed a relational database of invertebrates, including butterflies and diurnal moths, the records of which form the basis of this atlas. He also spent two years at Harvard University as a Fulbright Fellow (1999–2001). In 2001, Michael was awarded the Whitley Medal for the best book on the natural history of Australian animals, *Butterflies of Australia: Their identification, biology and distribution*.

**Dr Donald Franklin** is an ecologist and natural historian who has lived and worked in the Australian Monsoon Tropics for more than two decades, being based in Darwin for much of that time, but now living in far north Queensland. His diverse research interests include plants, birds and butterflies, with perspectives including species ecology and conservation biology, interactions between plants and fauna, phenology in seasonal and arid environments and community ecology. Eucalypts—the trees that dominate so many Australian forests and woodlands—are a special interest and he is currently working on a regional field guide to them. Don is the author of two books, 11 book chapters, 108 peer-reviewed papers and numerous technical reports and popular science articles. In his spare time during the past 13 years, he has been conducting butterfly surveys, cultivating a garden for butterflies, conducting surveys of eucalypts and photographing plants, butterflies and landscapes.

**Deborah Bisa** lived in the Northern Territory for 22 years and during that time developed strong connections to the local environment and culture through her study, work, volunteering efforts and publishing achievements. Between 2012 and 2017, she was the collections and facility manager at the Northern Territory Herbarium. She was editor and publication coordinator for Nic Smith's *Weeds of the Wet/Dry Tropics of Australia: A field guide* (2002), co-author with Don Franklin of the 'Field Key to the Lycaenid Butterflies of the Top End and Kimberley' (2008) and in 2016 she published five years of historical research, *Remember Me Kindly: A history of the Holtze family in the Northern Territory*. She spent five years working in the remote Arnhem Land coastal community of Maningrida (2005–10). In 2001, Deb pioneered the development of a relational database (c. 7,900 records) for capturing records from butterfly surveys in the Top End, undertaken in conjunction with Don Franklin and a small group of enthusiastic 'butterfliers';

these records form a key component of this atlas. Some of Deb's Northern Territory on-country experiences and notable findings were published in the *Northern Territory Naturalist* in 2013.

**Dr Matthew Williams** is a Senior Research Scientist at the Department of Biodiversity, Conservation and Attractions in Western Australia. He has studied the biology, ecology and taxonomy of Western Australia's butterflies and day-flying moths for more than 30 years, is actively involved in their conservation and has undertaken several major field expeditions in the Kimberley. His research aims to better understand the biogeography of Australia's butterfly fauna, particularly the processes that have shaped their distribution—both past and present. This work has important implications for preserving Australia's unique butterflies within the modern landscape of fragmented patches of native vegetation. Matthew is also the department's biometrician and in this role oversees the design and analytical robustness of all conservation research publications. Since 2010, he has coordinated and led the graceful sun-moth research program in south-western Western Australia, coordinating the skills of many enthusiastic nature-minded volunteers. Matthew's analytical expertise, coupled with firsthand experience of the Kimberley and western Northern Territory Lepidoptera fauna, makes him a key contributor to this atlas.

**Andrew Williams** grew up in Kenya, where he was introduced to wildlife by his father, John G. Williams, ornithologist at the Coryndon Museum in Nairobi. Andrew became interested in butterflies largely because of his father's passion for Lepidoptera. In his early career, Andrew conducted many faunal surveys in Africa for the Los Angeles County Museum of Natural History, the Smithsonian Institution in Washington, DC, and the Royal Ontario Museum in Canada. In 1969, Andrew joined the Western Foundation of Vertebrate Zoology in California, working primarily in Central America. In 1979, he immigrated to Australia to work for the West Australian Department of Fisheries and Wildlife (now the Department of Biodiversity, Conservation and Attractions), working in nature conservation for 37 years. Andrew conducted three major Lepidoptera field expeditions to the Kimberley; his data are an integral part of this atlas. Andrew has co-authored field guides on American and European orchids and written numerous research papers and wildlife articles. On retirement, Andrew became a departmental Research Associate. He is based at the Western Australia Wildlife Research Centre, where he works on threatened sun-moths and butterflies. His most recent co-authored paper on the sun-moths of Western Australia was published in *Records of the Western Australian Museum* in 2016.

**Dr Carly Bishop** is a landscape ecologist with a focus on the application of landscape ecology concepts to environmental management issues under state and Commonwealth legislation. Her research interests include conservation of mammals, butterflies and day-flying moths, particularly ensuring that scientific knowledge is incorporated into conservation policy and management.

**Rebecca Coppen** is an ecologist with expertise in invertebrate zoology and conducting botanical and Lepidoptera field surveys. Her research interests include the conservation of wetlands, their associated flora and fauna, particularly invertebrates, and the conservation of threatened butterflies and day-flying moths.

# Acknowledgements

The Northern Territory Department of Land Resource Management supplied the primary source of data analysed for this project. We are grateful to Alaric Fisher for providing access to these data. Damian Milne is especially thanked for building the relational MS Access invertebrate database, providing substantial expertise with data input and resolving numerous queries with data cleaning and curation. Yusuke Fukuda also helped with the initial development of the invertebrate database for the Northern Territory on which this atlas is based. We thank Sheryl Keates for extracting, georeferencing and databasing a substantial number of records from the scientific literature.

We are most grateful to the following people, who generously provided data records, assisted with fieldwork or clarified certain facts: Lynette Aitchison, Stacey Anderson, Jared Archibald, Glenn Bellis, Darryl Binns, Caroline Camilleri, Ashley Carlson, Neil Collier, Fabian Douglas, Kelvyn Dunn, Rod Eastwood, Ted Edwards, Matt Gillis, Phil Gilmour, Steven Gregg, Bertus Hanekom, Mark Heath, Ben Hoffmann, Peter Holbery, Andrea Hope, the late Lindsay Hunt, Sheryl and Arthur Keates, Vince Kessner, Michael Kessner, John and Anne Koeyers, David Liddle, Amanda Lilleyman, Kim McLachlan, Max Mace, Zig Madycki, Geoff Martin, Cliff Meyer, Bruce Michael, Ian Morris, Scott Morrison, Craig Nieminski, Kenji Nishida, Simon Normand, George Paras, John Peters, Frank Pierce, Robert Powell, Tissa Ratnayeke, Brian and Lynn Reid, Heather Ryan, Chris Sanderson, Ben Stuckey, George Swann, David Webb, Richard Willan, Eleanor Williams, David Wilson, Lyndsay Wilson, Phil Wise, Alison Worsnop, Andy Young and Stuart Young.

Gavin Dally helped enormously with access to the entomology collection in the Museum and Art Gallery of the Northern Territory (Darwin), as well as the provision of data held in that collection. Haidee Brown and Brian Thistleton similarly assisted with access to the entomology collection and provision of data held in their care at the Northern Territory Economic Insect Collection (Berrimah). Andras Szito (Western Australia Department of Agriculture, Perth), David Britton (Australian Museum, Sydney), Ted Edwards (Australian National Insect Collection, Canberra) and Catriona McPhee (Museum Victoria, Melbourne) kindly assisted with access to museum collections under their care. Alan Andersen and Ben Hoffmann provided considerable expertise with identification of ants, made available their substantial reference collection at the Tropical Ecosystems Research Centre (Commonwealth Scientific and Industrial Research Organisation, Darwin) for study and generously provided spatial data of ants for inclusion in this atlas. Special thanks go to Kym Brennan, Ian Cowie, Sarah Hirst, Peter Latz, John Westaway and Glenn Wightman for identifying numerous larval food plants and for answering many botanical queries; John Westaway also provided considerable assistance with information on the life forms and reproductive strategies of larval food plants. We are most grateful to John Woinarski for preparing the foreword and reviewing the text for the entire manuscript, and to Jan Borrie for meticulously copyediting the text. Ted Edwards and Axel Kallies are also thanked for checking over some sections of the text and for assistance with literature. Keith Willmott, Don Sands and an anonymous reviewer also provided helpful comments and constructive advice.

Most of the images of live butterflies and diurnal moths were taken by the authors. However, we are particularly grateful to the following people who provided superb images for inclusion in this book: Richard Allen (*Catopsilia pyranthe*), Justin Armstrong (*Radinocera* sp. 'Sandstone'), Roger Farrow (*Tirumala hamata*), Mark Golding (*Catopyrops florinda*, *Freyeria putli*), Stephanie Haygarth (*Catopsilia pyranthe*), Andrea Hope (*Euploea corinna*), Axel Kallies (*Pseudosesia oberthuri*), Rod Kennett (*Agarista agricola*), Frank Pierce (*Borbo cinnara*, *Euploea climena*, *Eurema brigitta*, *Hypolimnas anomala*, *Neohesperilla crocea*, *Neohesperilla xanthomera*, *Ocybadistes hypomeloma*, *Papilio aegeus*, *Parnara amalia*), Tissa Ratnayeke (*Arhopala micale*, *Danaus affinis*, *Danaus petilia*, *Liphyra brassolis*, *Ocybadistes walkeri*), Ian Morris (*Chaetocneme denitza*, *Ctimene* sp. 'Top End', *Euchromia creusa*, *Protographium leosthenes*), Kenji Nishida (*Hesperilla sexguttata*), Eleanor Williams (*Danaus plexippus*) and Alison Worsnop (*Liphyra brassolis*).

We are grateful to ANU Press for providing a subsidy to assist towards the copyediting and indexing costs for this publication.

Finally, this work would not have been possible without generous financial support from the Australian Biological Resources Study BushBlitz Applied Taxonomic Grants scheme through the Director of National Parks (grant number ATC214-09).

# Abbreviations

| | | | |
|---|---|---|---|
| * | introduced larval food plant | Qld | Queensland |
| ACT | Australian Capital Territory | SAM | South Australian Museum, Adelaide |
| AM | Australian Museum, Sydney | TERC | Tropical Ecosystems Research Centre, CSIRO, Berrimah |
| AMT | Australian Monsoon Tropics | | |
| ANIC | Australian National Insect Collection, Canberra | *TPWCA* | *Territory Parks and Wildlife Conservation Act* |
| AOO | area of occupancy | Vic | Victoria |
| BMNH | Natural History Museum, London | VU | Vulnerable |
| CRC | Cooperative Research Centre | WA | Western Australia |
| CSIRO | Commonwealth Scientific and Industrial Research Organisation | WADA | Western Australia Department of Agriculture collection, Perth |
| DD | Data Deficient | | |
| EOO | extent of occurrence | | |
| GIS | Geographical Information System | | |
| IBRA | Interim Biogeographic Regionalisation for Australia | | |
| IPA | Indigenous Protected Area | | |
| IUCN | International Union for Conservation of Nature | | |
| LC | Least Concern | | |
| NA | Not Applicable | | |
| NGO | non-governmental organisation | | |
| NMV | Museum Victoria, Melbourne | | |
| NRS | National Reserve System | | |
| NSW | New South Wales | | |
| NT | Northern Territory | | |
| NTEIC | Northern Territory Economic Insect Collection, Berrimah | | |
| NTM | Museum and Art Gallery of the Northern Territory, Darwin | | |

# Figures

# Maps

# Plates

Photo: *Taractrocera dolon*, M. F. Braby

# Tables

Photo: *Belenois java*, M. F. Braby

# Foreword

The monsoon tropics region of northern Australia is one of the world's few remaining vast natural landscapes. Much of it is still held as Indigenous estate, with traditional landowners applying management honed by an intricate knowledge of its nature acquired over countless generations. For most other Australians, it remains a poorly known frontier. Some see that frontier and naturalness as a backwardness, and seek instead the opportunity to engineer major development that will bend the land to commercial productivity. Whatever the future brings, it is important that decisions that are being made now and in the future are grounded in robust evidence. A critical component of that evidence concerns the significance of the natural values of the region, the way those values are embedded in and dependent on broader ecological processes and the extent to which those values may be subverted or compromised by changes in the way the land and waters are managed, used or transformed.

This book is an important piece of that evidence base, and it represents a major advance in knowledge of the nature of northern Australia. Intermittently over the past 200 or so years, collectors and scientists have skimmed the surface of the biology of Australia's monsoon tropics. Over the past 40 to 50 years, more detailed studies have been done, and this research has revealed that the area is an important conservation stronghold, particularly but not only for native mammal groups that have been lost or declined extensively elsewhere in Australia. Many studies over this period have described in increasing detail the inventory and ecological fit of plants, mammals and birds in monsoonal northern Australia. Collectively, these studies show clearly that the region is distinctive ecologically and has areas of endemicity that are significant at national and international levels. Over the past decade or so, there have also been novel landmark studies of less charismatic vertebrate groups—freshwater fish, frogs and reptiles—with the results of these studies showing a richness, endemicity and evolutionary antiquity that far surpassed previous estimates. Scientists have worked their way beyond the constraints of the frontier and found much that is fascinating and important; we are beginning to understand how the country works, how much of what is in it is special and how its nature is remarkably long established. Of course, much of this knowledge, and recognition of value, has been long held by Indigenous people living in, and caring for, this country.

With, to date, two notable exceptions, invertebrates have been largely neglected in this biological journey beyond the frontier. Those exceptions are somewhat quixotic and their study has been driven by the dedication of a few specialist invertebrate biologists. Since the 1980s, a series of surveys, initially by Alan Solem and colleagues, has documented the quite extraordinary radiation, species richness and narrow endemicity of land snails in the Australian Monsoon Tropics and particularly in the Kimberley—although much of this trove remains undiscovered. The ant fauna of monsoonal northern Australia is a much more conspicuous feature of these environments and self-evidently ecologically important. Studies by Alan Andersen and colleagues have also demonstrated a remarkably high richness for ant species, with local diversity among the richest in the world. These studies have also had major ecological underpinnings and have helped influence the environmental management of the region.

This book brings knowledge of a third major invertebrate group out of the shadows. Butterflies, and day-flying moths, would seem an obvious candidate for inventory and research. Surely, everybody loves butterflies. Most are readily observable, and their nuanced variations in colour, pattern, shape and behaviour represent a tractable but intriguing challenge for identification for many observers. But, as evident in the history of collection revealed in this book, their study in monsoonal northern Australia has long been limited and ephemeral. To aggregate these sparse fragments of previous information, and then to systematically seek to redress the deficiencies, is the quest described in this book. Michael Braby and co-authors have done a magnificent job in meeting this challenge. For the first time, the butterfly fauna (and, enmeshed with it, much of the day-flying moth fauna) is systematically catalogued, with painstaking care for precision and justification. This—the thorough and reliable documentation of what is there—is the grounding needed for any assessment of natural values.

But this book is much more than a checklist of butterfly (and diurnal moth) species. It places these species carefully within a biogeographical, ecological and conservation context. It describes, for every species, their phenology, likely movement patterns, diet and habitat relationships. It provides an atlas of known records and quantifies this distributional range. It illustrates and describes the species. It provides advice on management. It is an enticement and entrée for anyone with any passing interest in the wonderful world of butterflies in this country. For people across a broad spectrum, it will provide much knowledge on every relevant aspect, but also show that much remains to be discovered. Still, we live in a natural world with so much to discover.

The evolutionary and distributional patterns revealed here for butterflies are similar to but somewhat different from those recently revealed for other major taxonomic groups. Species richness tends to decline from higher rainfall to lower rainfall areas; some species have very extensive distributions across the broad arc of savannah woodland environments, whereas other species are far more narrowly confined, with some centres of endemism; and species vary markedly in their habitat associations. The ecological web in which butterflies live is more intimate than for many other taxonomic groups, with narrow host plant specificity (the plant species on which their larvae depend) for many butterfly species, and also remarkably intricate relationships of some butterfly species with particular ant species. Butterflies present archetypical examples of the life-giving links that connect different components of the environment. We cannot maintain butterflies in this system unless we also conserve those plants on which they depend. And, in many cases, the retention of those plants is dependent on appropriate fire regimes, the control of introduced pests and weeds, and appropriate constraints on habitat destruction or modification. This book provides a systematic and comprehensive body of evidence about values that we should respect and how these values should be managed.

The co-authors have invested decades of their lives in the research that underpins this book. Initially, those studies may have been disparate and anecdotal. However, as evident in this book, the research became orderly, directed, systematic and increasingly comprehensive. Increasingly, it had more strategic point and purpose. This book represents the culmination of that effort. It is a landmark achievement and a wonderful legacy for future generations of biologists, and indeed for anyone with an interest in nature in northern Australia and of the future of this country. Michael Braby and his co-authors know their study animals, they are passionate about them, they make that knowledge fascinating and hence this book allows its readers a remarkable opportunity and privilege to see and get a feel for this country from the perspectives of other lives.

— Professor John J. C. Z. Woinarski,
Charles Darwin University

# Preface

Northern Australia is a vast region that includes the Kimberley of northern Western Australia, the 'Top End' of the north of the Northern Territory, the 'Gulf Country' of central-eastern Northern Territory and western Queensland, and Cape York Peninsula of northern Queensland. Collectively, these areas make up the Australian Monsoon Tropics biome, a distinct geographical region renowned internationally for its large and relatively intact natural landscapes, high biodiversity and strong Indigenous culture. However, currently there is increasing pressure to exploit northern Australia's natural wealth, particularly through expansion of the pastoral, agricultural/horticultural and mining industries, all of which will inevitably result in substantial habitat loss and modification and erosion of biodiversity values. In addition, there is a multitude of other threats, including invasive species, inappropriate fire regimes, pastoral intensification and, of course, climate change. Hence, there is an urgent need to identify the region's biological assets, to inform policy and management agencies and to set priorities for biodiversity conservation.

This atlas deals specifically with the western portion of the Australian Monsoon Tropics—that is, the region west of the Gulf of Carpentaria (Kimberley, Top End, Northern Deserts and western Gulf Country). The main purpose is to compile a comprehensive inventory of the butterflies and diurnal moths (Insecta: Lepidoptera) of this region of northern Australia. In particular, we aim to answer the following questions regarding the entire fauna within the study region:

1. How many species occur in the region?
2. What kinds of species occur in the region?
3. What proportion of, and which, species are restricted to the region?
4. What is their geographical/breeding range?
5. What are their ecological requirements according to larval food plant specificity and habitat preferences?
6. What is the breeding status of those species?
7. When are they most abundant as adults, do they breed continuously or seasonally and what strategies have they evolved to cope with the adversity of the dry season?
8. What is their conservation status, and how well are they represented in the conservation reserve system?

Answers to such basic questions concerning the composition, distribution and abundance of the fauna are critical because they provide the baseline against which the extent and direction of change can be assessed in future.

Our dataset—comprising 23,885 records based on field observations (55 per cent), museum specimens (34 per cent) and literature (11 per cent)—represents 4,352 sites and spans more than 110 years of recording effort. These data indicate that 166 taxa representing 163 species (132 butterflies and 31 diurnal moths) have been recorded from the study region. The Top End appears to be substantially more diverse than the other subregions, with 150 species (122 butterflies and 28 diurnal moths, or 93 per cent of the total fauna) compared with 105 species (88 butterflies and 17 diurnal moths, or 63 per cent) from the

Kimberley, 82 species (74 butterflies and eight diurnal moths, or 51 per cent) from the western Gulf Country and 53 species (45 butterflies and eight diurnal moths, or 33 per cent) from the Northern Deserts (less than 700 mm mean annual rainfall). Surprisingly, a substantial number of species (37, or 23 per cent) recorded from the study region—including several that are yet to be scientifically described—have been detected only during the past four decades, highlighting that northern Australia is still a frontier for biodiversity discovery.

The fauna comprises seven (4 per cent) taxa of Papilionidae, 31 (19 per cent) Hesperiidae, 20 (12 per cent) Pieridae, 33 (20 per cent) Nymphalidae, 43 (26 per cent) Lycaenidae and 32 (19 per cent) diurnal moths representing eight different families. Thus, the lycaenids, nymphalids and hesperiids are the dominant components in the fauna. No genera are endemic to the study region. Available data indicate that 17 species (seven undescribed) and 35 subspecies (six undescribed) are endemic to the region. The level of endemism is relatively low at the species level (10 per cent), but rises to 31 per cent if all taxa (i.e. species and subspecies) are considered, suggesting differentiation of the fauna has been relatively recent. Most of the endemic species and subspecies have narrow ranges and are restricted to the Top End or to the Kimberley and Top End. Interestingly, all of the endemics, with one exception, are restricted to savannah habitats, especially woodland associated with sandstone or sandy soils derived from sandstone. Clearly, the Top End has been important as an area of endemism in the evolution of the butterfly and diurnal moth fauna.

Breeding habitats, or suspected breeding habitats, based on the presence of immature stages on larval food plants or other evidence have been recorded for 142 (87 per cent) species in the study region. More than 50 (31 per cent) species are dependent on various types of monsoon rainforest, of which 34 (21 per cent) are entirely restricted to these habitats, despite the fact that monsoon forests occupy less than 1 per cent of the landscape. About 60 (37 per cent) species are found only in savannah woodland (including eucalypt heathy woodland, eucalypt open woodland, *Acacia* woodland, riparian woodland and tropical grassland). At least 88 (54 per cent) species occur in savannah woodland, but a number of these species breed equally in both savannah woodland and monsoon forest or mixed monsoon forest, while others that are more typical of savannah woodland also breed along the edges of monsoon forest. Eleven of these woodland species are restricted to habitats associated with laterite or sandstone outcrops or sandy soil derived from sandstone/laterite, often with a heathy understorey or a hummock (spinifex) grass understorey. At least nine species are regularly associated with paperbark woodland, paperbark swampland or mixed paperbark–pandanus swampland and other damp areas, often adjacent to evergreen monsoon vine forest. Fewer species are associated with floodplain wetlands, mangroves and coastal saltmarsh.

Of the 166 taxa (i.e. species and subspecies), 151 (91 per cent) are resident (i.e. breeding regularly within the study region, with permanently established populations), three (2 per cent) are immigrant (i.e. breeding irregularly, with temporary populations) and 12 (7 per cent) are vagrant or infrequent visitors (i.e. not breeding, with nonresident populations). Although many species (at least 64, or 40 per cent) have been recorded in each month of the year, most show pronounced seasonal changes in relative abundance, with some appearing for just a few months of the year. We distinguished five broad seasonal patterns, with groups of species peaking at different times of the year: early wet season (late October – early January), mid wet season (late December – early March), late wet season and/or early dry season (March–May), mid dry season (May–July) and late dry season (August–October). Available data on the breeding phenology and seasonal abundance of adults indicate that about half of the species breed, or are suspected to breed, continuously throughout the year, while the other half breed on a more seasonal basis. The continuous breeders have multiple generations during the year, whereas those that breed seasonally usually have only one or a few generations annually. Both groups have evolved a range of life-history strategies that enable them to either breed continuously or survive the long harsh dry season when their food plants are not available.

Of the 166 taxa assessed for their conservation status according to the International Union for Conservation of Nature (IUCN) Red List criteria, one (1 per cent) is categorised as Vulnerable (VU), four (2 per cent) as Near Threatened (NT), 131 (79 per cent) as Least Concern (LC), 16 (9 per cent) as Data Deficient (DD) and 14 (8 per cent) as Not Applicable (NA). No species from the study region are known to have become extinct since European settlement, although one species (*Pollanisus* sp. 7) has not been detected since it was first recorded 110 years ago. The taxa of most conservation concern are *Ogyris iphis doddi* (VU), *Euploea alcathoe enastri* (NT), *Hypochrysops apelles* ssp. 'Arnhem Land' (NT), *Idalima* sp. 'Arnhem Land' (NT) and *Hecatesia* sp. 'Arnhem Land' (NT)—all of which are endemic to the Top End. However, at least nine of the Data Deficient taxa—*Hesperilla crypsigramma* ssp. 'Top End', *Suniana lascivia lasus*, *Acrodipsas myrmecophila*, *A. decima*, *Ogyris barnardi barnardi*, *Nesolycaena caesia*, *Theclinesthes albocinctus*, *Pollanisus* sp. 7 and *Agarista agricola agricola*—are of conservation interest because they may qualify as Near Threatened (NT) once adequate data are available.

In terms of representation in the National Reserve System (NRS), most taxa are adequately represented to varying degrees, but 14 lack adequate representation. Populations of six taxa (*Suniana lascivia lasus*, *Acrodipsas myrmecophila*, *A. decima*, *Ogyris barnardi barnardi*, *Jalmenus icilius* and *Theclinesthes albocinctus*) are currently not represented in any conservation reserve, while a further eight (*Ogyris oroetes oroetes*, *O. iphis doddi*, *Nesolycaena caesia*, *Petrelaea tombugensis*, *Nacaduba kurava felsina*, *Theclinesthes sulpitius*, *Synemon* sp. 'Roper River' and *Agarista agricola agricola*) are each currently known from only a single conservation reserve.

Overall, the fauna may be considered in reasonably good health because there are apparently few threatened taxa with narrow geographic ranges. Most species, particularly those associated with savannah woodland, have large geographic range sizes across the study region. However, for species with such widespread distributions over relatively uniform landscapes, the loss of any local area/population will not necessarily be inconsequential. Many of the ecological processes that underpin the health and heterogeneity of the landscape—such as fire, flooding, pollination and seed dispersal—operate across large spatial scales such that loss of a subset of the range and disruption of natural processes may have far-reaching and unforeseen consequences.

A key issue that is likely to adversely affect those species with relatively small geographic range sizes is decline of ecological resources (larval food plants and/or habitat) through inappropriate fire regimes, especially an increase in the frequency and scale of dry season burns. At present, the interval between fires in many tropical savannahs of northern Australia is far too short, such that relatively long unburnt habitat (more than five years) is now rare in the landscape. Habitat loss and fragmentation are also a concern, and other threatening processes may become significant in future, such as invasion of grassy weeds (particularly gamba grass and mission grasses) and the concomitant grass–fire cycle affecting habitat specialist species inhabiting savannah woodland, riparian woodland/open forest and the edges of riparian monsoon forest. The long-term viability of monsoon forest patches and the disproportionally rich butterfly and diurnal moth assemblages they support may ultimately depend on reducing the frequency and intensity of fire in the surrounding matrix. High fire frequency may reduce both patch size and connectivity of monsoon forest, but the extent of connectivity between patches may also be adversely affected by any loss or decline of essential pollinators and seed dispersers.

Northern Australia is one of few tropical places left on Earth in which biodiversity—and the ecological processes underpinning that biodiversity—is still relatively intact. Moreover, scientific knowledge of that biodiversity is still in its infancy and the region remains a frontier for biological discovery. The butterfly and diurnal moth assemblages of the area, and their intimate associations with vascular plants (and sometimes ants), exemplify these points. However, the opportunity to fill knowledge gaps is quickly closing: proposals for substantial development and exploitation of Australia's north will inevitably repeat the ecological devastation that has occurred in temperate southern Australia—loss of species, loss of ecological communities, fragmentation of populations, disruption of healthy ecosystem function and so on—all of which will diminish the value of the natural heritage of the region before it is fully understood and appreciated.

# Introduction

Invertebrates comprise about 80 per cent of the world's biodiversity, yet they rarely attract the same level of conservation attention as the more 'charismatic' vertebrates, such as mammals and birds (Cardoso et al. 2011a, 2011b). They are rarely incorporated into standard wildlife inventory and monitoring programs, let alone used to inform conservation management. Butterflies, however, are an exception. They are popular with the general public, respond rapidly to environmental change and are widely recognised as a key flagship group for insect conservation (Pollard and Yates 1993; New 1997; McGeoch 1998, 2007; Braby and Williams 2016). These are the reasons we have chosen them as our focal group.

However, it may come as a surprise to many readers that there are currently no national long-term recording schemes to determine the geographic range size or to evaluate broad changes in the spatial distribution and relative abundance of butterflies in Australia. This contrasts markedly with other developed countries, where butterfly inventory and/or monitoring schemes—such as 'Butterflies for the New Millennium' in Great Britain and Ireland (Asher et al. 2001), the 'Mapping European Butterflies' and related monitoring programs (van Swaay et al. 2008; Kudrna et al. 2011) and other collaborative initiatives such as the 'Tropical Andean Butterfly Diversity Project' in South America (Willmott et al. 2011; Merckx et al. 2013)—have been running

for many years or even decades. The vast size of the Australian continent and the comparatively low interest in insects have undoubtedly been impediments to mapping and monitoring the distribution and abundance of butterflies nationally.

There have been only three previous attempts in Australia to systematically catalogue the distribution of butterflies in the form of an atlas. The first atlas was produced more than 30 years ago when a set of maps for all species in Victoria was published by the Entomological Society of Victoria based on a pioneering study (ENTRECS: insect distribution data collection and recording scheme) using a grid cell resolution of 10° longitude x 10° latitude (Crosby 1986). These maps were updated and released 10 years later in CD format by Gullan et al. (1996). More recently, Field (2013) revised the distribution maps of Victorian butterflies, presenting the records as individual data points and distinguishing the records from before and after 1970. The second atlas was the popular book on Tasmanian butterflies by Virtue and McQuillan (1994), which included distribution maps for all 39 species presented as plots of actual records on a 10 km x 10 km grid based on a database maintained by the Tasmanian Parks and Wildlife Service. This work built on the substantial field inventory and base maps prepared earlier by Couchman and Couchman (1977). The species distribution maps presented in the books by Field (2013) and Virtue and McQuillan (1994), however, were limited by the exceedingly small map sizes and the relatively small spatial scale of the areas covered (Victoria and Tasmania together comprise only 3.9 per cent of the land area of Australia). The third major study was a useful set of maps for all Australian butterfly species compiled by Dunn and Dunn (1991). However, that report and the dataset on which it is based are now out of date and not widely accessible to the general public.

Diurnal moths have received substantially less attention in Australia; however, some progress has recently been made to systematically document the geographical distributions of sun-moths (Castniidae) in Western Australia (Williams et al. 2016). That work shows that the few known species from the northern Kimberley are vastly undersampled and highlights the need for further field surveys.

Sixteen years ago, Sands and New (2002: 29) called for an atlas of the butterflies of Australia, stating: 'We regard the establishment of a National Data Base for Australian Butterflies as a key tool for conservation planning'. Despite this vision and critical objective, little progress has been made in this direction. The online *Atlas of Living Australia* (2017) seeks to address this gap, but contains a relatively small amount of data based primarily on collections in most state museums; however, there is currently no process in place to verify or moderate the accuracy of these records, and therefore the error rate in data quality (whether it be taxonomic, spatial, temporal or observer related) is fairly high.

We thus attempt to fill some of these deficiencies by producing a dedicated atlas of butterflies and diurnal moths of northern Australia—a stepping stone towards a national atlas. Our focus is on the western portion of the Australian Monsoon Tropics (AMT)—that is, the region west of the Gulf of Carpentaria (i.e. Kimberley, Top End, Northern Deserts and western Gulf Country). Although we deal with only a small subset of the continent, the region is still a vast and remote area representing about 1,212,200 sq km (15.8 per cent of Australia). This coverage

is substantially larger than the combined areas of Victoria and Tasmania for which point data maps of butterflies have previously been produced (Virtue and McQuillan 1994; Field 2013). Until now, this region was arguably the most poorly known area of the continent for butterflies in terms of basic knowledge, such as taxonomy, distribution and biology. Moreover, it is hoped that this work will stimulate similar studies elsewhere in Australia and ultimately lead to the development of a national recording and monitoring scheme.

In this book, we have attempted to break new ground. Our study is unique in several respects: not only does it produce a set of maps for a remote and poorly known area of Australia, but also the distributional point data of each species are integrated and compared with the spatial distribution of their larval food plants. The geographic range of each species is then estimated using Geographical Information System (GIS) software. Using a novel approach, the geographic range is inferred using a set of explicit criteria that integrate the spatial records of each butterfly and diurnal moth with those of their larval food plants (based on online data in *Australia's Virtual Herbarium* and the *Atlas of Living Australia*). In addition, the book includes images of living butterflies, graphs of seasonal changes in relative abundance and phenology charts of the immature stages for each species—information that has been lacking in previous Australian butterfly atlases.

There is currently increasing pressure for the exploitation of northern Australia's natural resources, particularly expansion of the pastoral, agricultural/horticultural and mining industries, which will inevitably result in substantial habitat loss, fragmentation and degradation and loss of biodiversity (Garnett et al. 2010). In addition, there are a multitude of other threats to the region's biodiversity, including invasive species (e.g. weeds, feral animals, tramp ants), inappropriate fire regimes, intensification of pastoralism (impacts of cattle grazing) (Garnett et al. 2010) and, of course, climate change, particularly increased atmospheric carbon dioxide concentrations and its effect on vegetational change (Parr et al. 2014). Hence, there is an urgent need to identify the region's biological assets, to inform policy and management agencies in their decision-making processes and to set priorities for biodiversity conservation.

While a broad-scale inventory of the terrestrial vertebrates of northern Australia has been undertaken in relation to the extent of their representativeness in the conservation reserve system (Woinarski 1992), there has been little synthesis of the region's invertebrates. Such baseline data on the composition, distribution and abundance of this key component of biodiversity are critical because they allow us to identify areas of high conservation value, as well as determine the extent and direction of change in future. This atlas aims to collate and disseminate baseline information for one key group of invertebrates: the butterflies and diurnal moths of the insect order Lepidoptera.

## Australian Monsoon Tropics

The AMT biome is a distinct geographical region defined by a combination of climate, vegetation types and the kinds of animals and plants adapted to it (Bowman et al. 2010). The AMT covers a vast area (more than 1.5 million sq km, or 20 per cent of Australia) that includes the Kimberley of northern Western Australia, the 'Top End' of the Northern Territory, the 'Gulf Country' adjacent to the Gulf of Carpentaria and Cape York Peninsula of northern Queensland. The AMT is defined by areas that receive more than 85 per cent of their rainfall between November and April (Bowman et al. 2010); thus, the southern boundary of the region equates to approximately 20–21°S latitude and the northern boundary to the coastline. A characteristic feature of the AMT is the strong latitudinal rainfall gradient, with parts adjacent to the northern coastline having average annual rainfall exceeding 1,500 mm, but with much of the inland region experiencing averages of only 300–600 mm.

The AMT is of international significance because of its large and relatively intact natural landscapes, high biodiversity and strong Indigenous culture (Woinarski et al. 2007a;

Bowman et al. 2010; Garnett et al. 2010; Moritz et al. 2013) that spans 65,000 years (Clarkson et al. 2017). The geological landscape is mostly old, eroded and infertile with few marked topographic features. The climate is harsh, with extreme seasonality driven by monsoon rainfall, destructive cyclonic events and an intense dry period during which little or no rain falls. Disturbance by fire is an integral part of the ecosystem (Russell-Smith and Yates 2007; Andersen et al. 2012; Parr et al. 2014), particularly during the dry season. The biome supports a range of habitats and vegetation types not found in the temperate areas of southern Australia. Moreover, the AMT supports the largest single expanse of tropical savannah woodland in good ecological condition in the world, and contains more than 25 per cent of the world's remaining savannahs (Woinarski et al. 2007a). The high integrity of the tropical savannahs of the AMT is related to the comparatively low human population density, low density of livestock (cattle and sheep, excluding feral herbivores) and the small proportion of land cleared for agriculture and mining. Unlike the temperate woodlands of southern Australia—which have largely been cleared, fragmented or heavily degraded—the tropical woodlands of northern Australia are intact and remain in a relatively unmodified condition (Woinarski et al. 2007a).

Another striking feature of the AMT is the disjunct blocks of ancient Proterozoic sandstone, which are embedded within the extensive savannah lowland plains (Woinarski et al. 2005, 2007a; Bowman et al. 2010). These sandstone blocks form substantial plateaus associated with steep cliffs and escarpments in the northern Kimberley and Top End (especially in western Arnhem Land) and to some extent in the western Gulf Country and on Cape York Peninsula, and support high levels of species richness and narrow-range endemism with many relict species (Press et al. 1995; Crisp et al. 2001; Woinarski et al. 2006), which contrast markedly with the savannahs in which species typically have very wide distributions.

Two major biogeographic barriers that divide the AMT into three main subregions are the Carpentarian Gap in the east and the Joseph Bonaparte Gulf in the west (Bowman et al. 2010; Eldridge et al. 2012; Catullo et al. 2014; Edwards et al. 2017). The Carpentarian Gap or Carpentaria Basin (at the base of the Gulf of Carpentaria) comprises an extensive area of flat and seasonally dry alluvial plains, tidal estuaries and deltas, and, together with the Gulf of Carpentaria, it separates Cape York Peninsula from the Top End. The Joseph Bonaparte Gulf may actually form two barriers that separate the Top End from the Kimberley, according to recent investigations on the distribution of frogs (Catullo et al. 2014) and vascular plants (Edwards et al. 2017). Eldridge et al. (2012) referred to this broad area as the Ord Arid Intrusion–Victoria River Drainage Barrier. The Ord Arid Intrusion (or Ord region) is a region of lowland country immediately west of the Ord River Basin and Cambridge Gulf, whereas the Victoria River Drainage Barrier (also known as the Daly River Plains) comprises a large tract of lowland country formed by the Victoria River and the headwaters of the Roper River, devoid of any permanent streams (Edwards et al. 2017). The effect of these barriers is reflected in the distinctive elements of the flora and fauna associated with each major subregion.

The biodiversity of the AMT is exceptionally rich (McKay 2017), but only recently have scientific studies revealed the true extent of endemism and the historical assembly (evolutionary origins and adaptive radiation) of its biota. Recent systematic and molecular phylogenetic studies of invertebrates (land snails) (Cameron et al. 2005; Köhler 2010; Criscione et al. 2012; Criscione and Köhler 2013; Köhler and Criscione 2013) and vertebrates (for a review, see Moritz et al. 2013) in the Kimberley and Top End suggest a vastly underdescribed fauna with high levels of narrow-range endemism and a deep evolutionary legacy. The emerging picture is one in which the AMT has substantially higher biodiversity value than previously realised, with multiple 'hotspots' of endemism in the Kimberley and Top End (Crisp et al. 2001; Rosauer et al. 2009; Moritz

et al. 2013; Pepper and Keogh 2014; Rosauer et al. 2016; Oliver et al. 2017). These hotspots are areas that support high concentrations of taxa with restricted geographic ranges and typically coincide with evolutionary refugia, usually associated with high topographic variability enabling mesic species to persist and evolve during past climatic extremes.

A review of the biogeography of butterflies in the AMT in terms of patterns of species richness, endemism and historical area relationships (Braby 2008a) indicated that the AMT supports a rich fauna, comprising 265 species (62 per cent of the Australian fauna), but endemism is low, with only 15 endemic species (6 per cent). Most of the endemic species (13, or 87 per cent) are associated with savannah, eucalypt woodland and heathy woodland habitats, while most of the non-endemic range-restricted species (38 of 46 species for which the breeding habitat was known) are restricted to various rainforest habitats. Cape York Peninsula was identified as an area of exceptional biodiversity. However, while the biodiversity of the butterfly fauna of Cape York Peninsula has been relatively well explored and documented (Kikkawa et al. 1981), the Kimberley and Top End remain a biological frontier in which the taxonomy and distribution of species are substantially less well known. Hence, one of the goals of this work is to fill knowledge gaps concerning the composition and distribution of butterflies in this western part of the AMT.

Butterflies are also deeply interwoven in Aboriginal culture and are recognised by the traditional owners of northern Australia. For example, in the Top End, butterflies (and moths) are known to the Burarra people in central coastal Arnhem Land around Maningrida as *burnpa*, to the Yolngu people of eastern Arnhem Land as *bonba*, to the Mangarrayi and Yangman peoples of the Roper River district as *bardbarda*, to the Dalabon people of southern Arnhem Land as *merlemerleh*, to the Bilinarra, Gurindji and Malngin peoples of the Victoria River District as *marlimarli* and to the Warray people of Adelaide River as *mirli-mirli*. The Warray people even have a specific name, *langga-langga*, for the Common Crow, *Euploea corinna* (White et al. 2009)—a species known for its habit of forming spectacular aggregations of adult butterflies in sheltered gorges during the dry season. Traditional language names have been used as the scientific names for several taxa described from the Top End—for example, *Protographium leosthenes* **geimbia** (Tindale 1927), *Suniana lascivia* **larrakia** (Couchman 1951), *Synemon* **wulwulam** (Angel 1951) and *Candalides geminus* **gagadju** (Braby 2017). The application of such names for butterflies is unparalleled elsewhere in Australia, reflecting the rich traditional culture that still exists in the AMT.

## Study region

For the purposes of this work, the boundaries of the study region are similar to those adopted by Woinarski (1992): the western boundary was set to 120°E, the eastern boundary to 140°E, the northern boundary to 10°S (i.e. the northern coast including continental islands) and the southern boundary to 20°S. Within the study region, we recognise four subregions: the Kimberley, Top End, Northern Deserts and the western Gulf Country (Map 1). The four subregions have been delineated more for convenience for describing broad distribution patterns of butterflies and diurnal moths, and do not necessarily represent natural bioregions (c.f. Ebach 2012; Ebach et al. 2015), although the Kimberley Plateau, Top End and Northern Deserts are now generally recognised as distinct bioregions (Bowman et al. 2010; Eldridge et al. 2012; Catullo et al. 2014; González-Orozco et al. 2014; Edwards et al. 2017). In this work, the Kimberley includes almost all of the area of Western Australia north of the Great Sandy and Tanami deserts; the Top End includes areas in the Northern Territory north of approximately 15.5°S latitude; the Northern Deserts comprise the semi-arid and arid zones receiving less than 700 mm mean annual rainfall; whereas the western Gulf Country extends approximately from Limmen Bight in the Northern Territory to Burketown in Queensland and inland to the Barkly Tableland. The definition of the 'Top End' is somewhat arbitrary—for example, some

authors treat the whole area in the Northern Territory and western Queensland north of about 19°S latitude as the Top End (Eldridge et al. 2012); others limit it to the Northern Territory north of 18°S latitude (Russell-Smith and Bowman 1992; Woinarski et al. 2007a); while yet others have adopted a more stringent approach and just include the area in the Northern Territory north of approximately 15°S latitude, bounded by the mouth of the Victoria River in Joseph Bonaparte Gulf in the south-west and the mouth of the Roper River in Limmen Bight in the south-east (Rosauer et al. 2016).

Two useful biogeographical regionalisations within Australia have been recognised that have relevance to the AMT and hence our study region (Map 1). One is the widely used Interim Biogeographic Regionalisation for Australia (IBRA), in which 89 geographical areas are recognised, representing a landscape approach to classifying the land surface for protecting biological diversity (Thackway and Cresswell 1995). A boundary commonly used to delineate northern Australia is that adopted by the Tropical Savannas Cooperative Research Centre (CRC), which was based on the IBRA classification; the CRC's southern boundary essentially follows the northern boundary of the Great Sandy Desert, Tanami Desert and Davenport Murchison Ranges bioregions. The second is a set of five phytogeographical regions for Australia recently developed by González-Orozco et al. (2014) and Ebach et al. (2015) based on species turnover in plant distributions (i.e. the rate of change in species composition between sites) that is also correlated with climatic variables (rainfall and temperature seasonality). For northern Australia, two phytogeographical regions are recognised: the Northern and Northern Desert bioregions. The phytogeographical boundary between these two bioregions is shown in Map 1, together with the IBRA bioregional boundary adopted by the CRC. We have included these biogeographical boundaries in our maps because they provide a useful comparison with the delineation of the AMT based on rainfall patterns and with the geographical distributions of our faunal group of interest, butterflies and diurnal moths.

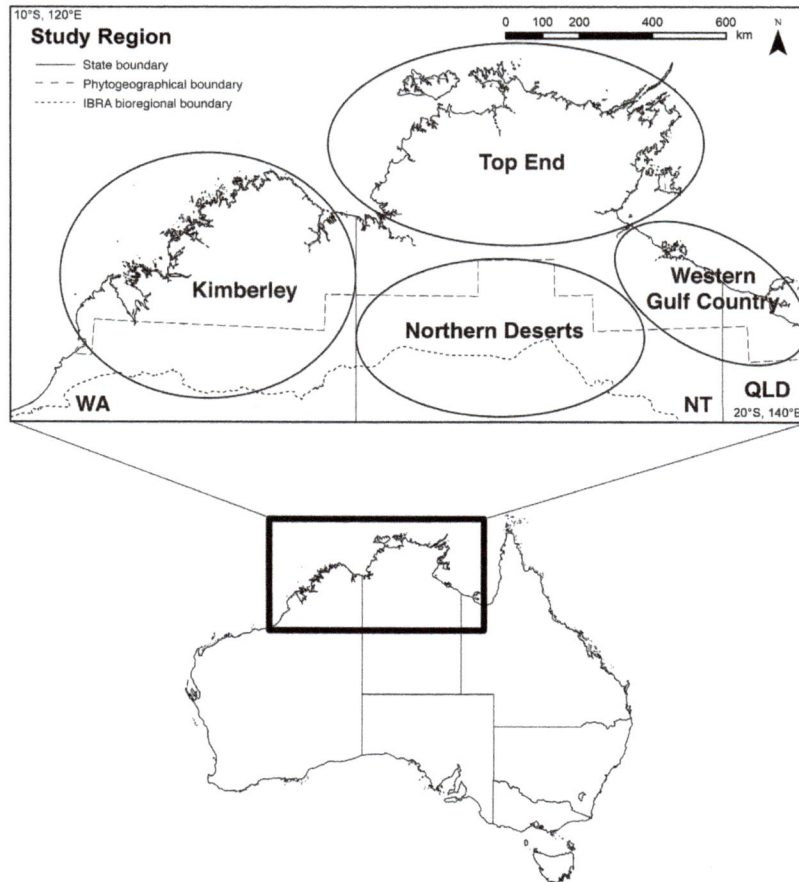

## Map 1 Location of the study region in northern Australia

Four component subregions are shown: the Kimberley, Top End, Northern Deserts and western Gulf Country. Two biogeographical boundaries are shown: the phytogeographical boundary that separates the Northern and Northern Desert bioregions (González-Orozco et al. 2014; Ebach et al. 2015), and the IBRA bioregional boundary (Thackway and Cresswell 1995) adopted by the Tropical Savannas CRC.

Source: Prepared by the authors.

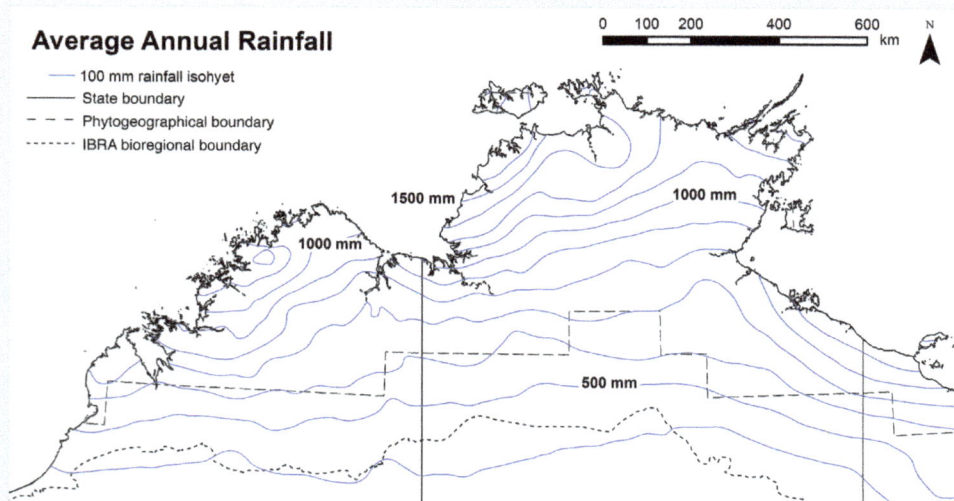

## Map 2 Map of the study region showing spatial variation in average annual precipitation

Both the phytogeographical and bioregional boundaries coincide closely with average rainfall, which shows a pronounced latitudinal gradient. Note: the 1,100 mm mean rainfall isohyet is not shown.

Source: Prepared by the authors.

## Climate

The climate in the AMT is one of extremes: it is highly seasonal or 'monsoonal', in which there are pronounced wet and dry seasons driven largely by seasonal fluctuations in rainfall (Bowman 2002; Bowman et al. 2010). The warmer months are characterised by high temperatures and high humidity, a pronounced wet period and north-westerly trade winds, typically from November to March or April (the wet season), while the cooler months are characterised by a long drought period and south-easterly trade winds, usually from May to October (the dry season). During the wet season, high moisture and torrential rainfall are associated with equatorial monsoon troughs and severe tropical storms, including cyclones, whereas the dry season is dominated by a subtropical high-pressure system and dry air. The Australian summer monsoon is a component of the large-scale Asian–Australian monsoon system, which appears to be of great antiquity, its origins dating back to the Late Eocene–Early Oligocene (for review, see Bowman et al. 2010). The onset and duration of the wet season, and the amount of rainfall and intraseasonal breaks between the monsoon troughs, are highly variable from year to year (Garnett and Williamson 2010; Drosdowsky and Wheeler 2014).

There is a strong latitudinal gradient in annual rainfall, which is a distinctive feature of the AMT (Map 2). The highest rainfall zones in the study region occur in the north-western corner of the Top End (> 1,600 mm mean annual rainfall), Gove Peninsula (> 1,400 mm) and the coastal areas of the north-western Kimberley (> 1,200 mm), with progressively lower rainfall in the inland areas. The highest mean annual rainfall occurs on the Tiwi Islands (> 2,000 mm) and the lowest in the Tanami and Great Sandy deserts (< 400 mm). For the purposes of this atlas, we divide the study region into three climatic zones according to rainfall pattern based on Gaffney (1971): the wetter tropics, with 700–2,200 mm mean annual rainfall; the semi-arid zone, with a mean annual rainfall of 350–700 mm; and the drier arid zone, with a mean annual rainfall of less than 350 mm (Map 5).

The seasonality in rainfall is a major driver of the ecology of the region, with species showing a range of adaptive responses to the alternating annual cycle of rainfall and flood, followed by drought and fire (Bowman 2002; Woinarski et al. 2007a; Garnett et al. 2010). Many animals, for instance, undertake local dispersal or seasonal migration at various spatial scales in response to changes in the availability of resources (Woinarski et al. 2005). The pronounced latitudinal rainfall gradients in the Kimberley and Top End also have a major influence in determining the vegetation type and habitats (see below) and hence the geographic distribution of animals, including butterflies and diurnal moths. Studies of vertebrates (Woinarski 1992; Woinarski et al. 1999) and vascular plants (Bowman 1996) have revealed substantial changes in species richness and composition, with high spatial turnover associated with this latitudinal rainfall gradient from the northern coast to the dry interior of arid Australia. In contrast, there is remarkably little longitudinal variation in species composition across the lowland savannah plains that extend across the region from west to east, with many species having extensive geographic ranges (Woinarski 1992; Woinarski et al. 2005).

## Habitats

The study region supports a variety of habitats or vegetation types used by butterflies and diurnal moths, including savannah woodland, rainforest, heathland, mangrove, saltmarsh and floodplain wetlands (Brock 2001; Woinarski et al. 2007a; McKay 2017) (Plates 1 and 2).

The overwhelmingly predominant habitat in the landscape is tropical savannah woodland, also known globally as 'tropical grassy biomes' (Parr et al. 2014). This habitat is distinguished by the ubiquitous presence of $C_4$ grasses in the understorey, shade-intolerant plant species, an overstorey dominated by eucalypts (*Eucalyptus* and *Corymbia*) (Plates 1a–c) and the prevalence of fire and herbivory. Tropical savannah woodlands in northern Australia occur extensively on the lowland plains comprising deep, well-drained sandy or lateritic soils.

In the higher rainfall areas, the savannahs comprise eucalypt woodland or eucalypt open forest with a tall grassy understorey (Plate 1a) and, depending on the frequency of fire and soil type, a woody shrub layer. In the lower rainfall areas, the savannahs typically comprise eucalypt open woodland (Plate 1b), tropical grassland or, in some areas, *Acacia* low open woodland (Plate 1d) or *Acacia* tall open shrubland. On the sandstone country that dominates much of the Kimberley and parts of the Top End, such as the Arnhem Land Plateau, tropical heathland, low shrubland and eucalypt heathy woodland (Plate 1c) persist on the dissected plateaus and steep breakaways with nutrient-poor, acidic soils that hold little moisture through the dry season (Russell-Smith et al. 2002). On wetter sites—such as seasonal swamps and the edges of permanent billabongs and creeks—various types of paperbark (*Melaleuca*) habitats prevail (Platess 2g–h), including paperbark open forest, paperbark woodland and paperbark swampland. Paperbark forests are the wetland analogue to the drier eucalypt forests and persist in wetter places at the expense of rainforest where the level of disturbance by fire and/or flood is high (Franklin et al. 2007b).

Where the landscape is protected from fire, forests with a closed canopy, buttressed roots and lianes may occur. These habitats are variously termed rainforest or 'monsoon forest' because of their adaptation to the pronounced dry season; they vary in stature and in the dominance of evergreen or deciduous species (Plates 2a, 2b and 2d). Monsoon forests typically occur as a mosaic of relatively small fragmented patches (< 5 ha), with an estimated 16,500 patches occurring in the study region, embedded in the vast savannah landscape (Russell-Smith 1991; Russell-Smith et al. 1992; Russell-Smith and Bowman 1992). The largest and most diverse patches of monsoon forest occur in the higher rainfall areas of the north-western Kimberley (Kenneally et al. 1991; McKenzie et al. 1991), northern coastal areas of the Top End and in western Arnhem Land, where they may exceed 100 ha in extent (Russell-Smith et al. 1992). Although in total they comprise a relatively small area of the study region (2,750 sq km or 0.4 per cent), they nevertheless support a disproportionately high number of plant species (13 per cent of the Northern Territory total) (Woinarski et al. 2005). They also support a rich diversity of butterflies (Hutchinson 1973, 1978; Bailey and Richards 1975; Kikkawa and Monteith 1980; Naumann et al. 1991) and diurnal moths, many of which do not occur in other habitats, while others use them as refuges for aggregation during the long dry season (Monteith 1982). The wetter monsoon forests comprise evergreen monsoon vine forest (Plate 2a) or riparian (gallery) forest and occur where water is available throughout the dry season, such as along perennial rivers, below springs associated with underground aquifers or in deep rocky sandstone gorges and escarpments. The drier monsoon forests frequently occur as semi-deciduous monsoon vine thicket (Plates 2b and 2d) on coastal laterite, coastal sand dunes, inland rocky outcrops composed of limestone, granite and basalt, and sandstone escarpment talus slopes where access to water is limited or unavailable during the dry season. These dry monsoon forests include both evergreen and deciduous tree species, hence the term 'semi-deciduous' monsoon vine thicket.

In the higher rainfall areas of the savannah landscape, both eucalypt and paperbark woodland/swampland may develop rainforest elements in the understorey if fire is excluded for a considerable period (Plate 2f). These habitats are important for butterflies and diurnal moths, particularly in riparian areas where mixed riparian monsoon forest (Plate 2c) or mixed riparian woodland develops, because of the high floristic diversity, longer growing season of the larval food plants and cooler, moist microclimatic conditions. Another important habitat for butterflies is rainforest edge (Plate 2e), which comprises the ecotone between monsoon forest and eucalypt woodland.

Other habitats that are used by a few ecologically specialised butterflies include mangrove, saltmarsh, floodplain wetland and open sandstone pavement (Plates 1e–h). Mangrove and saltmarsh occur along the coastline and adjacent areas and are subject to

**Plate 1 Examples of breeding habitats used by butterflies and diurnal moths in the study region**
(a) savannah woodland, in higher rainfall areas; (b) savannah woodland, in lower rainfall areas; (c) eucalypt heathy woodland on sandstone breakaway; (d) *Acacia* low open woodland; (e) open sandstone pavement; (f) floodplain wetland; (g) mangrove; (h) saltmarsh.

**Plate 2 Examples of breeding habitats used by butterflies and diurnal moths in the study region**
(a) wet monsoon forest (evergreen monsoon vine forest); (b) dry monsoon forest (semi-deciduous monsoon vine thicket) on coastal laterite; (c) mixed riparian monsoon forest; (d) dry monsoon forest (semi-deciduous monsoon vine thicket) on limestone; (e) rainforest edge (ecotone between monsoon forest and savannah woodland); (f) paperbark tall open forest with rainforest elements in the understorey; (g) paperbark swampland; (h) paperbark woodland.

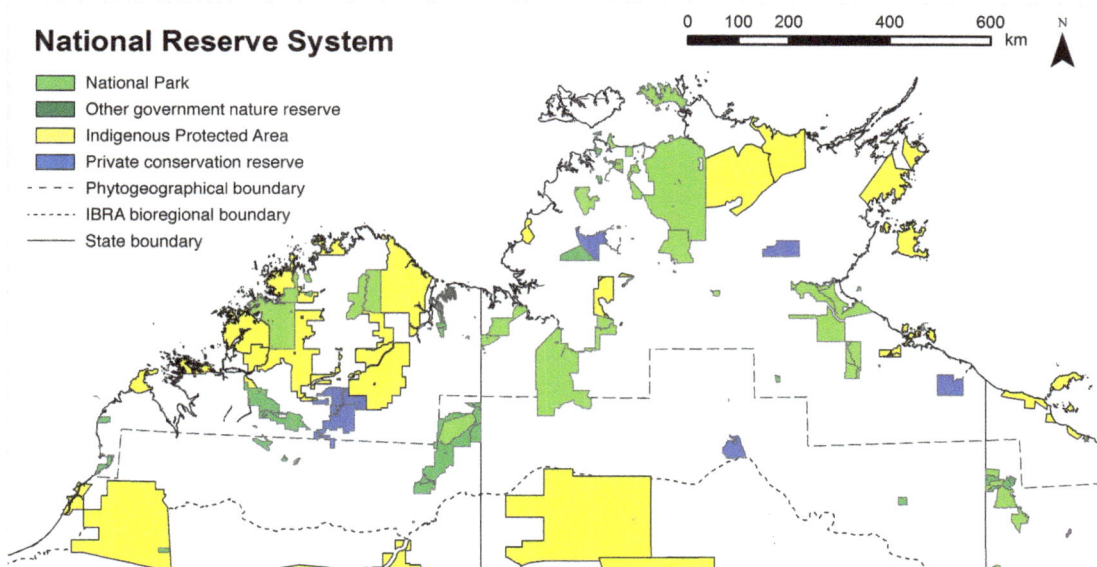

Map 3 Map of the study region showing the extent and land tenure of the National Reserve System
Source: Prepared by the authors.

## National Reserve System

periodic tidal inundation; saltmarsh typically occurs as extensive flats along the landward edge of mangroves. Floodplain wetlands are ephemeral or seasonal habitats reliant on major river systems flooding during the wet season; during the late dry season, the heavy clay soils crack and dry out. Open sandstone pavements support 'resurrection' grasses (*Micraira* spp.), the only plant genus that has adapted and radiated to any extent on this very specialised and harsh habitat.

## National Reserve System

The National Reserve System (NRS) in northern Australia, according to the *Collaborative Australian Protected Area Database* (2014), consists of three major categories of land tenure: government, Indigenous and private (Map 3). Government land includes world heritage areas, national parks and other nature reserves managed by the Commonwealth, state or Northern Territory governments. Indigenous Protected Areas (IPAs) are Aboriginal-owned land managed by traditional owners through dedicated ranger groups on country. Private conservation reserves include areas that have been purchased for conservation purposes by

non-governmental organisations (NGOs), such as the Australian Wildlife Conservancy, Bush Heritage Australia and the Nature Conservancy. In the study region, IPAs cover 154,575 sq km (61 per cent), government national parks and nature reserves 86,062 sq km (34 per cent) and NGO reserves represent 12,765 sq km (5 per cent) of the NRS. Collectively, these different reserve types in northern Australia, which constitute 253,402 sq km (21 per cent of the study region), aim to protect and manage unique landscapes and their biological diversity by forming a network of protected areas that is comprehensive, adequate and representative, with the added value that IPAs also protect and maintain cultural values and customary practices (Woinarski et al. 2007a; Garnett et al. 2010; Moritz et al. 2013).

Some of the more significant national parks in the study region include Prince Regent River, Drysdale River and Purnululu national parks in Western Australia; Keep River, Judbarra/ Gregory, Litchfield, Kakadu, Garig Gunak Barlu (on Cobourg Peninsula) and Limmen national parks in the Northern Territory; and Boodjamulla (Lawn Hill) National Park in Queensland. Both Purnululu and Kakadu national parks have been afforded World

Heritage status. Significant IPAs include Karajarri, Bardi-Jawi, Wilinggin, Uunguu, Dambimangari and Balanggarra in the Kimberley; Northern Tanami in the Northern Deserts; Wardaman, Warddeken, Djelk, Laynhapuy, Dhimurru and Anindilyakwa (on Groote Eylandt) in the Top End; and Yanyuwa (Barni-Wardimantha Awara), Thuwathu/Bujimulla (on Mornington Island) and Nijinda Durlga in the western Gulf Country. Examples of private conservation reserves are Mornington Wildlife Sanctuary in Western Australia and Fish River Station, Wongalara Sanctuary and Pungalina-Seven Emu Wildlife Sanctuary in the Northern Territory.

The land tenure outside the NRS consists predominantly of pastoral land (both freehold and leasehold) used for cattle production (c. 70 per cent of the total area) and private Indigenous lands, with smaller areas devoted to agriculture/horticulture and mining (Woinarski et al. 2007a; Garnett et al. 2010). Mining may also occur under permit in conservation reserves, on Indigenous lands (for example, Melville Island) and on pastoral leases. However, while protected areas are crucial for biodiversity conservation, management of the non-protected areas (i.e. the larger 'matrix' surrounding conservation reserves) is also necessary to ensure connectivity and continuity of ecological and evolutionary processes within the landscape (Woinarski 1992; Garnett et al. 2010; Moritz et al. 2013). In other words, pastoral lands and Indigenous lands that are not part of the NRS also have critical roles to play in biodiversity conservation.

## Early history of butterfly studies

The sparse population density of the study region and its remoteness from the major population centres of Australia, which are primarily concentrated in the south-east of the continent, has meant that not until recent times have there been regular and substantial field collecting and research activity. Prior to 1970, recording of the butterfly and diurnal moth fauna was spasmodic, undertaken largely by naturalists and entomologists visiting the Kimberley or Top End for brief periods and usually concentrated around the major ports such as Darwin, Wyndham and Derby, although some did make arduous explorations to areas that would have been extremely challenging at the time, and are still frontier places today! A brief account of some of these pioneers and their published studies is provided here.

It is not certain when the first specimens were collected from the study region. The subspecies *Euploea darchia darchia*, which is endemic to the Kimberley and Top End, was described by W. S. Macleay in 1826 (under the name *Danais darchia*) based on material collected during Captain P. P. King's voyages (Macleay 1826; Edwards et al. 2001), and may represent the first scientific collection of butterflies from the region. King made four voyages between 1817 and 1822 during which he explored and meticulously charted the northern coast of Australia. He was accompanied by the botanist A. Cunningham, who also collected insects on these voyages, but it is not clear where the specimens were collected, other than somewhere along the Northern Territory coast (Oberprieler et al. 2016). Unfortunately, the type material of *E. darchia* is lost and presumably destroyed (Waterhouse 1937b), and the exact locality for the type is unlikely to be established, as is the case for all type specimens of other insects described by Macleay (Oberprieler et al. 2016).

King was also the first European to visit Cobourg Peninsula and, in 1818, he mapped (and named) the inlet of Port Essington. A remote British military colony was subsequently established at Port Essington between 1838 and 1849, which at that time was the only European settlement in northern Australia. The colony allowed many early Victorian-era natural history collectors to visit the area, the most notable of whom was John Gilbert, in 1840 and 1841 (Fisher and Calaby 2009). Although Gilbert collected mainly birds and mammals, he also collected insects, which were sent back to England for taxonomic appraisal (Fisher and Calaby 2009). The insects almost certainly included samples of butterflies and moths, but this material has not yet been studied. However, the type of *Idalima leonora* was collected from Port Essington (Walker 1854); this fine diurnal moth species was described (as *Agarista leonora*) by Edward Doubleday in England in 1846 based on material (two specimens) in the John Gould collection. Doubleday subsequently described *Euploea sylvester pelor* in 1847, although it is not certain exactly where the type came from other than somewhere in 'north-western Australia' (Edwards et al. 2001). A few years earlier, *Papilio fuscus canopus* was described by J. O. Westwood, in 1842 (as *Papilio canopus*), based on type material from nearby Melville Island (Edwards et al. 2001). A small British military outpost was established on Melville Island between 1824 and 1829 (Fisher and Calaby 2009), and this may have been the time when the first specimens of *P. fuscus canopus* were collected.

In 1887, W. W. Froggatt—who a few years earlier was appointed the natural history collector for the Macleay Museum at the University of Sydney—was sent to the western Kimberley (Derby, Lennard River, Fitzroy River, 'Barrier Ranges') for almost a year to collect insects and other natural history objects (Froggatt 1934). Most of Froggatt's material is housed in the Macleay Museum and has not been examined, although a specimen of the rare and currently undescribed *Hecatesia* sp. 'Amata' on permanent loan in the Australian National Insect Collection (ANIC) has been examined.

In 1908–09, F. P. Dodd and his second son, W. D. Dodd, spent 10 months stationed at Darwin (Anonymous 1909; Dodd 1935a; Monteith 1991), where they made a substantial collection of butterflies and other insects. They were based at the old railway workshops and locomotive depot (now subsumed by the suburb of Parap) and collected extensively within a radius of a few kilometres and from the nearby East Point (Braby and Nielsen 2011). Most notable among the many discoveries made by the Dodds was *Ogyris iphis doddi*, which was formally described by G. A. Waterhouse and G. Lyell in 1914 (Waterhouse and Lyell 1914; see also Dodd 1935a; Braby 2015a). Waterhouse (1932) also later described

several hesperiids based on Dodd material from 'Port Darwin'. W. D. Dodd subsequently visited the Kimberley (Broome, Derby, Fitzroy River, Grant Range) and the Top End (Melville and Bathurst islands, Darwin, Pine Creek, western Arnhem Land) during an epic (18-month) solo field expedition in 1912–13 funded by the South Australian Museum (Monteith 1991). The butterfly and other natural history specimens collected on that field trip are currently registered in the South Australian Museum (SAM). Many years later, Dodd (1935b, 1935c, 1935d) recounted his experiences in the Kimberley and the Top End and referred to several butterflies—namely, *Ogyris amaryllis*, *Papilio fuscus*, *Liphyra brassolis* and *Candalides margarita*.

During 1921–22, N. B. Tindale, an anthropologist and entomologist at the South Australian Museum, undertook extensive field studies on Groote Eylandt and its adjacent islands and the Roper River area, in the Northern Territory, and published several key papers from this work (Tindale 1922, 1923, 1927), including the description of *Nesolycaena urumelia* (as *Adaluma urumelia*). Tindale subsequently visited Mornington Island, Queensland, in 1963 (Fisher 1992). T. G. Campbell visited the Northern Territory from 1929 to 1933; he mainly collected butterflies at Brocks Creek, but also on Melville Island and Wyndham, in Western Australia. Much of his material was new and subsequently described by Waterhouse (1933, 1938). J. O. Campbell visited Darwin in 1945 and published an account of the butterflies he recorded (Campbell 1947). F. M. Angel and F. E. Parsons made an expedition to the Northern Territory in 1948; they travelled along the Stuart Highway from Adelaide, in South Australia, and made thorough collections around the major towns between Elliott and Darwin, details of which they subsequently published (Angel 1951; Couchman 1951). Of particular interest were the discovery and description of two new taxa: *Synemon wulwulam* and *Suniana lascivia larrakia* (as *Suniana larrakia*).

Warham (1957) and Koch (1957) made early compilations of the butterfly fauna of the Kimberley. Koch subsequently published a list of species from Koolan Island in Yampi Sound, Western Australia (Koch and van Ingen 1969; Koch 1975), which was later updated by McKenzie et al. (1995). J. C. Le Souëf visited the Top End in 1969 and again in 1971 (Le Souëf 1971); he mainly collected between Darwin and Mataranka, and between Timber Creek in the Northern Territory and Wyndham in Western Australia. Of particular note was the discovery of *Deudorix smilis*, which was recorded for the first time from Australia and described as a distinct subspecies, *Deudorix smilis dalyensis* (as *Virachola smilis dalyensis*) (Le Souëf and Tindale 1970; see also Braby 2016c).

Other less well-known field workers operating during the early twentieth century were M. Lain and F. Omer-Cooper, both of whom collected in northern Arnhem Land (including King River and Maningrida) in 1915–16 and 1968, respectively (Peters 1969). L. D. Crawford collected many butterflies from Darwin in 1955, as did R. G. Byrnes, who also travelled widely in the north-western corner of the Top End and eastern Kimberley during 1969–70.

In the 1950s and 1960s, agricultural research stations were established at Kununurra in Western Australia and Humpty Doo and Katherine in the Northern Territory. This led to the appointment of permanent entomologists in these towns and, although their focus was mainly on insect pest species, butterflies were reared and collected, especially skippers (Hesperiidae) associated with rice. The study by Koch (1957), for example, was based primarily on material in the Kimberley Research Station and Ord River Station (near Kununurra) made by C. F. H. Jenkins in 1944, R. G. Lukins in 1953 and L. E. Koch in 1957, among others.

Between 1970 and 1980, the level of field exploration of northern Australia and resulting scientific publications on the butterflies of the region increased dramatically, particularly from remote areas that previously had been poorly surveyed, such as the western and northern Kimberley and western Arnhem Land (e.g. Common 1973, 1981; Hutchinson 1973,

1978; Bailey and Richards 1975; Hall 1976, 1981; Common and Upton 1977; Edwards 1977, 1980, 1987; Dunn 1980; Kikkawa and Monteith 1980; Monteith 1982). From the 1980s to the present there has been a more or less permanent presence of local naturalists, amateur lepidopterists and professional entomologists, mainly based in Darwin, with an interest in butterflies.

## Aims and purpose

This book has several aims. The main purpose is to compile a comprehensive inventory and atlas of the butterflies and diurnal moths of the western section of the AMT. In particular, we aimed to answer the following questions:

1. How many species occur in the region (i.e. what is the overall species richness)?

2. What kinds of species occur in the region (i.e. what is the composition)?

3. What proportion of and which taxa are restricted to the region (i.e. what is the level of endemism)?

4. What is their geographical/breeding range (i.e. what is their spatial distribution)?

5. What are their ecological requirements according to larval food plant specificity and habitat preferences?

6. What is the breeding status of those species (i.e. which taxa are resident, immigrant, visitor or vagrant)?

7. When are they most abundant as adults (i.e. what is their temporal distribution), do they breed continuously or seasonally and what strategies have they evolved to cope with the adversity of the dry season?

8. What is their conservation status (i.e. which taxa are under threat)?

The overall goal is to provide the scientific basis on which to ensure the future conservation of this diverse group of invertebrates. By answering such fundamental questions, analysis of distributional patterns can then be undertaken to provide a broader perspective of the historical biogeography and evolutionary history of the fauna as a whole (Bowman et al. 2010), as well as to identify centres of endemism or biodiversity 'hotspots' that enable policymakers and managers to set priorities for conservation management (Braby and Williams 2016). Although not strictly part of this work, these later goals will be the subject of a forthcoming study.

Although our main focus is on butterflies, we have included a set of diurnal moths (31 species) for several reasons. First, diurnal moths are usually ignored by butterfly workers because they are moths, yet they show many characteristics similar to, and are often confused with, butterflies—for example, they are day-flying, colourful, often conspicuous, may feed on flowers and many have similar life histories in that most are phytophagous folivores. Second, moth workers frequently ignore diurnal moths because they are not active, or rarely active, at night. However, our selection and inclusion of diurnal moths are somewhat subjective. We have generally included only the more conspicuous species representing the families Sesiidae, Castniidae, Zygaenidae, Immidae, Geometridae, Uraniidae, Erebidae and Noctuidae (Agaristinae) and have not dealt with the smaller, less conspicuous groups (e.g. Heliozelidae, Glyphipterigidae, Heliodinidae, Brachodidae, Choreutidae). Also, with the exception of one species, we have not included members of the diurnal Arctiinae (family Erebidae) (e.g. *Amata*, *Asura*) because the taxonomy and species boundaries of these genera are in such a parlous state. And we have not included the Sphingidae, several of which are day-flying (e.g. *Cephonodes*) or crepuscular (e.g. *Macroglossum*), because of the great difficulty of identifying the species in flight. Even specimens of *Cephonodes* are difficult to identify and the genus in northern Australia comprises a complex of at least four species, which are currently under taxonomic revision (M. S. Moulds, unpublished data). Two species of Agaristinae appear to be strictly nocturnal (*Leucogonia cosmopis*, *Ipanica cornigera*) while several others are crepuscular (*Periopta ardescens*, *Radinocera* spp., *Mimeusemia* spp.); however, all of these are included for completeness given that the subfamily is such a dominant group in the fauna, with 19 taxa.

## Using this book

This book is intended as a reference work for lepidopterists, butterfly naturalists, hobbyists and collectors, as well as professional natural resource managers, conservation biologists, entomologists, scientists and students interested in landscape ecology and tropical ecology. It should also provide an important resource for those people interested in butterflies who are visiting northern parts of Australia. Moreover, it will provide a baseline assessment of the biodiversity of butterflies and day-flying moths in the region, given that the AMT is undergoing substantial environmental change. Thus, it will undoubtedly be viewed as a valuable information resource for political debate on future land use.

Of interest to most readers will be the species accounts, which form the basis of this book. For each species, we have organised the information into eight topics: distribution, any excluded records, habitat, larval food plants, attendant ants (for lycaenid butterflies), seasonality, breeding status and conservation status. The distribution (and excluded data) section includes a map showing the point data, together with the spatial distribution of the relevant larval food plants, and a geographic range map based on those data points. The habitat and larval food plant (and attendant ants) sections provide a summary of the main ecological resources required for each species within their geographic distribution. The seasonality section includes phenology charts of the adult and immature stages and, where sufficient data are available, graphs showing seasonal changes in relative abundance—that is, their occurrence in time. The section on breeding status was derived by integrating information on distribution, habitat, larval food plants and seasonality. The section on conservation status was determined from an analysis of distribution and breeding status.

Photo: *Synemon phaseoptila*, M. F. Braby

# 2

# Methods

## Database

The data compiled for this atlas were stored and managed in a dedicated relational database, using Microsoft Access, developed by the Northern Territory Flora and Fauna Division, and is now publicly available through the *Atlas of Living Australia* (www.ala.org.au). The database was constructed specifically to manage invertebrate data collected from targeted field surveys of threatened species, faunal inventories and incidental observations. The structure of this database is shown in Figure 1. The fields were organised into six tables, of which the three most relevant are the taxon table, the site table and the incidental table. Descriptions of some of the more critical fields used to capture data are summarised in Table 1.

The total number of species records assembled for this project was 23,885, of which 13,146 (55 per cent) were observations, 8,110 (34 per cent) specimens and 2,629 (11 per cent) literature. The spatial and temporal data collected for each species consisted of five main sources: vouchered specimens (S), specimens netted and released (N), photographs (P), field observations (O) and scientific literature (L) (Table 1). For the purposes of this work, codes N, P and O were combined into a single category, 'Observation'. Vouchered specimens consisted of specimens in museum collections and in most cases these are registered with an accession number. The majority of specimens are registered in the Museum and Art Gallery of the Northern Territory (NTM), the Northern Territory Economic Insect Collection (NTEIC) and the Australian National Insect Collection (ANIC). Some additional museum data were sourced through the *Atlas of Living Australia*, particularly for material that has been registered and databased in the South Australian Museum (SAM), Museum Victoria (NMV) and Australian Museum (AM).

A 'species record' was defined as a unique site–date occurrence for a particular species. Sites were defined as point localities more than 1 km apart (i.e. the spatial precision from the centre of the point was 500 m) and locations were defined as areas more than 10 km apart. For example, if 10 species were observed at a site (e.g. Lameroo Beach, the Esplanade, Darwin, Northern Territory) and then the same 10 species were observed at another nearby site, more than 1 km away but less than 10 km distant (e.g. East Point Reserve, Darwin, Northern Territory), on the same day (e.g. 1 January 2000), that would comprise a total of 20 species records from one location representing two sites.

Geocoordinates of all spatial records were first transformed to decimal degrees if required and then plotted using ArcMap, ArcGIS 10.3 Esri. Our dataset consisted of 4,352 spatial records (i.e. unique sampling sites), which are shown in

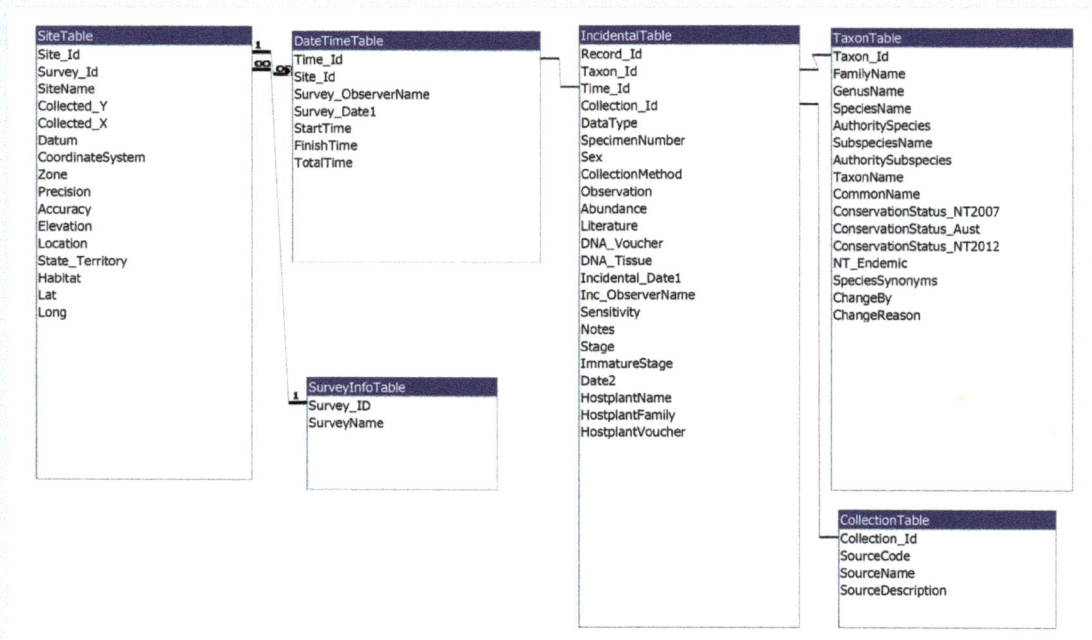

Figure 1 Structure and relations of the database used to manage species records
Source: Prepared by the authors.

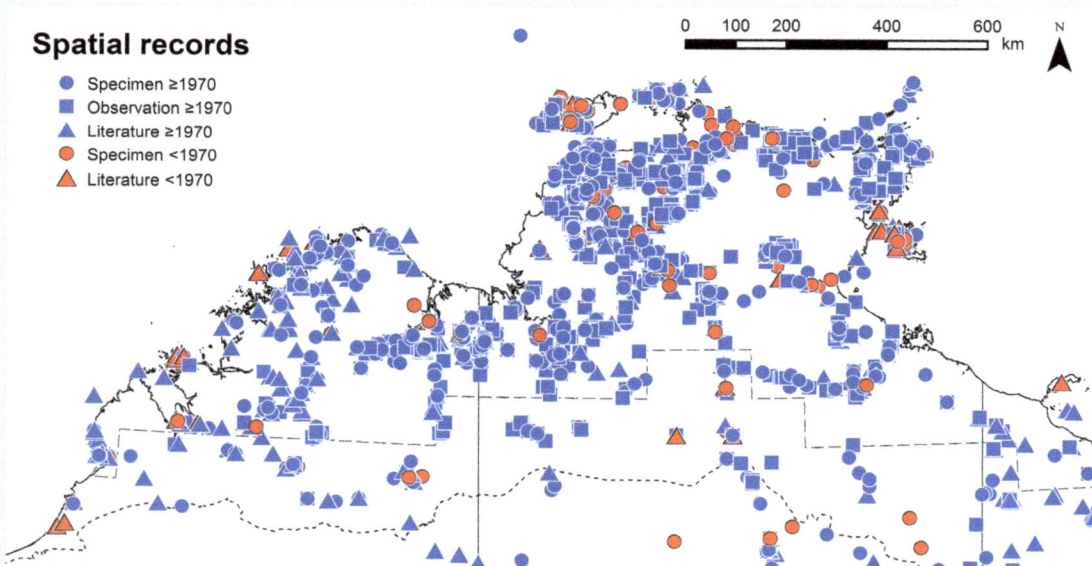

Map 4 Geographical distribution of spatial records (unique sampling sites) within the study region (n = 4,352)
Source: Prepared by the authors.

Map 4. The distribution of these spatial records shows that most areas within the study region have been sampled, but to varying extents. For instance, there is a strong bias in sampling towards urban areas, such as around the city of Darwin in the north-western corner of the Top End. In contrast, there are fewer sites from the semi-arid and arid zones towards the southern boundary of the study region. Some areas have been very sparsely sampled, such as eastern Arnhem Land, the Barkly Tableland, the Tanami Desert and the Great Sandy Desert.

## Excluded records

During the course of compiling an inventory of butterflies and diurnal moths for the study region, a number of records came to our attention that required scrutiny or clarification. Several of these records have already been removed from the inventory following detailed investigation and analysis (see Meyer et al. 2006; Braby 2008a, 2012b, 2014a). These records were excluded because they were deemed to be in error, comprising taxonomic misidentifications (determination errors) or mislabelling (transcription errors), or there was considerable doubt regarding their authenticity, with insufficient evidence provided to substantiate them. Thus, we do not recognise the following 10 species from the study region: *Pachliopta liris*, *Oriens augustulus*, *Pseudoborbo bevani*, *Pelopidas agna*, *Telicota ancilla*, *Telicota mesoptis*, *Deudorix diovis*, *Theclinesthes onycha*, *Theclinesthes serpentatus* and *Leptotes plinius*. In the case of *Telicota ancilla* (for the subspecies *T. ancilla baudina*), the taxon proved to be a junior synonym of *T. augias krefftii* (Braby 2012b).

Table 1 Descriptions of some of the critical fields used to capture species data, with an example provided to illustrate how data were coded for an individual species record

| Field name | Description | Example |
|---|---|---|
| **Taxon table** | | |
| FamilyName | Name of family | Papilionidae |
| GenusName | Name of genus | *Protographium* |
| SpeciesName | Name of species | *leosthenes* |
| AuthoritySpecies | Authority of species | (Doubleday 1846) |
| SubspeciesName | Name of subspecies | *geimbia* |
| AuthoritySubspecies | Authority of subspecies | (Tindale 1927) |
| TaxonName | Genus_species_subspecies | *Protographium leosthenes geimbia* |
| CommonName | Common name | Kakadu Swordtail |
| **Site table** | | |
| SiteName | Site code name or number | |
| Collected_Y | Y coordinates (latitude or northing) | 12.84262°S |
| Collected_X | X coordinates (longitude or easting) | 132.81895°E |
| Datum | WGS84, GDA94 or AGD66 | WGS84 |
| CoordinateSystem | Dd = decimal degrees, DMS = degrees, minutes, seconds, DMm = decimal minutes, or E_N (easting and northing) | Dd |
| Zone | Only when projected coordinate system is used (E_N) | |
| Precision | Radius from point: 5 m, 10 m, 50 m, 100 m, 250 m, 500 m | 250 m |
| Accuracy | Error in point reading | |
| Elevation | Elevation (m) | 50 m |
| Location | Description of location name | Kakadu National Park, Nourlangie Rock, Nanguluwur Art site |
| State_Territory | NT, WA, Qld | NT |

| Field name | Description | Example |
|---|---|---|
| Habitat | Vegetation community | Monsoon vine thicket at base of sandstone cliff |
| Lat | Mapping Y coordinates converted to latitude in Dd with GDA94 or WGS84 | −12.84262 |
| Long | Mapping X coordinates converted to longitude in Dd with GDA94 or WGS84 | 132.81895 |
| **Incidental table** | | |
| DataType | S = specimen, O = observation, L = literature, N = netted and released, P = photograph | S |
| SpecimenNumber | Museum repository/private collection and voucher number of specimen collected | NTM I005572 |
| Sex | M = male, F = female | F |
| CollectionMethod | Sweep net, hand, light trap, etc. | Hand |
| Literature | Author (year) | |
| Date1 | Day/month/year | 20 December 2009 |
| ObserverName | Initials and surname | M. F. Braby |
| Notes | Early stages, adult behaviour numbers collected etc. | 6 larvae, various instars, collected from LFP growing at base of escarpment |
| Stage | A = adult, I = immature | I |
| ImmatureStage | E = egg, L = larva, P = pupa | L |
| Date2 | Date of adult emergence for reared specimen | 8 January 2010 |
| HostplantName | Name of genus and species of larval food plant | *Melodorum rupestre* |
| HostplantFamily | Name of family of larval food plant | Annonaceae |
| HostplantVoucher | Herbarium repository and voucher number of plant specimen | M. F. Braby 200, DNA |

# Data analysis

## Geographic range

The geographic range (Maps 6–8) of each butterfly and diurnal moth species was estimated directly from the distribution of spatial records or inferred by combining these records with those of larval food plant(s) or attendant ant. The geographic ranges were generated using a rule set based on the minimum distances between distribution records and, where applicable, the larval food plant or attendant ant data. Records for larval food plants included both known and putative species; however, only native food plant records were plotted.

For each species, several criteria were used to create polygons to estimate the geographic range and determine whether ranges were continuous or disjunct. Based on the distances between spatial records, distributions were classified as continuous or disjunct, with continuous distributions incorporated into range polygons and disjunct distributions surrounded by an arbitrary buffer of 15 km so they could be clearly discerned in the range maps. The method is broadly similar to the α-hull recommended by the IUCN Standards and Petitions Subcommittee (2016) for estimating the extent of occurrence (EOO) of a taxon, but with some notable differences, as follows.

## Continuous distribution

Distribution records were included in the polygon only if they were located within a specified distance of one another or had intervening larval food plant records. Thus, the distribution was considered to be continuous when spatial records within the geographic range were 200 km apart, with or without intervening larval food plant records, or 200–500 km apart, but only with intervening

**Seasonal Rainfall Zones**

Summer dominant (more than 1200mm)
Summer dominant (650 - 1200mm)
Summer dominant (350 - 650mm)
Arid  (less than 350mm)

0  100  200  400  600 km

Summer dominant (more than 1200mm)

Summer dominant (650 - 1200mm)

Summer dominant (350 - 650mm)

Arid  (less than 350mm)

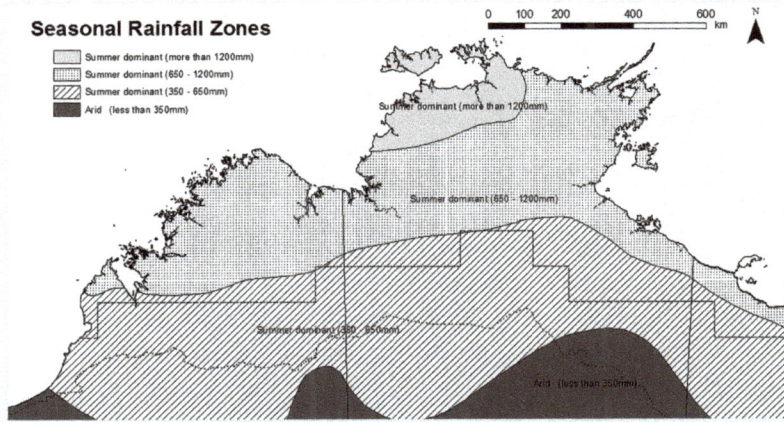

Map 5 Seasonal rainfall zones of northern Australia
Source: After Gaffney (1971).

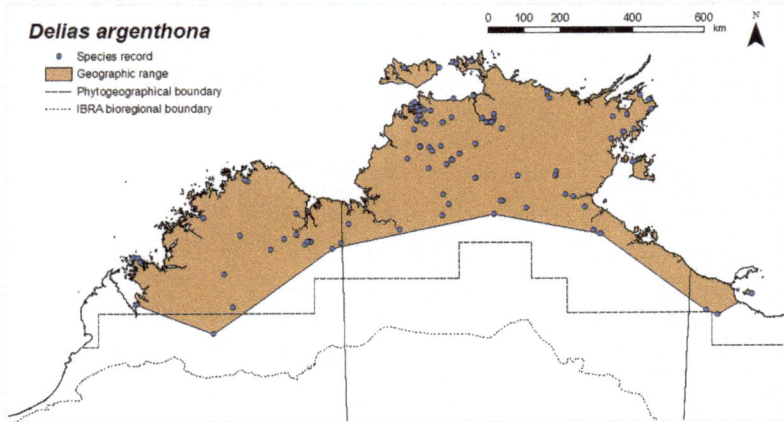

**Delias argenthona**

• Species record
Geographic range
— — — Phytogeographical boundary
· · · · · IBRA bioregional boundary

0  100  200  400  600 km

Map 6 Example of the method used to estimate geographic range: *Delias argenthona*, a species with a continuous distribution, as inferred from distribution records and larval food plant data
Source: Prepared by the authors.

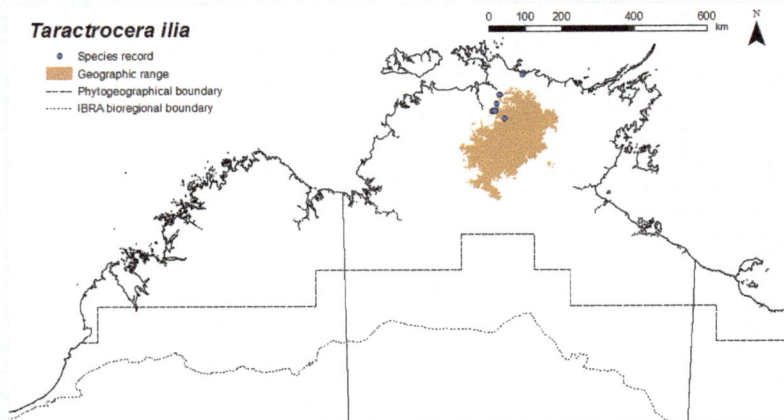

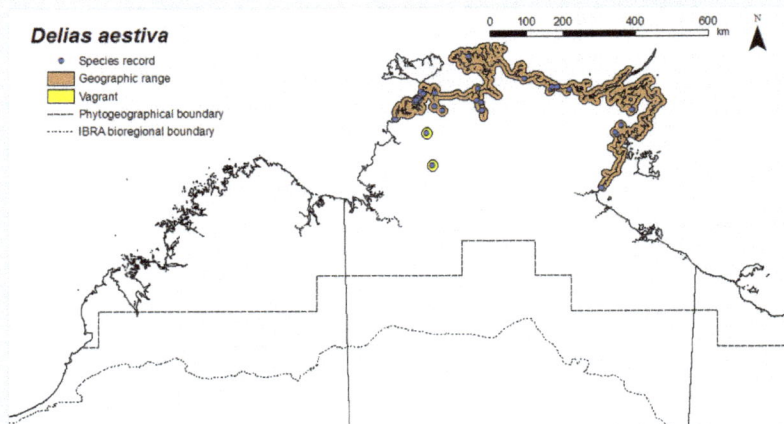

**Taractrocera ilia**

• Species record
Geographic range
— — — Phytogeographical boundary
· · · · · IBRA bioregional boundary

0  100  200  400  600 km

Map 7 Example of the method used to estimate geographic range: *Taractrocera ilia*, a narrow-range endemic restricted to the Arnhem Land Plateau
Source: Prepared by the authors.

**Delias aestiva**

• Species record
Geographic range
Vagrant
— — — Phytogeographical boundary
· · · · · IBRA bioregional boundary

0  100  200  400  600 km

Map 8 Example of the method used to estimate geographic range: *Delias aestiva*, a species normally restricted to coastal habitats (mangrove), but with two inland non-breeding records (vagrants)
Source: Prepared by the authors.

larval food plant records. These distances were chosen as a conservative estimate for how far a species may disperse in the presence or absence of larval food plants. Distribution records that fell outside these thresholds were considered to be disjunct (see below). However, there was a sampling bias in the study region (for butterflies and diurnal moths and their larval food plants) related both to the distance from major urban centres and to seasonal rainfall zones (Map 5), with comparatively few distribution records from the remote semi-arid and arid zones (< 700 mm mean annual rainfall). Thus, to adjust for this sampling bias, different distance rule sets were applied to records from the lower rainfall zones with intervening larval food plant data (Table 2).

In addition, seven other criteria based on particular idiosyncrasies of the data or geography were applied to estimate the geographic range for continuous distributions. First, for those species for which the larval food plant was unknown, distribution records were included in the polygon when the distance between them was less than 500 km. Second, marine areas were excluded from polygons and the distance rule sets were not applied across marine gaps, such as gulfs. Third, if a species occurred on any of the larger islands (Bathurst, Melville, Groote or Mornington), the whole island was included in the geographic range. However, if a species had not been recorded from an island, the island was excluded from the geographic range, even if the larval food plant was present and it fell within the polygon. Fourth, areas near the coast (i.e. within 150 km of the coastline) and nearby small islands that fell outside the line joining two distribution records were included in the geographic range, but only if the larval food plant was present or if the butterfly would be expected to occur in the intervening area based on expert opinion. Fifth, for species known to be restricted to particular geological elements—for example, the sandstone plateau of western Arnhem Land—the geological element was used to delimit the geographic range (Map 7). It should be noted that the eastern part of the sandstone plateau of western Arnhem Land is difficult to access and hence there is a paucity of larval food plant (and butterfly/diurnal moth) data from this remote area. Sixth, for species restricted to coastal habitats (e.g. mangroves, saltmarsh), a 10 km wide buffer was applied along the coastline and along mangrove/saltmarsh-lined watercourses between distribution records to estimate the geographic range (Map 8). Finally, for widespread species known to occur throughout the central arid zone, poorly sampled areas such as the south-eastern corner of the study region were included in the geographic range, despite the absence of distribution records.

## Disjunct distribution

Distribution records were considered to be disjunct or isolated when the closest points were separated by 200 km or more and there were no intervening larval food plant records, or when the closest points were separated by more than 500 km and there were intervening larval food plant records. However, as noted above, different distance rule sets were applied to the lower rainfall areas of the semi-arid and arid zones (Table 2) to account for the low sampling effort in the southern half of the study region.

Disjunct or isolated distribution records may represent: 1) resident breeding populations, 2) vagrant individuals or 3) temporary range expansions. Disjunct resident breeding populations comprising a single isolated point, or a small cluster of points, outside the main distribution were included in the geographic range with a 15 km buffer. However, often we were not able to distinguish whether the disjunction was a natural one due to an inhospitable area (e.g. absence of breeding habitat) between two resident breeding populations or simply due to low sampling effort in the intervening area. Vagrant individuals or temporary breeding populations outside the normal breeding range were not included in the geographic range size calculations, and were buffered in a contrasting colour (yellow) to indicate their exclusion from the natural geographic range (Map 8).

## Endemism

Taxa restricted to the study region are referred to as endemics. We distinguished three types of endemics according to their geographic

Table 2 Threshold criteria used to determine the inclusion of butterfly and diurnal moth distribution records in the continuous range polygon with intervening larval food plant records according to mean annual rainfall zones

| Seasonal rainfall zone | Distance between spatial records (km) |
|---|---|
| Summer dominant rainfall zone (≥ 700 mm) | ≤ 500 |
| Summer dominant rainfall, semi-arid zone (350–700 mm) | ≤ 750 |
| Arid zone (≤ 350 mm) | ≤ 1,000 |

Note: See also Map 5.

range size. Taxa that had small geographical range sizes (≤ 40,000 sq km) were categorised as 'narrow-range endemics', whereas those that had exceedingly small distributions (≤ 10,000 sq km) were classified as 'short-range endemics' following the definition of M. S. Harvey (Harvey 2002; Harvey et al. 2011). Taxa with slightly larger ranges (40,000–100,000 sq km) were considered 'restricted'. The 40,000 sq km threshold for narrow-range endemism was chosen because this is the maximum extent of the Arnhem Land Plateau—an area that supports numerous endemic plants (Woinarski et al. 2006) and invertebrates (Andersen et al. 2014).

## Relative abundance

Seasonal trends in relative abundance throughout the year were graphed for each species. Relative abundance was estimated from data pooled across the species' range based on the number of temporal records for each month. For our purposes, a temporal record was defined as the occurrence on a particular date (time) at a given site (space) irrespective of the number of adults recorded or time spent at the site. As noted above, the spatial precision of sites was set to a 500 m radius from its central point—that is, a minimum distance of 1 km was used to distinguish two adjacent sites sampled on the same day. Only species with more than 25 temporal records were plotted, along with two highly seasonal species, Genus 1 sp. 'Sandstone' and *Periopta ardescens*, which had 21 and 23 temporal records, respectively.

The dataset comprised 4,603 temporal records (i.e. unique sampling dates); however, there was substantial seasonal variation, with April having the most records and three times more than

September, which had the least (Figure 2). This seasonal difference reflects a bias in sampling effort when surveys or field collections are generally undertaken: April heralds the end of the wet season, when areas become increasingly accessible and insect activity for many species is at its greatest, whereas September coincides with the end of the dry season, when conditions are dry and hot and insect activity is considerably reduced. Hence, more surveys have been carried out in the late wet season than in the late dry season. Thus, to minimise the effect of this seasonal bias in sampling on interpretation of seasonal patterns, we applied a correction factor to adjust for the variation in sampling effort for each month, calculated as follows: correction factor = $616 / n$, where 616 is the maximum number of temporal records in April and $n$ is the total number of temporal records in the month (see Table 3).

Table 3 Correction factor used to estimate relative abundance for each month to account for seasonal variation in sampling effort (the number of temporal records)

| Month | Number of records | Correction factor |
|---|---|---|
| January | 392 | 1.571 |
| February | 394 | 1.563 |
| March | 484 | 1.273 |
| April | 616 | 1.000 |
| May | 524 | 1.176 |
| June | 450 | 1.369 |
| July | 388 | 1.588 |
| August | 327 | 1.884 |
| September | 206 | 2.990 |
| October | 312 | 1.974 |
| November | 270 | 2.282 |
| December | 240 | 2.567 |

Note: For seasonal variation in the number of temporal records, see Figure 2.

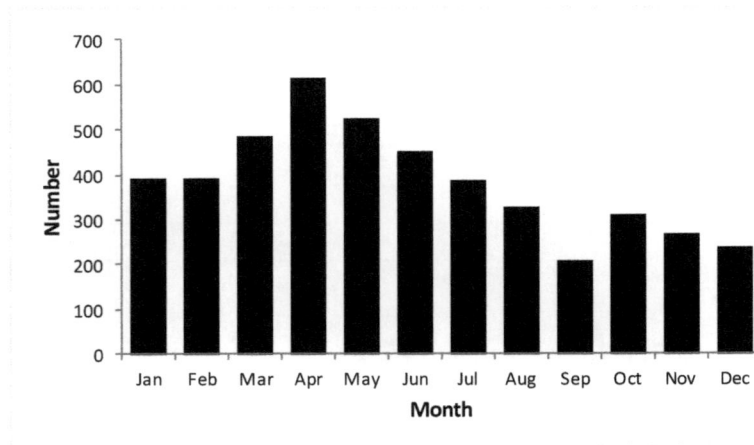

Figure 2 Seasonal variation in the number of temporal records (unique sampling dates) within the study region (*n* = 4,603)
Source: Prepared by the authors.

## Nomenclature

Scientific nomenclature for butterflies largely follows Braby (2010b, 2011b, 2016a), while that for moths follows Nielsen et al. (1996). The recent proposal to place the species *Mycalesis sirius* in the genus *Mydosama* (Kodandaramaiah et al. 2010; Aduse-Poku et al. 2015) is followed. Nomenclature for plants follows the *Australian Plant Census* (2017) and FloraBase (2017). We have also included nine undescribed species of diurnal moths (from the families Castniidae, Zygaenidae, Geometridae, Erebidae and Noctuidae) and five undescribed subspecies of butterflies (from the families Hesperiidae, Pieridae and Lycaenidae). Formal descriptions of these taxa are currently in preparation (M. F. Braby, unpublished data).

The taxonomic status of the pierid *Elodina tongura* (Pipeclay Pearl-white), which is endemic to the Top End, requires comment. It is closely related to *E. walkeri* and may be a junior synonym of that species. *Elodina tongura* was originally described by Tindale (1923) as a subspecies of *E. perdita*, based on 16 specimens collected from Groote Eylandt and Winchelsea and Woodah islands, in the Northern Territory, in the wet season between February and April. However, De Baar and Hancock (1993) and De Baar (2004) treated *E. tongura* as a separate species, considering it to be closely allied to *E. walkeri*, which was reinstated by De Baar and Hancock (1993) as a species distinct from *E. perdita*. The type locality of *E. walkeri* is Darwin (Edwards et al. 2001).

De Baar (2004) indicated that *E. walkeri* and *E. tongura* are sympatric in the Darwin region and possibly elsewhere in coastal areas of the Northern Territory, and alluded to differences in adult size, the shade of the yellow basal patch on the underside of the forewing and length of the vesica of the male phallus to distinguish the two species. In contrast, Braby (2000) regarded *E. tongura* as a junior synonym of *E. walkeri* based on examination of the syntypes, an extensive series of other material and an unpublished geographical study of the male genitalia. It was concluded that many of the character differences (e.g. adult size, yellow basal flash on the underside of the forewing, the presence of a dark apical patch on the underside of the forewing and the shape of the forewing apex) vary seasonally; thus, the validity of *E. tongura* as a distinct taxon was considered very doubtful. In the present work, we have therefore not recognised *E. tongura* until a thorough taxonomic study of the complex is undertaken.

Common names for butterflies follow Braby (2016a); however, most of the diurnal moths do not have standard common names, although some genera do have group names—for example, 'sun-moth' for *Synemon*. Hence, we have proposed common names for all species of moths, including the names for three species of sun-moths recently given by Williams et al. (2016).

# Species accounts

## Distribution

For each species, the geographical distribution of records, along with those of its native and/or putative larval food plant(s), is presented. For some species of Lycaenidae that are obligately attended by ants or are ant predators—notably, *Liphyra brassolis*, *Acrodipsas* spp., *Hypochrysops* spp., *Arhopala* spp. and *Hypolycaena phorbas*—the spatial distribution of the ant is also included.

Data for known and putative larval food plants were extracted from *Australia's Virtual Herbarium* (2017). For each plant species, records were vetted using the following procedure. Records were first removed if they were flagged as being of uncertain identity, cultivated, had more than 10 km uncertainty in the coordinates, had no coordinates or were outside the study region by more than one degree of latitude or longitude. The remaining records were then examined spatially and compared against benchmark publications—Liddle et al. (1994) for the Northern Territory and FloraBase (2017) for Western Australia—to identify obvious outliers. Where outliers were detected, the record was examined for potential errors such as mismatches between location and coordinates, and then either deleted or corrected if such an error was detected.

Data for ants were obtained either from the *Atlas of Living Australia* (for *Oecophylla smaragdina*) or from specimens curated in the Tropical Ecosystems Research Centre (TERC) and the ANIC (for *Froggattella kirbii* and *Papyrius* spp.).

For each butterfly and diurnal moth, spatial records were divided into two periods: historical (before 1970) and contemporary (1970 onwards). There were two reasons for this approach. First, prior to 1970 there was comparatively little collecting effort in the study region and a similarly low level of reporting in the scientific literature. Second, locality data for specimens collected before 1970 were rarely, if ever, georeferenced with latitude and longitude; it was not until detailed topographic maps (employing the AGD66 datum) became widely available in the late 1960s and early 1970s that professional entomologists started to routinely include geographic coordinates on their label data. Hence, plotting for historical records almost invariably has low precision and large spatial errors. For example, Tindale (1923) was stationed on Groote Eylandt—an island measuring approximately 70 km by 50 km—for almost a year (June 1921 – May 1922) and published a detailed list of the 52 species he recorded, which included temporal data on the months during which specimens were collected. However, Tindale (1923: 349) mentioned the actual place of capture in the text for only one species: 'female [*Liphyra brassolis*] flew to a light at 8.30 p.m. on a sultry night, in my camp at Yetiba'. For the others, precise locations were not given. Thus, plotting most of Tindale's records for Groote Eylandt with any level of precision is problematic. In this case, we have assigned and plotted all records to 'Yedikba' (Yetiba) (14.079ºS, 136.457ºE), approximately 25 km south of Alyangula, because this is the only site where we can be certain Tindale was stationed, but the spatial error of such records may well be more than 10 km.

Where multiple data points had the same geographic coordinate, we plotted the contemporary records (≥ 1970, blue symbols) over historic records (< 1970, red symbols) to emphasise more recent occurrences. Thus, for some species historical records may not always be clearly visible in the distribution maps.

Comparison of the spatial distribution of a butterfly or diurnal moth species with its larval food plant(s) required exploration of possible reasons for discrepancies where differences arose. For example, a species may be more widespread than its larval food plant because: 1) the adult regularly disperses outside its breeding area (i.e. the non-breeding distribution is wider than the breeding distribution); 2) the species has additional food plants that have not yet been reported (i.e. knowledge of the breeding area is incomplete); or 3) the food plant may be underreported and is in fact more widespread than herbarium records indicate. Conversely, a species may be less widespread than its larval food plant because: 1) the species' distribution may be limited by factors other than the food plant, such as climate; 2) breeding populations of the species may be limited to areas where the food plant occurs above a threshold density or where the extent of the breeding habitat is above a critical minimum size; or 3) the species' distribution may be underreported and thus our knowledge of it is incomplete.

## Excluded data

A number of spatial records were found to be erroneous or doubtful. These records are listed for the relevant species and the data are excluded from the distribution maps. Braby (2000, 2012b, 2014a) and Braby and Zwick (2015) have previously discussed the reliability of some of these records.

## Habitat

The habitats listed for each species refer to breeding habitats—that is, where the species completes its life cycle from egg to adult and where the particular larval food plants grow. Habitats were based largely on natural vegetation types and structural classification according to the *National Vegetation Information System* (Executive Steering Committee for Australian Vegetation Information 2003), together with topographic features and climatic factors. The adult stage of butterflies and diurnal moths may also use other habitats for feeding (e.g. nectar resources) and mating (e.g. landmarks such as hilltops); however, for most species in the study region, these habitats were not well known and thus are generally not reported in this work.

## Larval food plants

The known larval food plants, both native and non-native species, are summarised for each species based on records published in the scientific literature and our own unpublished observations. Where multiple food plants were used, comments on any apparent preferences are noted. If the larval food plant had not been

recorded from the study region, we listed the food plant reported from adjacent areas—usually northern Queensland. In most cases, these putative larval food plant records were derived from Braby (2000, 2016a), although in some cases we referred to the primary literature source.

## Attendant ants

The larvae, and to some extent pupae, of many Lycaenidae are attended by ants to various degrees. In this work, these ant–lycaenid associations are grouped into five categories: 1) those species not attended by ants; 2) those species usually unattended, but very occasionally attended by a few ants; 3) those species attended by a few ants from several genera in a facultative association; 4) those species constantly attended by many ants of a specific species or genus in an obligate association; and 5) those species that are dependent on ants as a food source in a myrmecophagous association. Ant associations are based primarily on information collated by Eastwood and Fraser (1999) and Braby (2000), combined with subsequent records published in the scientific literature and our own unpublished observations.

## Seasonality

For each species, the broad adult flight period and patterns of seasonal changes in relative abundance are given. Additional information on breeding, incidence of the immature stages (eggs, larvae and pupae), number of generations completed annually and incidence of dormancy are also provided where known. A phenology chart showing monthly temporal records of the immature stages is also presented for each species. These charts are necessarily incomplete, but are provided for two reasons: 1) to give an approximation of when the various stages are present during the year; and 2) to highlight knowledge gaps in recording of the life cycle stages. For instance, for some species there are very few or no available temporal data on the incidence of their eggs, larvae or pupae. Migration records are also summarised where this has been reported.

## Breeding status

The breeding status of each species was determined, noting whether it was: 1) a resident, in that the species breeds regularly and is permanently established in the study region; 2) an immigrant, in which the species breeds temporarily within the study region and then vacates it on a regular basis; 3) a visitor, in that the species does not breed within the study region, but regularly occurs within its boundaries; or 4) a vagrant, in which the species does not breed and only occasionally or rarely enters the study region, usually in very small numbers. Resident species include those that are nomadic—that is, the population occurs regularly within the study region, but at any given location breeding is temporary.

## Conservation status

The conservation status of each species was evaluated according to the IUCN Red List criteria (IUCN 2001; IUCN Standards and Petitions Subcommittee 2016). Although these criteria are designed for global taxon assessments, they can be applied to subsets of global data at different geographical scales (e.g. national, regional or local levels). In this work, we determined the conservation status at the regional level—that is, the status of each species within the study region. Only taxa (species, subspecies) that are endemic to the region qualify for global assessment. For taxa that occur outside the study region—either elsewhere in Australia or outside Australia—the Red List category may differ substantially from the global assessment because we have evaluated only a small subset of the geographic range. For these nonendemic, extralimital taxa we therefore followed IUCN guidelines for the application of Red List criteria to assess populations at the regional level (IUCN 2012). Under these regional criteria, taxa that are introduced, vagrants or rare immigrants (i.e. populations that breed occasionally during favourable conditions, but do not become permanently established) or have recently colonised the region and are currently expanding their range outside the region should not be assessed and are accordingly categorised as Not Applicable (NA).

It should be noted that taxa may be listed as threatened under IUCN criteria because they have small population sizes (criteria C and D) and/or marked population declines (criterion A). However, these parameters are generally unknown for butterflies and diurnal moths in northern Australia; thus, our conservation assessments focused mostly on those distributional parameters (criterion B) for which information is more readily available. Indeed, criterion B is the primary Red List criterion used for status evaluation of butterflies globally (Lewis and Senior 2011), and it has previously been used to assess the conservation status of butterflies in Australia (Braby and Williams 2016).

The EOO, a component of IUCN Criterion B1, was used to evaluate whether a taxon belonged in a threatened category (IUCN Standards and Petitions Subcommittee 2016). Taxa were first sorted according to their geographic range size (based on the calculations described above; see 'Data analysis') to determine whether they fell within or close to the threshold of 20,000 sq km to potentially qualify for a threatened category (i.e. Vulnerable). The initial sorting produced a short list of 34 taxa (i.e. with distributions of < 40,000 sq km), which were then scrutinised in more detail to determine the EOO, number of locations or extent of fragmentation and any other known conservation issues such as evidence of decline and/or threatening processes. The EOO was calculated (in square kilometres) using minimum convex polygons or convex hulls in ArcMap, ArcGIS 10.3 Esri Projected Coordinate System: GDA 1994 Australia Albers. Areas of unsuitable habitat (e.g. ocean) within the minimum convex polygon were also included in the area calculation. Based on this information, IUCN Red List categories were then allocated to each of the 34 taxa.

For taxa for which the EOO could not be estimated, because they were known from only one or two sites, the area of occupancy (AOO) was used as an alternative method of assessment. The AOO is a component of IUCN Criterion B2, and the most common approach is to calculate the area of distribution from range-wide occurrences (grid cells) (Gaston and Fuller 2009). However, this method proved unsatisfactory because of the low spatial resolution of the grid cells employed across the study region (100 km x 100 km = 10,000 sq km) (Braby et al., unpublished data). We therefore used spatial buffering for these locality records (with the buffer set to 15 km) in conjunction with habitat suitability where this was known. The spatial buffering is not an actual estimate of the AOO, but rather an indication that these species are currently known to have exceedingly small geographic range sizes within the study region. That is, these taxa are likely to fall below the threshold of 2,000 sq km to potentially qualify for a threatened category (i.e. Vulnerable) under IUCN Red List criteria.

For taxa assessed as Data Deficient (DD), but which are of conservation interest because they may qualify as Near Threatened (NT) once adequate data are available, we have followed Bland et al. (2017), who recommended that justification tags be assigned to identify knowledge gaps and help prioritise reassessments. For example, lack of information could arise because there are few records or historical records only, there is uncertainty about locations or distribution, or there are uncertain threats or uncertain taxonomy. The actions needed for these taxa are also provided.

Although listing undescribed taxa on the IUCN Red List is discouraged—for example, as Least Concern (LC) or DD (see IUCN Standards and Petitions Subcommittee 2016)—the guidelines do allow for taxa to be evaluated provided there is certainty that the species are being or are about to be described; this is certainly the case for the putative butterfly subspecies and agaristine diurnal moth species that we have assessed (M. F. Braby, unpublished data).

# Results and discussion

## Patterns of biodiversity

Four key questions we sought to answer relating to biodiversity were: 1) How many species occur in the study region (i.e. what is the overall species richness)? 2) What kinds of species occur in the region (i.e. what is the composition)? 3) What proportion of and which taxa are restricted to the region (i.e. what is the level of endemism)? 4) How are those species distributed spatially?

## Species richness

For the first question, our investigations have revealed that 163 species (132 butterflies and 31 diurnal moths), representing 166 taxa, have been recorded from the study region. Three species are each represented by two subspecies within the region: *Suniana lascivia* (with *S. lascivia larrakia* and *S. lascivia lasus*), *Appias albina* (with *A. albina albina* and *A. albina infuscata*) and *Agarista agricola* (with *A. agricola agricola* and *A. agricola biformis*). In terms of their representation within each of the four subregions, 150 species (122 butterflies and 28 diurnal moths) have been recorded from the Top End, 105 species (88 butterflies and 17 diurnal moths) from the Kimberley, 82 species (74 butterflies and eight diurnal moths) from the western Gulf Country and 53 species (45 butterflies and eight diurnal moths) from the Northern Deserts. Thus, the Top End is substantially more diverse than the other subregions and almost three times richer than the semi-arid and arid zones of the Northern Deserts.

Interestingly, a substantial number of the species known from the study region have only been detected during the past four decades. Thirty-seven taxa (22 per cent) have been discovered or recorded for the first time since 1970 (Table 4). Of these, 11 species represent their first recording for Australia. Some of these newly recorded species concern intrusions from South-East Asia (e.g. *Appias albina infuscata*, *Danaus chrysippus cratippus*, *Acraea terpsicore* and *Junonia erigone*), while others are species entirely new to science (e.g. *Taractrocera psammopetra*, *Acrodipsas decima* and *Nesolycaena caesia*) or new taxa that have not yet been formally described (e.g. *Leptosia nina* ssp. 'Kimberley', *Synemon* sp. 'Kimberley', *Idalima* sp. 'Arnhem Land', *Hecatesia* sp. 'Arnhem Land' and *Radinocera* sp. 'Sandstone'). Notably, during the course of this study, two taxa not recorded for more than 100 years were rediscovered: an extant breeding population of *Ogyris iphis doddi* was found in

## Table 4 Records of 'new' species from the study region during the past four decades (since 1970)

| Species | Year | Comments | Reference |
|---|---|---|---|
| *Zizula hylax attenuata* | 1971 | First record for study region | Le Souëf (1971) |
| *Alcides metaurus* | 1972 | First record for study region | Braby (2014a) |
| *Idalima* sp. 'Arnhem Land' | 1972 | First record for Australia | This volume |
| *Candalides geminus gagadju* | 1972 | First record of species for study region; first record of subspecies for Australia | Braby (2017) |
| *Hecatesia* sp. 'Arnhem Land' | 1973 | First record for Australia | This volume |
| *Hypochrysops apelles* ssp. 'Top End' | 1973 | First record of species for study region; first record of subspecies for Australia | This volume (see also Common and Waterhouse 1981) |
| *Yoma sabina sabina* | 1976 | First record for study region | This volume (see also Common and Waterhouse 1981) |
| *Eurema brigitta australis* | 1977 | First record for study region | Braby (2014a) |
| *Danaus chrysippus cratippus* | 1977 | First record for Australia | Common and Waterhouse (1981); Braby (2014a) |
| *Junonia erigone* | 1977 | First record for Australia | Edwards (1977) |
| *Petrelaea tombugensis* | 1977 | First record for study region | Common and Waterhouse (1981); Braby (2015b) |
| *Danaus plexippus* | 1979 | First record for study region | Dunn (1980) |
| *Leptosia nina* ssp. 'Kimberley' | 1980 | First record for Australia | Common and Waterhouse (1981); Naumann et al. (1991) |
| *Acrodipsas myrmecophila* | 1981 | First record for study region | Dunn and Dunn (1991) |
| *Euploea alcathoe enastri* | 1988 | First record of species for study region; first record of subspecies for Australia | Fenner (1991) |
| *Hesperilla crypsigramma* ssp. 'Top End' | 1989 | First record of species for study region; first record of subspecies for Australia | Field (1990a) |
| *Nesolycaena caesia* | 1990 | First record for Australia | d'Apice and Miller (1992) |
| *Papilio anactus* | 1991 | First record for study region | Puccetti (1991) |
| *Cephrenes augiades* ssp. 'Top End' | 1991 | First record of species for study region; first record of subspecies for Australia | Braby (2000) |
| *Vanessa kershawi* | 1991 | First record for study region | Puccetti (1991) |
| *Acrodipsas decima* | 1991 | First record for Australia | Miller and Lane (2004) |
| *Theclinesthes sulpitius* | 1991 | First record for study region | Meyer and Wilson (1995) |
| *Euchromia creusa* | 1993 | First record for study region | Braby (2014a) |
| *Jalmenus icilius* | 1995 | First record for study region | Braby (2000) |
| *Eurema alitha novaguineensis* | 1997* | First record for study region | Braby (1997) |
| *Theclinesthes albocinctus* | 1997 | First record for study region | Grund (1998) |
| *Synemon* sp. 'Kimberley' | 2000 | First record for Australia | This volume (see also Williams et al. 2016) |
| *Agarista agricola agricola* | 2005 | First record of subspecies for study region | This volume |
| *Mimeusemia centralis* | 2007 | First record for study region | Braby (2014a) |
| *Comocrus behri* | 2008 | First record for study region | Braby (2011a) |
| *Bindahara phocides* | 2009 | First record for study region | This volume |
| *Radinocera* sp. 'Sandstone' | 2009 | First record for Australia | Braby (2015e) |
| *Appias albina infuscata* | 2010 | First record of subspecies for Australia | Braby et al. (2010b) |
| *Taractrocera psammopetra* | 2010 | First record for Australia | Braby and Zwick (2015) |
| *Ogyris barnardi barnardi* | 2011 | First record for study region | Dunn (2013) |
| *Acraea terpsicore* | 2012 | First record for Australia | Braby et al. (2014a, 2014b) |
| *Mimeusemia econia* | 2017 | First record for study region | This volume |

\* The earliest specimens of this species date back to 1909; however, it was previously confused with *Eurema hecabe* such that *E. alitha* was not formally recognised in the fauna until 1997.

Note: Taxa are listed in chronological order according to the year they were first discovered or detected, not necessarily the year they were first reported in the literature.

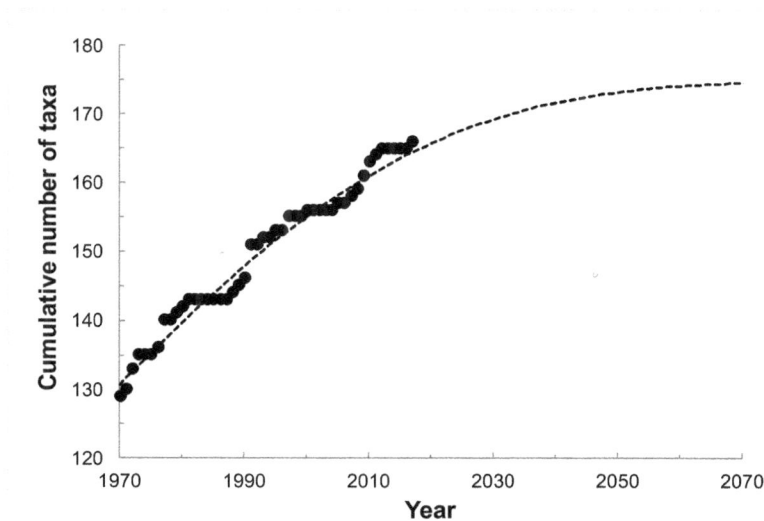

Figure 3 A species accumulation curve fitted to the number of new taxa discovered since 1970

Note: The curve is a cumulative extreme-value function (see Williams et al. 2009): $\log(-\log(1 - y/174.91)) = 0.01524 + 0.30394 * (x - 1969)$ ($r^2 = 0.975$), where y = total taxa known and x = year. The rate of species accumulation since 1970 suggests a predicted total richness of 175 taxa, which, at the current discovery rate, will be discovered progressively up to the year 2068.

Source: Prepared by the authors.

the Top End in 2014, 105 years after it was last recorded breeding, in 1908–09, by F. P. Dodd (Braby 2015a); and an extant breeding population of *Hecatesia* sp. 'Amata' was found in 2008, 121 years after it was last recorded. The latter species was previously known only from a single specimen collected in the Kimberley in 1887 by W. W. Froggatt (M. F. Braby, unpublished data). Also, the subspecies *Suniana lascivia lasus* was rediscovered on Bathurst Island in 2009, more than 70 years after it was last recorded on the island; it was previously known only from three specimens collected in 1933 by T. G. Campbell (Waterhouse 1937a). These rediscoveries and the large number of recent additions to the fauna highlight how poorly known the region is in terms of its butterfly and diurnal moth biodiversity, but also the substantial field effort undertaken during the past 45 years; northern Australia has clearly been a frontier for field discovery.

An obvious question, then, is how many species remain to be discovered? Modelling the rate of species accumulation since 1970 suggests a predicted total richness of 175 taxa, which will be discovered progressively up to the year 2068 (Figure 3). The uncertainty of this estimate (95 per cent confidence interval) is ± 9 species. Thus, it is estimated that approximately 10 ± 9 more taxa remain to be found, which at the current discovery rate may take another 50 years. In other words, the modelling suggests the inventory is close to completion, with an estimated 94 per cent of the species recorded.

## Composition

In terms of composition, the fauna comprises seven taxa of Papilionidae (4 per cent), 31 Hesperiidae (19 per cent), 20 Pieridae (12 per cent), 33 Nymphalidae (20 per cent), 43 Lycaenidae (26 per cent) and 32 diurnal moths (19 per cent) representing nine different families. Thus, the lycaenids, nymphalids and hesperiids are the dominant components in the fauna. For the butterflies, these proportions compare well nationally at the species level, although the Hesperiidae (30 species, or 23 per cent) are underrepresented (hesperiids comprise 29 per cent of the Australian fauna), whereas the Nymphalidae (33 species, or 25 per cent) and Pieridae (19 species, or 14 per cent) are slightly overrepresented (nymphalids and pierids comprise 21 per cent and 9 per cent of the Australian fauna, respectively) (Braby 2016a).

## Endemism

Our third question was to ascertain which species are found only within the study region. Although no genera are endemic to the study region, 17 species (including seven undescribed diurnal moths) and 35 subspecies (including five undescribed butterflies) are endemic to the region (Table 5). The poorer taxonomic resolution among the moths, with a high proportion of undescribed taxa (seven of 31 species, or 23 per cent), reflects the greater level of taxonomic interest and attention given to the butterflies. This level of endemism is relatively low at the species level (10 per cent), but rises to 31 per cent if all taxa (i.e. species and subspecies) are considered. Thus, endemism is more pronounced below the level of species, suggesting that differentiation of the fauna has been relatively recent. In addition, six species (*Leptosia nina*, *Appias albina*, *Danaus genutia*, *Cethosia penthesilea*, *Phalanta phalantha* and *Deudorix smilis*) have—within Australian limits—their geographic ranges restricted or largely restricted to the study region. That is, these species do not occur elsewhere on the Australian continent, but do occur more widely in South-East Asia.

Most of the endemic species and subspecies have narrow ranges and are restricted to the Top End or to the Top End and Kimberley subregions. Indeed, of the 52 taxa endemic to the study region, 49 (15 species and 34 subspecies, or 96 per cent) occur in the Top End and, of these, 28 taxa (10 species and 18 subspecies, or 55 per cent) are found only in the Top End (Table 5). The three remaining endemic taxa that do not occur in the Top End are *Leptosia nina* ssp. 'Kimberley', *Nesolycaena caesia* and *Synemon* sp. 'Kimberley'—all of which are restricted to the Kimberley. No taxa are endemic to the western Gulf Country or the semi-arid and arid areas of the Northern Deserts. Despite the high variation in collecting effort between the subregions, available data indicate that the Top End has been important in the evolution of the butterfly and diurnal moth fauna, which supports the conclusion of Cracraft (1991) and Bowman et al. (2010) that the Top End is an area of endemism.

At the species level, all the endemics with one exception are restricted to woodland or open woodland habitats, especially those associated with sandstone or sandy soils derived from sandstone. The single exception is the diurnal moth *Ctimene* sp. 'Top End', which is restricted to monsoon forest. Braby (2008a) first noted this striking association between endemism and broad habitat type for butterflies across the AMT as a whole. He suggested this pattern may reflect differences in the origin and historical assembly of the biome, with the savannah fauna possibly comprising an older autochthonous element compared with those taxa associated with monsoon forest, which may be a more recent element from South-East Asia.

Table 5 Taxa (species and subspecies, including undescribed taxa) endemic to the study region and their occurrence within the four major subregions

| Taxon | Distribution | | | |
|---|---|---|---|---|
| | Kimberley | Top End | Northern Deserts | Western Gulf Country |
| **Endemic species** | | | | |
| *Mesodina gracillima* E. D. Edwards, 1987 | | ++ | | |
| *Taractrocera ilia* Waterhouse, 1932 | | ++ | | |
| *Taractrocera psammopetra* Braby, 2015 | + | + | | |
| *Acrodipsas decima* Miller & Lane, 2004 | | ++ | | |
| *Nesolycaena urumelia* (Tindale, 1922) | | + | | + |
| *Nesolycaena caesia* d'Apice & Miller, 1992 | ++ | | | |
| *Pseudosesia oberthuri* (Le Cerf, 1916) | | ++ | | |
| *Synemon* sp. 'Kimberley' | ++ | | | |
| *Synemon* sp. 'Roper River' | + | + | + | + |
| *Pollanisus* sp. 7 | | ++ | | |
| *Ctimene* sp. 'Top End' | | ++ | | |
| *Radinocera* sp. 'Sandstone' | + | + | | |
| *Idalima metasticta* Hampson, 1910 | | ++ | | |
| *Idalima leonora* (Doubleday, 1846) | + | + | | |
| *Idalima* sp. 'Arnhem Land' | | ++ | | |
| *Hecatesia* sp. 'Arnhem Land' | | ++ | | |
| *Cruria darwiniensis* (Butler, 1884) | | ++ | | |
| **Endemic subspecies** | | | | |
| *Protographium leosthenes geimbia* (Tindale, 1927) | | ++ | | |
| *Graphium eurypylus nyctimus* (Waterhouse & Lyell, 1914) | + | + | | + |
| *Papilio fuscus canopus* Westwood, 1842 | + | + | | |
| *Hasora hurama territorialis* Meyer et al., 2015 | | ++ | | |
| *Hesperilla crypsigramma* (Meyrick & Lower, 1902) ssp. 'Top End' | | ++ | | |
| *Borbo impar lavinia* (Waterhouse, 1932) | | ++ | | |
| *Taractrocera dolon diomedes* Waterhouse, 1933 | + | + | | + |
| *Ocybadistes flavovittatus vesta* (Waterhouse, 1932) | + | + | | + |
| *Ocybadistes walkeri olivia* Waterhouse, 1933 | + | + | | |
| *Ocybadistes hypomeloma vaga* (Waterhouse, 1932) | + | + | | + |
| *Suniana sunias sauda* Waterhouse, 1937 | | ++ | | |
| *Suniana lascivia lasus* Waterhouse, 1937 | | ++ | | |
| *Suniana lascivia larrakia* L. E. Couchman, 1951 | + | + | | + |
| *Cephrenes augiades* (C. Felder, 1860) ssp. 'Top End' | | ++ | | |
| *Leptosia nina* (Fabricius, 1793) ssp. 'Kimberley' | ++ | | | |
| *Delias aestiva aestiva* Butler, 1897 | | ++ | | |
| *Delias argenthona fragalactea* (Butler, 1869) | + | + | | + |
| *Libythea geoffroyi genia* Waterhouse, 1938 | + | + | | |
| *Danaus genutia alexis* (Waterhouse & Lyell, 1914) | + | + | | |
| *Euploea sylvester pelor* Doubleday, 1847 | + | + | | + |
| *Euploea darchia darchia* W. S. Macleay, 1826 | + | + | | |

| Taxon | Distribution | | | |
|---|---|---|---|---|
| | Kimberley | Top End | Northern Deserts | Western Gulf Country |
| *Euploea alcathoe enastri* Fenner, 1991 | | ++ | | |
| *Phalanta phalantha araca* (Waterhouse & Lyell, 1914) | | ++ | | |
| *Hypolimnas alimena darwinensis* Waterhouse & Lyell, 1914 | | ++ | | |
| *Hypocysta adiante antirius* Butler, 1868 | + | + | + | |
| *Hypochrysops apelles* (Fabricius, 1775) ssp. 'Top End' | | ++ | | |
| *Hypochrysops ignitus erythrina* (Waterhouse & Lyell, 1909) | + | + | | + |
| *Arhopala eupolis asopus* Waterhouse & Lyell, 1914 | + | + | | |
| *Arhopala micale* Boisduval, 1853 ssp. 'Top End' | | ++ | | |
| *Ogyris iphis doddi* (Waterhouse & Lyell, 1914) | | ++ | | |
| *Deudorix smilis dalyensis* (Le Souëf & Tindale, 1970) | | ++ | | |
| *Candalides margarita gilberti* Waterhouse, 1903 | + | + | | + |
| *Candalides geminus gagadju* Braby, 2017 | | ++ | | |
| *Nacaduba kurava felsina* Waterhouse & Lyell, 1914 | | ++ | | |
| *Agarista agricola biformis* Butler, 1884 | | ++ | | |
| **Total endemic species** | **6** | **15** | **1** | **2** |
| **Total endemic subspecies** | **17** | **34** | **1** | **9** |
| **Total endemic taxa** | **23** | **49** | **2** | **11** |

+ occurs in subregion

++ endemic to subregion

## Distribution

The fourth question was to estimate how each species was distributed in space across the landscape and to determine whether there were congruent patterns in geographical range. The geographic range maps suggest there are at least 10 broad patterns (Maps 9a–h and 10a–h), which are briefly summarised as follows.

### Very wide ranges

These species occur throughout all or most of the study region, from coastal areas of high rainfall to inland areas of low rainfall, and often extend to the arid areas of central Australia (Map 9a). They include *Papilio demoleus, Catopsilia pomona, C. scylla, Belenois java, Danaus petilia, Euploea corinna, Acraea andromacha, Hypolimnas bolina, Junonia villida, Ogyris zosine, O. amaryllis, Theclinesthes miskini, Nacaduba biocellata, Lampides boeticus, Zizina otis* and *Famegana alsulus*.

### Wide ranges

These species are widely distributed throughout the Kimberley, Top End and western Gulf Country (Maps 9b and 9c). Two subgroups can be distinguished: 1) those that extend well into the semi-arid zone (< 500 mm mean annual rainfall), either permanently or on a seasonal basis (*Eurema herla, E. hecabe, Junonia orithya, Charaxes sempronius, Melanitis leda, Catopyrops florinda, Catochrysops panormus, Zizeeria karsandra* and *Freyeria putli*) (Map 9b); and 2) those that are restricted to the higher rainfall areas, generally above 700 mm annually (*Hesperilla sexguttata, Pelopidas lyelli, Ocybadistes flavovittatus, Telicota colon, Cephrenes trichopepla, Eurema laeta, Cepora perimale, Delias argenthona, Tirumala hamata, Danaus affinis, Euploea sylvester, Acraea terpsicore, Junonia hedonia, Ypthima arctous, Candalides margarita, Jamides phaseli, Euchrysops cnejus* and *Zizula hylax*) (Map 9c).

## Kimberley and Top End

These species are restricted to the Kimberley and Top End and fall into three subgroups (Maps 9d–f): 1) those that are distributed continuously between the two areas across the Ord Arid Intrusion and Victoria River Drainage Basin barrier (*Graphium eurypylus*, *Papilio fuscus*, *Telicota augias*, *Elodina walkeri*, *Hypocysta adiante*, *Arhopala eupolis*, *Anthene lycaenoides*, *Prosotas dubiosa* and *Comocrus behri*) (Map 9d); 2) those that have disjunct populations on either side of the barrier (*Cressida cressida*, *Chaetocneme denitza*, *Taractrocera dolon*, *Eurema alitha*, *Appias paulina*, *Euploea darchia*, *Liphyra brassolis*, *Hypolycaena phorbas*, *Everes lacturnus*, *Birthana cleis*, *Dysphania numana* and *Idalima leonora*) (Map 9e); and 3) those that have patchy and highly disjunct ranges in the two areas (*Petrelaea tombugensis*, *Taractrocera psammopetra* and *Radinocera* sp. 'Sandstone') (Map 9f).

## Top End and western Gulf Country

These species are restricted to the Top End and western Gulf Country and include *Neohesperilla xiphiphora*, *Mycalesis perseus* and *Nesolycaena urumelia* (Map 9g).

## Top End (broad ranges)

These species are restricted to the Top End, but have relatively wide ranges throughout all or most of the subregion and include *Neohesperilla crocea*, *Ocybadistes walkeri*, *Cethosia penthesilea*, *Hypolimnas alimena*, *Mydosama sirius*, *Arhopala micale*, *Deudorix smilis*, *Anthene seltuttus*, *Idalima metasticta* and *Cruria darwiniensis* (Map 9h).

## Top End (narrow ranges)

These species are restricted to the Top End, but have fairly narrow ranges within the subregion. At least four subsets can be distinguished (Maps 10a–d): 1) those restricted to the northern half of the Top End (*Mesodina gracillima*, *Cephrenes augiades*, *Appias albina*, *Sahulana scintillata* and *Ctimene* sp. 'Top End') (Map 10a); 2) those restricted to the higher rainfall areas of the north-western corner of the Top End (*Hesperilla crypsigramma*, *Parnara amalia*, *Phalanta phalantha*, *Acrodipsas decima*, *Ogyris iphis*, *Nacaduba kurava*, *Idalima aethrias* and *Agarista agricola biformis*) (Map 10b); 3) those restricted to the Arnhem Land Plateau (*Protographium leosthenes*, *Taractrocera ilia*, *Candalides geminus*, *Idalima* sp. 'Arnhem Land' and *Hecatesia* sp. 'Arnhem Land') (Map 10c); and 4) those restricted to Gove Peninsula and adjacent islands (*Euploea alcathoe*, *Yoma sabina* and *Agarista agricola agricola*) (Map 10d).

## Kimberley

These species are restricted to the Kimberley and include *Leptosia nina*, *Nesolycaena caesia*, *Synemon* sp. 'Kimberley' and *Mimeusemia econia* (Map 10e).

## Low rainfall areas

These species breed in areas with below 900 mm mean annual rainfall and occur predominantly in the Northern Deserts of the semi-arid zone (< 700 mm) (Maps 10f and 10g). They include species that are very widespread (*Catopsilia pyranthe*, *Eurema smilax*, *Elodina padusa*, *Candalides delospila* and *Synemon wulwulam*) (Map 10f) and those that apparently have more restricted ranges (*Ogyris barnardi*, *Jalmenus icilius*, *Synemon* sp. 'Roper River' and *Hecatesia* sp. 'Amata') (Map 10g).

## Coastal/estuarine areas

These species are restricted to coastal/estuarine areas and include *Hasora hurama*, *Delias aestiva*, *Hypochrysops apelles* and *Theclinesthes sulpitius* (Map 10h).

## Idiosyncratic ranges

These species show no apparent pattern, and include *Papilio aegeus*, *Libythea geoffroyi*, *Danaus genutia* and Genus 1 sp. 'Sandstone'.

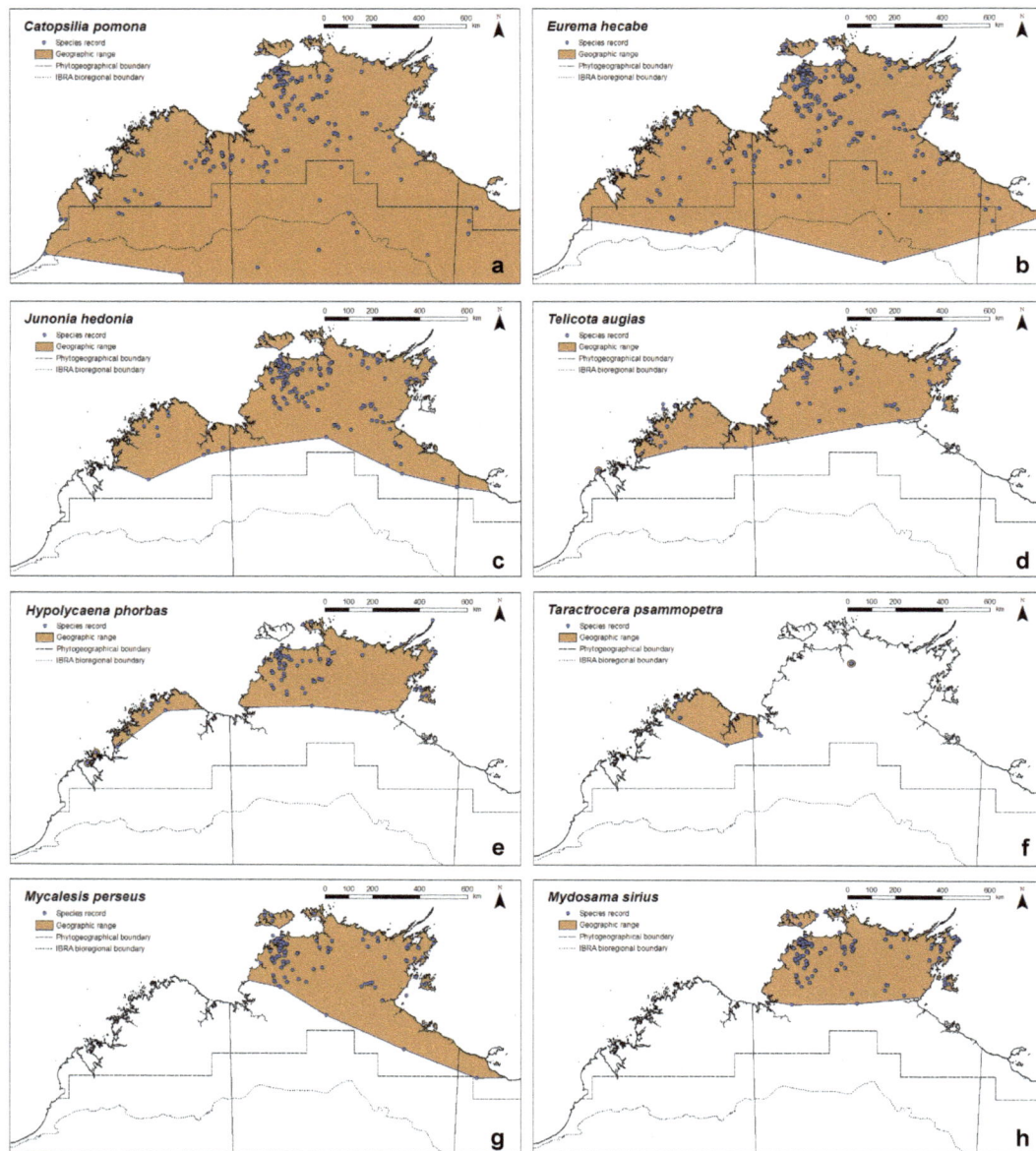

**Map 9 Geographic ranges of butterflies and diurnal moths in the study region, showing examples of species**

(a) very widely distributed (*Catopsilia pomona*); (b) widely distributed across the Kimberley, Top End and western Gulf Country, but also extending well into lower rainfall areas of the semi-arid zone (< 700 mm mean annual rainfall) (*Eurema hecabe*); (c) widely distributed across the Kimberley, Top End and western Gulf Country, but restricted to higher rainfall areas (> 700 mm) (*Junonia hedonia*); (d) restricted to the Kimberley and Top End, but distributed continuously between the two areas (*Telicota augias*); (e) restricted to the Kimberley and Top End, but with a disjunct distribution (*Hypolycaena phorbas*); (f) restricted to the Kimberley and Top End, but with a highly disjunct and limited range (*Taractrocera psammopetra*); (g) restricted to the Top End and western Gulf Country (*Mycalesis perseus*); (h) restricted to the Top End, but widespread in area (*Mydosama sirius*).

Source: Prepared by the authors.

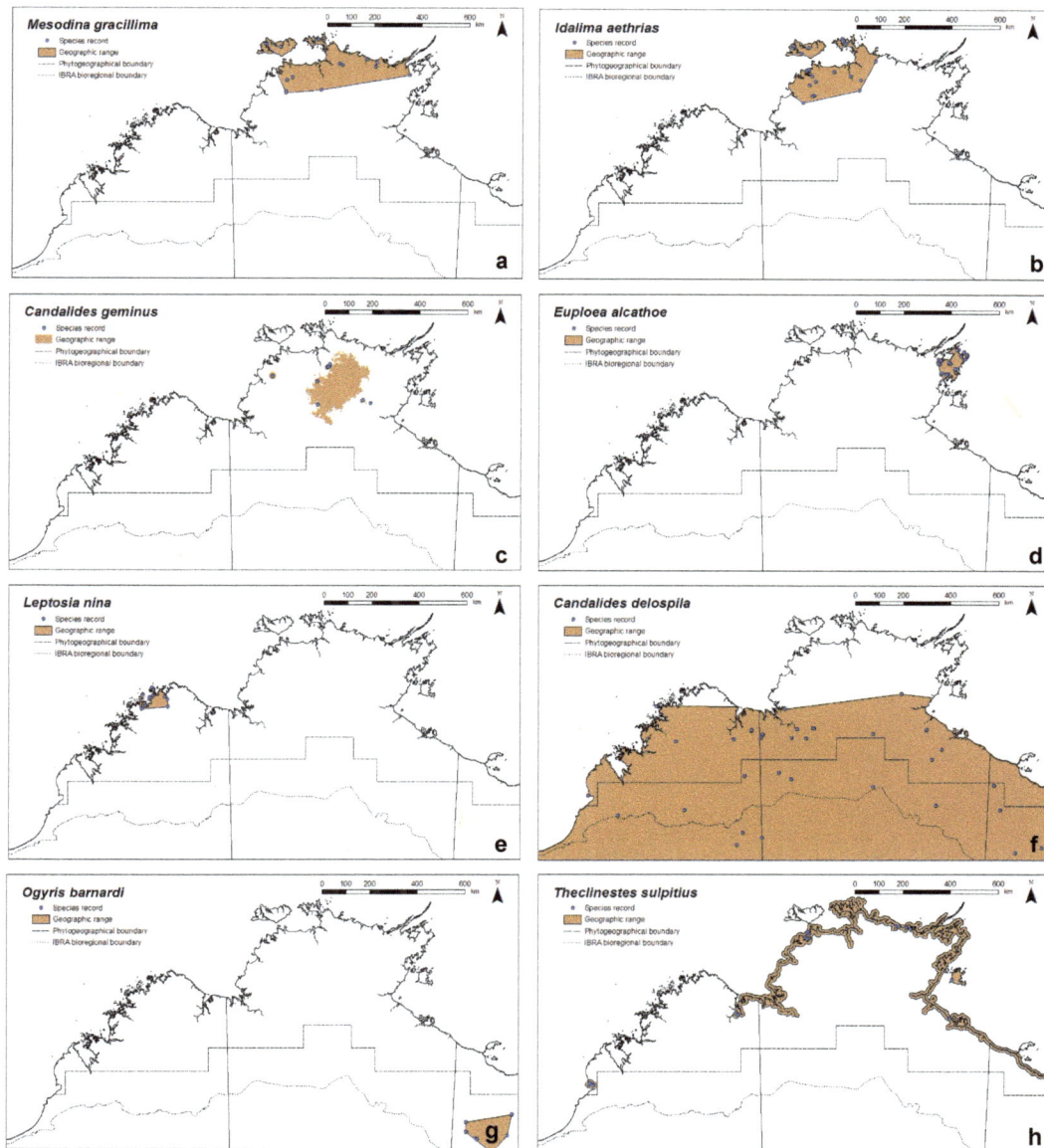

**Map 10 Geographic ranges of butterflies and diurnal moths in the study region, showing examples of species**

(a) restricted to the Top End, but limited to the northern half (*Mesodina gracillima*); (b) restricted to the Top End, but limited to the north-western corner (*Idalima aethrias*); (c) restricted to the Top End, but limited to or mainly limited to the Arnhem Land Plateau (*Candalides geminus*); (d) restricted to the Top End, but limited to Gove Peninsula (*Euploea alcathoe*); (e) restricted to the Kimberley (*Leptosia nina*); (f) restricted to low rainfall areas (< 900 mm mean annual rainfall), but widespread in area (*Candalides delospila*); (g) restricted to low rainfall areas of the semi-arid zone, but limited in range (*Ogyris barnardi*); (h) restricted to coastal/estuarine areas (*Theclinesthes sulpitius*).

Source: Prepared by the authors.

# Critical habitats and larval food plants

Of the 163 species recorded in the study region, 152 are resident or immigrant (see 'Breeding status' below). Of these resident and immigrant species, breeding habitats or suspected breeding habitats have been recorded for 141 (93 per cent). The 11 resident species for which the natural breeding habitat has not been recorded are: *Papilio aegeus*, *Acrodipsas myrmecophila*, *A. decima*, *Bindahara phocides*, *Sahulana scintillata*, *Synemon* sp. 'Kimberley', *Pollanisus* sp. 7, *Hestiochora xanthocoma*, *Euchromia creusa*, *Leucogonia cosmopis* and *Ipanica cornigera*. Some species breed in only one type of habitat, particularly those with very specialised ecological requirements in which the larvae feed on just one plant species or genus, whereas others breed in several different types of habitat and may utilise many different kinds of plants.

## Monsoon forest

More than 50 (31 per cent) species depend on various types of monsoon forest, of which 34 (21 per cent) are obligatorily restricted to these habitats in that they do not breed in other habitats. Among these monsoon forest specialists, at least one (*Cephrenes augiades*) is restricted to the wetter monsoon forests (i.e. evergreen monsoon vine forest) associated with permanent springs, creeks and other riparian areas, whereas 12 others occur only in drier monsoon forests (i.e. semi-deciduous monsoon vine thicket) on rocky outcrops, along seasonal gullies or on coastal laterite cliffs or dunes adjacent to the beach (*Protographium leosthenes*, *Graphium eurypylus*, *Papilio fuscus*, *Badamia exclamationis*, *Hasora chromus*, *Leptosia nina*, *Elodina walkeri*, *Appias albina*, *Libythea geoffroyi*, *Hypolycaena phorbas*, *Mimeusemia econia* and *M. centralis*). At least 14 species occur in both wet and dry monsoon forest (*Telicota augias*, *Appias paulina*, *Tirumala hamata*, *Phalanta phalantha*, *Euploea sylvester*, *E. darchia*, *Cethosia penthesilea*, *Arhopala micale*, *Deudorix smilis*, *Anthene lycaenoides*, *Petrelaea tombugensis*, *Nacaduba kurava*, *Ctimene* sp. 'Top End' and

*Dysphania numana*), although in the wetter monsoon forests they usually breed only along the drier edges of the forest. Seven species breed only in light gaps within, or along the edge of, the rainforest, especially in the ecotone between monsoon forest and woodland (*Suniana sunias*, *Euploea alcathoe*, *Yoma sabina*, *Hypolimnas alimena*, *Anthene seltuttus*, *Everes lacturnus* and *Agarista agricola*). These last mentioned species often also occur in mixed open forest with rainforest elements in the understorey, particularly in riparian areas.

Sixteen (10 per cent) species use monsoon forest facultatively—that is, they breed equally in both monsoon forest or mixed monsoon forest and savannah woodland (*Neohesperilla crocea*, *Pelopidas lyelli*, *Taractrocera ina*, *Cephrenes trichopepla*, *Catopsilia scylla*, *Eurema alitha*, *E. hecabe*, *Elodina padusa*, *Belenois java*, *Cepora perimale*, *Acraea andromacha*, *Charaxes sempronius*, *Melanitis leda*, *Candalides margarita*, *Prosotas dubiosa* and *Catopyrops florinda*). A further nine that are more typical of savannah woodland also breed along the edges of monsoon forest (*Telicota colon*, *Euploea corinna*, *Hypocysta adiante*, *Ypthima arctous*, *Arhopala eupolis*, *Candalides erinus*, *Jamides phaseli*, *Pseudosesia oberthuri* and *Radinocera vagata*).

Monsoon forests make up only a relatively small area (less than 1 per cent) of the landscape in the study region (Russell-Smith 1991; Russell-Smith and Bowman 1992; Russell-Smith et al. 1992), yet they support a disproportionally high number of butterfly and diurnal moth species (21–31 per cent), highlighting the high biodiversity value of this habitat type. These findings parallel similar trends for plants (Woinarski et al. 2005), but contrast markedly with those for ants (Andersen and Reichel 1994; Reichel and Andersen 1996; Andersen et al. 2006, 2007b). Interestingly, despite the high proportion of rainforest specialists, few taxa appear to be endemic to monsoon forests—a pattern also evident in ants (Reichel and Andersen 1996). Of the 34 taxa obligatorily associated with monsoon forest, only one species (*Ctimene* sp. 'Top End') and 16 subspecies (*Protographium leosthenes geimbia*, *Graphium eurypylus nyctimus*, *Papilio fuscus canopus*, *Suniana sunias sauda*, *Cephrenes augiades* ssp. 'Top End', *Leptosia nina* ssp. 'Kimberley', *Libythea geoffroyi genia*, *Euploea sylvester pelor*, *E. darchia darchia*, *E. alcathoe enastri*, *Phalanta phalantha araca*, *Hypolimnas alimena darwinensis*, *Arhopala micale* ssp. 'Top End', *Deudorix smilis dalyensis*, *Nacaduba kurava felsina* and *Agarista agricola biformis*) are endemic to monsoon forests within the study region. Four monsoon forest specialists (*Leptosia nina*, *Cethosia penthesilea*, *Phalanta phalantha* and *Deudorix smilis*) represent the only known Australian occurrences of predominantly South-East Asian species. A fifth monsoon specialist, the predominantly South-East Asian pierid *Appias albina*, also has its main occurrence within Australia in the study region.

## Savannah woodland

At least 88 (54 per cent) species occur in savannah woodland, but a number of these breed in other habitats. About 60 (37 per cent) species are found only in savannah woodland in the broad sense—that is, eucalypt heathy woodland, open woodland, *Acacia* woodland, riparian woodland and tropical grassland. Some of these species favour open disturbed areas for breeding (*Papilio demoleus*, *Acraea terpsicore*, *Hypolimnas bolina*, *H. misippus*, *Theclinesthes miskini*, *Lampides boeticus*, *Zizeeria karsandra*, *Euchrysops cnejus*, *Freyeria putli* and *Periopta diversa*). Eleven of these savannah species are restricted to habitats associated with laterite or sandstone outcrops or sandy soils derived from sandstone/laterite, often with a heathy understorey or a hummock (spinifex) grass understorey (*Hesperilla crypsigramma*, *Mesodina gracillima*, *Proeidosa polysema*, *Candalides geminus*, *C. delospila*, *Nesolycaena urumelia*, *N. caesia*, Genus 1 sp. 'Sandstone', *Idalima* sp. 'Arnhem Land', *Hecatesia* sp. 'Arnhem Land' and *Hecatesia* sp. 'Amata').

## Other habitats

At least nine species are regularly associated with paperbark woodland, paperbark swampland or mixed paperbark–pandanus swampland and other damp areas (*Hesperilla sexguttata*, *Parnara amalia*, *Suniana lascivia*, *Danaus affinis*, *D. genutia*, *Junonia hedonia*, *Mydosama sirius*, *Zizula hylax* and *Cruria darwiniensis*)—often adjacent to evergreen monsoon vine forest. Another species (*Borbo impar*) inhabits floodplain wetlands and swamps where the native larval food plant grows in standing water, and at least three others (*Papilio demoleus*, *Junonia hedonia* and *Mydosama sirius*) breed along the margins of these habitats; most of these species also breed in other habitats. Three species are restricted to the taller closed forest mangrove communities in the Top End (*Hasora hurama*, *Delias aestiva* and *Hypochrysops apelles*). Interestingly, on Cape York Peninsula, these species (or their nearest relatives) are typically associated with tropical forest, and Braby (2012c) hypothesised that historical contraction of rainforest habitat in the Top End during periods of aridification in the Miocene–Pleistocene may have led to pronounced habitat/food plant shifts, ultimately contributing to the evolution and adaptation of these insects to nonrainforest habitats, such as mangroves.

Other species occur in very specialised habitats. For example, *Radinocera* sp. 'Sandstone' breeds only on its food plant growing among foot-slope boulders at the base of sandstone escarpments. Open sandstone pavements, rock ledges and the base of rock overhangs are interesting habitats because they support 'resurrection' grasses (*Micraira* spp.), the only plant genus that has adapted and radiated to any extent in this very specialised habitat (Lazarides 2005); in turn, these grasses support two ecologically specialised butterflies endemic to the study region (*Taractrocera ilia* and *T. psammopetra*) (Braby and Zwick 2015). *Theclinesthes sulpitius* is the only species that breeds in coastal saltmarsh—a specialised habitat that is highly saline and periodically inundated by incoming tides.

## Larval food plants

A complete list of known and putative larval food plants is provided in Appendix I. A total of 373 known butterfly/diurnal moth–plant associations, and a further 55 putative associations that require confirmation, have been recorded from the study region.

The putative associations are based on records from adjacent areas (e.g. Queensland), on the spatial distribution of the relevant species or on other evidence (e.g. female pre-oviposition behaviour) within the study region.

While larval food plants have been recorded for most species in the study region, data are still lacking for 26 species, and for nine species only fragmentary information, such as associations with introduced weedy taxa, is available. Notable taxa for which there are currently no published native food plant associations are the grass-feeding hesperiids, satyrines and sun-moths (*Neohesperilla senta*, *N. crocea*, *N. xanthomera*, *Parnara amalia*, *Borbo cinnara*, *Taractrocera dolon*, *Ocybadistes flavovittata*, *O. walkeri*, *O. hypomeloma*, *Suniana sunias*, *Mycalesis perseus*, *Ypthima arctous* and *Synemon* sp. 'Kimberley'), the legume-feeding pierids and lycaenids (*Catopsilia pyranthe*, *C. pomona*, *Eurema laeta*, *E. smilax*, *Zizina otis*, *Zizula hylax* and *Everes lacturnus*), three nymphalids (*Danaus chrysippus*, *Junonia villida* and *Hypolimnas bolina*), several other lycaenids (*Bindahara phocides*, *Petrelaea tombugensis* and *Sahulana scintillata*) and a number of diurnal moths (*Pollanisus* sp. 7, *Hestiochora xanthocoma*, *Ctimene* sp. 'Top End', *Euchromia creusa*, Genus 1 sp. 'Sandstone', *Leucogonia cosmopis*, *Ipanica cornigera*, *Mimeusemia econia* and *M. centralis*).

# Patterns of breeding and seasonal abundance

## Breeding status

Of the 166 taxa, 151 (91 per cent) are resident within the study region (i.e. breeding regularly, with permanently established populations), three (2 per cent) are immigrant (i.e. breeding irregularly, with temporary populations) and 12 (7 per cent) are vagrant or infrequent visitors (i.e. not breeding, with non-resident populations). The three immigrants are *Badamia exclamationis*, *Nacaduba biocellata* and *Lampides boeticus*, and these species appear to enter and vacate the study region on a seasonal basis. The 12 vagrant or infrequent visitor taxa are listed in Table 6. These species rarely enter the study region—and usually in very small numbers—or they may enter on a more regular basis. Two of these species (*Borbo cinnara* and *Danaus chrysippus*) may actually be rare immigrants in that breeding possibly occurs, but the colonising populations are temporary and they fail to establish permanently; however, only circumstantial evidence is currently available on the breeding status of these species. The majority of the vagrant or infrequent visitor species belong to the family Nymphalidae (Table 6).

Table 6 List of taxa that are vagrants and/or infrequent visitors

| Scientific name | Common name | Family |
| --- | --- | --- |
| *Borbo cinnara* (Wallace, 1866) | Rice Swift | Hesperiidae |
| *Eurema brigitta australis* (Wallace, 1867) | No-brand Grass-yellow | Pieridae |
| *Appias albina infuscata* Fruhstorfer, 1910 | White Albatross | Pieridae |
| *Delias mysis mysis* (Fabricius, 1775) | Red-banded Jezebel | Pieridae |
| *Danaus chrysippus cratippus* (C. Felder, 1860) | Plain Tiger | Nymphalidae |
| *Danaus plexippus* (Linnaeus, 1758) | Monarch | Nymphalidae |
| *Euploea climena macleari* Butler, 1887 | Climena Crow | Nymphalidae |
| *Vanessa kershawi* (McCoy, 1868) | Australian Painted Lady | Nymphalidae |
| *Vanessa itea* (Fabricius, 1775) | Yellow Admiral | Nymphalidae |
| *Junonia erigone* (Cramer, [1775]) | Northern Argus | Nymphalidae |
| *Hypolimnas anomala albula* (Wallace, 1869) | Crow Eggfly | Nymphalidae |
| *Alcides metaurus* (Hopffer, 1856) | Zodiac Moth | Uraniidae |

Note: These taxa probably do not breed in the study region (i.e. are non-resident).

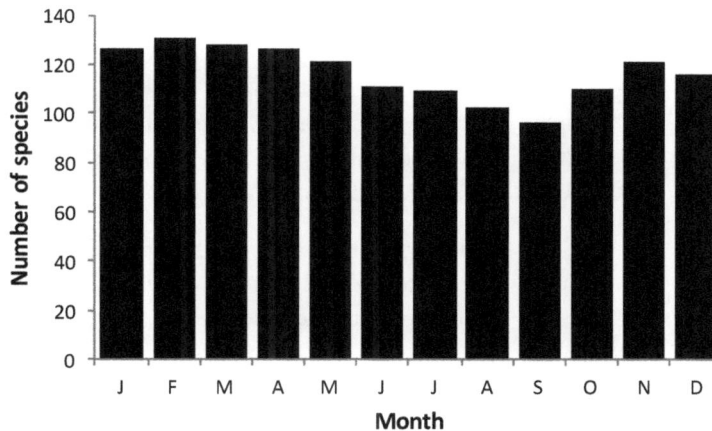

Figure 4 Temporal distribution of species richness in the study region

Note: Data are based on monthly flight period records for each of the 152 resident and immigrant species (i.e. those that breed in the region).

Source: Prepared by the authors.

## Seasonality

A summary of the monthly incidence of the resident and immigrant species in the study region is shown in Figure 4. Although the phenology data for some species are incomplete or based on limited sampling, available records suggest a seasonal component in species richness. Species richness is apparently highest during the mid to late wet season (January–April), with a peak in February, steadily declining as the dry season progresses and is lowest in the late dry season (September), before increasing slightly during the 'build-up' (October–December). Although many species (at least 64) have been recorded in every month of the year, some are strictly seasonal and appear for just a few months (see below). For example, many species of Hesperiidae and most Agaristinae fly mainly during the wet season, but a few Lycaenidae and Castniidae occur predominantly during the dry season (*Acrodipsas myrmecophila*, *Ogyris barnardi*, *Sahulana scintillata*, *Theclinesthes albocinctus*, *Synemon wulwulam* and *Synemon* sp. 'Roper River').

Available data on the breeding phenology (i.e. temporal incidence of the immature stages) and seasonal abundance of adults indicate that about half of the region's species breed, or are suspected to breed, continuously throughout the year, whereas the other half breed seasonally. The continuous breeders have multiple generations during the year and

they achieve this in one of several ways. First, some species specialise on larval food plants that are perennial and thus available all year round (e.g. non-deciduous vines, mistletoes, palms or grasses that grow where conditions are permanently moist and buffered from climatic extremes). Second, other species (e.g. some lycaenids) feed on a range of plants, but switch their food plants on a seasonal basis according to the availability of new leaf shoots, flowers or fruits. Some ecologically specialised species (e.g. *Hasora hurama*, *H. chromus* and *Deudorix smilis*) that feed only on a single plant species have a similar strategy but appear to switch between individual plants that are seasonally available. Third, a strategy adopted by some species (e.g. several satyrines and *Acraea* spp.) is to breed throughout the year, but as the dry season progresses their reproductive activity declines and/or they contract to moist refuges. Fourth, at least four species (*Eurema smilax*, *E. hecabe*, *Belenois java* and *Zizeeria karsandra*) are highly nomadic and/or migratory, with populations shifting across a wide area within the study region, presumably tracking resources that are ephemeral or unpredictable in space and time. Several other highly mobile species (e.g. *Papilio demoleus* and *Catopsilia* spp.) may adopt this strategy, but it is not clear whether they breed all year round. Another—and perhaps the most bizarre—tactic of a continuous breeder is displayed by *Taractrocera ilia* (Braby and Zwick 2015). The adults emerge and fly all year round, but the larvae (all instars) enter diapause during the dry season when their food plants (resurrection grasses) become completely desiccated. However, diapause in final instar larvae is frequently terminated before the onset of the wet season such that the larvae pupate and emerge as adults shortly afterwards in the mid to late dry season. These adults then mate and the females lay eggs on the dry plants; the eggs develop and hatch soon after, but the first instar larvae then enter diapause.

Species that breed seasonally usually have only one or a few generations annually, or one generation followed by a partial second generation. The larvae of these species generally specialise on plants that are either short-lived annuals or seasonal perennials (e.g. deciduous vines) or they exploit plant parts (e.g. new leaf shoots) that are only available for a few months of the year, typically during the wet season. These species have developed an array of life history strategies to survive the non-breeding season when the food plants are not available, which is usually during the long dry season. We identified at least four strategies. First, several grass-feeding hesperiids (e.g. *Taractrocera* spp., *Neohesperilla* spp. and *Proeidosa polysema*) and possibly a few lycaenids (e.g. *Zizula hylax*, *Everes lacturnus*, *Euchrysops cnejus* and *Freyeria putli*) remain, or are suspected to remain, in larval diapause. Second, several swallowtails (e.g. *Protographium leosthenes*, *Graphium eurypylus* and *Papilio fuscus*), at least two lycaenids (*Nesolycaena urumelia* and *N. caesia*) and most of the agaristine moths remain in pupal diapause, often for many months or even years. The agaristines typically pupate underground or deep under bark at the base of tree trunks, where they are also protected from dry season fires. Four other species (*Appias albina*, *Delias aestiva*, *Cethosia penthesilea* and *Phalanta phalantha*) are suspected to have capacity for pupal diapause during periods of food shortfall, but to our knowledge this has not been confirmed. Third, a number of pierids (e.g. *Catopsilia pomona*, *Eurema laeta* and *E. herla*) and nymphalids (most danaines, *Junonia* spp., *Yoma sabina*, *Hypolimnas* spp., *Melanitis leda* and *Mycalesis perseus*) remain, or are suspected to remain, in adult reproductive diapause. The adults of these species contract to moist refuges, where they remain relatively inactive for many months, sometimes in immense numbers (e.g. *Euploea corinna*). Finally, three immigrant species (*Badamia exclamationis*, *Nacaduba biocellata* and *Lampides boeticus*) enter the study region on a seasonal basis and appear to breed only for a short period before vacating.

Although some caution must be exercised when interpreting patterns of seasonal changes in relative abundance based on collection and observation records, our data suggest that, where sufficient records are available, at least six broad patterns may be discerned (Figures 5a–h), as follows.

## Early wet season

These species peak in adult abundance during the 'build-up' of the early wet season when conditions are very humid and hot—typically in November and December, but sometimes in late October or early January, depending on the arrival of the pre-monsoon storms (Figure 5a). All of these species are seasonal breeders and include three swallowtails (*Protographium leosthenes*, *Graphium eurypylus* and *Papilio fuscus*), several agaristines (*Periopta* spp., *Radinocera* spp., *Idalima* spp. and *Cruria donowani*) and the undescribed noctuid Genus 1 sp. 'Sandstone'.

## Mid wet season

These species peak in adult abundance during the monsoon rains of the mid wet season— typically in January and February, but sometimes in late December or early March, depending on the timing of the monsoon(s) (Figures 5b and 5c). Continuous breeders that show this pattern are *Hasora chromus*, *Taractrocera ilia*, *Telicota augias*, *Appias paulina* and *Acraea terpsicore*. Seasonal breeders that show this pattern include *Badamia exclamationis*, *Taractrocera ina*, *Neohesperilla xiphiphora*, *Appias albina*, *Synemon phaeoptila* and *Ipanica cornigera*.

## Late wet season – early dry season

These species peak in adult abundance during the late wet season and/or early dry season (March–May) (Figures 5d and 5e). A large number of species show this pattern,

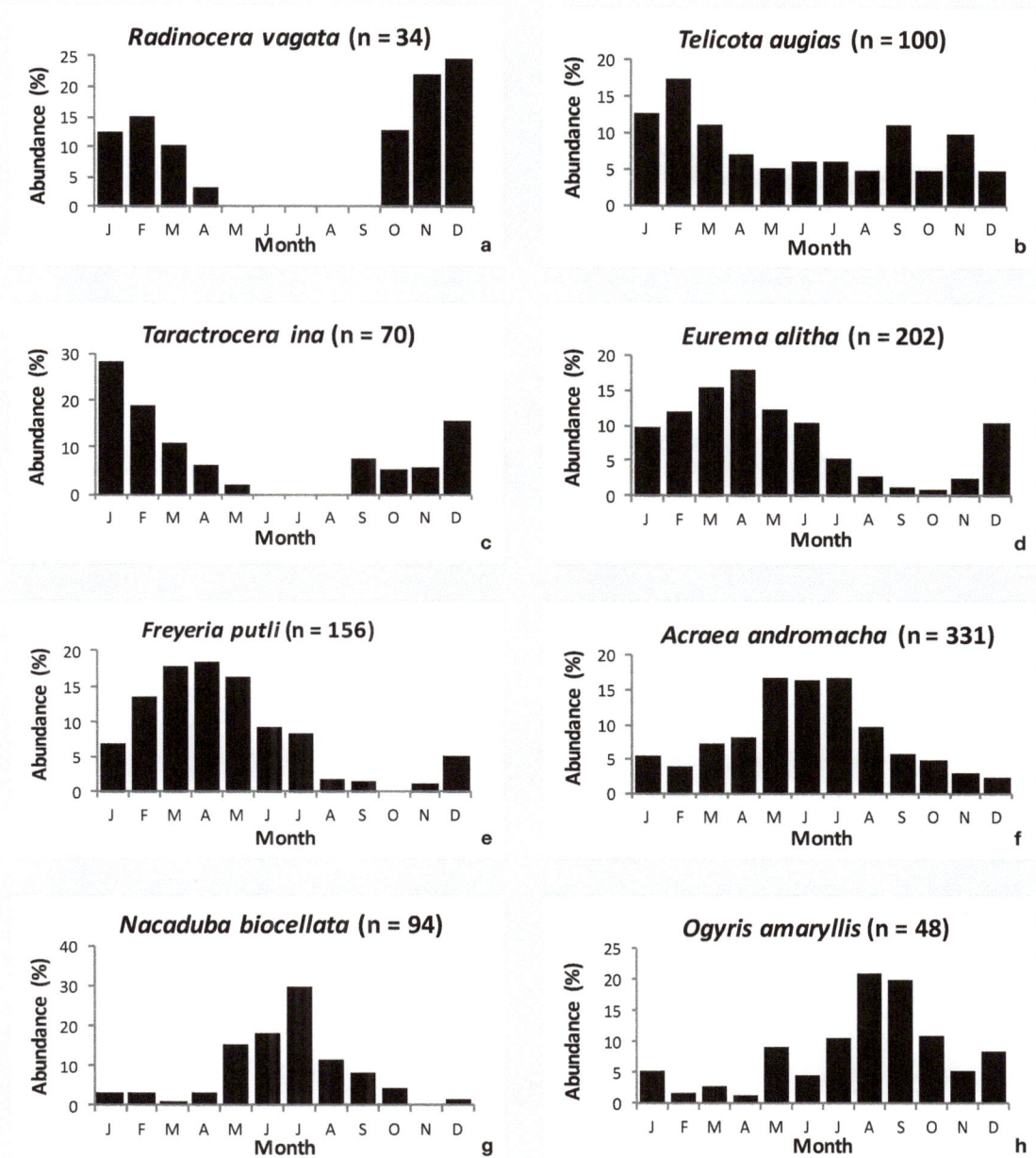

**Figure 5 Seasonal variation in adult abundance of butterflies and diurnal moths in the study region, showing examples of species peaking at different times of the year**

(a) early wet season (*Radinocera vagata*, a seasonal breeder); (b, c) mid wet season (*Telicota augias*, a continuous breeder, and *Taractrocera ina*, a seasonal breeder); (d, e) late wet season – early dry season (*Eurema alitha*, a continuous breeder, and *Freyeria putli*, a seasonal breeder); (f, g) mid dry season (*Acraea andromacha*, a continuous breeder, and *Nacaduba biocellata*, a seasonal breeder); (h) late dry season (*Ogyris amaryllis*, a continuous breeder).

Note: The relative abundance of each species (per cent) is based on the number of temporal records, defined as the occurrence on a particular date (time) at a given site (space), for the study region.

including both continuous breeders or likely continuous breeders (e.g. *Papilio demoleus, Cressida cressida, Borbo impar, Parnara amalia, Pelopidas lyelli, Ocybadistes flavovittatus, Telicota colon, Cephrenes trichopepla, Catopsilia pyranthe, C. scylla, Eurema alitha, E. hecabe, Cepora perimale, Mydosama sirius, Catochrysops panormus, Jamides phaseli, Famegana alsulus, Zizina otis* and *Zizeeria karsandra*) and seasonal or putative seasonal breeders (e.g. *Hesperilla crypsigramma, Taractrocera anisomorpha, T. dolon, Eurema herla, Libythea geoffroyi, Junonia villida, J. orithya, Hypolimnas misippus, H. bolina, Melanitis leda, Nesolycaena urumelia, Everes lacturnus, Euchrysops cnejus, Freyeria putli* and *Synemon wulwulam*).

### Mid dry season

These species peak in adult abundance during the mid dry season (May/June and July), when conditions are dry and cool, especially at night (Figures 5f and 5g). This group includes both continuous or putative continuous breeders (e.g. *Ocybadistes walkeri, Suniana* spp., *Eurema smilax, Elodina padusa, Acraea andromacha, Ypthima arctous, Arhopala micale, Deudorix smilis, Candalides margarita, C. erinus* and *Catopyrops florinda*) and seasonal breeders (e.g. *Eurema laeta, Delias aestiva, Cethosia penthesilea, Junonia hedonia, Mycalesis perseus, Sahulana scintillata, Nacaduba biocellata, Zizula hylax* and *Lampides boeticus*).

### Late dry season

These species peak in adult abundance during the late dry season, when conditions are dry and hot to very hot—typically during the period August–October (Figure 5h). All of these species are continuous breeders and include *Belenois java, Charaxes sempronius* and several taxa associated with mistletoes (*Delias argenthona, Ogyris* spp. and *Comocrus behri*).

### Nonseasonal

These species show little seasonal variation in abundance or no clear or consistent pattern throughout the year (e.g. *Catopsilia pomona, Elodina walkeri, Tirumala hamata, Danaus affinis, Euploea sylvester, E. darchia,* *Hypolimnas alimena, Liphyra brassolis, Arhopala eupolis, Hypochrysops ignitus, Anthene seltuttus, A. lycaenoides* and *Dysphania numana*).

## Conservation status

Of the 166 taxa assessed for their conservation status according to the IUCN Red List criteria, 14 were ranked as Not Applicable (NA) and therefore were not evaluated. For the remaining 152 taxa, the data for 16 were inadequate to make an evaluation on the extent of extinction risk and thus were categorised as Data Deficient (DD). For the 136 taxa for which there were adequate data, one species was ranked as Vulnerable (VU), four as Near Threatened (NT) and 131 were assessed as Least Concern (LC) (Table 7). No species from the study region are known to have become extinct since European settlement, although one species (*Pollanisus* sp. 7) has not been detected since it was first recorded 110 years ago. However, the paucity of records of *Pollanisus* sp. 7 probably reflects a combination of lack of targeted survey, seasonality, likely occurrence in remote areas and the inconspicuous nature of adults of this species. The putative larval food plant (*Pipturus argenteus*) occurs mainly in eastern Arnhem Land, with a very sporadic occurrence closer to Darwin, in the Northern Territory.

The threatened (VU), Near Threatened (NT) and DD taxa are also listed in Appendix II, which provides a summary of the criteria, justifications and actions needed. The taxa of most conservation concern are *Ogyris iphis doddi, Euploea alcathoe enastri, Hypochrysops apelles* ssp. 'Top End', *Idalima* sp. 'Arnhem Land' and *Hecatesia* sp. 'Arnhem Land'—all of which are endemic to the Top End. However, at least nine of the DD species (*Hesperilla crypsigramma* ssp. 'Top End', *Suniana lascivia lasus, Acrodipsas myrmecophila, A. decima, Ogyris barnardi barnardi, Nesolycaena caesia, Theclinesthes albocinctus, Pollanisus* sp. 7 and *Agarista agricola agricola*) are of conservation interest because they may qualify as Near Threatened (NT) once adequate data are available. That is, these taxa are possibly threatened (Appendix II).

Table 7 Summary of conservation status of taxa based on IUCN Red List categories and criteria

| Red List category | Number of taxa | Taxa |
|---|---:|---|
| Vulnerable (VU) | 1 | †*Ogyris iphis doddi* |
| Near Threatened (NT) | 4 | †*Euploea alcathoe enastri*, †*Hypochrysops apelles* ssp. 'Top End', †*Idalima* sp. 'Arnhem Land', †*Hecatesia* sp. 'Arnhem Land' |
| Least Concern (LC) | 131 | |
| Data Deficient (DD) | 16 | *Badamia exclamationis*, †*Hesperilla crypsigramma* ssp. 'Top End', †*Suniana lascivia lasus*, *Acrodipsas myrmecophila*, †*A. decima*, *Ogyris barnardi barnardi*, *Bindahara phocides*, †*Nesolycaena caesia*, *Petrelaea tombugensis*, *Theclinesthes albocinctus*, †*Synemon* sp. 'Kimberley', †*Pollanisus* sp. 7, *Euchromia creusa*, *Mimeusemia econia*, *M. centralis*, *Agarista agricola agricola* |
| Not Applicable (NA) | 14 | *Papilio anactus*, *Borbo cinnara*, *Eurema brigitta australis*, *Appias albina infuscata*, *Delias mysis mysis*, *Danaus chrysippus cratippus*, *Danaus plexippus*, *Euploea climena macleari*, *Acraea terpsicore*, *Vanessa kershawi*, *Vanessa itea*, *Junonia erigone*, *Hypolimnas anomala albula*, *Alcides metaurus* |
| Total | 166 | |

† taxa endemic to the study region

Note: See Appendix II for justifications and actions needed for threatened (VU), NT and DD taxa.
Taxa assessed as LC are not listed.

Very few species have restricted distributions (Appendix II). Of the 34 resident and immigrant taxa identified as having small geographic range sizes (< 40,000 sq km), nine (*Protographium leosthenes geimbia*, *Taractrocera ilia*, *Leptosia nina*, *Hypochrysops apelles*, *Ogyris barnardi barnardi*, *Euploea alcathoe enastri*, *Yoma sabina*, *Nesolycaena caesia* and *Agarista agricola biformis*) had an estimated EOO of less than the threshold of 20,000 sq km to qualify for a possible Red List category under criterion B1. A further 17 taxa (*Badamia exclamationis*, *Suniana lascivia lasus*, *Acrodipsas myrmecophila*, *A. decima*, *Ogyris iphis doddi*, *Bindahara phocides*, *Petrelaea tombugensis*, *Theclinesthes albocinctus*, *Synemon* sp. 'Kimberley', *Pollanisus* sp. 7, *Euchromia creusa*, *Idalima* sp. 'Arnhem Land', *Radinocera* sp. 'Sandstone', *Hecatesia* sp. 'Arnhem Land', *Mimeusemia econia*, *M. centralis* and *Agarista agricola agricola*) are known only from one or two spatial records or breeding sites (i.e. their AOO is likely to be < 2,000 sq km) and thus they may qualify for a Red List category under criterion B2. However, most of these taxa were evaluated to be LC or DD when two of three other subcriteria were considered—specifically, the number of locations and evidence of decline (Appendix II).

In terms of representation in the NRS, most taxa are adequately represented to varying degrees. However, 14 taxa are poorly represented in the conservation reserve system: *Suniana lascivia lasus*, *Acrodipsas myrmecophila*, *A. decima*, *Ogyris barnardi barnardi*, *Jalmenus icilius* and *Theclinesthes albocinctus* are currently not known to be represented in any conservation reserve. *Ogyris oroetes oroetes*, *O. iphis doddi*, *Nesolycaena caesia*, *Petrelaea tombugensis*, *Nacaduba kurava felsina*, *Theclinesthes sulpitius*, *Synemon* sp. 'Roper River' and *Agarista agricola agricola*, which have been recorded from two or more locations, are each currently known from only a single conservation reserve. Of these 14 taxa, four (*Ogyris barnardi barnardi*, *O. oroetes oroetes*, *Jalmenus icilius* and *Synemon* sp. 'Roper River') occur predominantly in the semi-arid zone (in *Acacia* low open woodland or eucalypt open woodland habitats) of the Northern Deserts and western Gulf Country, while two (*Suniana lascivia lasus* and *Ogyris iphis doddi*) occur on the Tiwi Islands.

Overall, our assessment indicates that the fauna is in reasonably good health; no species are known to have become extinct and there are few threatened taxa. The relatively low frequency of threatened taxa in large part reflects two standout features of northern Australia: 1) the landscapes it supports are still relatively intact; and 2) many species have large geographic range sizes across the study region, particularly those associated with savannah woodland. Although the region is not immune from threats, the geographic ranges of most species are large enough to buffer against local impacts. However, as Woinarski et al. (2005) point out, having species with such widespread distributions over a relatively uniform landscape does not necessarily imply that loss of any local area/population may be inconsequential. Many of the ecological processes underpinning the health and heterogeneity of the landscape, such as fire and flooding, must also operate over vast areas, such that loss of a subset of the range and disruption of natural processes may have far-reaching (and unforeseen) consequences.

For species that have relatively small geographic range sizes, the key threatening process likely to adversely affect butterfly populations at present is decline of ecological resources (larval food plants and/or habitat) through inappropriate fire regimes, especially an increase in the frequency and extent of dry season burns. At present, the fire frequency in many tropical savannahs of northern Australia, particularly in the higher rainfall areas, is far too frequent, such that relatively long unburnt habitat (more than five years) is now rare in the landscape (less than 3 per cent of the total area) (Andersen et al. 2005, 2012; Russell-Smith and Yates 2007). Habitat loss and fragmentation are also a concern, particularly for *Ogyris iphis doddi* and *Hypochrysops apelles* ssp. 'Arnhem Land'. However, other threats may become significant in future, such as the ongoing invasion of grassy weeds—particularly gamba grass (*Andropogon gayanus*) and mission grasses (*Cenchrus pedicellatus* and *C. polystachios*)—and the concomitant grass–fire cycle (Rossiter et al. 2003; Douglas and Setterfield 2005; Setterfield et al. 2010, 2013), especially for specialist species inhabiting savannah woodland, riparian woodland/open forest and the edges of riparian monsoon forest. The African gamba grass currently poses the greatest invasion threat due to substantial changes in community structure, fuel loads and impacts on fire regimes, and has been listed as a key threatening process under Commonwealth legislation. Habitat loss due to the expanding pastoral, agricultural/horticultural and mining industries is likely to have a detrimental effect on savannah woodland specialists in future (Garnett et al. 2010), and impacts on butterfly populations (their abundance and/or occupancy) will need to be carefully monitored. The long-term viability of patches of monsoon forest and the disproportionally rich butterfly assemblages they support may ultimately depend on reducing the frequency and intensity of fire in the surrounding matrix (Bowman 2000), as well as maintaining connectivity; any loss or decline of essential pollinators and seed dispersers or impediments that reduce their ability to move effectively between patches may have detrimental consequences (Russell-Smith and Bowman 1992).

The effects of global climate change and the response or resilience of tropical butterflies to such change are presently unclear. Predictions are that atmospheric carbon dioxide concentrations will continue to increase, leading to promotion of woody vegetation (monsoon forest) over grass (savannah woodland); sea level will rise, which will most likely impact the extent of coastal floodplains; and cyclones will become more intense, which will likely render patches of monsoon forest more susceptible to weed invasion. The impacts of increased temperature and rainfall, however, are less certain (Garnett et al. 2010). The recent catastrophic loss of large areas of mangrove communities in the Gulf of Carpentaria—due possibly to a combination of above-average temperatures and successive poor wet seasons—is indicative of the rapid and extensive environmental change that may be triggered more frequently in the future by escalating global climate change.

# Conclusion

4

Collectively, the Kimberley, Top End, Northern Deserts and western Gulf Country make up a vast and remote area of tropical northern Australia (about 16 per cent of the continent). Until now, this region was arguably the most poorly known area of the Australian continent for butterflies and diurnal moths in terms of basic natural history. Thus, a major goal of this atlas was to address this knowledge gap by compiling a detailed inventory of the species known to occur in the region based on review of the scientific literature, examination of material in museum collections, field surveys and incidental observations.

Our data indicate that 166 taxa representing 163 species (132 butterflies and 31 diurnal moths) have been recorded from the study region, of which 151 (91 per cent) are resident, three (2 per cent) are immigrant and 12 (7 per cent) are vagrant or infrequent visitors. Overall, the fauna has a relatively low level of endemism: 17 species (10 per cent, including seven undescribed diurnal moths) and 35 subspecies (21 per cent, including six undescribed butterflies) are endemic to the region. Most of the endemic species are restricted to the higher rainfall areas of the Top End, or the Top End and northern Kimberley, where they occur predominantly in savannah landscapes, especially woodland or open woodland associated with sandstone. Within the study region, the Top End is substantially richer than the other subregions, with 150 species (93 per cent of the total fauna), compared with the Kimberley (105 species), western Gulf Country (82 species) and Northern Deserts (53 species). These broad patterns of species richness and endemism suggest the Top End has been important in the evolution and historical assembly of the butterfly and diurnal moth fauna.

Within the Top End, available data indicate the north-western corner, which includes the Arnhem Land Plateau, is a biodiversity 'hotspot' based on the concentration of mesic-adapted taxa or lineages with restricted geographic ranges. The Arnhem Land Plateau supports five taxa (*Taractrocera ilia*, *Protographium leosthenes geimbia*, *Candalides geminus gagadju*, *Idalima* sp. 'Arnhem Land' and *Hecatesia* sp. 'Arnhem Land') that are endemic to it, plus a further two species (*Taractrocera psammopetra* and *Radinocera* sp. 'Sandstone') that occur elsewhere only in the northern Kimberley. All of these endemics are associated with sandstone plateaus and escarpments. Further analysis of fine-scale distribution patterns and population genetics (phylogeographic structure) is needed to determine whether putative biodiversity hotspots occur elsewhere within the Top End and Kimberley. These hotspots are likely to coincide with evolutionary refugia, enabling species to persist during past (and potentially future) climatic extremes, and thus represent important areas for biodiversity conservation (Pepper and Keogh 2014; Rosauer et al. 2016; Oliver et al. 2017).

Monsoon forests comprise a very small fraction (less than 1 per cent) of the landscape in the study region, yet they support a disproportionally high number of butterfly and diurnal moth species. More than 50 species (31 per cent) breed, or are suspected to breed, in various types of monsoon forest, and 34 (21 per cent) of these species are obligatorily dependent on these habitats, in that they do not breed in other habitats. Interestingly, despite the high proportion of monsoon forest specialists, few of these taxa are actually endemic to the study region; only one species (*Ctimene* sp. 'Top End') and 16 subspecies are endemic to northern Australia. Four of these monsoon forest specialists (*Leptosia nina*, *Cethosia penthesilea*, *Phalanta phalantha* and *Deudorix smilis*) represent the only known Australian occurrences of species that are predominantly distributed in South-East Asia. A fifth monsoon specialist, the predominantly South-East Asian pierid *Appias albina*, also has its main occurrence within Australia in the study region. In addition, several species (particularly Danainae) use monsoon forest as refuges during the dry season, further highlighting the high biodiversity value of this habitat type.

Northern Australia is one of the few tropical places left on Earth in which biodiversity and the ecological processes underpinning that biodiversity are still relatively intact. Yet, paradoxically, at a time when the region is under increasing threat from development (pastoral, agricultural/horticultural and mining industries), invasive species (feral animals and weeds), inappropriate fire regimes (especially an increase in the frequency and scale of fires) and climate change (especially elevated carbon dioxide levels), scientific knowledge of the invertebrate biodiversity is still in its infancy. Even for a popular insect group such as butterflies and diurnal moths, there are still, for some species, substantial knowledge gaps in taxonomic status, spatial distribution and ecology, such as larval food plant associations. It is hoped this work—particularly the

geographic range maps, relative abundance charts and conservation status assessments— will not only provide the foundation for further research, but also provide the baseline against which the extent and direction of change can be assessed in future. It should also serve to help identify the region's biological assets to set priorities for biodiversity conservation. The fact that 79 per cent of the butterfly and diurnal moth fauna is presently evaluated as LC is encouraging; it tells us not only that the fauna is in relatively good health, but also what stands to be lost if the north is opened up for wholesale development.

While the fauna, overall, may be considered to be in reasonably good condition, with more than three-quarters of the species presently secure (LC), the need for further survey and monitoring will be crucial to reevaluate the conservation status of those species that are threatened (VU), Near Threatened (NT) or Data Deficient (DD), as well as those that are currently inadequately represented in the National Reserve System. Further research on these species should focus on clarifying the extent of their geographic distribution, determining their ecological resources (e.g. larval food plants), monitoring adult abundance and occupancy of critical habitats and managing threats from fire, as well as determining any other key threatening processes. More generally, in addition to understanding broad-scale distributional patterns, there is a complementary need to understand abundance and population trends for the fauna as a whole. Long-term monitoring of butterfly populations (and/or their larval food plants) is needed to determine whether species are stable or declining, and to help identify threats and management priorities and their effectiveness. The impact on the butterfly and diurnal moth fauna as a whole from developmental and management processes such as fire management, the intensification of pastoralism and selective clearing of native vegetation from the most fertile and productive land systems should be assessed as a high priority.

# Swallowtails

## (Papilionidae)

5

# Four-barred Swordtail, Kakadu Swordtail
## *Protographium leosthenes* (Doubleday, 1846)

Plate 3 Kakadu National Park, NT
Photo: Ian Morris

Plate 4 Kakadu National Park, NT
Photo: Ian Morris

## Distribution

This species is represented by the subspecies *P. leosthenes geimbia* (Tindale, 1927), which is endemic to the study region. It occurs in the Top End, where it is restricted to western Arnhem Land, extending from Ubirr Rock (A. Carlson) south to Deaf Adder Gorge (M. B. Malipatil) in Kakadu National Park, NT. Its geographic range closely corresponds with the spatial distribution of its larval food plant, which is also endemic to western Arnhem Land, although the food plant extends slightly further south (to the upper Gimbat Creek area). Outside the study region, *P. leosthenes* occurs in north-eastern and eastern Australia.

## Habitat

*Protographium leosthenes* breeds in patches of monsoon vine thicket on sandstone escarpments in steep rocky hill-slopes and the base of rock overhangs and boulders where the larval food plant grows as a scrambling vine. Adults also fly in open woodland and males congregate on hilltops to locate females, but they do not breed in these habitats.

## Larval food plant

*Melodorum rupestre* (Annonaceae).

## Seasonality

Adults are seasonal, occurring from October to May. They are most abundant during the 'build-up' and early wet season (November–January), when humidity and temperatures are high before the onset of the first monsoon rains, when the larval food plant produces flushes of new leaf growth. The species breeds mainly during the pre-monsoon period, with the immature stages (eggs or larvae) found in December and January. There is possibly only a single generation in most years, with a partial second generation during the late wet season and early dry season (March–May), when some adults may emerge during or after the monsoon. The butterfly survives the long dry season in pupal diapause (Sands and New 2002)—a strategy also adopted by the nominate subspecies in Queensland.

## Breeding status

This species is resident in the study region.

## Conservation status

LC. The subspecies *P. leosthenes geimbia* is a short-range endemic (EOO = 1,700 sq km) and its entire range occurs in two conservation reserves: Kakadu National Park and Warddeken IPA. Despite its restricted occurrence, there are no known threats facing the taxon (Sands and New 2002). The larval food plant is currently listed as LC under the *Territory Parks and Wildlife Conservation Act (2014) (TPWCA)*.

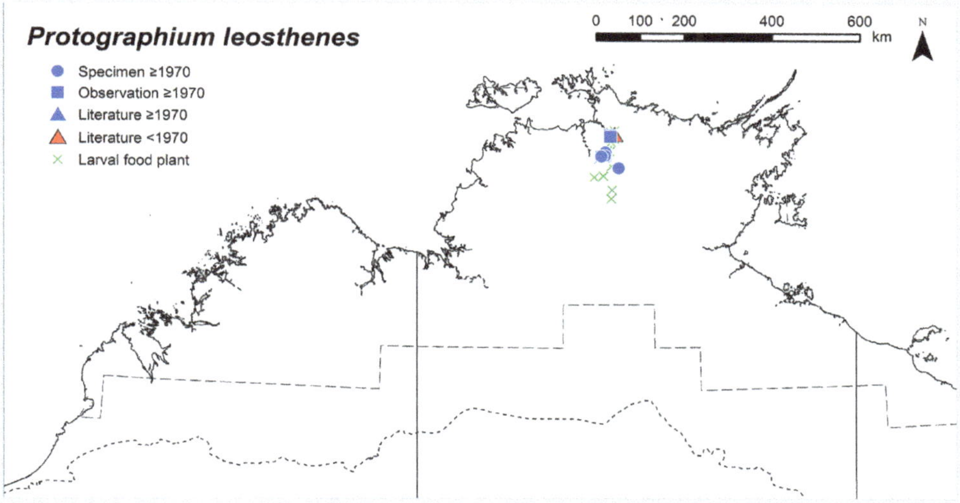

Protographium leosthenes
0 100 200 400 600
km
N
Specimen ≥1970
Observation ≥1970
Literature ≥1970
Literature <1970
Larval food plant

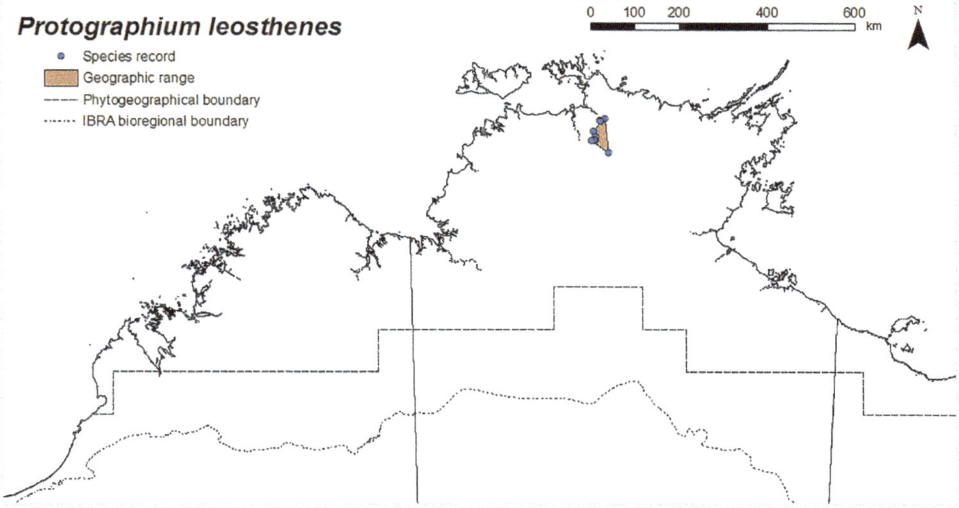

Protographium leosthenes
0 100 200 400 600
km
N
Species record
Geographic range
Phytogeographical boundary
IBRA bioregional boundary

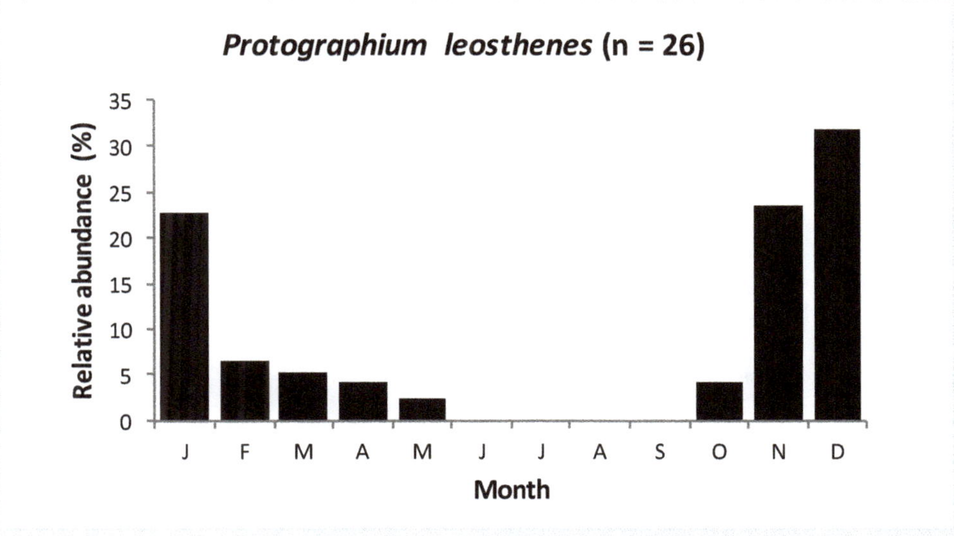

Protographium leosthenes (n = 26)
Relative abundance (%)
35
30
25
20
15
10
5
0
J F M A M J J A S O N D
Month

| Month | J | F | M | A | M | J | J | A | S | O | N | D |
|---|---|---|---|---|---|---|---|---|---|---|---|---|
| Egg | | | | | | | | | | | | |
| Larva | | | | | | | | | | | | |
| Pupa | | | | | | | | | | | | |
| Adult | | | | | | | | | | | | |

# Pale Triangle
## *Graphium eurypylus* (Linnaeus, 1758)

Plate 5 Wanguri, Darwin, NT
Photo: M. F. Braby

## Distribution

This species is represented by the subspecies *G. eurypylus nyctimus* (Waterhouse & Lyell, 1914), which is endemic to the study region. It occurs in the Kimberley and throughout the Top End. The geographic range closely corresponds with the spatial distribution of its larval food plants. The native food plants are absent from the western Gulf Country south of Groote Eylandt; hence, it is uncertain whether records from near Mataranka, Ngukurr and Limmen National Park, NT, and Doomadgee, Qld, represent vagrants from further north or localised populations breeding on ornamental or cultivated food plants. At Doomadgee, Puccetti (1991: 144) noted that only '[o]ne very worn specimen was observed but not taken', which suggests the species may not be established in the western Gulf Country. Similarly, in the south-western Kimberley, it was recorded on several occasions in the township of Broome, WA, during the mid to late 1990s, when it may have bred temporarily on ornamental *Annona* (G. Swann), but the population did not establish. More recently, G. Swann observed the species in Broome, in January 2018. Outside the study region, *G. eurypylus* occurs widely from India, southern China, Japan and South-East Asia through mainland New Guinea and north-eastern and eastern Australia to the Bismarck Archipelago.

## Habitat

*Graphium eurypylus* breeds mainly in semi-deciduous monsoon vine thicket and mixed eucalypt woodland–vine thicket in coastal, riparian and inland areas where the larval food plants grow, some as semi-deciduous shrubs or small trees. It also occurs in suburban parks and gardens where the ornamental food plants are propagated.

## Larval food plants

*Meiogyne cylindrocarpa, Melodorum rupestre, Miliusa brahei, Miliusa traceyi, Monoon australe, Hubera nitidissima* (Annonaceae), *Cryptocarya cunninghamii* (Lauraceae), *Diospyros maritima* (Ebenaceae); also \*Annona muricata, \*Polyalthia longifolia* (Annonaceae).

## Seasonality

Adults occur throughout the year, but they are more abundant during the wet season. They are particularly numerous during the 'build-up' and pre-monsoon storms of the early wet season (October–January), when the larval food plants start to produce new leaf growth. Adults are generally absent during the cooler dry season (May–August), although a few have been recorded at this time of year. Several generations are completed annually; the immature stages (eggs or larvae) have been recorded from September to April, indicating that breeding occurs over an extended period. The population survives the dry season in the pupal stage, which may remain in diapause for up to seven months. In the Darwin area, adults start to emerge in September or October following the first heavy downpours.

## Breeding status

This species is resident in the study region. It is not known whether it migrates or disperses outside the normal breeding range.

## Conservation status

LC.

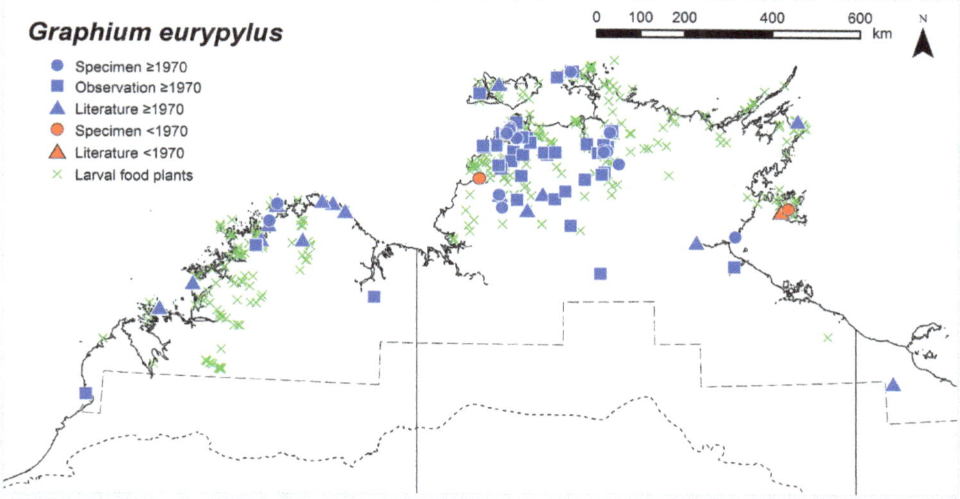

Graphium eurypylus
0 100 200 400 600 km
N
Specimen ≥1970
Observation ≥1970
Literature ≥1970
Specimen <1970
Literature <1970
Larval food plants

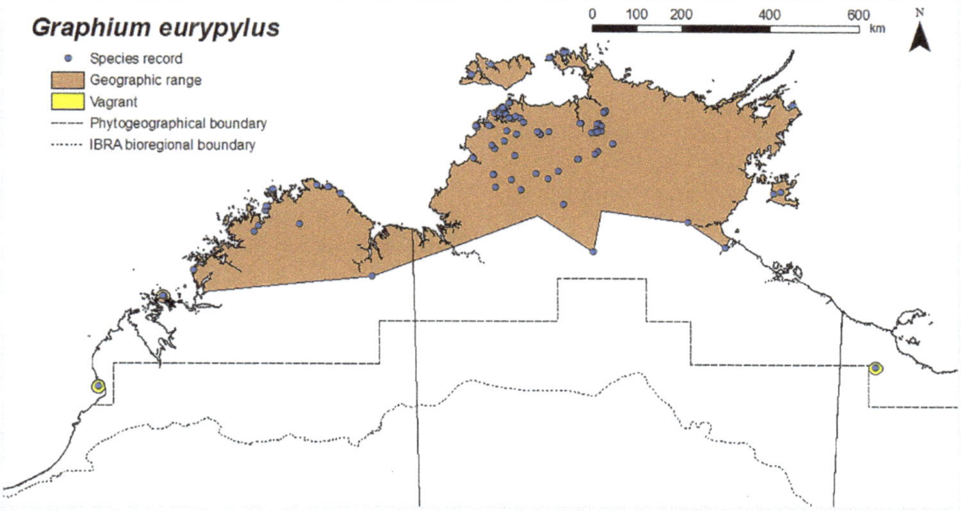

Graphium eurypylus
0 100 200 400 600 km
N
Species record
Geographic range
Vagrant
Phytogeographical boundary
IBRA bioregional boundary

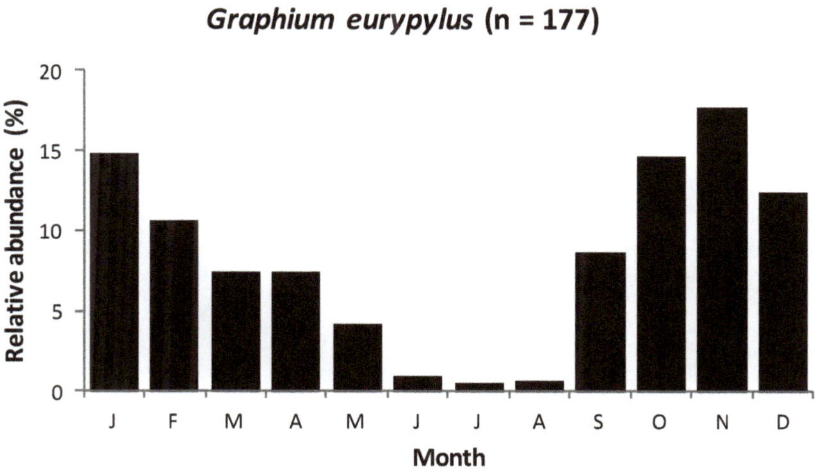

Graphium eurypylus (n = 177)
Relative abundance (%)
20
15
10
5
0
J F M A M J J A S O N D
Month

| Month | J | F | M | A | M | J | J | A | S | O | N | D |
|---|---|---|---|---|---|---|---|---|---|---|---|---|
| Egg | ■ | ■ | ■ | ■ | | | | | | | ■ | ■ |
| Larva | ■ | | ■ | ■ | | | | | ■ | ■ | ■ | ■ |
| Pupa | | | ■ | ■ | ■ | ■ | ■ | ■ | ■ | ■ | | |
| Adult | ■ | ■ | ■ | ■ | ■ | ■ | ■ | ■ | ■ | ■ | ■ | ■ |

# Dainty Swallowtail
## *Papilio anactus* W. S. Macleay, 1826

Plate 6 Mt Piper, Broadford, Vic
Photo: M. F. Braby

## Distribution

This species has been recorded only from the western Gulf Country of the study region, at Doomadgee, Qld (Puccetti 1991). Outside the study region, *P. anactus* occurs widely in eastern and south-eastern Australia.

## Habitat

*Papilio anactus* has been recorded only from urban areas where its non-native larval food plant is propagated in suburban gardens; it does not appear to be established in natural areas outside this habitat.

## Larval food plants

*\*Citrus* sp. (Rutaceae). The native food plant has not been recorded in the study region, but in eastern Queensland the species is known to feed on various native species of *Citrus* (Braby 2000).

## Seasonality

The seasonal abundance and breeding phenology of this species are not well understood. Adults have been recorded during most months of the year, but there are too few records (n = 8) to assess any seasonal changes in abundance. Puccetti (1991: 144) noted: 'This species has been common at times … the species appears to be well-established on local citrus'. Presumably, the species breeds throughout the year.

## Breeding status

The breeding status of *P. anactus* is uncertain. It appears to be resident, but it is possible the species does not occur naturally in the region and is introduced, having become established recently with the cultivation of its ornamental larval food plant.

## Conservation status

NA.

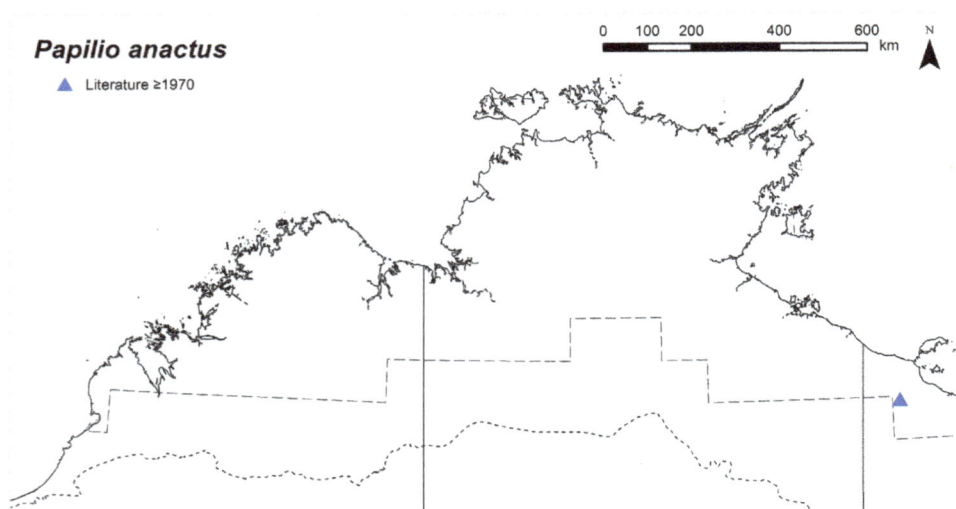

*Papilio anactus*

▲   Literature ≥1970

| Month | J | F | M | A | M | J | J | A | S | O | N | D |
|-------|---|---|---|---|---|---|---|---|---|---|---|---|
| Egg   |   |   |   |   |   |   |   |   |   |   |   |   |
| Larva |   |   |   |   |   |   |   |   |   |   |   |   |
| Pupa  |   |   |   |   |   |   |   |   |   |   |   |   |
| Adult |   |   |   |   |   |   |   |   |   |   |   |   |

# Orchard Swallowtail

*Papilio aegeus* Donovan, 1805

Plate 7 Mallacoota, Vic
Photo: Frank Pierce

Plate 8 Crystal Cascades, Qld
Photo: Frank Pierce

## Distribution

This species is represented in the study region by the subspecies *P. aegeus aegeus* Donovan, 1805. It is restricted to eastern Arnhem Land (Wessel Islands, Gove Peninsula and Groote Eylandt, NT), Limmen Bight and the western Gulf Country, where it occurs sporadically in coastal and near-coastal areas. Despite the widespread distribution of its native larval food plant, *P. aegeus* is not permanently established further west in the Top End. The species is normally absent from Kakadu National Park, NT, but a male was observed at West Alligator Head in March 1991 and sightings of a second male for several weeks were subsequently made at the same location (K. McLachlan); these records are considered to be vagrants from further east. Outside the study region, *P. aegeus* occurs from Tanimbar, the Kai and Aru islands, through mainland New Guinea and adjacent islands and eastern Australia to the Bismarck Archipelago and the Santa Cruz Islands east of the Solomon Islands.

## Excluded data

A previous record of *P. aegeus* from Darwin, NT (Dunn and Dunn 1991), based on a pair of specimens collected in January 1978 by J. T. Moss, appears to represent an accidental introduction that failed to establish following Tropical Cyclone Tracy (Braby 2014a).

## Habitat

*Papilio aegeus* breeds in urban areas where its non-native larval food plant is propagated in suburban gardens (Braby 2011a); however, the natural breeding habitat is not well documented. Fenner (1991) recorded *P. aegeus* breeding on Marchinbar Island, NT, on the native food plant (*Micromelum minutum*), which, in coastal areas, typically grows in monsoon vine thicket on sand dunes and low lateritic cliffs above the beach. Males have been observed in savannah woodland, patrolling encounter sites to locate females.

## Larval food plants

*Micromelum minutum* (Rutaceae); also *Citrus* sp. (Rutaceae).

## Seasonality

The seasonal abundance and breeding phenology of this species are not well understood. Adults have been recorded during most months of the year, usually in low numbers, but we have too few records (n = 16) to assess any seasonal changes in abundance. The immature stages (larvae or pupae) have been recorded during the dry season (July–September). Presumably, the species breeds throughout the year.

## Breeding status

This species is resident in the study region. It is possible that *P. aegeus* does not occur naturally in the region, and has become established only within the past three decades, perhaps facilitated by the cultivation of its ornamental larval food plant in suburban areas on the mainland. The species did not become established on Gove Peninsula in north-eastern Arnhem Land until about 2000 (Braby 2011a). The only record prior to 1986 is a historical specimen (in SAM) from the Roper River, NT, by N. B. Tindale.

## Conservation status

LC. Although the species *P. aegeus* has a restricted range in the study region, there are no known threats facing the taxon.

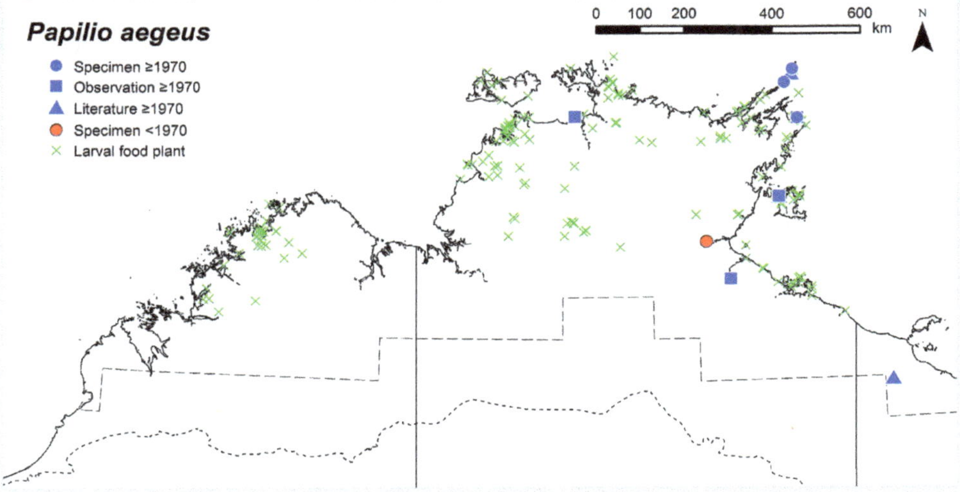

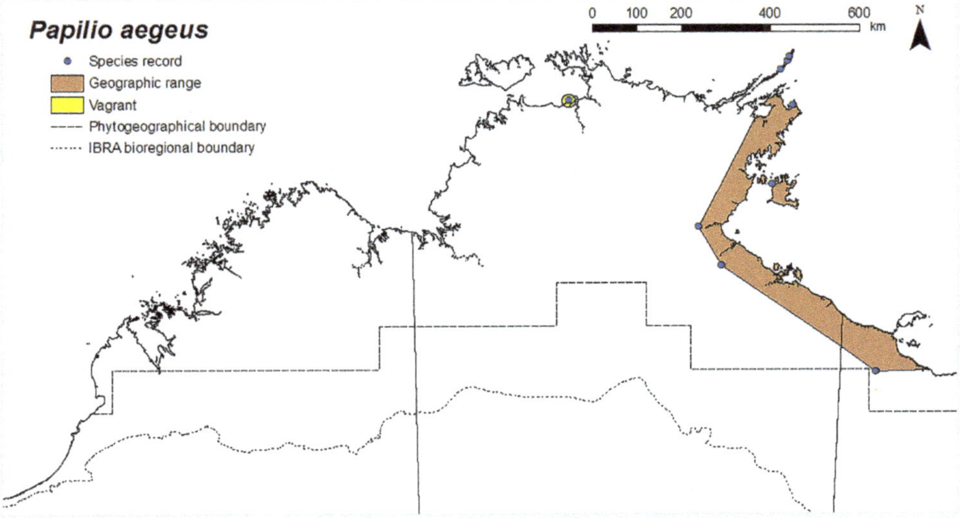

| Month | J | F | M | A | M | J | J | A | S | O | N | D |
|-------|---|---|---|---|---|---|---|---|---|---|---|---|
| Egg   |   |   |   |   |   |   |   |   |   |   |   |   |
| Larva |   |   |   |   |   | ▓ |   |   | ▓ |   |   |   |
| Pupa  |   |   |   |   |   | ▓ | ▓ |   | ▓ |   |   |   |
| Adult | ▓ | ▓ | ▓ | ▓ | ▓ |   | ▓ |   | ▓ | ▓ |   |   |

# Fuscous Swallowtail
## *Papilio fuscus* Goeze, 1779

Plate 9 Lee Point, Darwin, NT
Photo: M. F. Braby

Plate 10 Wanguri, Darwin, NT
Photo: M. F. Braby

## Distribution

This species is represented by the subspecies *P. fuscus canopus* Westwood, 1842, which is endemic to the study region. It occurs widely in coastal areas of the western and northern Kimberley and throughout the Top End, extending as far south as Judbarra/Gregory National Park (Limestone Gorge) (Braby and Archibald 2016), near Mataranka and Bing Bong, NT, in the Gulf of Carpentaria. Its geographic range closely corresponds with the spatial distribution of its native larval food plants. Outside the study region, *P. fuscus* occurs from the Andaman Islands, the Malay Peninsula and Indonesia, through mainland New Guinea and adjacent islands and north-eastern and eastern Australia to the Solomon Islands and Vanuatu.

## Habitat

*Papilio fuscus* breeds mainly in semi-deciduous monsoon vine thicket in both coastal and inland areas where the native larval food plants grow as tall shrubs (Hall 1976). It also occurs in suburban gardens where ornamental citrus trees are cultivated. Adults sometimes disperse into savannah woodland, but they do not breed in this habitat.

## Larval food plants

*Glycosmis trifoliata*, *Micromelum minutum*, *Zanthoxylum parviflorum* (Rutaceae); also *Citrus* sp. (Rutaceae). The main food plants are *M. minutum* and *G. trifoliata* (Hall 1976; Meyer 1996a), but occasionally the species also breeds on *Z. parviflorum* (Braby 2015e) and cultivated *Citrus* (Braby 2011a).

## Seasonality

Adults occur throughout the year, but they are most abundant during the pre-monsoon 'build-up' and throughout the wet season. Franklin (2011) found similar trends near Darwin, NT, in that adults were most abundant during the wet season, based on quantitative studies conducted over 14 months during 2008–09. Very few adults occur during the winter dry season (June–August), when the species does not breed and remains in pupal diapause. Hall (1976: 41) noted that 'the pupal duration … is extremely variable, ranging from 14 days to a little over 24 months'. In the higher rainfall areas, the immature stages (eggs or larvae) have been recorded from October to June, indicating that breeding occurs over an extended period during which several generations are completed.

## Breeding status

This species is resident in the study region.

## Conservation status

LC.

## Papilio fuscus

- ● Specimen ≥1970
- ■ Observation ≥1970
- ▲ Literature ≥1970
- ● Specimen <1970
- ▲ Literature <1970
- × Larval food plants

0  100  200  400  600 km

N

## Papilio fuscus

- • Species record
- ▨ Geographic range
- — — Phytogeographical boundary
- ······ IBRA bioregional boundary

0  100  200  400  600 km

N

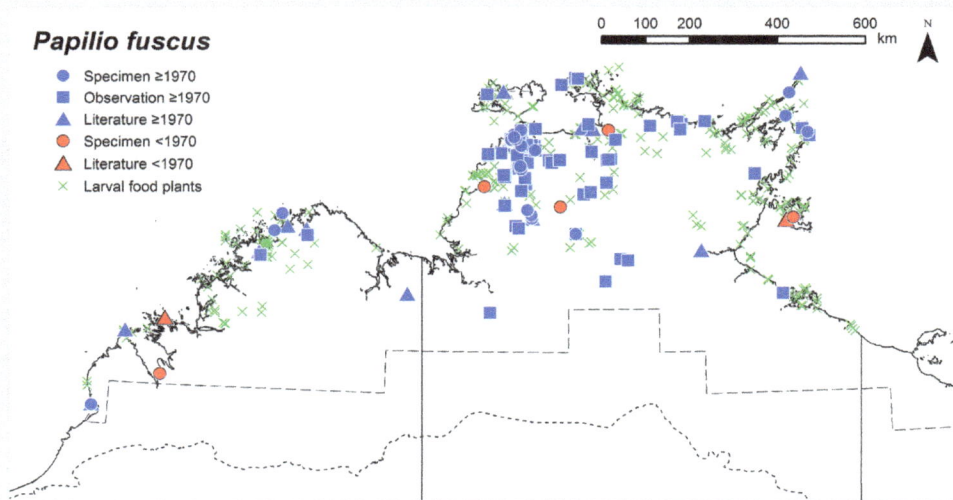

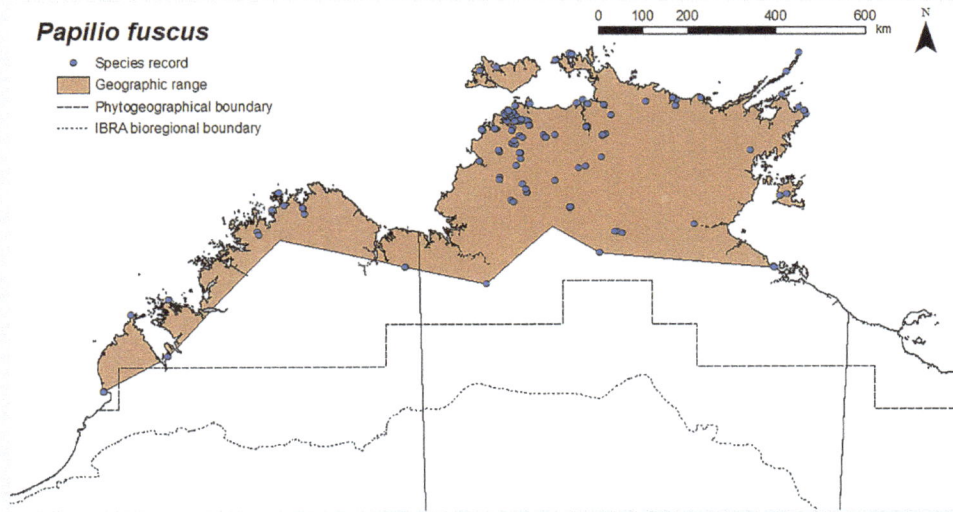

## *Papilio fuscus* (n = 232)

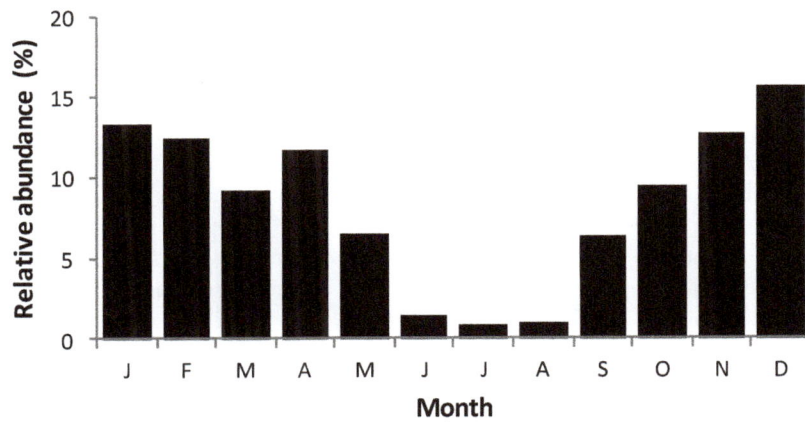

Relative abundance (%)

| Month | J | F | M | A | M | J | J | A | S | O | N | D |
|-------|---|---|---|---|---|---|---|---|---|---|---|---|
| Egg   | | | | | | | | | | | | |
| Larva | | | | | | | | | | | | |
| Pupa  | | | | | | | | | | | | |
| Adult | | | | | | | | | | | | |

# Chequered Swallowtail
## *Papilio demoleus* Linnaeus, 1758

Plate 11 Cardwell, Qld
Photo: M. F. Braby

Plate 12 Cardwell, Qld
Photo: M. F. Braby

## Distribution

This species is represented in the study region by the subspecies *P. demoleus sthenelus* W. S. Macleay, 1826. It occurs throughout almost the entire region, extending from moist coastal areas to the arid zone of central Australia beyond the southern boundary of the study region. The broad geographic range closely corresponds with the spatial distribution of its larval food plants, of which two (*Cullen balsamicum* and *C. cinereum*) occur widely in the drier inland areas. There are no records of either the butterfly or the food plants from the Tiwi Islands, so *P. demoleus* may be absent from this area. Outside the study region, *P. demoleus* occurs widely from the Middle East, India, southern China and South-East Asia, through mainland New Guinea to Australia, where it occurs throughout the continent, as well as in the Dominican Republic, where it has been accidently introduced (Eastwood et al. 2006; Morgun and Wiemers 2012).

## Habitat

*Papilio demoleus* breeds mainly in savannah woodland and open, low-lying grassy areas and floodplains where the larval food plants grow, either as perennial shrubs (*Cullen badocanum* and *C. balsamicum*) or as a seasonal annual (*C. cinereum*). Males also commonly fly on prominent hills, ridges and other landmarks, which are used as encounter sites to locate females for mating, but they do not breed in these habitats.

## Larval food plants

*Cullen badocanum, C. balsamicum, C. cinereum* (Fabaceae).

## Seasonality

Adults have been recorded throughout the year, but they are generally more abundant during the early dry season (April–June) following good wet seasons of average or above average rainfall. In some seasons or months, immense numbers of adults have been observed, particularly in semi-arid areas. The species becomes scarce as the dry season progresses, and there are very few records in the early wet season (November and December). The breeding phenology is not well understood. The immature stages have been recorded from February to June, which broadly coincides with the late wet season and early dry season and is when adults are more abundant, but it is not clear whether the species breeds at other times of the year. *Papilio demoleus* is a well-known migrant, but there are few published details of movement patterns (Smithers 1978). Smithers and McArtney (1970) recorded hundreds of specimens flying south-east over a distance of 25 km across the Stuart Highway between Elliott and Renner Springs, NT, in May 1969. In Darwin suburban and rural areas, a large-scale population movement, which lasted for about two weeks, was recorded in February 2015 (Braby 2016b). Adults flew rapidly between mid morning and mid afternoon, but the direction of flight progressively shifted from easterly, through southerly to westerly over the migration period.

## Breeding status

This species is assumed to be resident in the study region, but populations appear to be nomadic and are possibly temporary in many areas.

## Conservation status

LC.

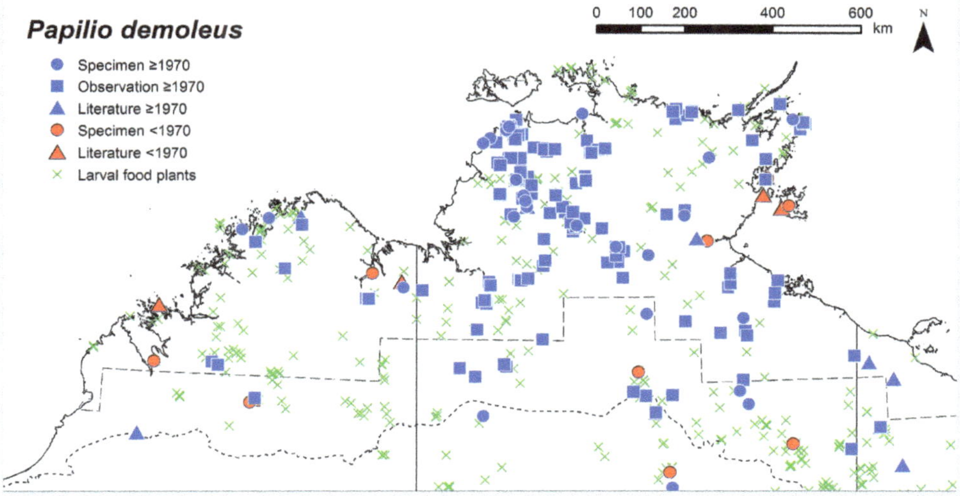

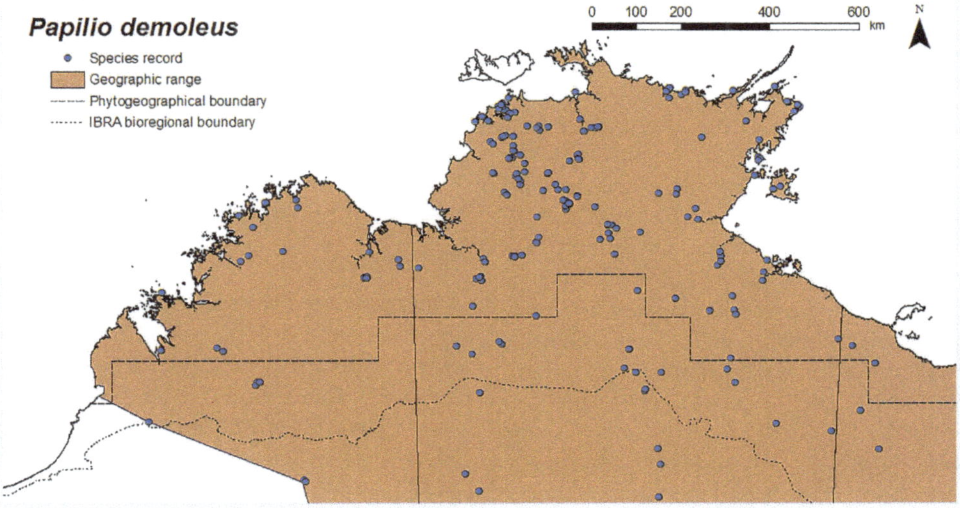

## Papilio demoleus

Legend:
- Specimen ≥1970 (blue circle)
- Observation ≥1970 (blue square)
- Literature ≥1970 (blue triangle)
- Specimen <1970 (orange circle)
- Literature <1970 (orange triangle)
- Larval food plants (green cross)

0  100  200  400  600 km

## Papilio demoleus

- Species record
- Geographic range
- Phytogeographical boundary
- IBRA bioregional boundary

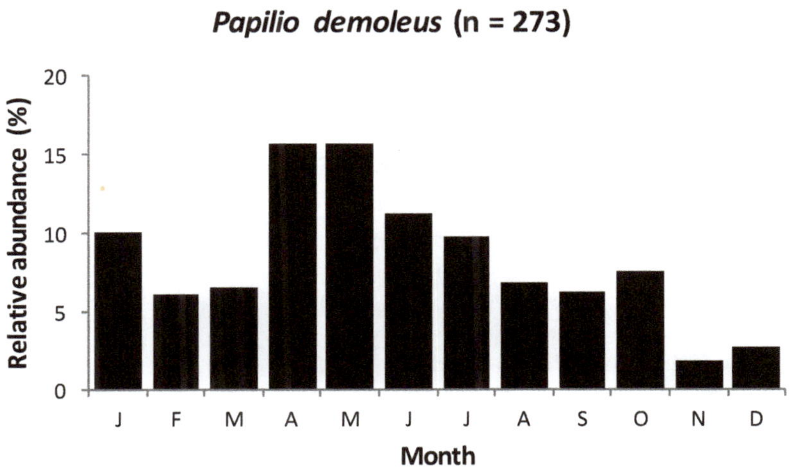

0  100  200  400  600 km

### Papilio demoleus (n = 273)

Relative abundance (%) vs Month (J F M A M J J A S O N D)

| Month | J | F | M | A | M | J | J | A | S | O | N | D |
|-------|---|---|---|---|---|---|---|---|---|---|---|---|
| Egg   |   |   | yellow | yellow | | yellow | | | | | | |
| Larva |   | green | green | green | green | | | | | | | |
| Pupa  |   |   |   | blue | | | | | | | | |
| Adult | orange | orange | orange | orange | orange | orange | orange | orange | orange | orange | orange | orange |

# Clearwing Swallowtail
## *Cressida cressida* (Fabricius, 1775)

Plate 13 Irvinebank, Qld
Photo: Don Franklin

## Distribution

This species is represented in the study region by the subspecies *C. cressida cressida* (Fabricius, 1775). It has a disjunct distribution, occurring in the western Kimberley and in the Top End; there are also a few records from the western Gulf Country. The geographic range is substantially broader than the spatial distribution of its known native larval food plant (*Aristolochia holtzei*), which is limited to the north-western corner of the Top End, indicating that several other (as yet unreported) food plants are used. In particular, *A. acuminata*, which occurs in the coastal areas of the western Kimberley, and *A. pubera*, which occurs commonly in north-eastern Arnhem Land, Wessel Islands and Groote Eylandt, NT, are known food plants of *C. cressida* in Queensland (Braby 2016a). However, in northern Australia, the native *Aristolochia* spp. are restricted to the higher rainfall areas (generally > 1,300 mm mean annual rainfall) and, in the Northern Territory, are not known to extend south of latitude 14°S. Hence, scattered occurrences of *C. cressida* in the southern areas of the Top End and western Gulf Country represent either vagrants that have extended beyond the normal breeding range or resident populations that have become established on naturalised food plants. Outside the study region, *C. cressida* occurs from the Lesser Sunda Islands, through south-eastern mainland New Guinea and adjacent islands to north-eastern and eastern Australia.

## Habitat

*Cressida cressida* breeds in a variety of savannah woodland and eucalypt open woodland habitats where the larval food plants grow—typically as herbs or small vines in the ground layer.

## Larval food plants

*Aristolochia holtzei* (Aristolochiaceae); also *Aristolochia indica*; probably *A. acuminata*, *A. pubera*, *A. thozetii*. On Groote Eylandt, Tindale (1923: 351) noted: 'The female was discovered laying eggs on one of the Aristolochia vines'. He listed the food plant as '*A. indica*', but this record refers to either *A. pubera* or *A. thozetii*, both of which have been collected from the island. Meyer (1996a) listed the food plant from Channel Island near Darwin as *Aristolochia* sp. 'Channel Island', and this record now refers to *A. indica*, which appears to have been introduced to the island from Asia. The food plant in the western Kimberley is likely to be *A. acuminata*.

## Seasonality

Adults have been recorded throughout the year, but they are generally more abundant during the wet season (January–April) and less common during the cooler dry season (May–August). We have few breeding records of this species, with the immature stages (eggs or larvae) recorded in February, June and November. Presumably, the species breeds throughout the year.

## Breeding status

This species is resident in the study region, but it is not known whether it disperses outside the normal breeding range.

## Conservation status

LC.

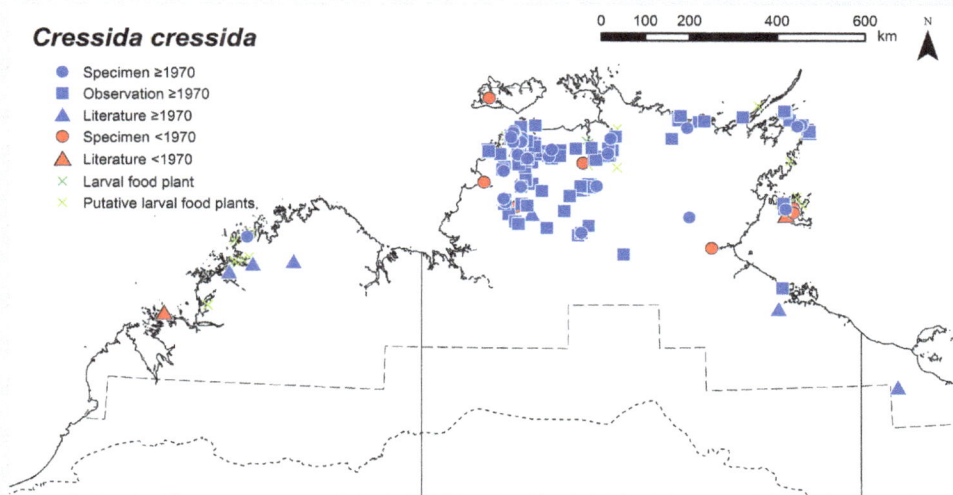

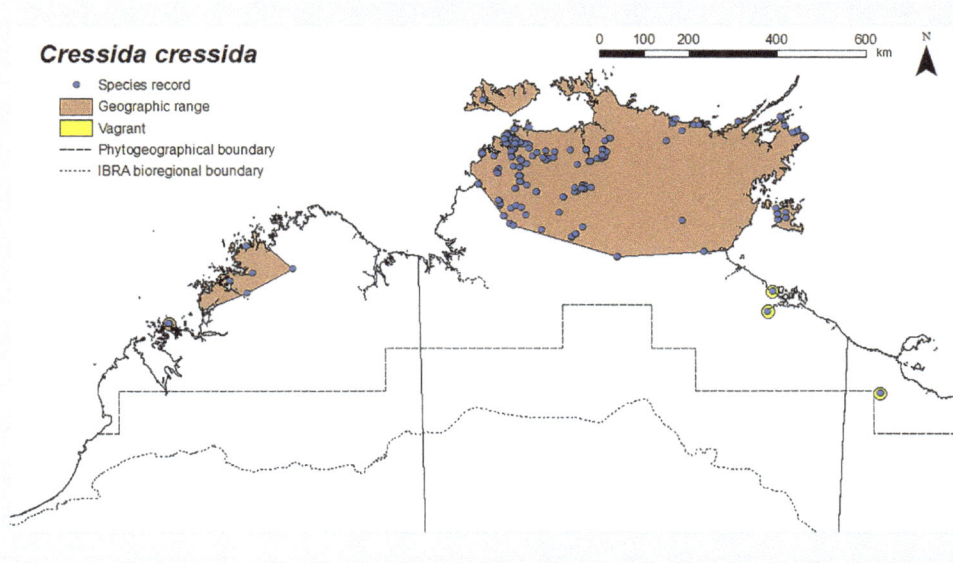

## Cressida cressida

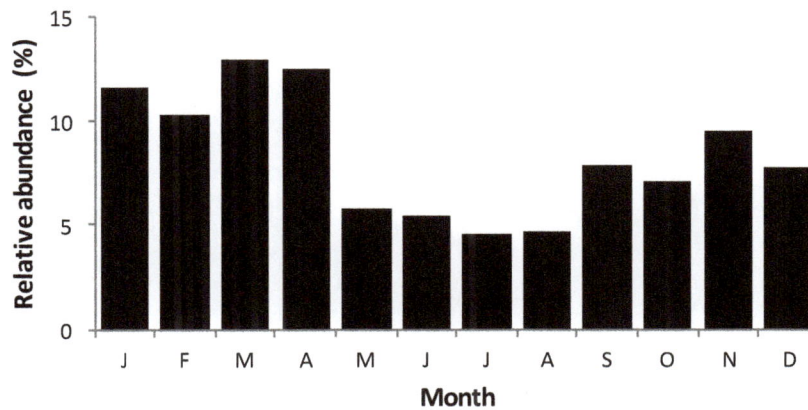

| Month | J | F | M | A | M | J | J | A | S | O | N | D |
|-------|---|---|---|---|---|---|---|---|---|---|---|---|
| Egg | | | | | | | | | | | | |
| Larva | | | | | | | | | | | | |
| Pupa | | | | | | | | | | | | |
| Adult | | | | | | | | | | | | |

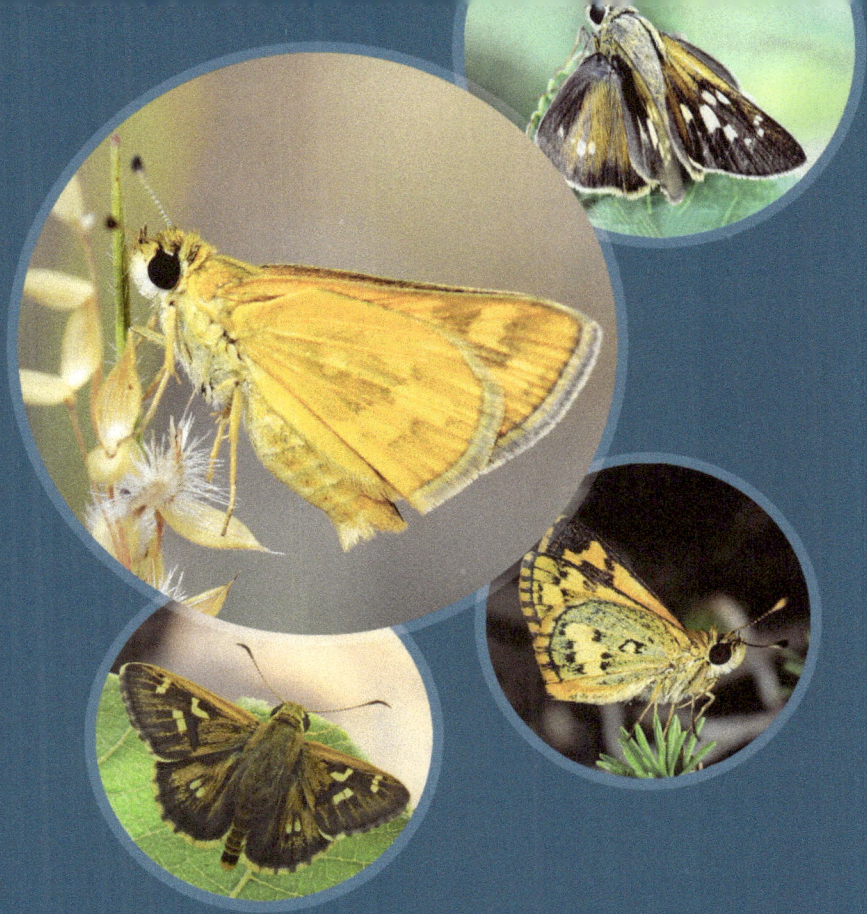

# Skippers

(Hesperiidae)

# 6

# Narrow-winged Awl
## *Badamia exclamationis* (Fabricius, 1775)

Plate 14 Near Borroloola, NT
Photo: Deb Bisa

## Distribution

This species is distributed widely throughout the study region. It is also recorded from remote areas offshore in the Timor Sea, including Ashmore Reef, WA (D. C. Binns). The breeding range is not well understood; the only documented breeding record is from the lower rainfall areas of the eastern Kimberley near Kununurra, WA (Meyer 1996a). Comparison of the geographic range with the spatial distribution of its known larval food plant indicates that *B. exclamationis* extends slightly further inland, with the food plant also widespread but restricted to coastal and near-coastal areas in which the mean annual rainfall is generally above 800 mm. It is not known whether other food plants are utilised throughout its broad range, but it is considered unlikely that *B. exclamationis* breeds in the drier inland areas (600–800 mm mean annual rainfall). Outside the study region, *B. exclamationis* occurs widely from India, southern China and South-East Asia, through mainland New Guinea and adjacent islands and north-eastern and eastern Australia to the Solomon Islands, New Caledonia and Fiji.

## Habitat

*Badamia exclamationis* has been recorded breeding in riparian monsoon vine thicket in the eastern Kimberley (at Black Rock Falls track near Kununurra, WA), where the larval food plant grew in abundance as a small tree (C. E. Meyer, pers. comm.).

## Larval food plant

*Terminalia microcarpa* (Combretaceae).

## Seasonality

Adults are seasonal, being most abundant during the pre-monsoon 'build-up' (October and November) and then again during the wet season (January–March). They are generally absent during the dry season (May–September). The breeding phenology is poorly known; the immature stages have been found only in December (C. E. Meyer). *Badamia exclamationis* is a well-known migrant (Smithers 1978; Common and Waterhouse 1981), but few details have been reported for northern Australia. On several occasions in the Top End, migratory flights comprising small to moderate numbers have been observed between late January and early April, particularly in March and early April (Braby 2016b). In general, adults fly rapidly between mid morning and early afternoon in a northerly direction (with the direction of flight varying from north-west, north–north-east to east–north-east). The only exception to this general pattern was a southerly flight (south–south-west) recorded in late January 2011, which suggests the arrival of an immigrant population. The timing of these migrations probably varies depending on the season and the start of the monsoon rains.

## Breeding status

This species appears to be a regular immigrant in the study region, breeding temporarily during the wet season and then vacating before the onset of the dry.

## Conservation status

DD. The geographical extent of the breeding habitat of *B. exclamationis* is very poorly understood, and it is currently known from only one site, but it may be very restricted—for example, confined to the lower rainfall areas of the monsoon tropics. Targeted field surveys to clarify the extent of the breeding distribution, determine its critical habitat and identify key threatening processes are required for this species.

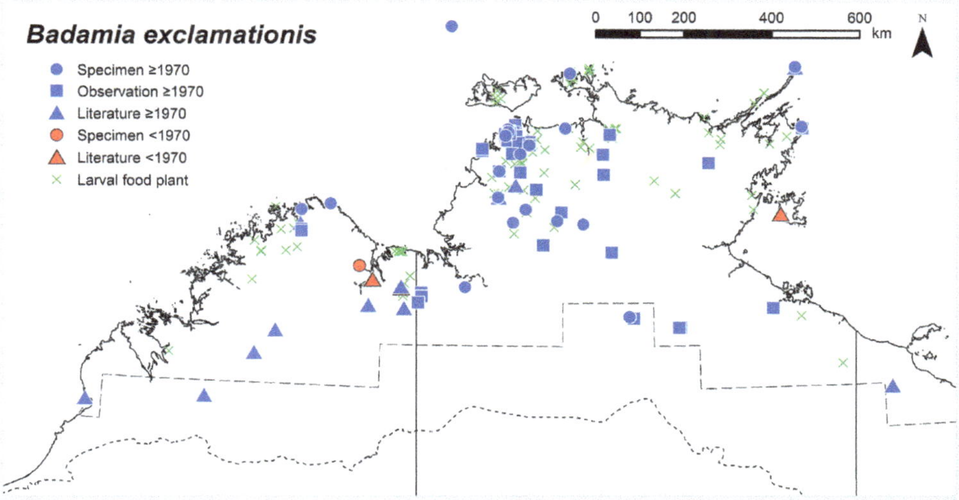

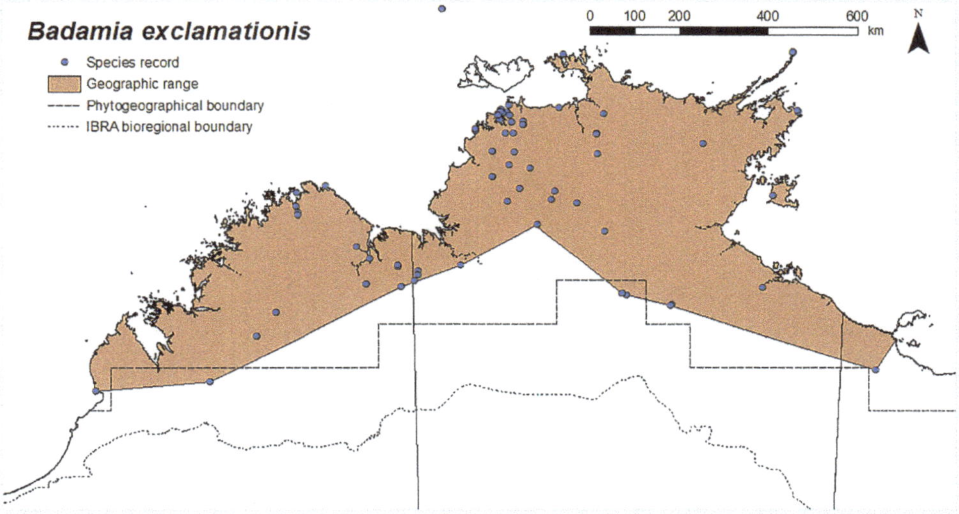

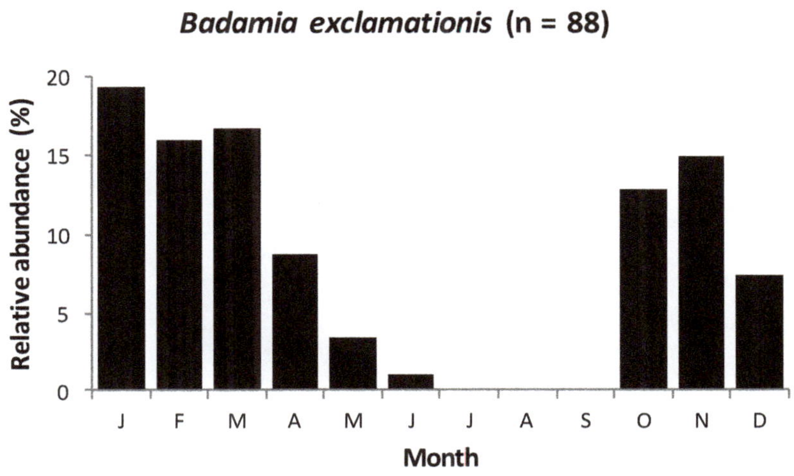

*Badamia exclamationis*

| Month | J | F | M | A | M | J | J | A | S | O | N | D |
|-------|---|---|---|---|---|---|---|---|---|---|---|---|
| Egg | | | | | | | | | | | | |
| Larva | | | | | | | | | | | | |
| Pupa | | | | | | | | | | | | |
| Adult | | | | | | | | | | | | |

# Broad-banded Awl
## *Hasora hurama* (Butler, 1870)

Plate 15 Adelaide River crossing,
Arnhem Highway, NT
Photo: M. F. Braby

## Distribution

This species is represented by the subspecies *H. hurama territorialis* Meyer, Weir & Brown, 2015, which is endemic to the study region. It occurs in the north of the Northern Territory, where it is restricted to the higher rainfall areas (> 1,300 mm mean annual rainfall) in the coastal and near-coastal areas of the Top End. Its geographic range closely corresponds with the spatial distribution of its larval food plant. The food plant, however, is wider in extent, occurring slightly further north (Bathurst and Croker islands) and further east (Gove Peninsula, Groote Eylandt), suggesting *H. hurama* is likely to have a larger range than present records indicate. Further field surveys are thus required to determine whether *H. hurama* is present on the Tiwi Islands and on Gove Peninsula, NT. Outside the study region, *H. hurama* occurs from Maluku, through mainland New Guinea and adjacent islands and north-eastern Australia to the Bismarck Archipelago and the Solomon Islands.

## Habitat

*Hasora hurama* breeds in the edge of mangroves and mixed mangrove–monsoon forest associations along the banks of rivers and estuaries in floodplains where the larval food plant grows as a vine (Meyer et al. 2015).

## Larval food plant

*Derris trifoliata* (Fabaceae).

## Seasonality

Adults have been recorded, or reared, during most months of the year, except August, but we have too few records (n = 12) to assess any seasonal changes in abundance. Similarly, the immature stages have been recorded throughout the year. The larvae feed on new leaf growth and the life cycle is completed relatively quickly (within a few weeks), with no evidence of diapause in any of the life history stages (Meyer et al. 2015). This fact, together with available phenological data, suggests *H. hurama* breeds continuously throughout the year.

## Breeding status

This species is resident in the study region.

## Conservation status

LC. The subspecies *H. hurama territorialis* is a narrow-range endemic (geographic range = 28,050 sq km) and it occurs in several conservation reserves, including Fogg Dam Conservation Reserve, Djukbinj National Park, Kakadu National Park and Djelk IPA.

## Hasora hurama

- ● Specimen ≥1970
- ■ Observation ≥1970
- ▲ Literature ≥1970
- ▲ Literature <1970
- × Larval food plant

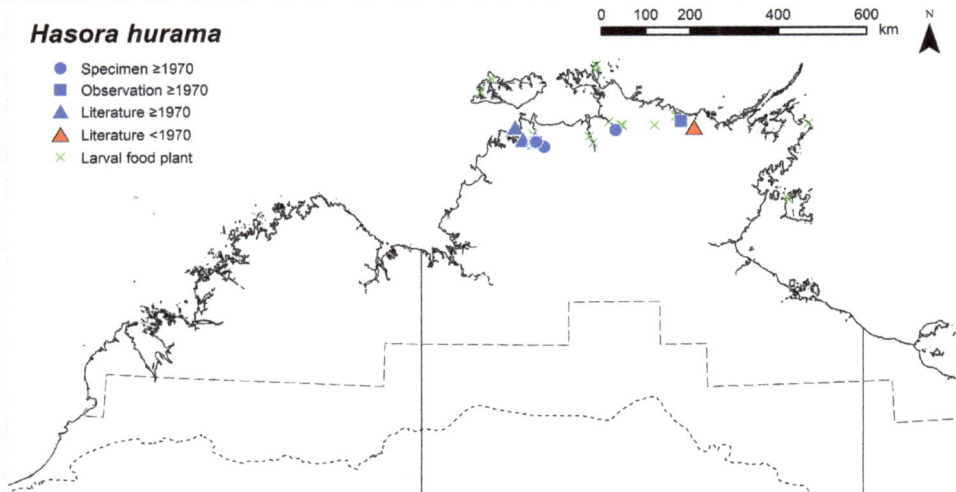

## Hasora hurama

- ● Species record
- ▬ Geographic range
- — — Phytogeographical boundary
- ········ IBRA bioregional boundary

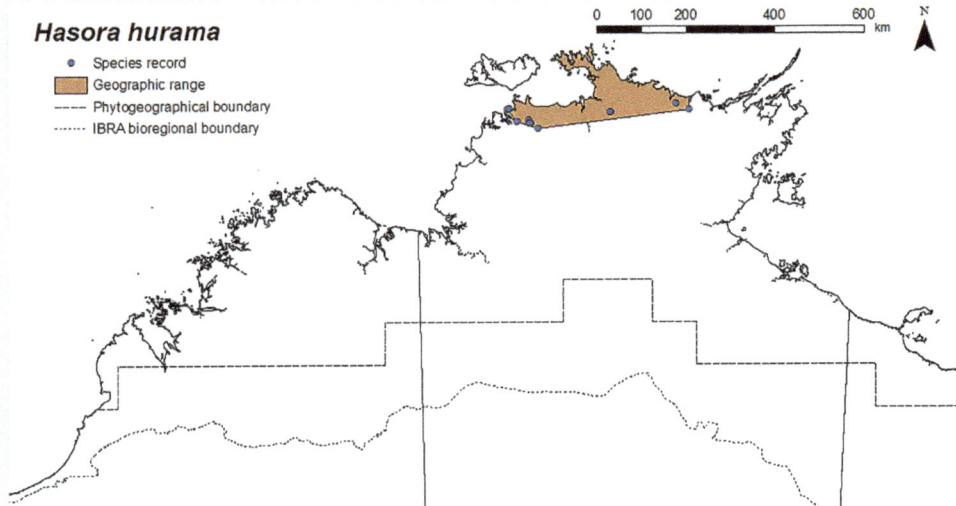

| Month | J | F | M | A | M | J | J | A | S | O | N | D |
|-------|---|---|---|---|---|---|---|---|---|---|---|---|
| Egg | | | | | | | | | | | | |
| Larva | | | | | | | | | | | | |
| Pupa | | | | | | | | | | | | |
| Adult | | | | | | | | | | | | |

# Chrome Awl
## *Hasora chromus* (Cramer, [1780])

Plate 16 Wanguri, Darwin, NT
Photo: M. F. Braby

## Distribution

This species is represented in the study region by the subspecies *H. chromus chromus* (Cramer, [1780]). It occurs at Broome (G. Swann) and the eastern Kimberley (Williams 2014) in Western Australia, and the Victoria River District and the northern coastal and near-coastal areas of the Top End. However, the natural distribution of the larval food plant has a much narrower range, restricted to the north-western corner of the Top End, where it extends no further east than Cobourg Peninsula and as far south as Hardies Creek, NT (10 km north of Marrakai Plains). The food plant is often cultivated as an ornamental tree in parks and roadside nature strips, and this most likely accounts for the wider distribution of *H. chromus*. Indeed, this is certainly the case in the Victoria River District (Timber Creek, NT) and Kimberley (Broome, WA), where *H. chromus* has been found breeding only on planted street trees. There is little historical data for *H. chromus* in the Northern Territory; it was first recorded from Darwin by Waterhouse and Lyell (1914), but there are few other published records before 1970. Thus, it is difficult to determine how the geographic range has expanded during the past century. However, it is noteworthy that most of the peripheral records outside the natural range of the food plant—that is, Arnhem Land (Mann River crossing, 2005; Maningrida, 2007), Gove Peninsula (Nhulunbuy, 2008), Victoria River District (Timber Creek, 2013), eastern Kimberley (Home Valley

Station, 2014) and western Kimberley (Broome, 2017)—are all within the past decade, suggesting the expansion may be recent. Moreover, the chronological records also suggest a progressive eastern expansion to Nhulunbuy and a similar south-western expansion to Broome. Outside the study region, *H. chromus* occurs widely from India, southern China and South-East Asia, through mainland New Guinea and adjacent islands and north-eastern Australia to Vanuatu, New Caledonia and Fiji.

## Habitat

*Hasora chromus* breeds naturally in coastal semi-deciduous monsoon vine thicket where the larval food plant grows as a tree on sand dunes or low lateritic cliffs above the beach.

## Larval food plant

*Millettia pinnata* (Fabaceae).

## Seasonality

Adults have been recorded during most months of the year, but they appear to be more abundant during the 'build-up' and wet season, when humidity is higher (October–March). The immature stages occur throughout the year, whenever the food plant produces flushes of new foliage. The larvae, like *Hasora hurama*, feed on new leaf growth and the life cycle is completed relatively quickly (within a few weeks), with no evidence of diapause in any of the life history stages. This fact, together with available phenological data, suggests *H. chromus* breeds continuously throughout the year.

## Breeding status

This species is resident in the study region.

## Conservation status

LC. Although the core area of the breeding range of *H. chromus* is narrowly restricted to the north-western corner of the Top End, its extent has now expanded well beyond this area.

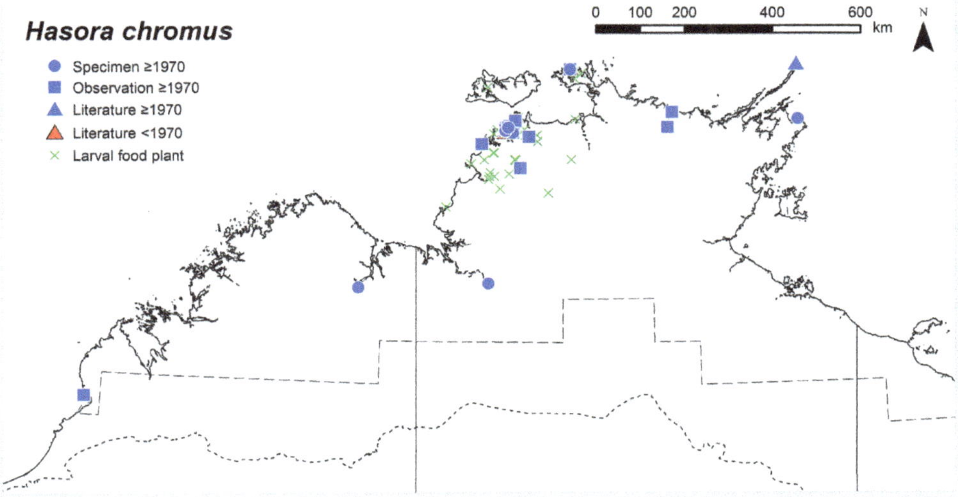

## Hasora chromus

Legend:
- ● Specimen ≥1970
- ■ Observation ≥1970
- ▲ Literature ≥1970
- ▲ Literature <1970
- × Larval food plant

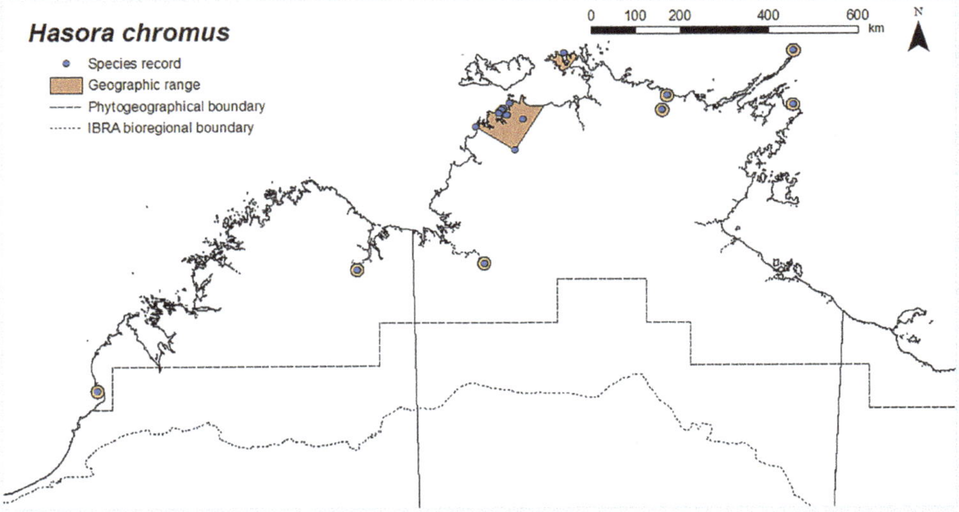

## Hasora chromus

Legend:
- • Species record
- Geographic range
- – – Phytogeographical boundary
- ······ IBRA bioregional boundary

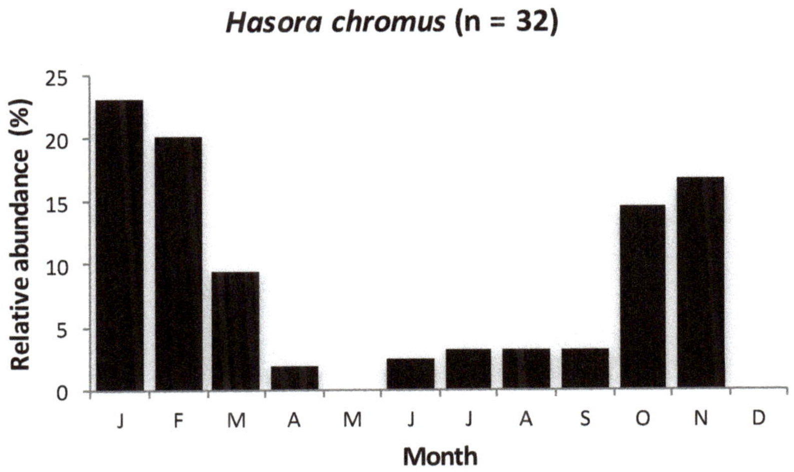

## Hasora chromus (n = 32)

| Month | J | F | M | A | M | J | J | A | S | O | N | D |
|-------|---|---|---|---|---|---|---|---|---|---|---|---|
| Egg | | | | | | | | ■ | | ■ | | |
| Larva | ■ | ■ | ■ | | ■ | ■ | ■ | ■ | ■ | ■ | ■ | ■ |
| Pupa | ■ | ■ | ■ | | | | | ■ | | ■ | ■ | ■ |
| Adult | ■ | ■ | ■ | ■ | | ■ | ■ | ■ | ■ | ■ | ■ | |

# Ornate Dusk-flat
## *Chaetocneme denitza* (Hewitson, 1867)

Plate 17 Elcho Island, NT
Photo: Ian Morris

Plate 18 Palmerston, NT
Photo: M. F. Braby

## Distribution

This species has a sporadic and possibly disjunct distribution in the study region. It has been recorded from the Kimberley in the Buccaneer Archipelago at Koolan Island in Yampi Sound (Koch and van Ingen 1969; McKenzie et al. 1995) and Drysdale River Station (J. E. and A. Koeyers), WA; and across the Top End, generally in the higher rainfall areas (> 900 mm mean annual rainfall, but mostly > 1,200 mm). The distribution of the known larval food plant is much more widespread than that of *C. denitza*; thus, further field surveys are required to determine whether *C. denitza* extends further south into the lower rainfall areas of the Top End. Outside the study region, *C. denitza* occurs in north-eastern and eastern Australia.

## Habitat

*Chaetocneme denitza* breeds in savannah woodland where the larval food plant commonly grows as a shrub or small tree (Braby 2011a). Freshly emerged adults have been collected in riparian woodland/open forest, suggesting they may also breed in this habitat.

## Larval food plants

*Planchonia careya* (Lecythidaceae). A number of other food plants have been recorded for the species in north-eastern Queensland (Braby 2016a), some of which may be used by *C. denitza* in northern Australia.

## Seasonality

The seasonal abundance and breeding phenology of this rarely seen crepuscular species are not well understood. Adults have been recorded during most months of the year, but we have too few records (n = 18) to assess any seasonal changes in abundance. The immature stages have been recorded from February to May, but undoubtedly occur at other times of the year. Presumably, the species breeds continuously throughout the year, with at least two or three generations completed annually.

## Breeding status

This species is resident in the study region.

## Conservation status

LC.

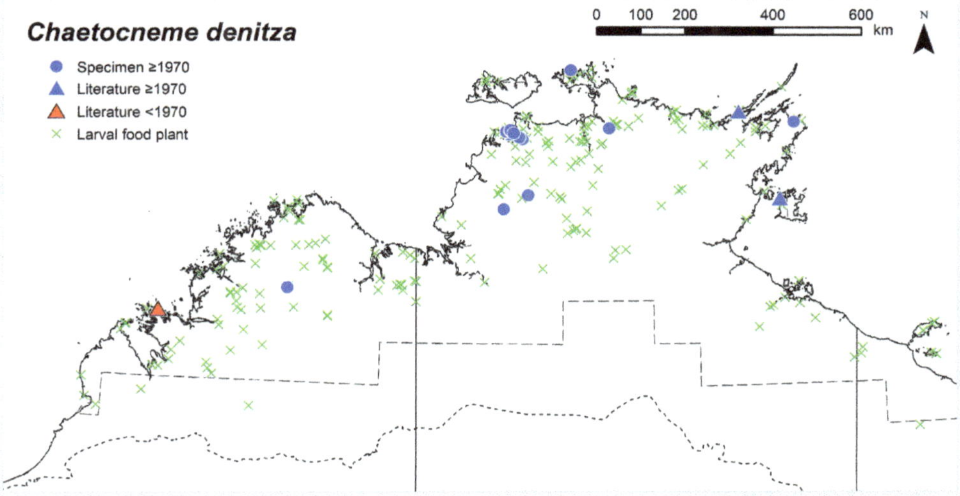

## Chaetocneme denitza

- ● Specimen ≥1970
- ▲ Literature ≥1970
- ▲ Literature <1970
- ✕ Larval food plant

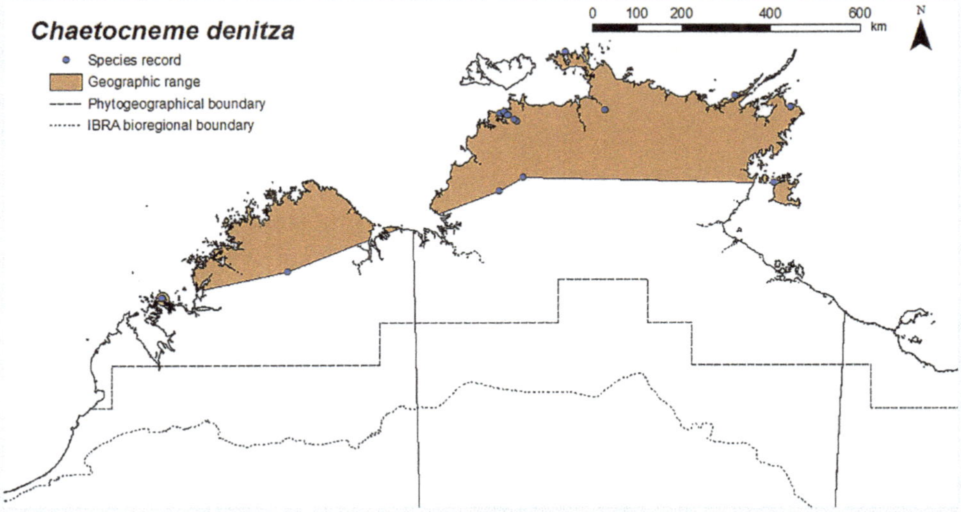

## Chaetocneme denitza

- ● Species record
- ▨ Geographic range
- --- Phytogeographical boundary
- ⋯ IBRA bioregional boundary

| Month | J | F | M | A | M | J | J | A | S | O | N | D |
|-------|---|---|---|---|---|---|---|---|---|---|---|---|
| Egg   |   |   |   |   |   |   |   |   |   |   |   |   |
| Larva |   |   |   |   |   |   |   |   |   |   |   |   |
| Pupa  |   |   |   |   |   |   |   |   |   |   |   |   |
| Adult |   |   |   |   |   |   |   |   |   |   |   |   |

# Spotted Grass-skipper
## *Neohesperilla senta* (Miskin, 1891)

Plate 19 Cardwell, Qld
Photo: M. F. Braby

## Distribution

This species has a disjunct distribution in the study region. It has been recorded from the northern Kimberley and from the Northern Territory, where it is restricted to higher rainfall areas (> 1,200 mm mean annual rainfall) of the north-western corner of the Top End. In the Kimberley, it has been recorded at Kalumburu (Johnson 1993) and Drysdale River Station (S. Craswell), WA; and in the Top End, it extends from Darwin (Berrimah) north-east to Cobourg Peninsula (27 km south–south-east of Black Point) and south-east to Eureka (Common and Waterhouse 1981), NT. The putative larval food plant (*Themeda triandra*) has a very broad distribution in northern Australia; if *N. senta* is found to use this species in the Kimberley and Top End then it may be more widespread in the study region than present records indicate. In particular, further field surveys in eastern Arnhem Land are required to determine whether *N. senta* is present in the eastern half of the Top End. Outside the study region, *N. senta* occurs in north-eastern Queensland.

## Habitat

The breeding habitat of *N. senta* has not been recorded in the study region. Adults have been collected in savannah woodland and eucalypt woodland, often very locally in disturbed open grassy areas, and they undoubtedly breed in these habitats.

## Larval food plants

Not recorded in the study region; probably *Themeda triandra* (Poaceae), which is the food plant in north-eastern Queensland (Braby 2000).

## Seasonality

Adults are seasonal, occurring only during the wet season (December–March), but there are too few records (n = 14) to assess temporal changes in abundance. In general, adults tend to be more numerous and freshly emerged during January and February. The breeding phenology and seasonal history of the immature stages have not been recorded, but it is possible there is only a single generation annually.

## Breeding status

This species is resident in the study region.

## Conservation status

LC. Available data suggest the species *N. senta* has a restricted range in the study region within which it occurs in at least two conservation reserves, Litchfield National Park and Garig Gunak Barlu National Park. Despite its restricted occurrence, there are no known threats facing the taxon.

## *Neohesperilla senta*

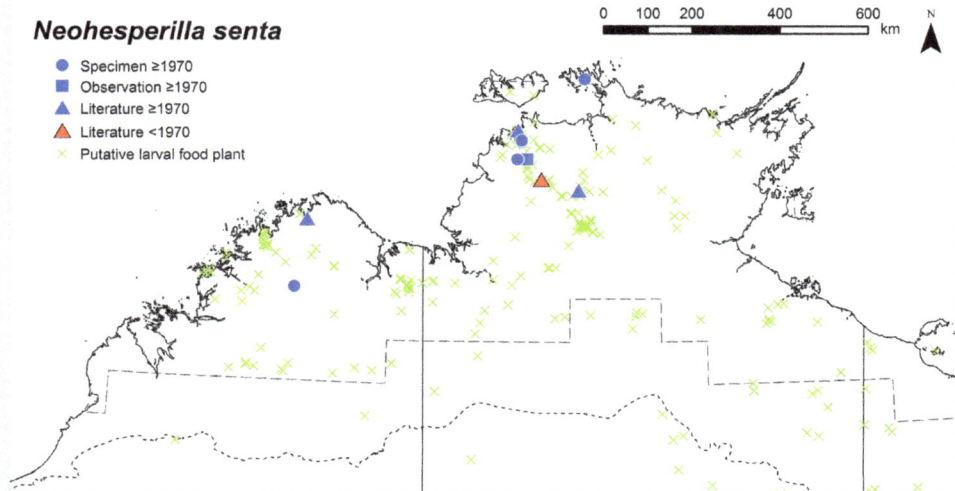

Legend:
- ● Specimen ≥1970
- ■ Observation ≥1970
- ▲ Literature ≥1970
- ▲ Literature <1970
- ✕ Putative larval food plant

0   100   200        400        600 km    N

## *Neohesperilla senta*

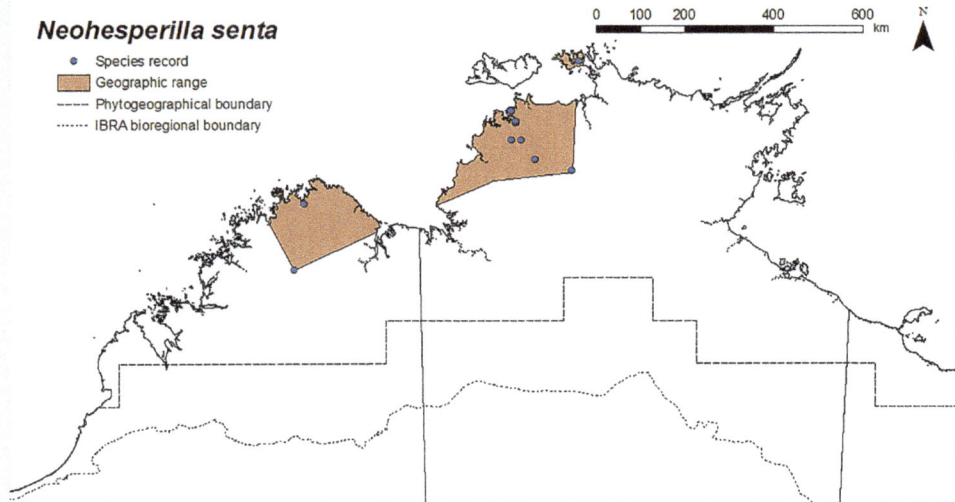

Legend:
- ● Species record
- ▨ Geographic range
- — — Phytogeographical boundary
- ·········· IBRA bioregional boundary

0   100   200        400        600 km    N

| Month | J | F | M | A | M | J | J | A | S | O | N | D |
|-------|---|---|---|---|---|---|---|---|---|---|---|---|
| Egg   |   |   |   |   |   |   |   |   |   |   |   |   |
| Larva |   |   |   |   |   |   |   |   |   |   |   |   |
| Pupa  |   |   |   |   |   |   |   |   |   |   |   |   |
| Adult |   |   |   |   |   |   |   |   |   |   |   |   |

# Sword-brand Grass-skipper
## *Neohesperilla xiphiphora* (Lower, 1911)

Plate 20 Mt Burrell, NT
Photo: M. F. Braby

Plate 21 Mt Burrell, NT
Photo: M. F. Braby

## Distribution

This species occurs in the north of the Northern Territory of the study region. It has been recorded mainly from the higher rainfall areas (> 1,000 mm mean annual rainfall) of the north-western corner of the Top End, but it has also been recorded from Groote Eylandt (Common and Waterhouse 1981) and in the western Gulf Country, at Caranbirini Creek, NT (Dunn and Dunn 1991). Its geographic range corresponds well with the spatial distribution of its known and putative larval food plants. The food plants, however, are more widely distributed than *N. xiphiphora*, with records from the Victoria River District and Tiwi Islands. Further field surveys are thus required to determine whether *N. xiphiphora* is established in these areas, particularly the drier areas of the Victoria River District. Outside the study region, *N. xiphiphora* occurs in north-eastern Queensland.

## Habitat

*Neohesperilla xiphiphora* breeds in savannah woodland and eucalypt woodland, particularly along the edges or bases of rocky outcrops, where the larval food plant grows as an annual grass in the open shaded understorey beneath eucalypt trees (Braby 2015e). Areas regenerating after dry season burns following wet season rains seem to be favoured for breeding. Males also congregate at the summit of steep hills, ridges and other landmarks, which are used as encounter sites to locate females for mating, but they do not breed in these habitats.

## Larval food plants

*Sorghum intrans* (Poaceae); probably *Schizachyrium perplexum* (Poaceae), which is a food plant in north-eastern Queensland (Braby 2000).

## Seasonality

Adults are seasonal, being most abundant during the wet season (December–March). The immature stages (eggs or larvae) have been recorded in February and March after the larval food plant has germinated and produced new soft leaf growth. The breeding phenology and seasonal history of the immature stages are not well understood, but it is possible there is only a single generation annually, with the final instar larvae remaining in diapause during the long dry season.

## Breeding status

This species is resident in the study region.

## Conservation status

LC.

## Neohesperilla xiphiphora

- ● Specimen ≥1970
- ■ Observation ≥1970
- ▲ Literature ≥1970
- ▲ Literature <1970
- ✕ Larval food plant
- ✕ Putative larval food plant

## Neohesperilla xiphiphora

- • Species record
- ▨ Geographic range
- — — Phytogeographical boundary
- ······· IBRA bioregional boundary

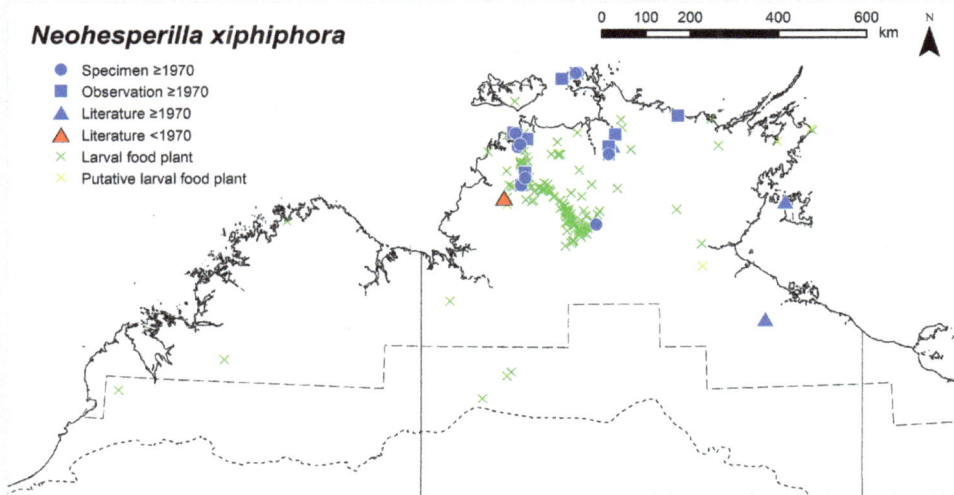

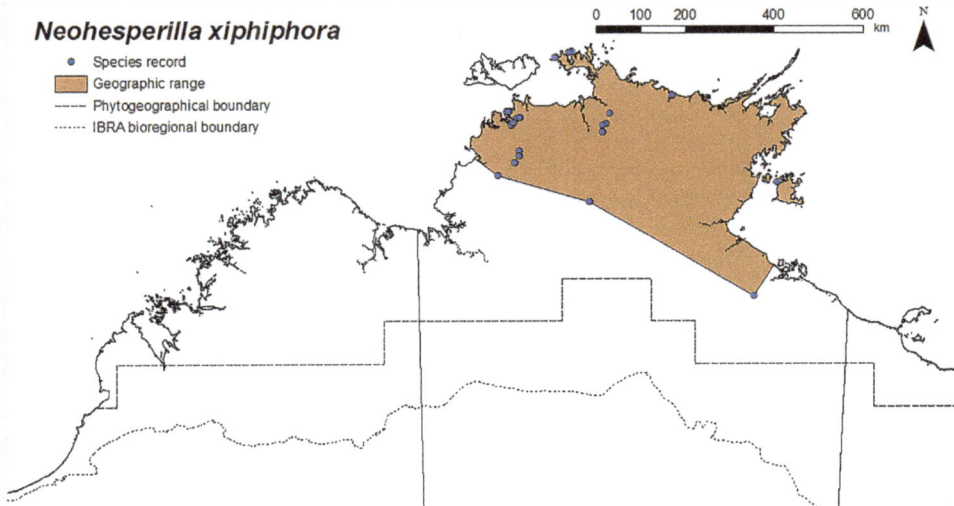

### Neohesperilla xiphiphora (n = 32)

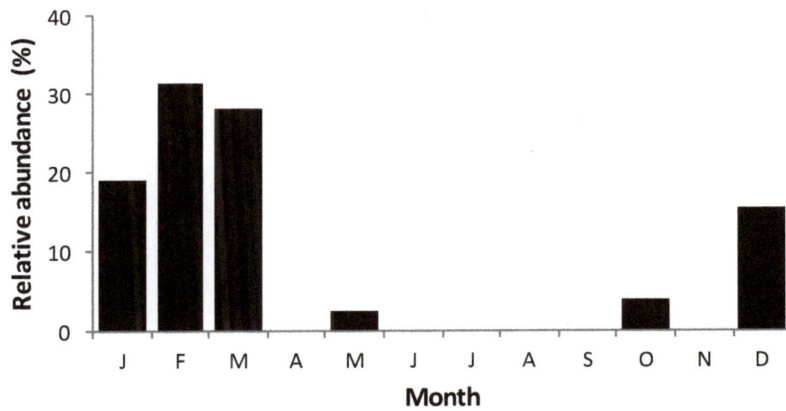

| Month | J | F | M | A | M | J | J | A | S | O | N | D |
|-------|---|---|---|---|---|---|---|---|---|---|---|---|
| Egg   |   |   |   |   |   |   |   |   |   |   |   |   |
| Larva |   |   |   |   |   |   |   |   |   |   |   |   |
| Pupa  |   |   |   |   |   |   |   |   |   |   |   |   |
| Adult |   |   |   |   |   |   |   |   |   |   |   |   |

# Narrow-brand Grass-skipper
*Neohesperilla crocea* (Miskin, 1889)

Plate 22 Endeavour Falls Tourist Park, Qld
Photo: Frank Pierce

## Distribution

This species has a disjunct distribution, occurring in the northern Kimberley and north of the Northern Territory of the study region. It is restricted to the higher rainfall areas (> 1,000 mm mean annual rainfall), with most records from the north-western corner of the Top End, extending as far south as Nitmiluk National Park (Edith Falls), but it has also been recorded on Groote Eylandt (Tindale 1923). In the Kimberley, it has recently been recorded at Drysdale River Station, WA (S. Craswell). The putative larval food plants (*Chrysopogon aciculatus* and *Schizachyrium pachyarthron*) occur widely in northern Australia; if *N. crocea* is found to use these species in the Kimberley and Northern Territory then it may have a greater geographic range within the study region than present records indicate. In particular, further field surveys are required to determine whether *N. crocea* is present on the Wessel Islands and Gove Peninsula of north-eastern Arnhem Land. Outside the study region, *N. crocea* occurs in Papua New Guinea and north-eastern Queensland.

## Habitat

The breeding habitat of *N. crocea* has not been recorded in the study region. Adults have been collected mainly in moister habitats than those of other species of *Neohesperilla*, including the edges of riparian forest and monsoon vine thicket, eucalypt woodland with a monsoon forest understorey and paperbark swampland, as well as savannah woodland, and no doubt they breed in these habitats.

## Larval food plants

Not recorded in the study region; probably *Chrysopogon aciculatus* and *Schizachyrium pachyarthron* (Poaceae), which are the food plants in north-eastern Queensland (Braby 2000).

## Seasonality

Adults are seasonal, being most abundant during the wet season (December–April), with an apparent peak in abundance in March. The flight season is somewhat protracted compared with other species of *Neohesperilla*, with adults also occurring in the early dry season (May–July). The breeding phenology and seasonal history of the immature stages have not been recorded, but it is possible only one or two generations are completed annually.

## Breeding status

This species is resident in the study region.

## Conservation status

LC.

## Neohesperilla crocea

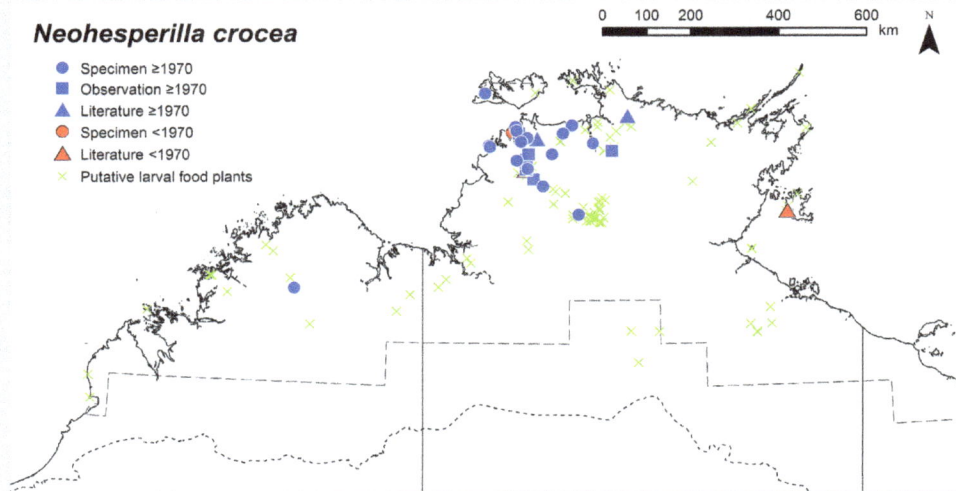

Legend:
- Specimen ≥1970
- Observation ≥1970
- Literature ≥1970
- Specimen <1970
- Literature <1970
- Putative larval food plants

## Neohesperilla crocea

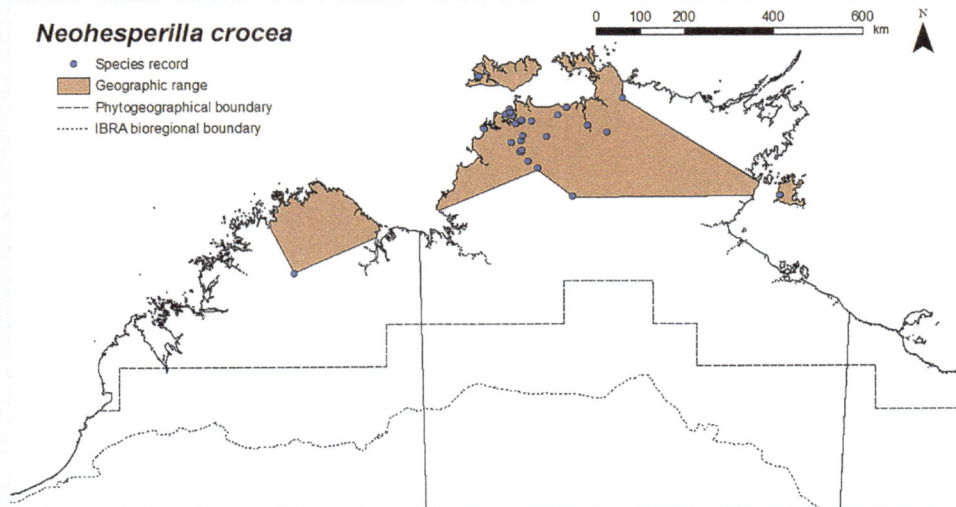

Legend:
- Species record
- Geographic range
- Phytogeographical boundary
- IBRA bioregional boundary

## Neohesperilla crocea (n = 34)

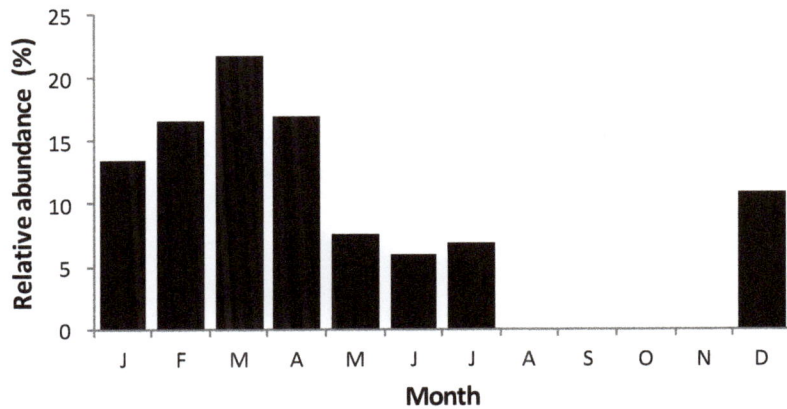

| Month | J | F | M | A | M | J | J | A | S | O | N | D |
|-------|---|---|---|---|---|---|---|---|---|---|---|---|
| Egg   |   |   |   |   |   |   |   |   |   |   |   |   |
| Larva |   |   |   |   |   |   |   |   |   |   |   |   |
| Pupa  |   |   |   |   |   |   |   |   |   |   |   |   |
| Adult |   |   |   |   |   |   |   |   |   |   |   |   |

# Yellow Grass-skipper

## *Neohesperilla xanthomera* (Meyrick & Lower, 1902)

Plate 23 Mt Burrell, NT
Photo: Frank Pierce

## Distribution

This species occurs in the north of the Northern Territory of the study region. It is restricted to the higher rainfall areas (> 1,000 mm mean annual rainfall, but mostly > 1,200 mm) of the north-western corner of the Top End. It has been recorded as far south as Daly River and as far east as Maningrida, NT (Braby 2000). Further field surveys are required to determine whether *N. xanthomera* occurs in the eastern half of the Top End, particularly in eastern Arnhem Land and the Limmen Bight area in the Gulf of Carpentaria. Outside the study region, *N. xanthomera* occurs in north-eastern and eastern Australia.

## Habitat

The breeding habitat of *N. xanthomera* has not been recorded in the study region. Adults have been collected mainly in savannah woodland and they undoubtedly breed in this habitat.

## Larval food plants

Not recorded in the study region. The food plants in north-eastern Queensland comprise various native grasses (Poaceae).

## Seasonality

Adults are seasonal, occurring only during the 'build-up' and wet season (October–March). They appear to be particularly abundant and freshly emerged early in the season (October) following the first substantial wet season rains, but we have too few records (n = 19) to adequately assess seasonal changes in abundance. The breeding phenology and seasonal history of the immature stages have not been recorded, but it is possible only one or two generations are completed annually.

## Breeding status

This species is resident in the study region.

## Conservation status

LC. Although the species *N. xanthomera* has a restricted range in the study region, there are no known threats facing the taxon.

## *Neohesperilla xanthomera*

- ● Specimen ≥1970
- ■ Observation ≥1970
- ▲ Literature ≥1970
- ▲ Literature <1970

## *Neohesperilla xanthomera*

- Species record
- Geographic range
- ‒ ‒ ‒ Phytogeographical boundary
- ······ IBRA bioregional boundary

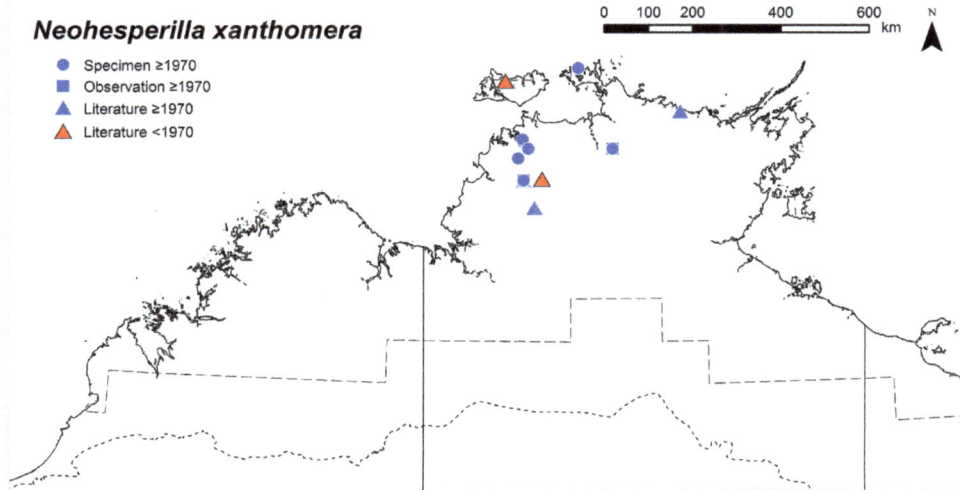

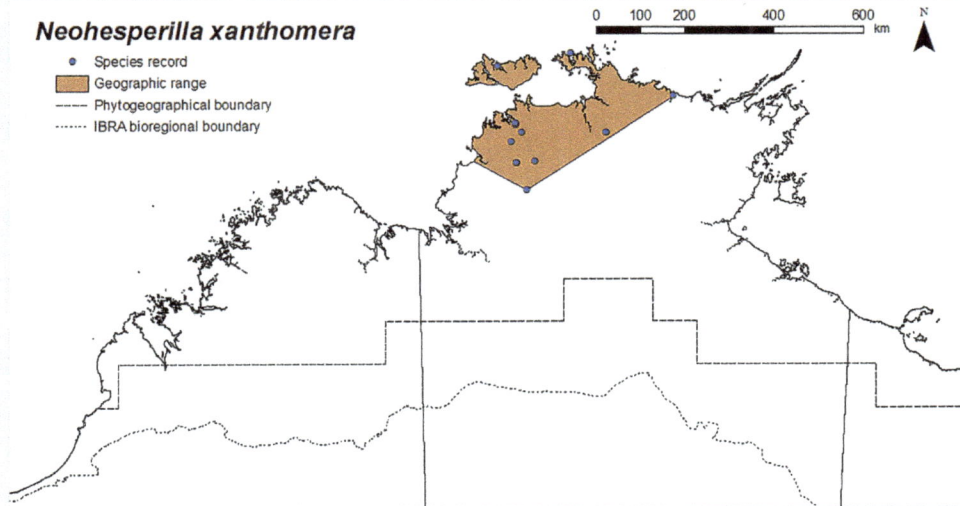

| Month | J | F | M | A | M | J | J | A | S | O | N | D |
|-------|---|---|---|---|---|---|---|---|---|---|---|---|
| Egg   |   |   |   |   |   |   |   |   |   |   |   |   |
| Larva |   |   |   |   |   |   |   |   |   |   |   |   |
| Pupa  |   |   |   |   |   |   |   |   |   |   |   |   |
| Adult |   |   |   |   |   |   |   |   |   |   |   |   |

# Wide-brand Sedge-skipper
## *Hesperilla crypsigramma* (Meyrick & Lower, 1902)

Plate 24 Mt Burrell, NT
Photo: M. F. Braby

## Distribution

This species is represented by an undescribed subspecies, which is endemic to the study region. It occurs in the north of the Northern Territory, its presence detected only as recently as 1989 (Field 1990a). It is restricted to the centre and north-western corner of the Top End, with all records from the higher rainfall areas (> 900 mm mean annual rainfall). The geographic range of *H. crypsigramma* closely corresponds with the spatial distribution of its larval food plant. The food plant, however, has a slightly wider range, occurring also on Melville Island, the Arnhem Land Plateau east of Jabiru and the Port Keats area (Macadam Range), NT. Further field surveys are thus required to determine whether *H. crypsigramma* also occurs in these areas. Outside the study region, *H. crypsigramma* occurs in north-eastern and eastern Australia.

## Excluded data

The record from the western Gulf Country at Boodjamulla/Lawn Hill National Park, Qld (Daniels and Edwards 1998), is excluded. The record was based on '[a] single female taken at Holts Creek' (Daniels and Edwards 1998: 89), but most likely refers to *Hesperilla sexguttata*. Females of the two species are difficult to distinguish, and *Scleria sphacelata*—the larval food plant of *H. crypsigramma*—does not occur in the park or adjacent areas.

## Habitat

*Hesperilla crypsigramma* breeds in eucalypt woodland and low open woodland on laterite or quartz–sandstone rocky outcrops, such as ridges and steep upper slopes and summits of hills, where the larval food plant grows as a sedge in high density.

## Larval food plant

*Scleria sphacelata* (Cyperaceae).

## Seasonality

Adults are seasonal, occurring mainly in the warmer and wetter periods of October and again from February to April. The immature stages (eggs or pupae) have also been detected in these months, indicating that breeding occurs at the time when adults are most abundant. It is not clear whether the few records of adults during the intervening months (November–January) reflect a period of reduced activity or simply a lack of recording. The species appears to survive the dry season (May–September) in the larval stage. Presumably, at least two generations are completed annually.

## Breeding status

This species is resident in the study region.

## Conservation status

DD. The putative subspecies *H. crypsigramma* ssp. 'Top End' is a narrow-range endemic (EOO = 20,980 sq km) and it occurs in several conservation reserves, including Litchfield National Park, Robin Falls and Fish River Station. However, the effect of inappropriate fire regimes requires further investigation. The immature stages are killed by fire and much of its habitat is subjected to landscape fire, with some areas of the range experiencing an increase in frequency of dry season burns, with short interfire intervals. The species may survive as larvae on sedges growing on steep rocky slopes/cliffs protected from fire or in unburnt patches. Presumably, the resulting adults from these larvae then recolonise food plants regenerating in burnt areas during the wet season. Although the larval food plant is currently listed as LC under the *TPWCA*, *H. crypsigramma* may qualify as Near Threatened (NT) once adequate data are available.

## Hesperilla crypsigramma

- Specimen ≥1970
- Observation ≥1970
- Literature ≥1970
- Larval food plant

## Hesperilla crypsigramma

- Species record
- Geographic distribution
- Phytogeographical boundary
- IBRA bioregional boundary

### Hesperilla crypsigramma (n = 27)

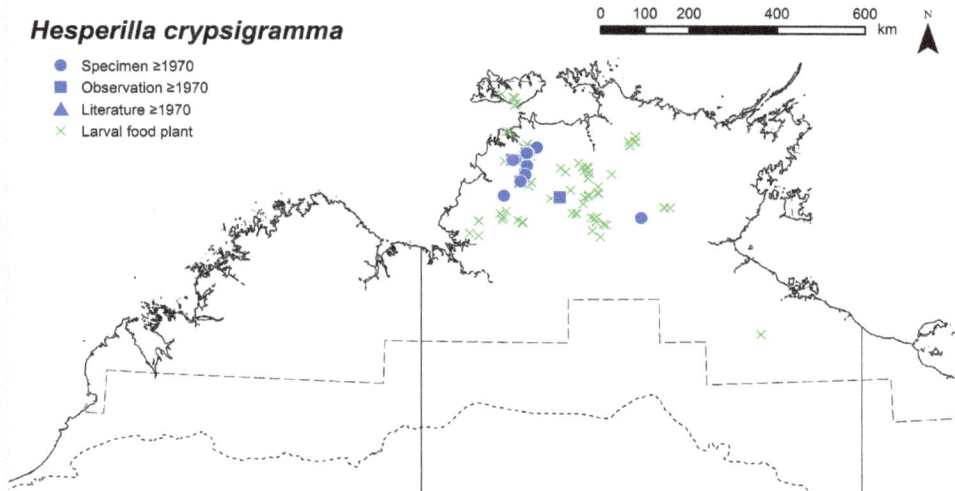

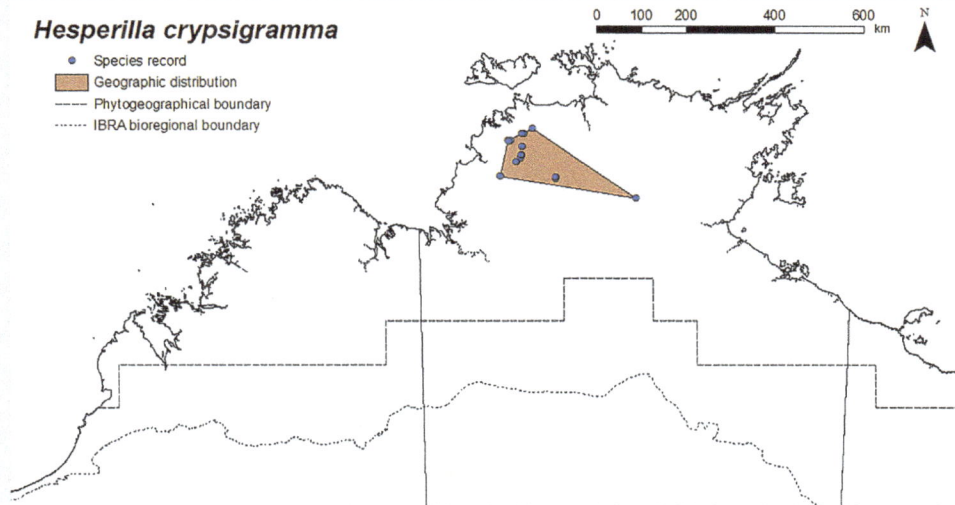

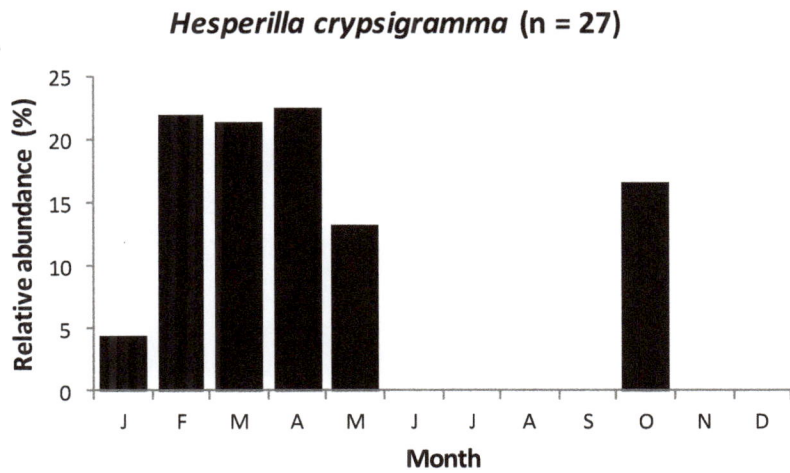

| Month | J | F | M | A | M | J | J | A | S | O | N | D |
|-------|---|---|---|---|---|---|---|---|---|---|---|---|
| Egg   |   |   |   |   |   |   |   |   |   |   |   |   |
| Larva |   |   |   |   |   |   |   |   |   |   |   |   |
| Pupa  |   |   |   |   |   |   |   |   |   |   |   |   |
| Adult |   |   |   |   |   |   |   |   |   |   |   |   |

# Riverine Sedge-skipper

## *Hesperilla sexguttata* Herrich-Schäffer, 1869

Plate 25 Kakadu National Park, NT
Photo: Kenji Nishida

## Distribution

This species occurs widely in the study region, extending from the Kimberley through the Top End to the western Gulf Country and its offshore islands. The geographic range corresponds well with the spatial distribution of its larval food plant, indicating that *H. sexguttata* has been well sampled across the region. The food plant occurs on the Tiwi Islands, and further field surveys are required to determine whether *H. sexguttata* also occurs on Bathurst and Melville islands. Outside the study region, *H. sexguttata* occurs in north-eastern and eastern Australia.

## Habitat

*Hesperilla sexguttata* usually breeds in paperbark swampland, mixed riparian paperbark woodland along creeks and riverine paperbark tall woodland, often with rainforest elements in the understorey or adjacent to monsoon forest, where the larval food plant grows as a sedge. In the drier inland areas, the breeding habitat consists of riverine corridors in which the food plant grows on sandy banks and riverbeds that are inundated during wet season floods. Males also fly in open rocky areas on the summit of steep hills, which are used as encounter sites to locate females for mating, but they do not breed in this habitat.

## Larval food plant

*Cyperus javanicus* (Cyperaceae).

## Seasonality

Adults have been recorded during most months of the year, but we have too few records (n = 17) to assess any seasonal changes in abundance. The immature stages have been recorded mainly from May to October, indicating that the species breeds during the dry season. However, it is very likely *H. sexguttata* breeds continuously throughout the year; the lack of data during the wet season (November–February) is probably because the breeding habitats are frequently inaccessible at that time of year.

## Breeding status

This species is resident in the study region.

## Conservation status

LC.

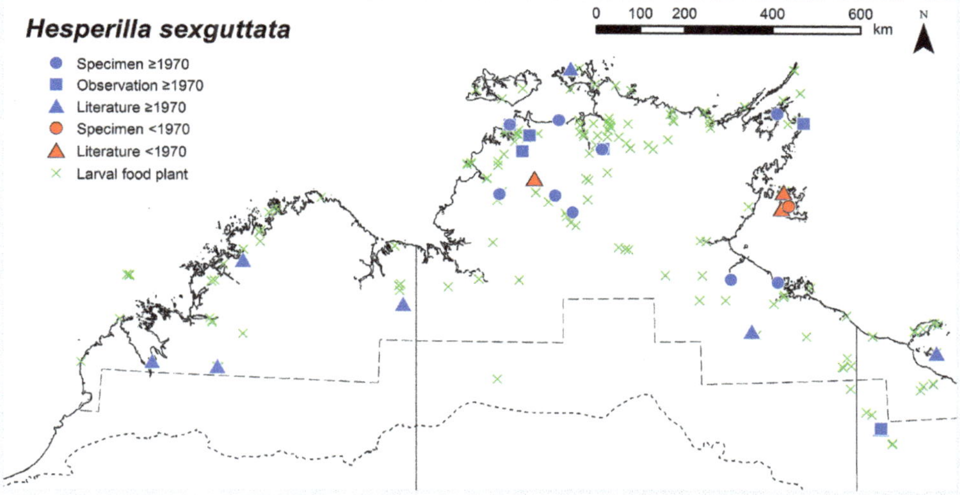

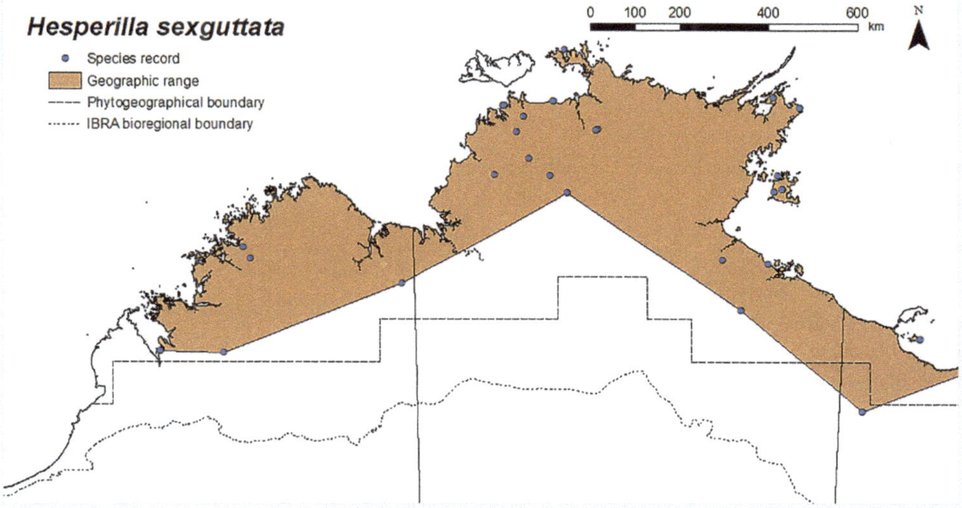

## Hesperilla sexguttata

| Month | J | F | M | A | M | J | J | A | S | O | N | D |
|-------|---|---|---|---|---|---|---|---|---|---|---|---|
| Egg | | | | | | | | | | | | |
| Larva | | | | | | | | | | | | |
| Pupa | | | | | | | | | | | | |
| Adult | | | | | | | | | | | | |

# Spinifex Sand-skipper
## *Proeidosa polysema* (Lower, 1908)

Plate 26 Curtin Springs, NT
Photo: M. F. Braby

## Distribution

This species occurs very widely in the study region. The northernmost limit is in the Noonamah–Berry Springs area, approximately 28 km southeast of Darwin, NT (Braby and Westaway 2016). The broad geographic range corresponds with the spatial distribution of its larval food plants (*Triodia* spp.). However, there are few records of *P. polysema* from the southern Kimberley and Great Sandy Desert, WA, and none from the Tiwi Islands, northern coastal areas and north-eastern Arnhem Land, despite the presence of the food plants. Further field surveys are therefore required to determine whether *P. polysema* occurs in these areas. Outside the study region, *P. polysema* occurs widely in the semi-arid and arid zones of central Australia and in northern and central Queensland.

## Habitat

*Proeidosa polysema* breeds mainly in eucalypt open woodland with a hummock/tussock grass understorey on sand and dry rocky sandstone, favouring shallow gullies, hill-slopes and escarpments where the larval food plants grow as perennial 'soft' resinous spinifex tussock-forming grasses (Braby 2015e). It also occurs in low open woodland on sandstone pavements and hummock grassland on sand dunes, but near Darwin it breeds in eucalypt open woodland on sandy soil derived from laterite (Braby and Westaway 2016).

## Larval food plants

*Triodia bitextura*, *T. microstachya*, *T. pungens* (Poaceae). The main food plants are *T. bitextura* and *T. microstachya*, but *T. pungens* is used in the drier inland areas of lower rainfall (< 900 mm mean annual rainfall).

## Seasonality

Adults are seasonal, occurring only during the wet season (November–April), but there are too few records (n = 19) to assess temporal changes in abundance. There are limited data on the phenology of the immature stages, and the number of generations completed annually is not known. The long dry season is passed in the larval stage—usually the final instar larva, which may remain in diapause inside its shelter for many months.

## Breeding status

This species is resident in the study region.

## Conservation status

LC. Although the species *P. polysema* has a very wide geographical range, the effect of inappropriate fire regimes requires further investigation. The immature stages are killed by fire and much of its habitat is subjected to landscape fire, with some areas of the range experiencing an increase in the frequency of dry season burns. The species may survive as larvae on grasses growing on steep rocky slopes/cliffs protected from fire or in unburnt patches. Presumably, the resulting adults from these larvae then recolonise food plants regenerating in burnt areas during the wet season.

## Proeidosa polysema

- ● Specimen ≥1970
- ■ Observation ≥1970
- ▲ Literature ≥1970
- ▲ Literature <1970
- × Larval food plants

## Proeidosa polysema

- ● Species record
- ▨ Geographic range
- – – Phytogeographical boundary
- ······ IBRA bioregional boundary

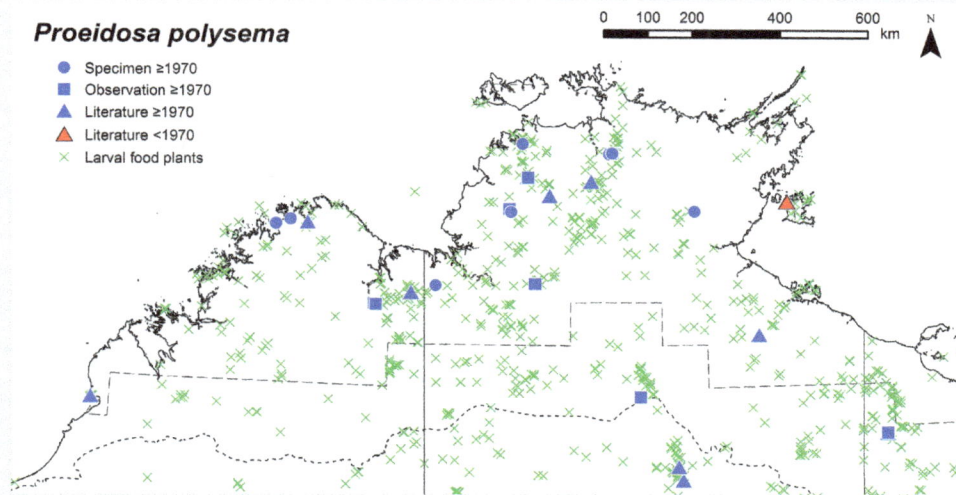

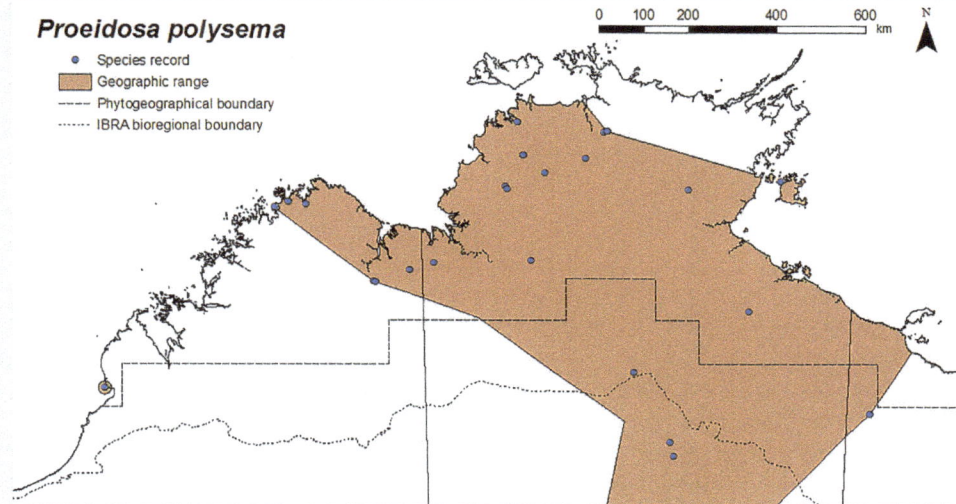

| Month | J | F | M | A | M | J | J | A | S | O | N | D |
|-------|---|---|---|---|---|---|---|---|---|---|---|---|
| Egg   |   |   |   |   |   |   |   |   |   |   |   |   |
| Larva |   |   |   | ▨ | ▨ | ▨ |   | ▨ | ▨ |   |   | ▨ |
| Pupa  |   |   |   |   |   |   |   |   |   |   |   |   |
| Adult | ▨ | ▨ | ▨ |   |   |   |   |   |   |   | ▨ | ▨ |

# Northern Iris-skipper
*Mesodina gracillima* E. D. Edwards, 1987

Plate 27 Marrakai Road, NT
Photo: M. F. Braby

Plate 28 Marrakai Road, NT
Photo: M. F. Braby

## Distribution

This species is endemic to the study region. It is restricted to the north of the Northern Territory, where it occurs in the higher rainfall areas (> 1,200 mm mean annual rainfall) of the Top End. Its geographic range closely corresponds with the spatial distribution of its larval food plant. The food plant, however, has a slightly broader range, extending further south to the Katherine district, including Nitmiluk (Katherine Gorge) National Park, NT. Further field surveys are therefore required to determine whether *M. gracillima* occurs in this area.

## Habitat

*Mesodina gracillima* breeds in eucalypt woodland and open woodland with a sparse understorey of grasses and herbs, including the larval food plant, which grows on well-drained sandy soils, often on gently sloping terrain or ridges (Edwards 1987).

## Larval food plant

*Patersonia macrantha* (Iridaceae).

## Seasonality

Adults have been recorded during most months of the year. They appear to be more abundant towards the end of the wet season (February and March), but we have too few records (n = 17) to adequately assess any seasonal changes in abundance. The immature stages have also been recorded during most months of the year, particularly during the late wet season and early to mid dry season (February–August). Larval development is protracted, but it is not certain how many generations are completed annually. It is likely *M. gracillima* breeds continuously throughout the year, except perhaps in the late dry season, when the food plants may become stressed from lack of water.

## Breeding status

This species is resident in the study region.

## Conservation status

LC. The species *M. gracillima* has a restricted range within which it occurs in several conservation reserves, including Garig Gunak Barlu National Park, Kakadu National Park and Djelk IPA. However, the effect of inappropriate fire regimes as a potential threat requires further investigation (Young 2005). The immature stages are killed by fire and much of its habitat is subjected to landscape fire, with some areas of the range experiencing an increase in the frequency of dry season burns. Presumably, some larvae survive on clumps growing in unburnt patches and the resulting adults then recolonise food plants regenerating in burnt areas during the wet season. It is not clear what the long-term effects of increased fire frequency are for the species and its larval food plant. The larval food plant is currently listed as LC under the *TPWCA*. Monitoring of the abundance or occupancy of the butterfly and its food plant is required to clarify the effect of short interfire intervals as a key threatening process.

## Mesodina gracillima

- ● Specimen ≥1970
- ■ Observation ≥1970
- ▲ Literature ≥1970
- ● Specimen <1970
- ▲ Literature <1970
- ✕ Larval food plant

## Mesodina gracillima

- ● Species record
- ▬ Geographic range
- — — Phytogeographical boundary
- ········ IBRA bioregional boundary

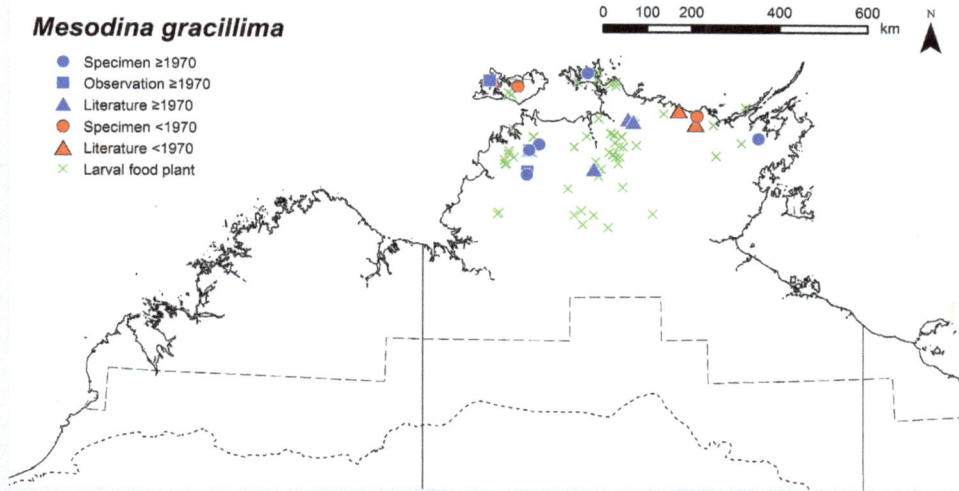

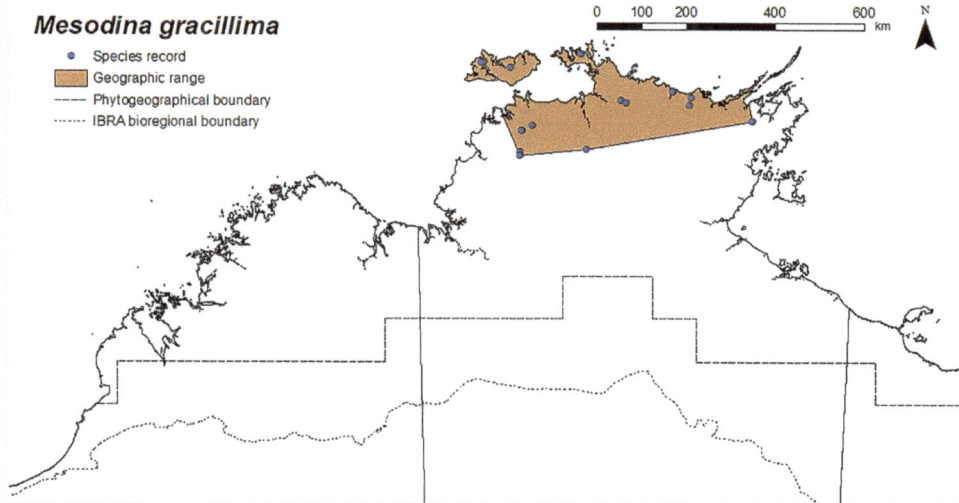

| Month | J | F | M | A | M | J | J | A | S | O | N | D |
|-------|---|---|---|---|---|---|---|---|---|---|---|---|
| Egg   |   |   |   |   |   |   |   |   |   |   |   |   |
| Larva |   |   |   |   |   |   |   |   |   |   |   |   |
| Pupa  |   |   |   |   |   |   |   |   |   |   |   |   |
| Adult |   |   |   |   |   |   |   |   |   |   |   |   |

# Orange Swift

## *Parnara amalia* (Semper, [1879])

Plate 29 Lake Mitchell, north of Mareeba, Qld
Photo: Frank Pierce

Plate 30 Lake Mitchell, north of Mareeba, Qld
Photo: Frank Pierce

## Distribution

This species occurs in the north of the Northern Territory of the study region. It is restricted to the north-western corner of the Top End, where it occurs in the higher rainfall areas (generally > 1,300 mm mean annual rainfall). It extends from Darwin south to the Daly River crossing (Oolloo crossing) (Hutchinson 1978) and east to Cooper Creek (Oenpelli–Murgenella Road crossing) (Dunn and Dunn 1991), NT. Further field surveys are required to determine whether *P. amalia* occurs in the eastern half of the Top End, particularly in north-eastern Arnhem Land. Outside the study region, *P. amalia* occurs in mainland New Guinea and eastern Australia.

## Habitat

The breeding habitat of *P. amalia* has not been recorded in the study region. Adults have been collected in a wide range of habitats, but they are generally more prevalent in paperbark swampland in juxtaposition to monsoon forest or mangroves. Presumably, the species breeds in this habitat and other moist low-lying areas.

## Larval food plants

*Oryza sativa* (Poaceae). The native food plant has not been recorded in the study region, but larvae have been reared on cultivated rice at Humpty Doo (C. S. Li) and Tortilla Flats (C. Wilson).

## Seasonality

Adults occur throughout the year, but they are most abundant during the wet season and early dry season, particularly in May following good wet seasons of average or above average rainfall. The breeding phenology and seasonal history of the immature stages have not been recorded, but it is likely the species breeds throughout the year.

## Breeding status

This species is resident in the study region.

## Conservation status

LC. The species *P. amalia* has a restricted range in the study region within which it occurs in several conservation reserves, including Howard Springs Nature Reserve, Fogg Dam Conservation Reserve, Litchfield National Park and Kakadu National Park. Despite its restricted occurrence, there are no known threats facing the taxon.

## Parnara amalia

- ● Specimen ≥1970
- ■ Observation ≥1970
- ▲ Literature ≥1970
- ● Specimen <1970

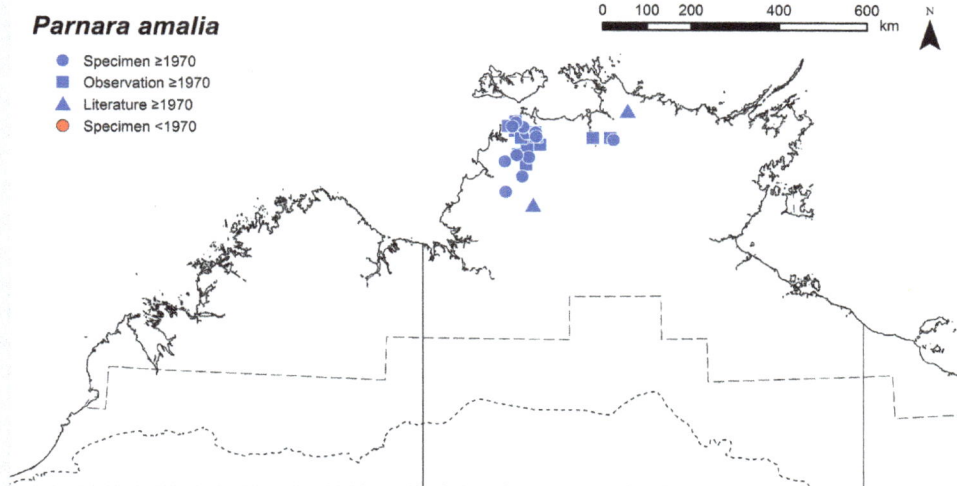

## Parnara amalia

- • Species record
- Geographic range
- — — Phytogeographical boundary
- ········ IBRA bioregional boundary

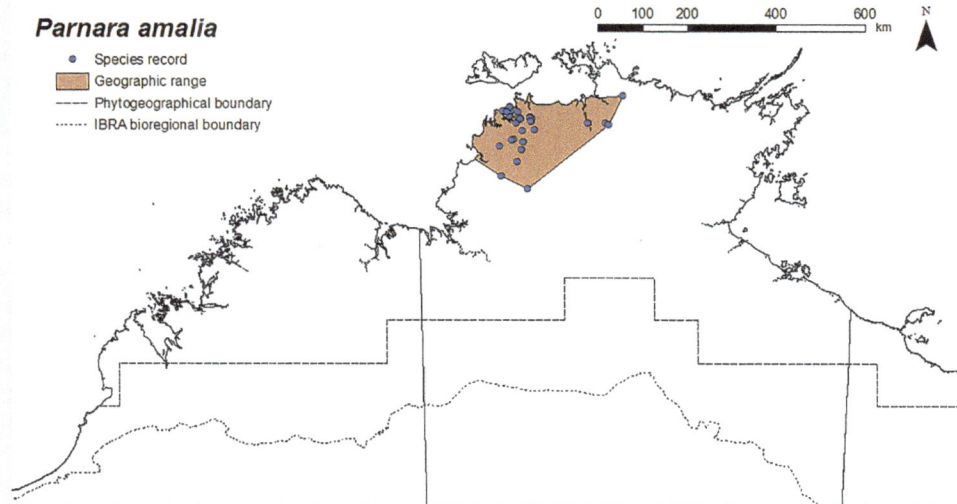

### Parnara amalia (n = 43)

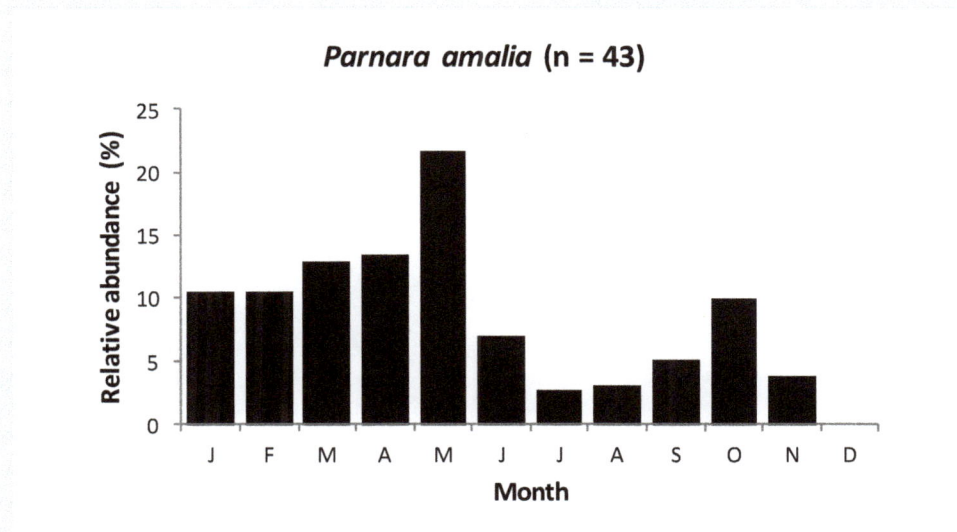

| Month | J | F | M | A | M | J | J | A | S | O | N | D |
|-------|---|---|---|---|---|---|---|---|---|---|---|---|
| Egg   |   |   |   |   |   |   |   |   |   |   |   |   |
| Larva |   |   |   |   |   |   |   |   |   |   |   |   |
| Pupa  | ▨ |   | ▨ |   |   |   |   |   |   |   |   |   |
| Adult | ▨ | ▨ | ▨ | ▨ | ▨ | ▨ | ▨ | ▨ | ▨ | ▨ | ▨ |   |

# Rice Swift
## Borbo cinnara (Wallace, 1866)

Plate 31 Christmas Island, WA
Photo: Frank Pierce

Plate 32 Christmas Island, WA
Photo: Frank Pierce

## Distribution

This species occurs in the north of the Northern Territory of the study region, where it is restricted to the north-western corner of the Top End. Historically, it was recorded from Darwin, Berry Springs and Adelaide River, NT. However, we are not aware of any additional records since it was first collected in 1948 by F. M. Angel and F. E. Parsons (Angel 1951)—a period of 70 years. Further field surveys are thus required to determine whether the species is still extant in the region. The putative larval food plant (*Rottboellia cochinchinensis*) occurs in the northern coastal areas of the Top End. Outside the study region, *B. cinnara* occurs widely from India, southern China and South-East Asia, through mainland New Guinea and adjacent islands and north-eastern Australia to the Solomon Islands, Vanuatu and New Caledonia.

## Habitat

The breeding habitat of *B. cinnara* has not been recorded in the study region. Sands and New (2002) suggested the species was likely to breed in wetland habitats.

## Larval food plants

Not recorded in the study region; possibly *Rottboellia cochinchinensis* (Poaceae), which is the food plant in the Torres Strait Islands, Qld (Braby 2000).

## Seasonality

The seasonal abundance and breeding phenology of this species are not well understood. Adults have been recorded in April and May, but there are too few records (n = 4) to assess any seasonal changes in abundance.

## Breeding status

*Borbo cinnara* does not appear to have become permanently established in the study region. Adults of *B. cinnara* resemble those of *B. impar* and *Pelopidas lyelli* and thus it may have been overlooked. However, at the time of its discovery in the Northern Territory, Angel (1951: 13) remarked that *B. cinnara* 'was more plentiful than the preceding species' (*Borbo impar lavinia*), which suggests the species was breeding locally. Currently, *B. impar* is seasonally abundant in the study region, whereas *B. cinnara* now appears to be absent. Angel's early observations were made before attempts to develop the rice industry in the Top End in the late 1950s to early 1960s (cultivated rice, *Oryza sativa*, is a common larval food plant in South-East Asia), which indicates that the subsequent collapse of the rice industry does not account for the disappearance of this species. It is therefore possible that *B. cinnara* is a vagrant or a rare immigrant from South-East Asia, occasionally colonising the region and breeding temporarily during favourable conditions.

## Conservation status

NA.

## *Borbo cinnara*

● Specimen <1970
▲ Literature <1970

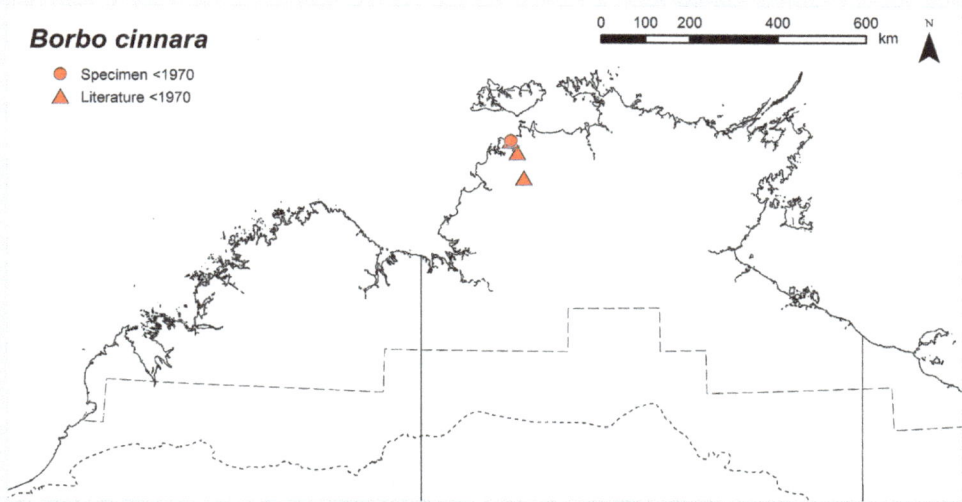

| Month | J | F | M | A | M | J | J | A | S | O | N | D |
|-------|---|---|---|---|---|---|---|---|---|---|---|---|
| Egg   |   |   |   |   |   |   |   |   |   |   |   |   |
| Larva |   |   |   |   |   |   |   |   |   |   |   |   |
| Pupa  |   |   |   |   |   |   |   |   |   |   |   |   |
| Adult |   |   |   | ■ | ■ |   |   |   |   |   |   |   |

Photo: Piccaninny, Bungle Bungles, WA, M.F. Braby

# Yellow Swift
## *Borbo impar* (Mabille, 1883)

Plate 33 Wanguri, Darwin, NT
Photo: M. F. Braby

Plate 34 Fogg Dam, NT
Photo: M. F. Braby

## Distribution

This species is represented by the subspecies *B. impar lavinia* (Waterhouse 1932), which is endemic to the study region. It occurs in the north of the Northern Territory, where it is restricted to the higher rainfall areas (> 1,300 mm mean annual rainfall) of the north-western corner of the Top End and also on Groote Eylandt, NT (Dunn and Dunn 1991). The geographic range closely corresponds with the spatial distribution of its preferred native larval food plant (*Hymenachne acutigluma*). The food plant, however, has a broader range, occurring in coastal areas near Port Keats (Palumpa Billabong) and in north-eastern Arnhem Land (Arafura Swamp, Goromuru River floodplain), NT. Further field surveys are thus required to determine whether *B. impar* also occurs in these areas. Outside the study region, *B. impar* occurs from Maluku, through mainland New Guinea and adjacent islands and the Torres Strait Islands, Qld, to the Solomon Islands and New Caledonia.

## Habitat

*Borbo impar* breeds mainly in floodplain wetlands and swamps where the native larval food plant (*Hymenachne acutigluma*) grows as a grass in standing water (Braby 2011a). It also breeds along the edges of monsoon forest where the introduced food plants *Cenchrus pedicellatus* and *Megathyrsus maximus* grow as tall grassy weeds, particularly in urban or disturbed areas.

## Larval food plants

*Hymenachne acutigluma*, *Whiteochloa airoides* (Poaceae); also *\*Cenchrus pedicellatus*, *\*Megathyrsus maximus*, *\*Oryza sativa*, *\*Zea mays* (Poaceae).

The main native food plant appears to be *H. acutigluma* (Braby 2011a), but the species also breeds seasonally on introduced *C. pedicellatus* and *M. maximus* (Meyer 1996a, 1997b).

## Seasonality

Adults occur throughout the year, but they are most abundant towards the end of the wet season (March and April). Few adults have been recorded during the mid to late dry season (June–September). The immature stages have been recorded from February to June and during the 'build-up' (October and November), and are not known to undergo diapause during the dry season. Larvae and pupae are particularly abundant in February–April, when the invasive annual mission grass (*Cenchrus pedicellatus*) germinates and produces new leaf growth following the monsoon rains. Presumably, the species breeds throughout the year on the perennial grasses and several generations are completed annually.

## Breeding status

This species is resident in the study region.

## Conservation status

LC. The subspecies *B. impar lavinia* is a narrow-range endemic (geographic range = 35,360 sq km) and it occurs in several conservation reserves, including Charles Darwin National Park, Black Jungle Conservation Reserve, Fogg Dam Conservation Reserve, Manton Dam Recreation Area and Kakadu National Park. Despite its restricted occurrence, there are no known threats facing the taxon.

## Borbo impar

- ● Specimen ≥1970
- ■ Observation ≥1970
- ▲ Literature ≥1970
- ● Specimen <1970
- ▲ Literature <1970
- ✕ Larval food plant

## Borbo impar

- ● Species record
- ▨ Geographic range
- — — Phytogeographical boundary
- ········· IBRA bioregional boundary

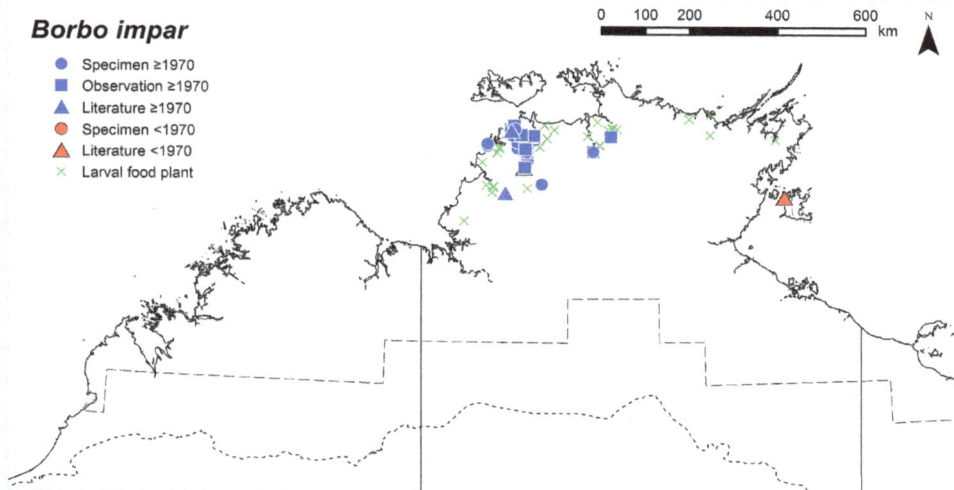

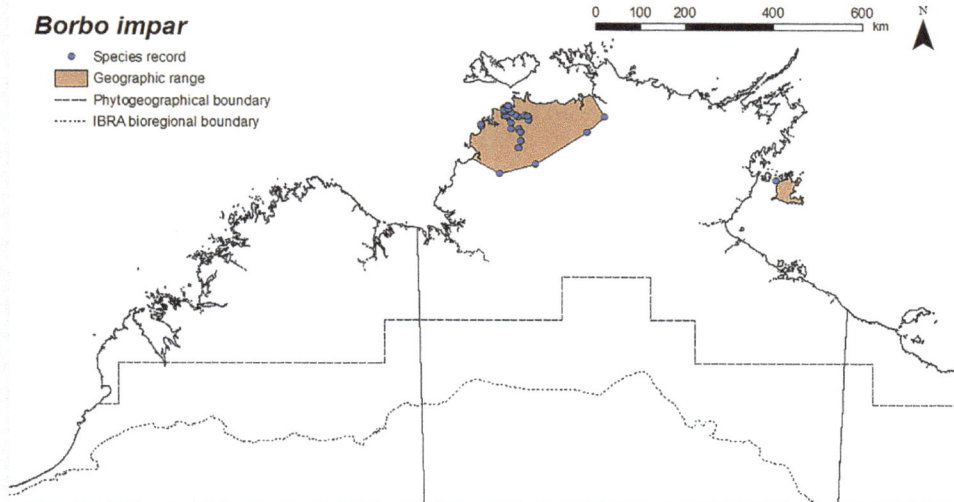

## Borbo impar (n = 57)

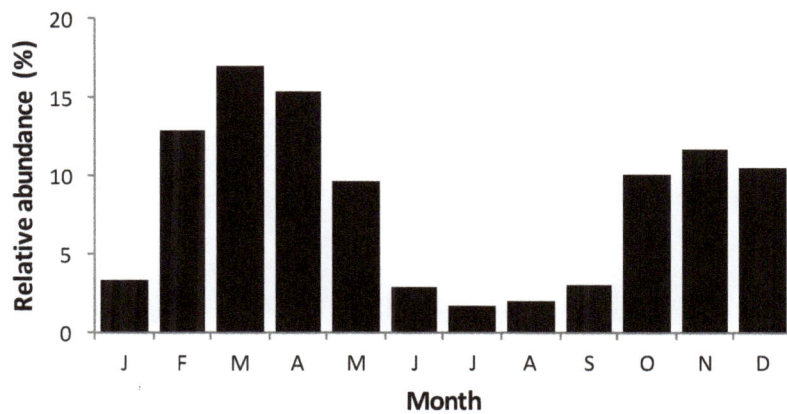

| Month | J | F | M | A | M | J | J | A | S | O | N | D |
|-------|---|---|---|---|---|---|---|---|---|---|---|---|
| Egg   |   | ▨ |   | ▨ |   | ▨ |   |   |   |   |   |   |
| Larva |   | ▨ | ▨ | ▨ | ▨ | ▨ |   |   |   | ▨ |   |   |
| Pupa  |   | ▨ | ▨ | ▨ |   |   |   |   |   |   | ▨ |   |
| Adult | ▨ | ▨ | ▨ | ▨ | ▨ | ▨ | ▨ | ▨ | ▨ | ▨ | ▨ | ▨ |

# Lyell's Swift
## *Pelopidas lyelli* (Rothschild, 1915)

Plate 35 Atherton, Qld
Photo: Don Franklin

## Distribution

This species is represented in the study region by the subspecies *P. lyelli lyelli* (Rothschild, 1915). It occurs widely throughout the region, extending from the Kimberley, through the Top End and Northern Deserts to the western Gulf Country. Its geographic range closely corresponds with the spatial distribution of its native larval food plants, indicating that *P. lyelli* has been well sampled in the region. Outside the study region, *P. lyelli* occurs from Maluku, through mainland New Guinea and adjacent islands and north-eastern and eastern Australia to the Solomon Islands and Vanuatu.

Previous records of the dingy swift, *Pelopidas agna* (Moore, 1866), refer to this species (see Braby 2012b).

## Habitat

*Pelopidas lyelli* occurs in a wide range of habitats, but it breeds mainly in riparian habitats, including woodland, open forest and the edges of mixed monsoon forest along seasonal creeks and riverbanks where the native larval food plants grow as dense perennial grasses in moist open areas (Braby 2015e).

## Larval food plants

*Chrysopogon elongatus*, *Eriachne triodioides*, *Mnesithea rottboellioides* (Poaceae); also *Cenchrus pedicellatus*, *Megathyrsus maximus*, *Oryza sativa* (Poaceae).

## Seasonality

Adults occur throughout the year and, like *Parnara amalia* and *Borbo impar*, they are most abundant during the late wet season and early dry season (April and May) after the monsoon rains have fallen and the larval food plants are green and luxuriant. The immature stages have been recorded sporadically from January to June, and are not known to undergo diapause during the dry season. Presumably, the species breeds continuously throughout the year.

## Breeding status

This species is resident in the study region.

## Conservation status

LC.

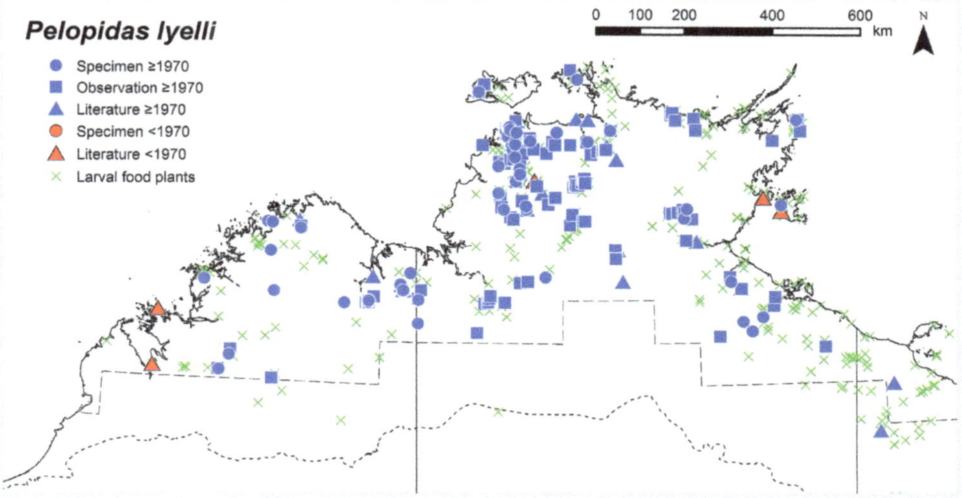

## Pelopidas lyelli

- ● Specimen ≥1970
- ■ Observation ≥1970
- ▲ Literature ≥1970
- ● Specimen <1970
- ▲ Literature <1970
- ✕ Larval food plants

0 100 200 400 600 km

N

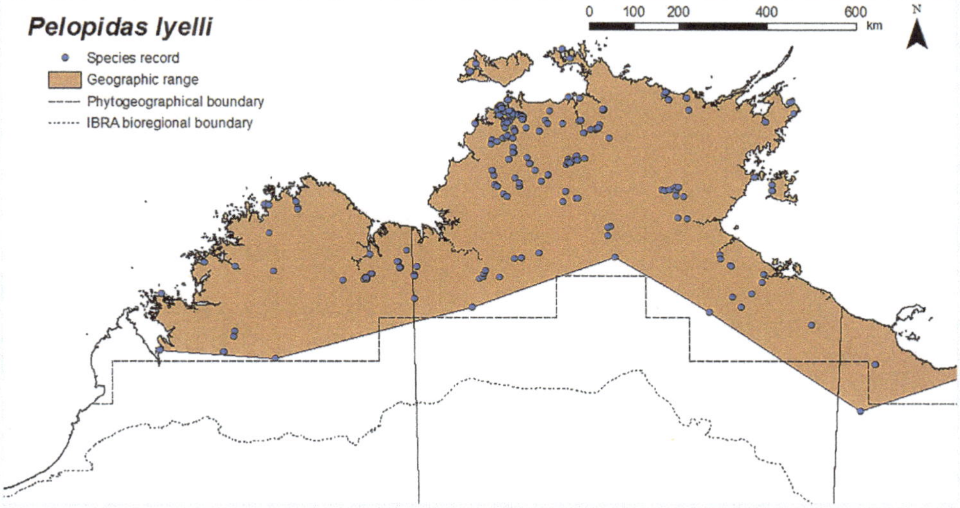

## Pelopidas lyelli

- ● Species record
- ▨ Geographic range
- — — Phytogeographical boundary
- ······ IBRA bioregional boundary

0 100 200 400 600 km

N

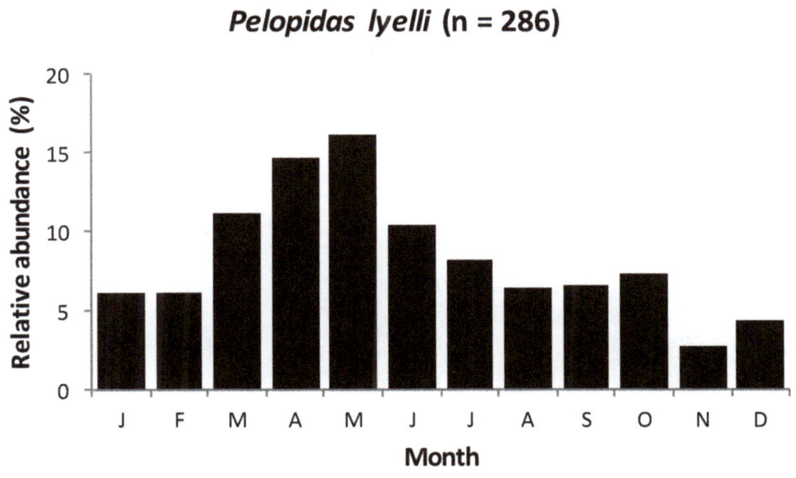

### Pelopidas lyelli (n = 286)

| Month | J | F | M | A | M | J | J | A | S | O | N | D |
|-------|---|---|---|---|---|---|---|---|---|---|---|---|
| Egg   |   |   |   |   |   |   |   |   |   |   |   |   |
| Larva | ██ |   |   |   | ██ | ██ |   |   |   |   |   |   |
| Pupa  | ██ | ██ | ██ |   | ██ | ██ |   |   |   |   |   |   |
| Adult | ██ | ██ | ██ | ██ | ██ | ██ | ██ | ██ | ██ | ██ | ██ | ██ |

# Large Yellow Grass-dart
## *Taractrocera anisomorpha* (Lower, 1911)

Plate 36 Standley Chasm, West MacDonnell
Ranges, NT
Photo: M. F. Braby

## Distribution

This species occurs widely in the study region,
extending from moist coastal areas to drier inland
areas of the semi-arid zone. Most records are from
areas that receive above 600 mm mean annual rainfall,
but it has also been recorded from Tennant Creek
(< 400 mm) (Waterhouse and Lyell 1914) and the
arid zone of central Australia beyond the southern
boundary of the study region. The distribution of
the native larval food plant is considerably broader
than the geographic range of *T. anisomorpha*,
occurring throughout northern Australia, especially
in semi-arid areas. *Taractrocera anisomorpha* has
a similar geographic range to the closely related
*Taractrocera ina*; however, there are no records of
*T. anisomorpha* from eastern Arnhem Land, the
Limmen Bight area in the Gulf of Carpentaria or
western Queensland in the Gulf Country, despite
the presence of *T. ina* in these areas, suggesting it
may have been overlooked. Further field surveys are
thus required to determine whether *T. anisomorpha*
occurs in these areas and to clarify the extent to
which it occurs in the drier inland areas of the study
region. Outside the study region, *T. anisomorpha*
occurs in the Lesser Sunda Islands, the Pilbara of
Western Australia, the southern Northern Territory
and northern and eastern Queensland.

## Habitat

The breeding habitat of *T. anisomorpha* has been
recorded only from the arid zone just outside the
study region, where the species is found associated
with tussocks of the native larval food plant growing
along dry watercourses and sand banks of rivers in
central Australia (Atkins 1991). Elsewhere, adults
have been collected in savannah woodland, often
near creeks, and they undoubtedly breed in this
habitat.

## Larval food plants

*Eulalia aurea* (Poaceae); also *\*Sorghum bicolor*
(Poaceae).

## Seasonality

Adults are seasonal, occurring only during the wet
season and early dry season (November–May),
with a peak in abundance in March. The breeding
phenology and seasonal history of the immature
stages have not been recorded, but it is possible
there are only one or two generations annually.
Presumably, the species remains in larval diapause
for long periods (possibly up to nine months) during
the drier months, as has been recorded elsewhere in
Australia (Atkins 1991).

## Breeding status

This species is resident in the study region.

## Conservation status

LC.

## *Taractrocera anisomorpha*

- ● Specimen ≥1970
- ■ Observation ≥1970
- ▲ Literature ≥1970
- ● Specimen <1970
- ▲ Literature <1970
- ✕ Larval food plant

## *Taractrocera anisomorpha*

- ● Species record
- ▨ Geographic range
- --- Phytogeographical boundary
- ⋯ IBRA bioregional boundary

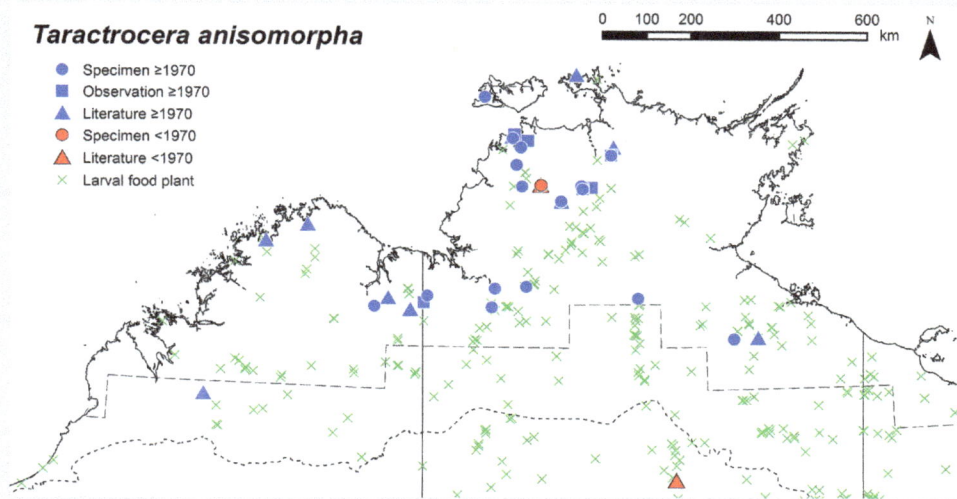

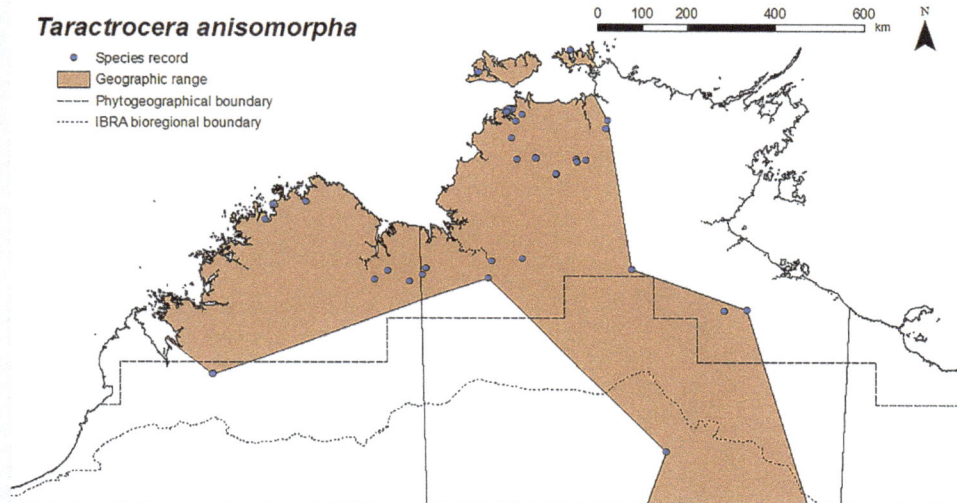

### *Taractrocera anisomorpha* (n = 38)

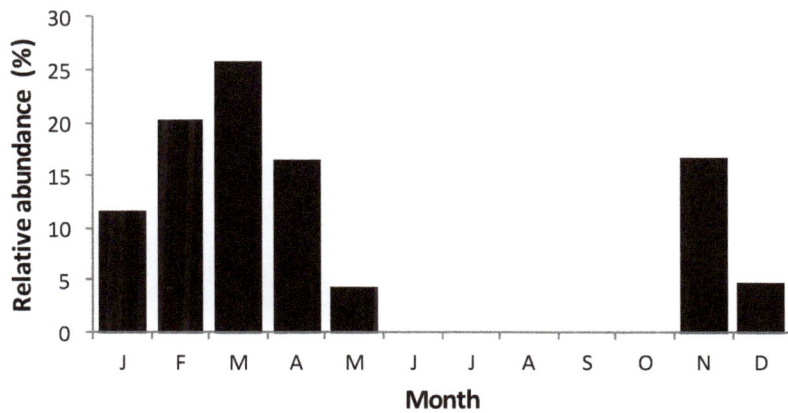

| Month | J | F | M | A | M | J | J | A | S | O | N | D |
|-------|---|---|---|---|---|---|---|---|---|---|---|---|
| Egg | | | | | | | | | | | | |
| Larva | | | | | | | | | | | | |
| Pupa | | | | | | | | | | | | |
| Adult | ▨ | ▨ | ▨ | ▨ | ▨ | | | | | | ▨ | ▨ |

# No-brand Grass-dart

## *Taractrocera ina* Waterhouse, 1932

Plate 37 Dundee Beach, NT
Photo: M. F. Braby

## Distribution

This species occurs widely in the study region. It occurs in the northern Kimberley and throughout much of the Northern Territory, where it extends to the arid zone of central Australia beyond the southern boundary of the study region. It has a similar geographic range to the closely related *Taractrocera anisomorpha*, although there are fewer records from the Kimberley and Northern Deserts. The spatial distribution of one of its main larval food plants (*Cymbopogon procerus*) suggests *T. ina* may be more widespread in semi-arid areas of the southern Kimberley and Great Sandy and Tanami deserts; thus, further field surveys are required to determine whether *T. ina* is established in these areas. Outside the study region, *T. ina* occurs in mainland New Guinea and central and eastern Australia.

## Habitat

*Taractrocera ina* breeds in savannah woodland and mixed monsoon vine thicket, favouring open or disturbed areas with a grassy understorey, where the native and introduced larval food plants grow as annual or perennial grasses (Braby 2011a, 2015e).

## Larval food plants

*Cymbopogon procerus*, *Sorghum macrospermum*, *Sorghum* sp. (Poaceae); also *Cenchrus pedicellatus*, *Cymbopogon citratus*, *Megathyrsus maximus* (Poaceae); possibly *Sacciolepis indica* (Poaceae).

## Seasonality

Adults are seasonal, occurring mainly during the warmer months, when humidity is higher, with a peak in abundance in the mid wet season (January). They are generally absent during the cooler dry season (May–August). The immature stages (larvae or pupae) have been recorded from January to May, during which several generations are completed. However, as the dry season progresses and the food plants dry out or decline in quality, the final instar larvae stop feeding and remain in diapause inside their tubular shelters for up to six months. Presumably, the larvae pupate and emerge as adults a few weeks later in response to the first pre-monsoon storms towards the end of the dry season and start of the wet season (September–November or December).

## Breeding status

This species is resident in the study region.

## Conservation status

LC.

## Taractrocera ina

- ● Specimen ≥1970
- ■ Observation ≥1970
- ▲ Literature ≥1970
- ● Specimen <1970
- ▲ Literature <1970
- × Larval food plants

0 100 200 400 600 km

N

## Taractrocera ina

- ● Species record
- ▬ Geographic range
- — — Phytogeographical boundary
- ......... IBRA bioregional boundary

0 100 200 400 600 km

N

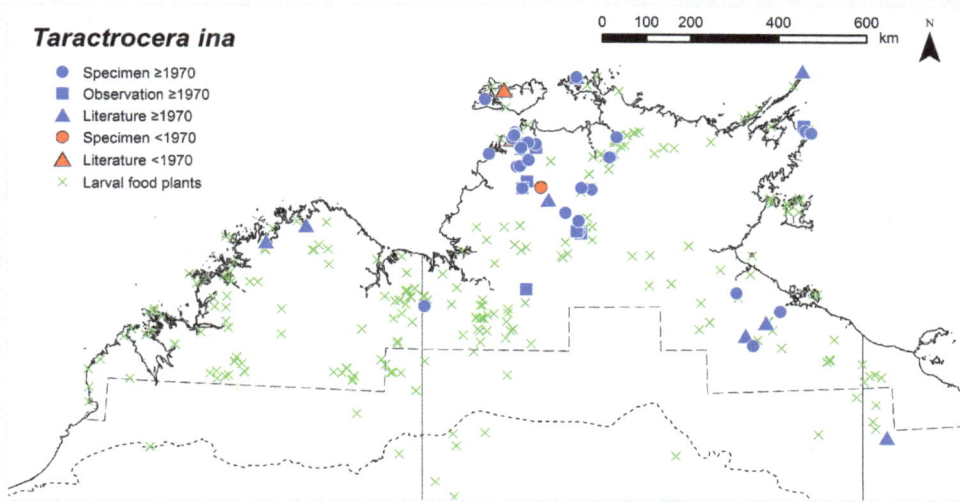

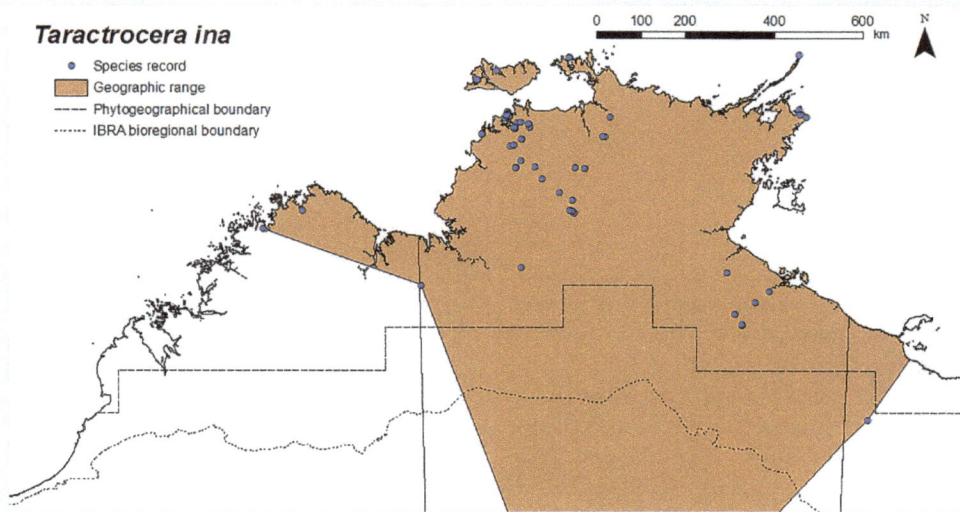

### Taractrocera ina (n = 70)

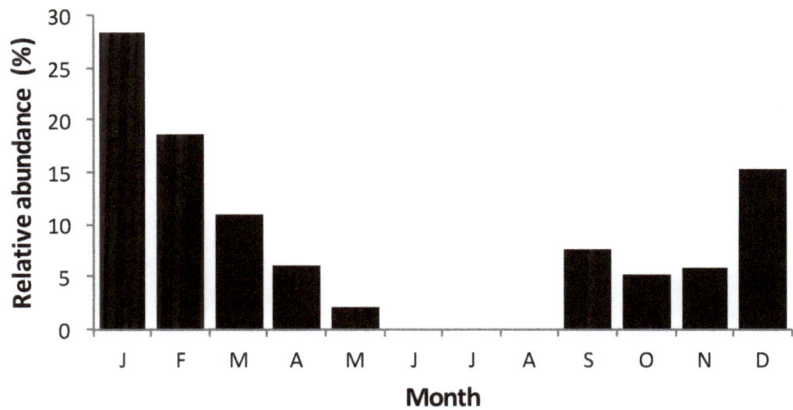

| Month | J | F | M | A | M | J | J | A | S | O | N | D |
|-------|---|---|---|---|---|---|---|---|---|---|---|---|
| Egg   |   |   |   |   |   |   |   |   |   |   |   |   |
| Larva |   |   |   |   |   |   |   |   |   |   |   |   |
| Pupa  |   |   |   |   |   |   |   |   |   |   |   |   |
| Adult |   |   |   |   |   |   |   |   |   |   |   |   |

# River-sand Grass-dart
## *Taractrocera dolon* (Plötze, 1884)

Plate 38 Mt Burrell, NT
Photo: M. F. Braby

## Distribution

This species is represented by the subspecies *T. dolon diomedes* Waterhouse, 1933, which is endemic to the study region. It has a disjunct distribution, occurring in the northern and western Kimberley at Middle Osborn Island and Samson Inlet, WA (J. E. and A. Koeyers); and the Top End, where it has been recorded mostly from the higher rainfall areas in the north-western corner (> 1,000 mm mean annual rainfall). It also occurs on the Wessel Islands (Rimbija Island) in the north-east (Dunn and Dunn 1991) and extends into drier areas of lower rainfall (< 800 mm), where it has been recorded as far south as Coolibah Station and Leila Creek, NT (Dunn and Dunn 1991). It has not been recorded from the Tiwi Islands, Cobourg Peninsula or Groote Eylandt; thus, further field surveys are required to determine whether *T. dolon* is present in these areas. Outside the study region, *T. dolon* occurs in eastern Australia and Tagula Island in the Louisade Archipelago.

## Habitat

The breeding habitat of *T. dolon* has not been recorded in the study region. Adults have been collected mainly in savannah woodland and riparian woodland, favouring open disturbed areas with patches of bare ground, and no doubt they breed on native grasses in these habitats.

## Larval food plants

Not recorded in the study region. The food plants in Queensland consist of various grasses (Poaceae), the identity of which have not been documented (Braby 2000).

## Seasonality

Adults are seasonal, being most abundant during the wet season (January–April). Very few adults have been recorded during the long dry season (May–October), and none has been recorded during the cooler winter months (June–August). A few adults have been collected after pre-monsoon storms in the early wet season (November and December). The breeding phenology and seasonal history of the immature stages have not been recorded, but it is possible only one or two generations are completed annually. Presumably, the species remains in larval diapause during the dry season.

## Breeding status

This species is resident in the study region.

## Conservation status

LC.

## Taractrocera dolon

- ● Specimen ≥1970
- ■ Observation ≥1970
- ▲ Literature ≥1970
- ● Specimen <1970
- ▲ Literature <1970

## Taractrocera dolon

- ● Species record
- Geographic range
- — — Phytogeographical boundary
- ········· IBRA bioregional boundary

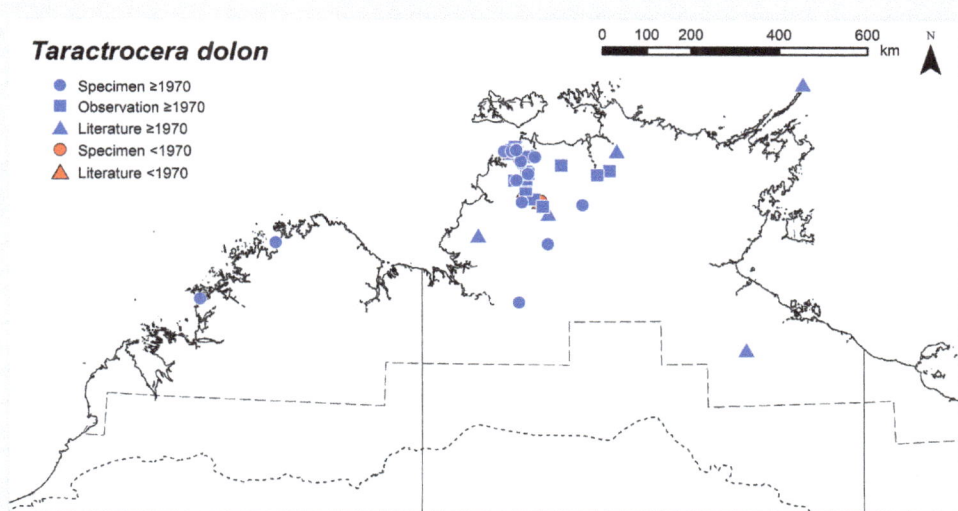

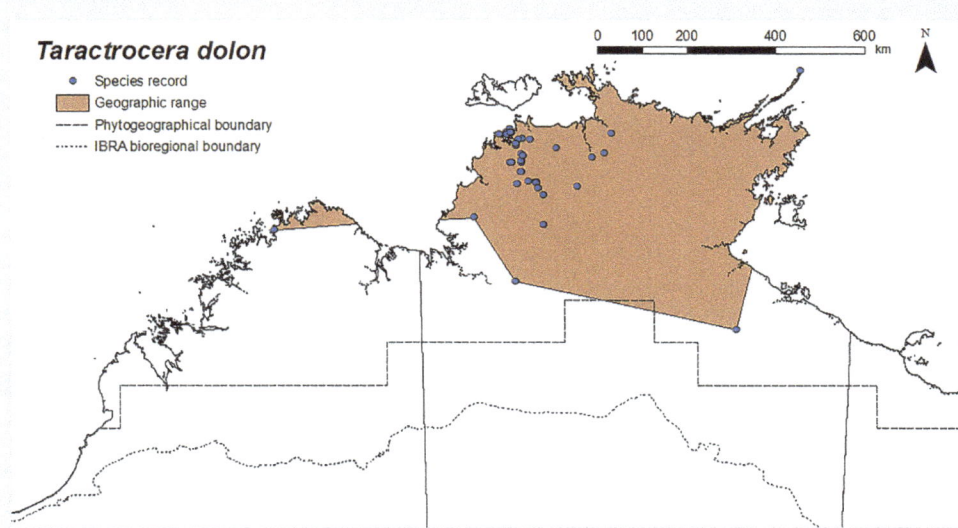

## Taractrocera dolon (n = 60)

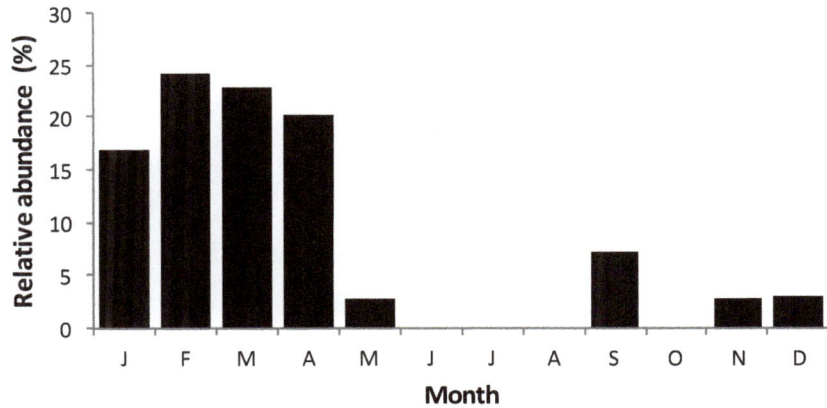

| Month | J | F | M | A | M | J | J | A | S | O | N | D |
|-------|---|---|---|---|---|---|---|---|---|---|---|---|
| Egg |  |  |  |  |  |  |  |  |  |  |  |  |
| Larva |  |  |  |  |  |  |  |  |  |  |  |  |
| Pupa |  |  |  |  |  |  |  |  |  |  |  |  |
| Adult |  |  |  |  |  |  |  |  |  |  |  |  |

# Rock Grass-dart
## *Taractrocera ilia* Waterhouse, 1932

Plate 39 Kakadu National Park, NT
Photo: M. F. Braby

Plate 40 Kakadu National Park, NT
Photo: M. F. Braby

## Distribution

This species is endemic to the study region. It is known only from the Top End, where it is restricted to western Arnhem Land. The known and putative larval food plants (*Micraira* spp.) are all restricted to the sandstone plateau of Arnhem Land, and their spatial distribution corresponds with the geographic range of *T. ilia*. The food plants, however, are more widely distributed, extending further south to the Katherine Gorge area, NT, and further east to the edge of the sandstone plateau—a remote area that is difficult to access. It is very likely *T. ilia* also occurs in these areas.

## Excluded data

Historical records from Melville Island and Darwin based on two specimens in the SAM (Waterhouse 1932) are erroneous (Braby and Zwick 2015).

## Habitat

*Taractrocera ilia* breeds in open sandstone pavements where the larval food plants grow as prostrate, dense mat-forming or hummock 'resurrection' grasses on gravelly rock scree, shallow depressions, seepage depressions and moist seepages at the base of rock overhangs (Braby and Zwick 2015).

## Larval food plants

*Micraira adamsii, M. compacta, M. multinervia, M. spinifera, M. tenuis* (Poaceae); probably *Micraira dentata, M. inserta, M. pungens, M. subspicata, M. viscidula.*

## Seasonality

Adults occur throughout the year, but they are most abundant during the wet season (particularly in February), when the food plants rehydrate, are green and produce new leaf growth. During the dry season—or during hot, dry periods between the monsoon troughs—the plants typically desiccate, are brown and stop growing. Despite the desiccated condition of the food plants during the dry season, breeding occurs throughout the year, but during the dry season the larvae remain in diapause inside their silken shelters, usually as final instar larvae (for up to six months) or as first instar larvae (at least for a month towards the end of the dry season).

## Breeding status

This species is resident in the study region.

## Conservation status

LC. The species *T. ilia* is a narrow-range endemic and its entire range occurs within two conservation reserves: Kakadu National Park and Warddeken IPA. Despite its restricted occurrence, there are no known threats facing the taxon.

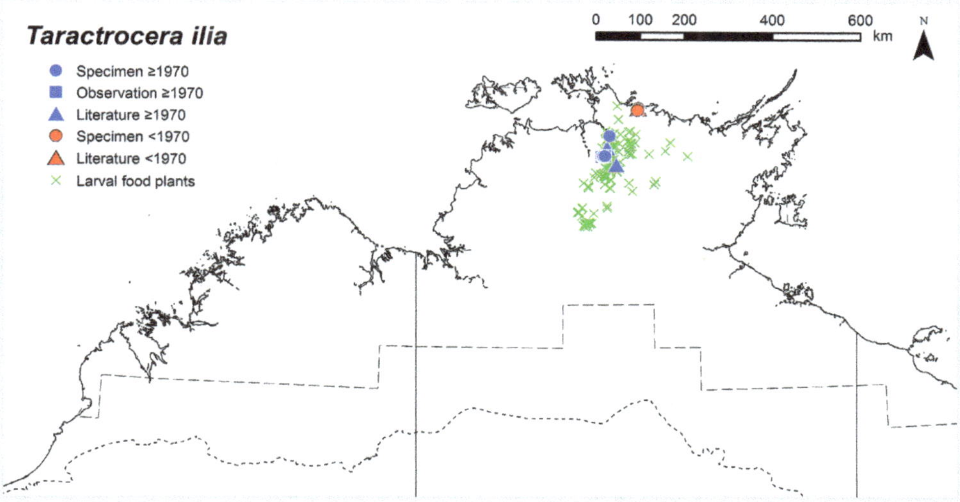

Taractrocera ilia
0 100 200 400 600
km
N
Specimen ≥1970
Observation ≥1970
Literature ≥1970
Specimen <1970
Literature <1970
Larval food plants

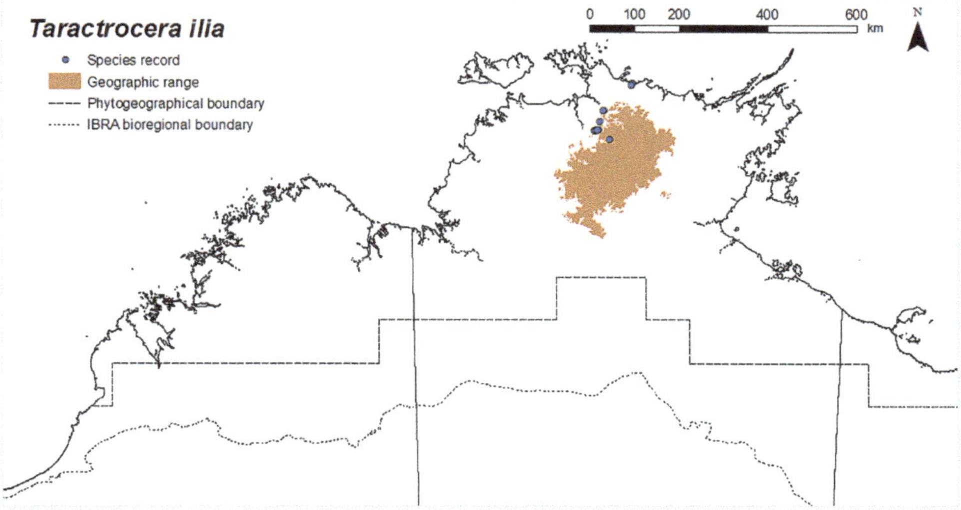

Taractrocera ilia
0 100 200 400 600
km
N
Species record
Geographic range
Phytogeographical boundary
IBRA bioregional boundary

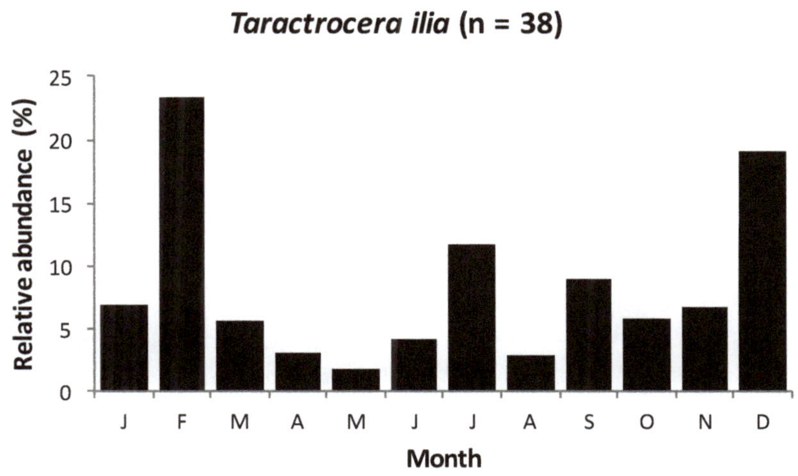

Taractrocera ilia (n = 38)
Relative abundance (%)
25
20
15
10
5
0
J F M A M J J A S O N D
Month

| Month | J | F | M | A | M | J | J | A | S | O | N | D |
|---|---|---|---|---|---|---|---|---|---|---|---|---|
| Egg | ● | ● | | | | | | | ● | | | ● |
| Larva | ● | ● | ● | ● | ● | ● | ● | ● | ● | ● | ● | ● |
| Pupa | | ● | ● | | ● | ● | ● | | | | ● | ● |
| Adult | ● | ● | ● | ● | ● | ● | ● | ● | ● | ● | ● | ● |

# Sandstone Grass-dart
## *Taractrocera psammopetra* Braby, 2015

Plate 41 Spirit Hills, Keep River National Park, NT
Photo: M. F. Braby

Plate 42 Spirit Hills, Keep River National Park, NT
Photo: M. F. Braby

## Distribution

This species is endemic to the study region. It was described only as recently as 2015, five years after it was first discovered, in 2010. It has a disjunct distribution, occurring in the Kimberley and in western Arnhem Land in the Top End. It is less common and has a much smaller extent in the north of the Northern Territory, where it occurs sympatrically with *Taractrocera ilia*. In Western Australia, it occurs in the higher rainfall areas of the northern Kimberley (> 1,200 mm mean annual rainfall), as well as in semi-arid areas of the eastern Kimberley (c. 700 mm). The known and putative larval food plants (*Micraira* spp.), however, are more widely distributed, extending as far south as the Bungle Bungle Range (Purnululu National Park) and King Leopold Ranges, WA. Further field surveys are thus required to determine whether *T. psammopetra* occurs in the southern Kimberley.

## Habitat

*Taractrocera psammopetra* breeds in open sandstone pavements, typically in shallow depressions or moist seepages on sandstone plateaus or on skeletal soils adjacent to watercourses, or on rock ledges of sandstone cliffs, where the larval food plants grow as prostrate, dense mat-forming or hummock 'resurrection' grasses (Braby and Zwick 2015).

## Larval food plants

*Micraira brevis*, *M. lazaridis*, *M. spiciforma*, *Micraira* sp. 'Purnululu' (Poaceae); probably *Micraira dunlopii*. The food plant for the population in western Arnhem Land in the Top End has not been determined.

## Seasonality

Adults are seasonal, occurring only during the wet season (November–February), but there are too few records (n = 12) to assess temporal changes in abundance. No adults have been recorded during the dry season despite targeted searches in April, May and August. The immature stages (eggs and pupae) have been recorded in February, when the food plants rehydrate, are green and produce new leaf growth. During the long dry season, the plants desiccate, are brown and stop growing, and the larvae remain in diapause as final instars inside their silken shelters. Presumably, only one or possibly two generations are completed annually.

## Breeding status

This species is resident in the study region.

## Conservation status

LC. Despite the wide geographical range of *T. psammopetra*, the available data suggest it has a disjunct distribution with a limited AOO. The species occurs in several conservation reserves, including Mitchell River National Park, El Questro Wilderness Park, Keep River National Park (Spirit Hills extension) and Kakadu National Park. Despite its restricted occurrence, there are no known threats facing the taxon.

## Taractrocera psammopetra

- ● Specimen ≥1970
- ■ Observation ≥1970
- ✕ Larval food plants

## Taractrocera psammopetra

- ● Species record
- ▨ Geographic range
- — — Phytogeographical boundary
- ······ IBRA bioregional boundary

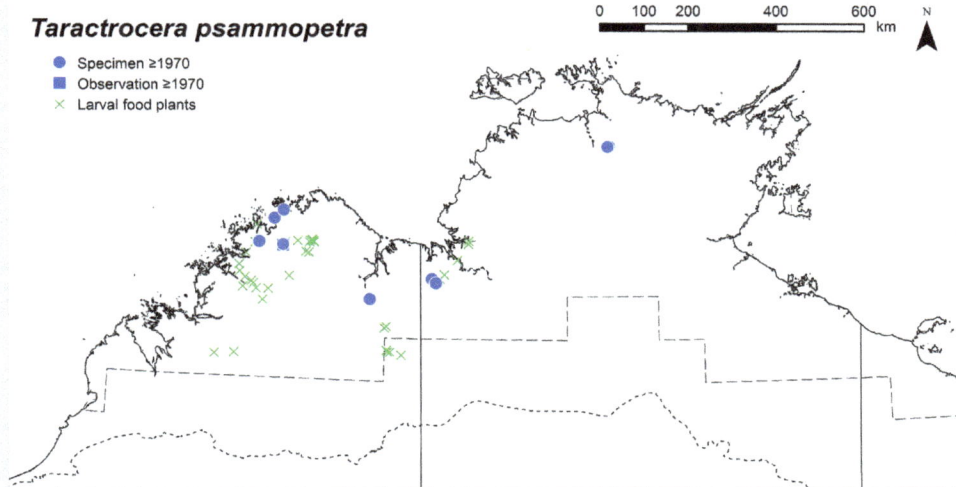

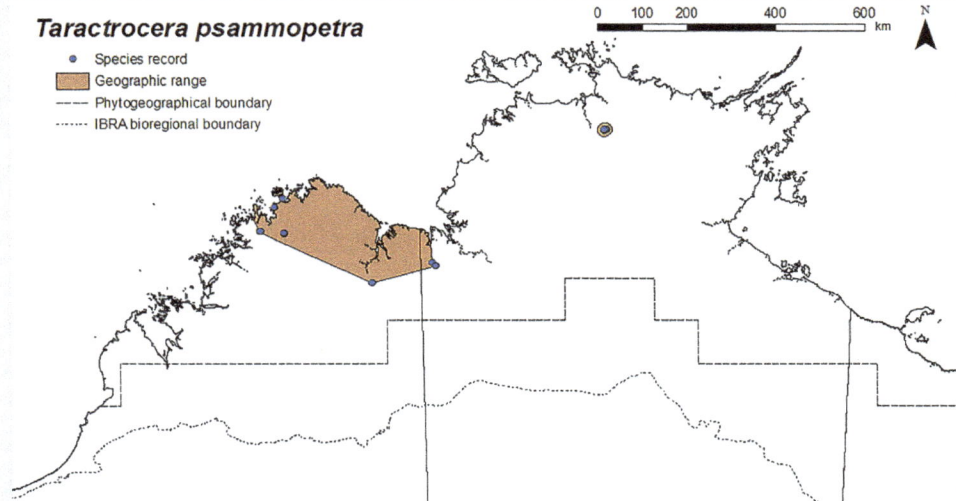

| Month | J | F | M | A | M | J | J | A | S | O | N | D |
|-------|---|---|---|---|---|---|---|---|---|---|---|---|
| Egg   |   | ▨ |   |   |   |   |   |   |   |   |   |   |
| Larva |   | ▨ |   | ▨ | ▨ |   | ▨ | ▨ |   |   |   |   |
| Pupa  |   | ▨ |   |   |   |   |   |   |   |   |   |   |
| Adult | ▨ | ▨ |   |   |   |   |   |   |   |   | ▨ | ▨ |

# Narrow-brand Grass-dart
## *Ocybadistes flavovittatus* (Latrielle, [1824])

Plate 43 Mornington Wildlife Sanctuary, WA
Photo: M. F. Braby

Plate 44 Mornington Wildlife Sanctuary, WA
Photo: M. F. Braby

## Distribution

This species is represented by the subspecies *O. flavovittatus vesta* (Waterhouse, 1932), which is endemic to the study region. It has a broad distribution, occurring in the Kimberley, Top End and western Gulf Country, extending from moist coastal areas to drier inland areas of the semi-arid zone (< 700 mm mean annual rainfall). It has been recorded as far south as the central Kimberley (Mornington Wildlife Sanctuary, WA) and western Gulf Country, at Cape Crawford and McArthur River Homestead, NT (Common and Waterhouse 1981; Braby 2000), and Doomadgee, Qld (Puccetti 1991). Outside the study region, *O. flavovittatus* occurs in mainland New Guinea and adjacent islands and widely in eastern Australia.

## Habitat

*Ocybadistes flavovittatus* has been recorded breeding in riparian woodland in the Kimberley (at Annie Creek, Mornington Wildlife Sanctuary), where the introduced larval food plant grew as an extensive hummock over the ground in a localised, damp open area with some shade provided by overstorey trees (Braby 2015e). Elsewhere, adults have been collected in a wide variety of habitats, including savannah woodland, paperbark woodland, paperbark open forest and mixed monsoon forest, but usually along or adjacent to riparian areas such as perennial creeks. Presumably, the species breeds in all of these habitats.

## Larval food plants

*Cynodon dactylon* (Poaceae). The native food plants have not been recorded in the study region, but in eastern Australia they consist of various grasses (Poaceae), the identity of which have not been documented (Braby 2000).

## Seasonality

Adults occur throughout the year, but they appear to be more abundant from the late wet season (March and April) to the mid dry season (July). We have few data on the phenology of the immature stages (eggs and larvae), which have been recorded in August. Presumably, the species breeds continuously throughout the year and several generations are completed annually.

## Breeding status

This species is resident in the study region.

## Conservation status

LC.

## Ocybadistes flavovittatus

- ● Specimen ≥1970
- ■ Observation ≥1970
- ▲ Literature ≥1970
- ▲ Literature <1970
- ✕ Larval food plant

## Ocybadistes flavovittatus

- ● Species record
- ▨ Geographic range
- — — Phytogeographical boundary
- ········ IBRA bioregional boundary

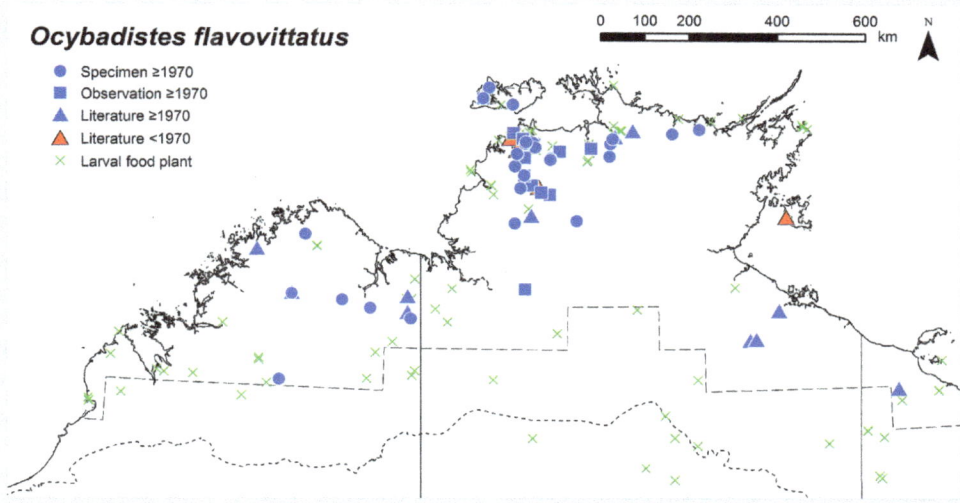

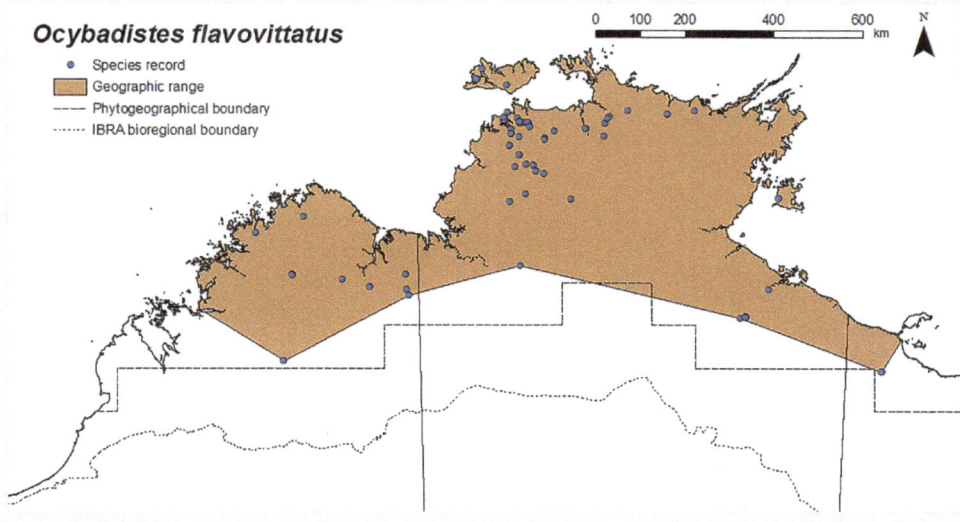

## Ocybadistes flavovittatus (n = 79)

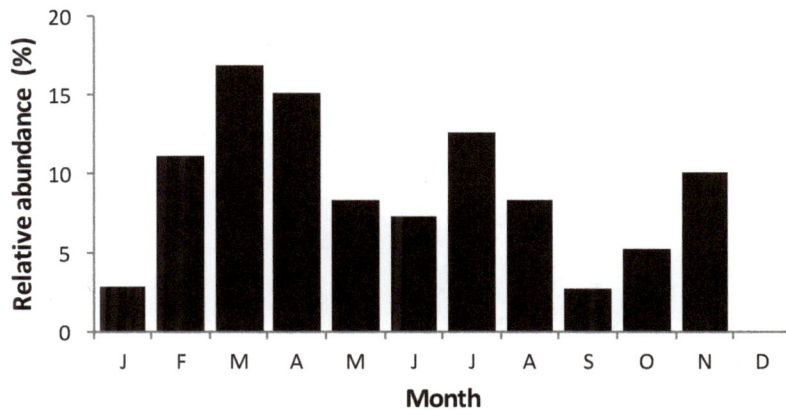

| Month | J | F | M | A | M | J | J | A | S | O | N | D |
|-------|---|---|---|---|---|---|---|---|---|---|---|---|
| Egg   |   |   |   |   |   |   |   | ▨ |   |   |   |   |
| Larva |   |   |   |   |   |   |   | ▨ |   |   |   |   |
| Pupa  |   |   |   |   |   |   |   |   |   |   |   |   |
| Adult |   |   |   |   |   |   |   |   |   |   |   |   |

# Green Grass-dart

## *Ocybadistes walkeri* Heron, 1894

Plate 45 Holmes Jungle, NT
Photo: Tissa Ratnayeke

## Distribution

This species is represented by the subspecies *O. walkeri olivia* Waterhouse, 1933, which is endemic to the study region. It is restricted mainly to the Top End (> 800 mm mean annual rainfall), reaching its southern limits at Victoria River Roadhouse and Mataranka, NT. It has also been recorded in the eastern Kimberley, based on a historical male specimen from Wyndham, WA, and a more recent observation at Keep River National Park (Cockatoo Lagoon), NT. The extent to which the species occurs in the Kimberley requires further scrutiny. It often occurs together with the closely related *Ocybadistes flavovittatus*, but it does not extend as far inland as that species. The putative native larval food plant (*Imperata cylindrica*) occurs widely in northern Australia. Outside the study region, *O. walkeri* occurs from the Lesser Sunda Islands, through mainland New Guinea and adjacent islands to eastern and south-eastern Australia.

## Excluded data

Yeates (1990) listed this species from the northern Kimberley based on one male and three females collected at Kalumburu Mission on 4–6 May 1989. However, examination of this material in the Western Australian Department of Agriculture collection (WADA) indicated that all were misidentified and were in fact *Ocybadistes flavovittatus*, a species known to occur in the northern Kimberley. Thus, the record of *O. walkeri* is excluded from the northern Kimberley.

## Habitat

The natural breeding habitat of *O. walkeri* has not been recorded in the study region. The species breeds commonly in urban areas where the naturalised larval food plants thrive, typically in well-watered lawns in suburban parks and gardens (Braby 2015e). Elsewhere, adults have been collected in a wide variety of habitats, including savannah woodland, paperbark swampland, woodland and open forest and mixed woodland–monsoon forest associations, usually along or adjacent to riparian areas such as perennial creeks and rivers. They also occur in moist open grassy areas within or along the edge of semi-deciduous monsoon vine thicket and evergreen monsoon vine forest. Presumably, the species breeds in all of these habitats.

## Larval food plants

*Axonopus compressus*, *Melinis repens* (Poaceae). The native food plants have not been recorded in the study region, but in eastern Australia they consist of various grasses, including *Imperata cylindrica* (Poaceae) (Braby 2000).

## Seasonality

Adults occur throughout the year, but they appear to be more abundant in the mid dry season (June and July) and least numerous in the late dry season (September and October). We have few data on the phenology of the immature stages, with eggs recorded in March and August. Presumably, the species breeds continuously throughout the year and several generations are completed annually.

## Breeding status

This species is resident in the study region.

## Conservation status

LC.

## Ocybadistes walkeri

Legend:
- ● Specimen ≥1970
- ■ Observation ≥1970
- ▲ Literature ≥1970
- ● Specimen <1970
- ▲ Literature <1970
- ✕ Putative larval food plant

## Ocybadistes walkeri

Legend:
- • Species record
- Geographic range
- — — Phytogeographical boundary
- ········ IBRA bioregional boundary

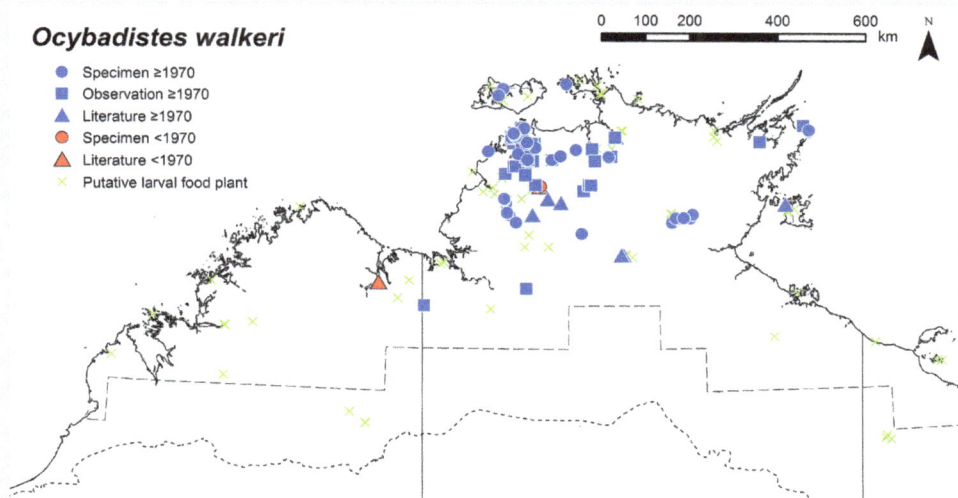

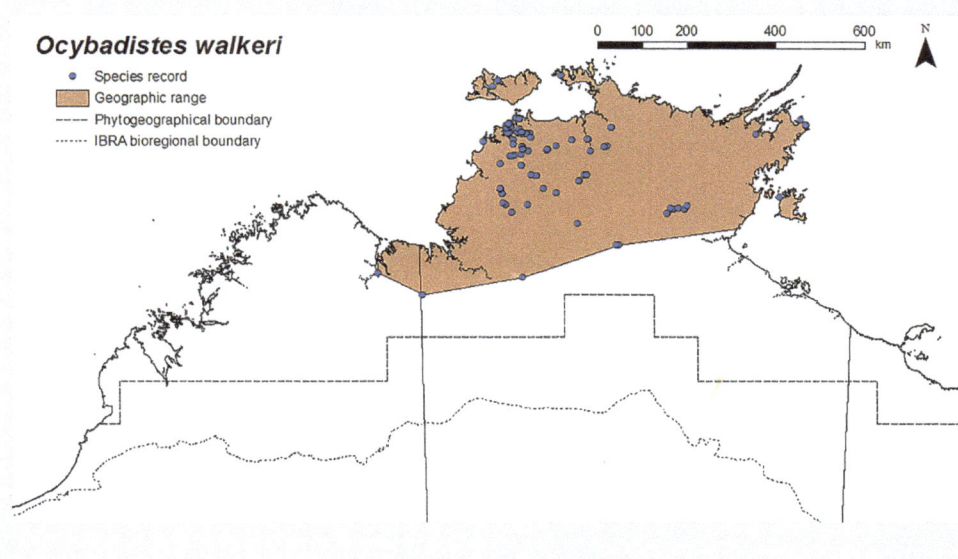

## Ocybadistes walkeri (n = 134)

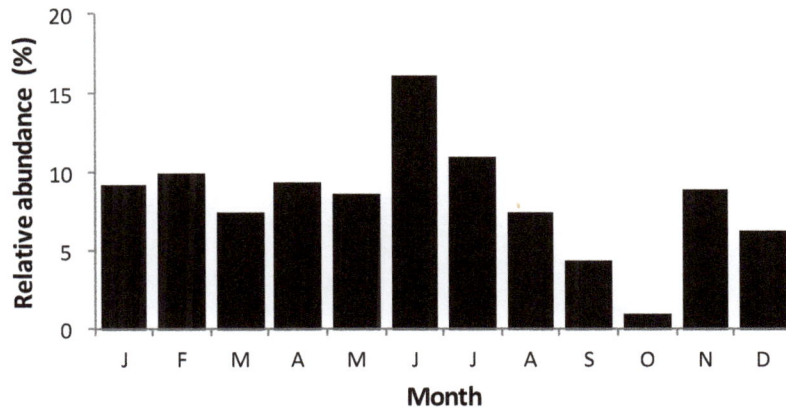

| Month | J | F | M | A | M | J | J | A | S | O | N | D |
|-------|---|---|---|---|---|---|---|---|---|---|---|---|
| Egg   |   |   | ▓ |   |   |   |   | ▓ |   |   |   |   |
| Larva |   |   |   |   |   |   |   |   |   |   |   |   |
| Pupa  |   |   |   |   |   |   |   |   |   |   |   |   |
| Adult | ▓ | ▓ | ▓ | ▓ | ▓ | ▓ | ▓ | ▓ | ▓ | ▓ | ▓ | ▓ |

# White-margined Grass-dart
## *Ocybadistes hypomeloma* Lower, 1911

Plate 46 West of Paluma, Qld
Photo: Frank Pierce

## Distribution

This species occurs in the Kimberley and western half of the Top End, with more sporadic occurrences in the western Gulf Country. It extends from the higher rainfall areas at Darwin to drier inland areas of the semi-arid zone (< 700 mm mean annual rainfall), where it has been recorded as far inland as El Questro Wilderness Park (Amalia Gorge, El Questro Gorge), WA (Braby 2012b). In the western Gulf Country, it has been recorded at McArthur River Homestead, NT (Common and Waterhouse 1981), and Boodjamulla/Lawn Hill National Park (Lawn Hill Gorge), Qld (Daniels and Edwards 1998). The putative larval food plants (*Ischaemum australe* and *Themeda triandra*) occur widely in northern Australia. Further field surveys are thus required to determine the geographic range of *O. hypomeloma* more precisely, particularly whether it occurs in the eastern half of the Top End and coastal areas of the Gulf of Carpentaria. Outside the study region, *O. hypomeloma* occurs in the Pilbara of Western Australia and widely in eastern Australia.

The subspecific status has not been determined with certainty: it may be *O. hypomeloma vaga* (Waterhouse, 1932), otherwise known only from the Torres Strait islands, or an undescribed subspecies.

## Habitat

The breeding habitat of *O. hypomeloma* has not been recorded in the study region. Adults have been collected mainly in savannah woodland and they undoubtedly breed on native grasses in this habitat.

## Larval food plants

Not recorded in the study region; probably *Ischaemum australe* and *Themeda triandra* (Poaceae), which are the food plants in eastern Australia (Braby 2000).

## Seasonality

Adults have been recorded sporadically throughout the year, with most records in the wet season (February and April). The breeding phenology and seasonal history of the immature stages have not been recorded, but presumably the species breeds continuously throughout the year.

## Breeding status

This species is resident in the study region.

## Conservation status

LC.

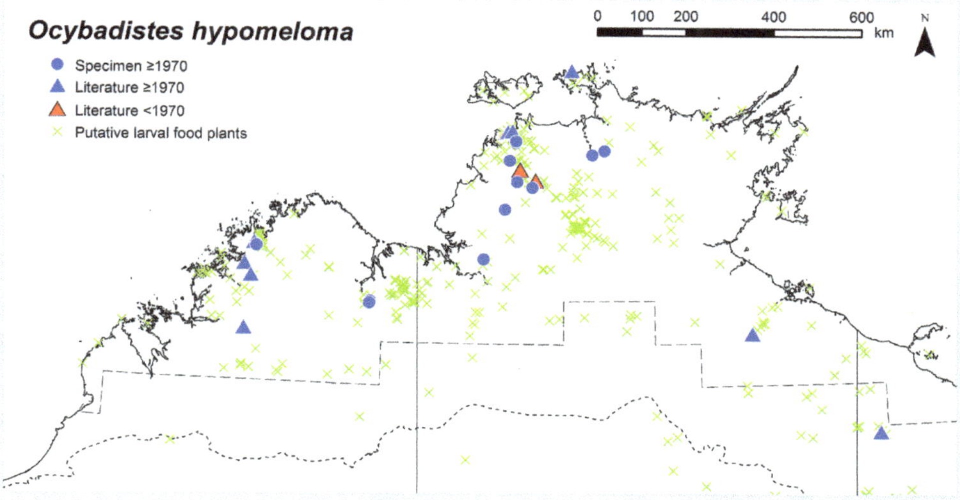

Ocybadistes hypomeloma
0 100 200 400 600 km
N
Specimen ≥1970
Literature ≥1970
Literature <1970
Putative larval food plants


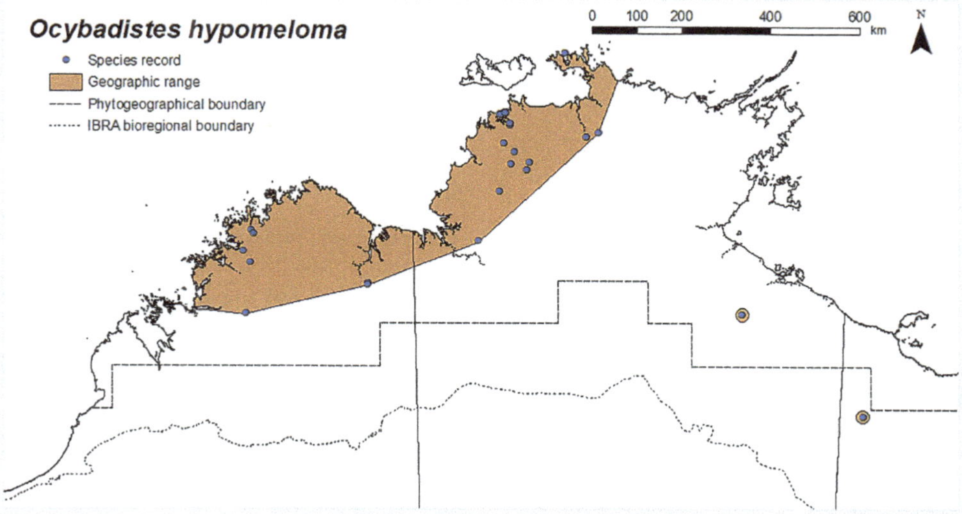

Ocybadistes hypomeloma
0 100 200 400 600 km
N
Species record
Geographic range
Phytogeographical boundary
IBRA bioregional boundary


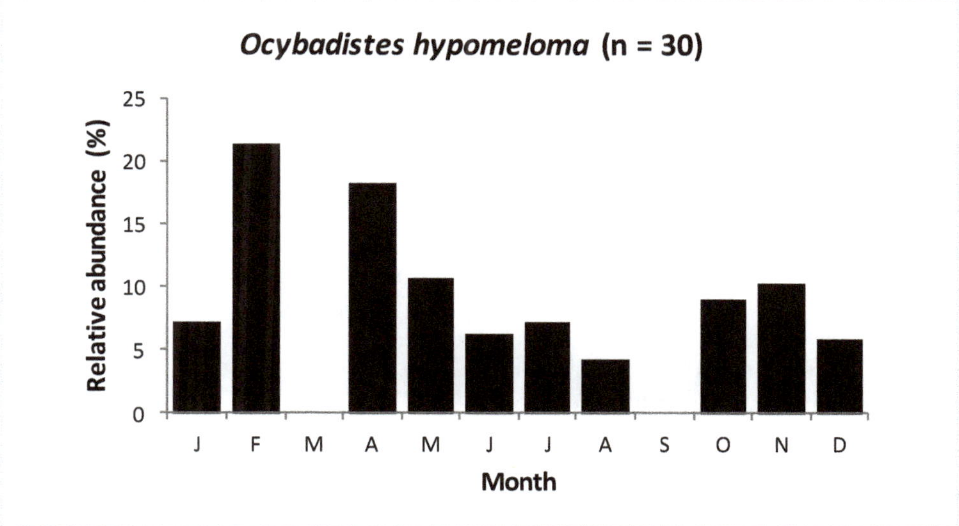

Ocybadistes hypomeloma (n = 30)
Relative abundance (%)
25
20
15
10
5
0
J F M A M J J A S O N D
Month


| Month | J | F | M | A | M | J | J | A | S | O | N | D |
|---|---|---|---|---|---|---|---|---|---|---|---|---|
| Egg | | | | | | | | | | | | |
| Larva | | | | | | | | | | | | |
| Pupa | | | | | | | | | | | | |
| Adult | | | | | | | | | | | | |

# Wide-brand Grass-dart
## *Suniana sunias* (C. Felder, 1860)

Plate 47 Mary River Reserve, NT
Photo: M. F. Braby

Plate 48 Marrakai Road, NT
Photo: M. F. Braby

## Distribution

This species is represented by the subspecies *S. sunias sauda* Waterhouse, 1937, which is endemic to the study region. It is restricted to the north of the Northern Territory, where it occurs in the higher rainfall areas (> 1,000 mm mean annual rainfall). Most records are from the north-western corner of the Top End, where it extends as far south-east as Nitmiluk (Katherine Gorge) National Park (Sweetwater Pool), but it has been recorded from Groote Eylandt in the Gulf of Carpentaria (Tindale 1923). It has not been recorded from the Tiwi Islands, Cobourg Peninsula or north-eastern Arnhem Land; thus, further field surveys are required to determine whether *S. sunias* is present in these northern areas. Outside the study region, *S. sunias* occurs from Maluku and the Lesser Sunda Islands, through mainland New Guinea and adjacent islands and north-eastern and eastern Australia to the Solomon Islands.

## Excluded data

Dunn (1985) listed this species from the eastern Kimberley, but Braby (2000: 212) cast doubt over this record, noting that '[a] single female collected at "Wyndham, WA" by J. C. Le Souëf (Dunn 1985) has not been traced, but one male each of *Ocybadistes walkeri* and *Suniana lascivia* are similarly labelled'. K. L. Dunn (pers. comm.) has subsequently advised that the specimen referred to was almost certainly the male *S. lascivia*, contributing to the mistaken identification.

## Habitat

The breeding habitat of *S. sunias* has not been recorded in the study region. Adults have been collected only in the grassy edges of riparian evergreen monsoon vine forest along creeks or the banks of rivers with permanent water, and in light gaps comprising grassy patches within coastal semi-deciduous monsoon vine thicket, and no doubt they breed in these habitats.

## Larval food plants

*\*Megathyrsus maximus* (Poaceae). The native food plants have not been recorded in the study region, but undoubtedly consist of native grasses (Poaceae).

## Seasonality

Adults have been recorded during most months of the year, but they are more abundant from the late wet season to the mid dry season (March–July). The breeding phenology and seasonal history of the immature stages have not been recorded, but it is likely the species breeds continuously throughout the year.

## Breeding status

This species is resident in the study region.

## Conservation status

LC.

## Suniana sunias

- ● Specimen ≥1970
- ■ Observation ≥1970
- ▲ Literature ≥1970
- ▲ Literature <1970

## Suniana sunias

- ● Species record
- ▨ Geographic range
- — — — Phytogeographical boundary
- ········· IBRA bioregional boundary

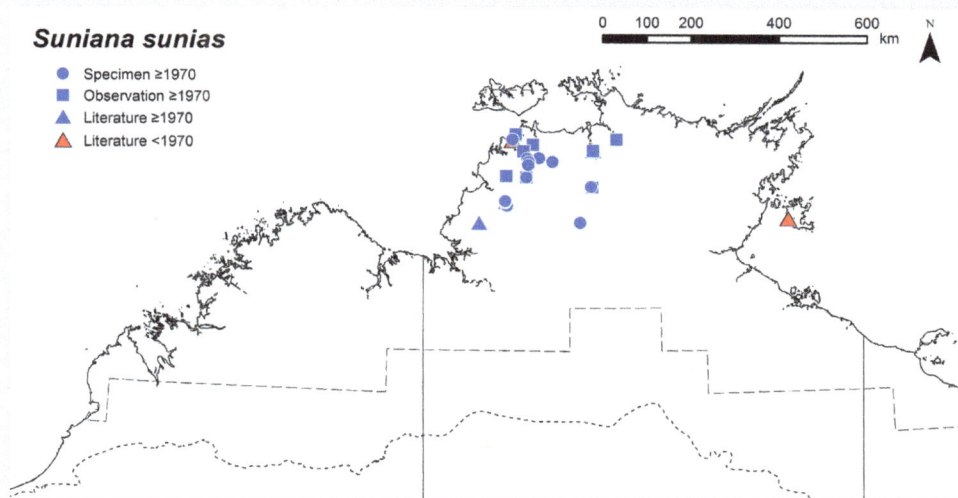

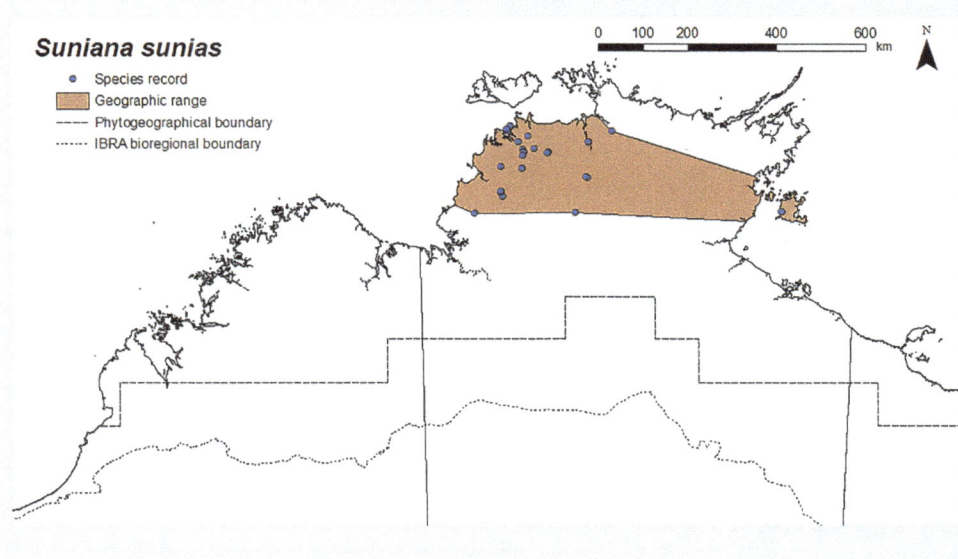

### Suniana sunias (n = 40)

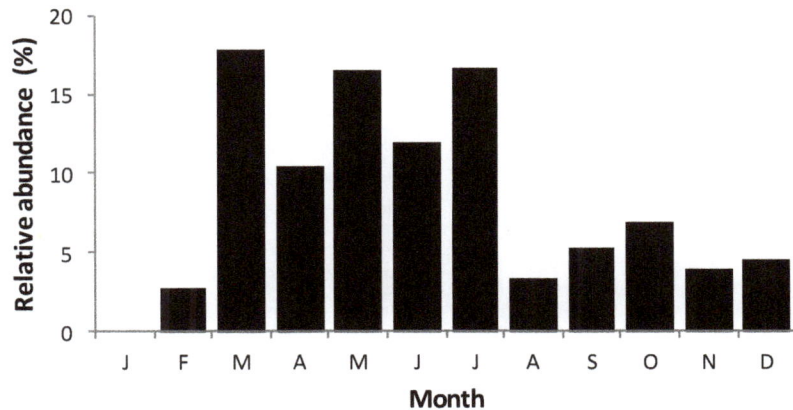

| Month | J | F | M | A | M | J | J | A | S | O | N | D |
|-------|---|---|---|---|---|---|---|---|---|---|---|---|
| Egg   |   |   |   |   |   |   |   |   |   |   |   |   |
| Larva |   |   |   |   |   |   |   |   |   |   |   |   |
| Pupa  |   |   |   |   |   |   |   |   |   |   |   |   |
| Adult |   |   |   |   |   |   |   |   |   |   |   |   |

# Dark Grass-dart
## *Suniana lascivia* (Rosenstock, 1885)

Plate 49 Wanguri, Darwin, NT
Photo: M. F. Braby

## Distribution

This species is represented by two subspecies: *S. lascivia larrakia* L. E. Couchman, 1951 and *S. lascivia lasus* Waterhouse, 1937, both of which are endemic to the study region. The latter subspecies is restricted to Bathurst Island, NT, but *S. lascivia larrakia* is much more widespread, occurring in the Kimberley and Top End, with more sporadic occurrences in the western Gulf Country. *S. lascivia larrakia* extends from moist coastal areas to drier inland areas of the semi-arid zone (c. 700 mm mean annual rainfall). The subspecies *S. lascivia lasus* was previously known only from three historical specimens (Waterhouse 1937a); however, it was rediscovered at Bathurst Island in March 2009 (D. A. Young). The geographical range of *S. lascivia* broadly corresponds with the spatial distribution of its two native larval food plants. The food plants, however, have a slightly wider extent, occurring also on Dampier Peninsula, WA, and in western Queensland in the Gulf Country. Further field surveys are therefore required to determine whether *S. lascivia* is present in these areas, particularly in the Limmen Bight area and elsewhere in the Gulf of Carpentaria. Further field surveys are also required to determine whether the subspecies *S. lascivia lasus* occurs on Melville Island, NT. Outside the study region, *S. lascivia* occurs in Timor, mainland New Guinea and adjacent islands and throughout eastern Australia.

## Habitat

*Suniana lascivia* occurs in a variety of habitats, but it breeds mainly in swampy and riparian areas, including paperbark–pandanus swampland adjacent to evergreen monsoon vine forest and the edges of mixed riparian woodland or open forest with rainforest elements in the understorey where the native larval food plants grow as perennial grasses in abundance in the more open areas (Braby 2011a, 2015e). In the drier inland areas, the breeding habitats are confined to perennial creeks along gorges.

## Larval food plants

*Ischaemum australe*, *Imperata cylindrica* (Poaceae); also *\*Megathyrsus maximus* (Poaceae).

## Seasonality

Adults have been recorded during most months of the year, but they are more abundant from the late wet season to the mid dry season (March–July). The apparent peak in abundance in May is a sampling bias due to targeted surveys in the eastern Kimberley (eight of 13 records for May are all from El Questro Wilderness Park in 2011). The immature stages have been recorded sporadically during the year, with no evidence of diapause in any of the life history stages. Presumably, the species breeds continuously throughout the year.

## Breeding status

This species is resident in the study region.

## Conservation status

*Suniana lascivia larrakia*: LC. *Suniana lascivia lasus*: DD. The subspecies *S. lascivia lasus* is a short-range endemic (AOO is likely to be < 2,000 sq km, with spatial buffering of records providing a first approximation of 700 sq km), and is currently known only from one extant location where it occurs on private Aboriginal land in which it has an uncertain future due to potential development. Thus, the taxon may qualify as Near Threatened (NT) once adequate data are available. Targeted field surveys to determine the extent of its distribution, critical habitat and key threatening processes on the Tiwi Islands should be a high priority for this subspecies.

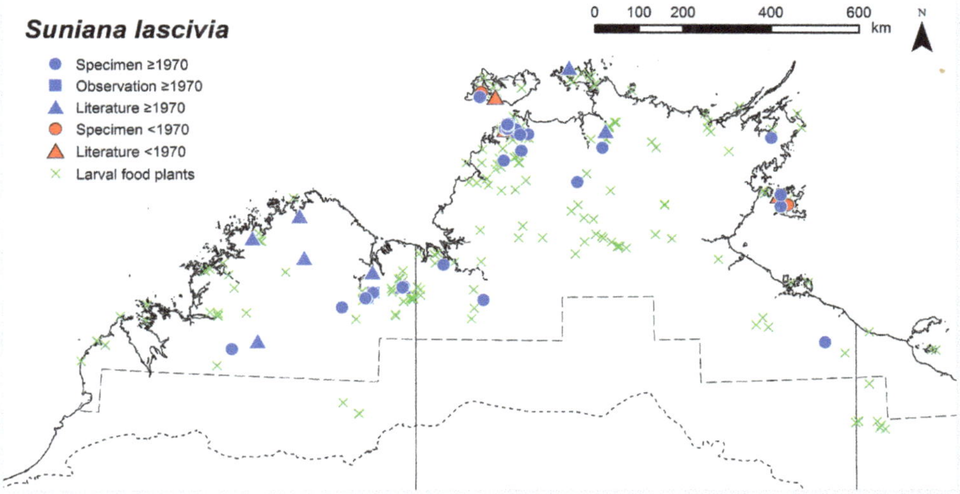

Suniana lascivia
0 100 200 400 600 km
N
Specimen ≥1970
Observation ≥1970
Literature ≥1970
Specimen <1970
Literature <1970
Larval food plants

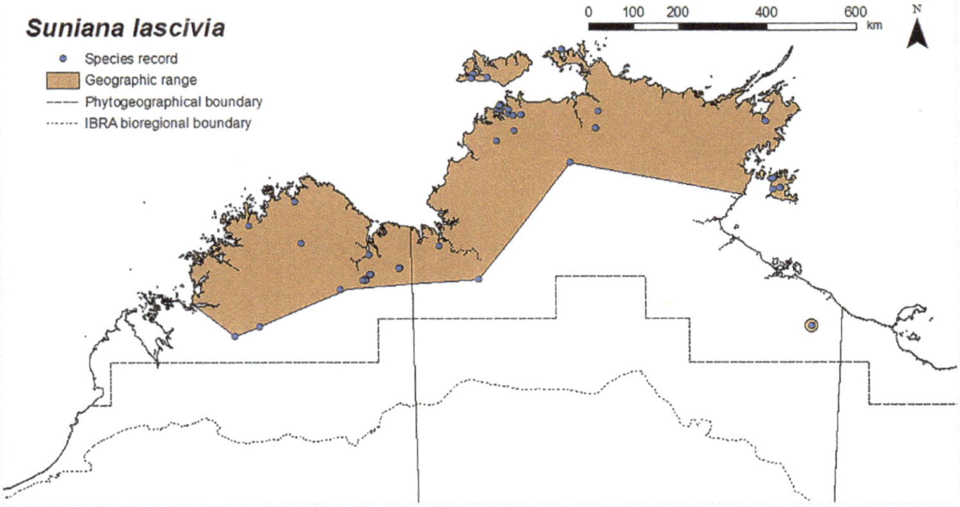

Suniana lascivia
0 100 200 400 600 km
N
Species record
Geographic range
Phytogeographical boundary
IBRA bioregional boundary

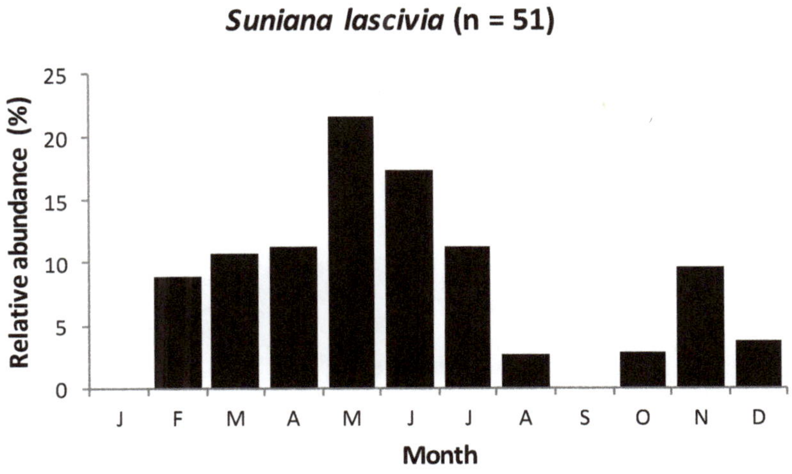

Suniana lascivia (n = 51)
Relative abundance (%)
0
5
10
15
20
25
J F M A M J J A S O N D
Month

| Month | J | F | M | A | M | J | J | A | S | O | N | D |
|---|---|---|---|---|---|---|---|---|---|---|---|---|
| Egg | | | | | | | ■ | | ■ | | | |
| Larva | | ■ | | | | | | | ■ | ■ | | |
| Pupa | | ■ | | | | | | | | ■ | | |
| Adult | | ■ | ■ | ■ | ■ | ■ | ■ | ■ | | ■ | ■ | ■ |

# Pale-orange Darter
## *Telicota colon* (Fabricius, 1775)

Plate 50 Adelaide River crossing,
Arnhem Highway, NT
Photo: Deb Bisa

## Distribution

This species is represented in the study region by the subspecies *T. colon argea* (Plötz, 1883). It occurs widely in the region, from the Kimberley, through the Top End to the western Gulf Country, extending from moist coastal areas to drier inland areas of the semi-arid zone (c. 700 mm mean annual rainfall). The geographical range corresponds well with the spatial distribution of the three known native larval food plants. The food plants, however, are more widely distributed, occurring also on Dampier Peninsula and the southern Kimberley, WA; Groote Eylandt, NT; and western Queensland in the Gulf Country. It is likely *T. colon* occurs in most of these areas and further field surveys are thus required to confirm this. Outside the study region, *T. colon* occurs widely from India and South-East Asia, through mainland New Guinea and north-eastern and eastern Australia to the Solomon Islands. It also occurs in the Pilbara of Western Australia.

## Habitat

*Telicota colon* occurs in a wide range of habitats, but it breeds mainly in open areas where the larval food plants commonly grow, including savannah woodland, mixed paperbark swampland and the edges of evergreen monsoon vine forest (Braby 2011a, 2015e). In the drier inland areas, the breeding habitats are confined to gorges with permanent water and other riparian areas, such as woodland along perennial creeks and the edges of open forest with rainforest elements in the understorey along riverbanks.

## Larval food plants

*Imperata cylindrica*, *Ischaemum australe*, *Mnesithea rottboellioides* (Poaceae); also *\*Oryza sativa*, *\*Paspalum scrobiculatum*, *\*Andropogon gayanus* (Poaceae).

## Seasonality

Adults occur during most months of the year, but they are more abundant during the late wet season (March and April). The immature stages have been recorded in January and from March to August, but undoubtedly occur at other times of the year. Presumably, the species breeds continuously throughout the year.

## Breeding status

This species is resident in the study region.

## Conservation status

LC.

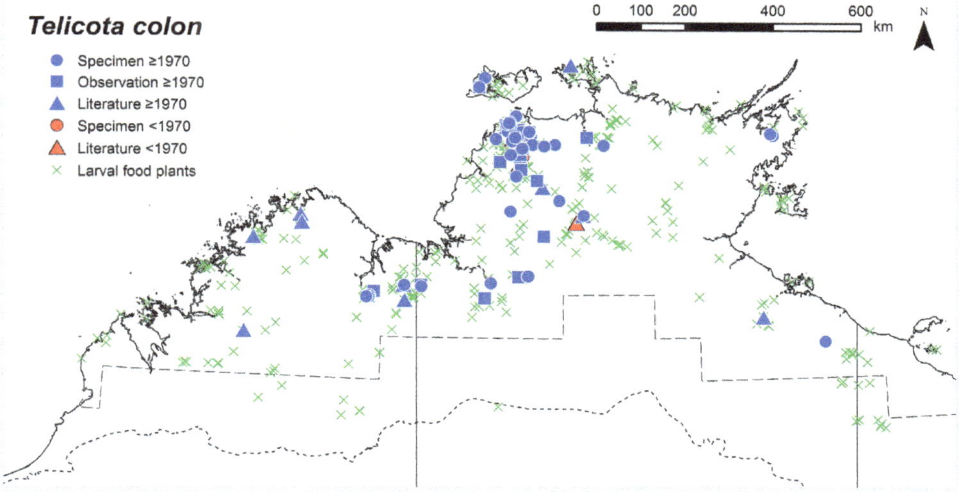

## Telicota colon

Specimen ≥1970
Observation ≥1970
Literature ≥1970
Specimen <1970
Literature <1970
Larval food plants

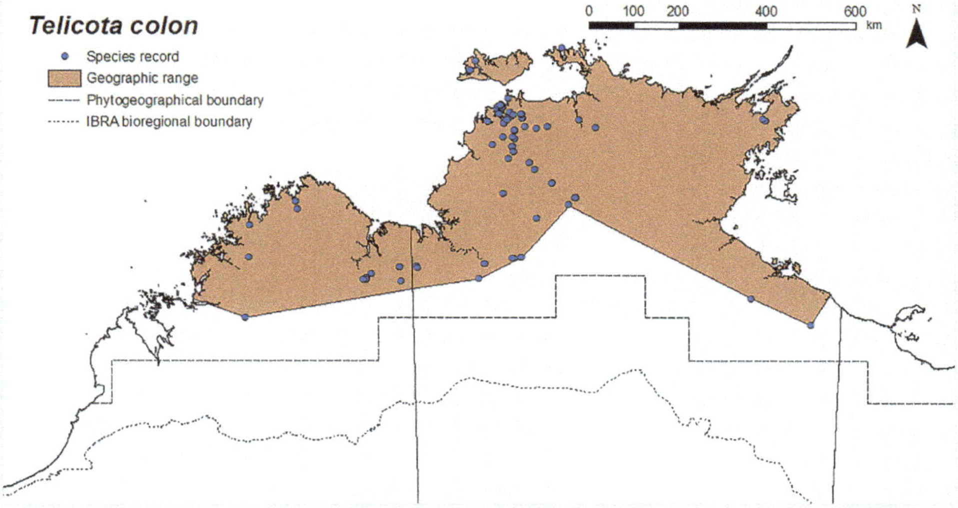

## Telicota colon

Species record
Geographic range
Phytogeographical boundary
IBRA bioregional boundary

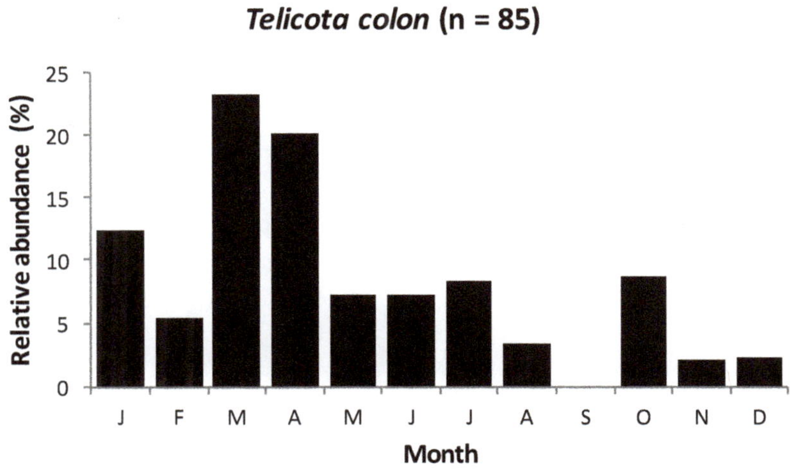

### Telicota colon (n = 85)

| Month | J | F | M | A | M | J | J | A | S | O | N | D |
|-------|---|---|---|---|---|---|---|---|---|---|---|---|
| Egg   |   |   |   |   |   |   | ▨ |   |   |   |   |   |
| Larva | █ |   | █ | █ | █ | █ | █ | █ |   |   |   |   |
| Pupa  |   |   |   | █ | █ | █ | █ | █ |   |   |   |   |
| Adult | █ | █ | █ | █ | █ | █ | █ | █ |   | █ | █ | █ |

# Bright-orange Darter
## *Telicota augias* (Linnaeus, 1763)

Plate 51 Dundee Beach, NT
Photo: M. F. Braby

Plate 52 Dundee Beach, NT
Photo: M. F. Braby

## Distribution

This species is represented in the study region by the subspecies *T. augias krefftii* (W. J. Macleay, 1866). It occurs widely in the Kimberley and Top End and does not intrude into the drier inland areas to any great extent (mainly > 800 mm mean annual rainfall). The geographical range closely corresponds with the spatial distribution of its larval food plant, indicating that *T. augias* has been well sampled in the region. The larval food plant, however, extends further south to the western Gulf Country at Limmen National Park (Nathan River Station) and Sir Edward Pellew Group (Vanderlin Island), NT. Further field surveys are therefore required to determine whether *T. augias* also occurs in these coastal and near-coastal areas. Outside the study region, *T. augias* occurs in South-East Asia, mainland New Guinea and north-eastern Australia.

Previous records of *Telicota ancilla baudina* Evans, 1949 refer to this species (Braby 2012b).

## Habitat

*Telicota augias* breeds in a variety of wet and dry monsoon forests, including evergreen monsoon vine forest associated with permanent water, coastal semi-deciduous monsoon vine thicket and mixed woodland with rainforest elements in the understorey associated with sandstone cliffs, creeks and gorges, where the larval food plant grows into the canopy as a tall scrambling vine.

## Larval food plant

*Flagellaria indica* (Flagellariaceae).

## Seasonality

Adults occur throughout the year, but they are more abundant during the mid wet season (February). The immature stages have been recorded during most months of the year, with no evidence of diapause in any of the life history stages. Presumably, the species breeds continuously throughout the year.

## Breeding status

This species is resident in the study region.

## Conservation status

LC.

## Telicota augias

- ● Specimen ≥1970
- ■ Observation ≥1970
- ▲ Literature ≥1970
- ● Specimen <1970
- ▲ Literature <1970
- ✕ Larval food plant

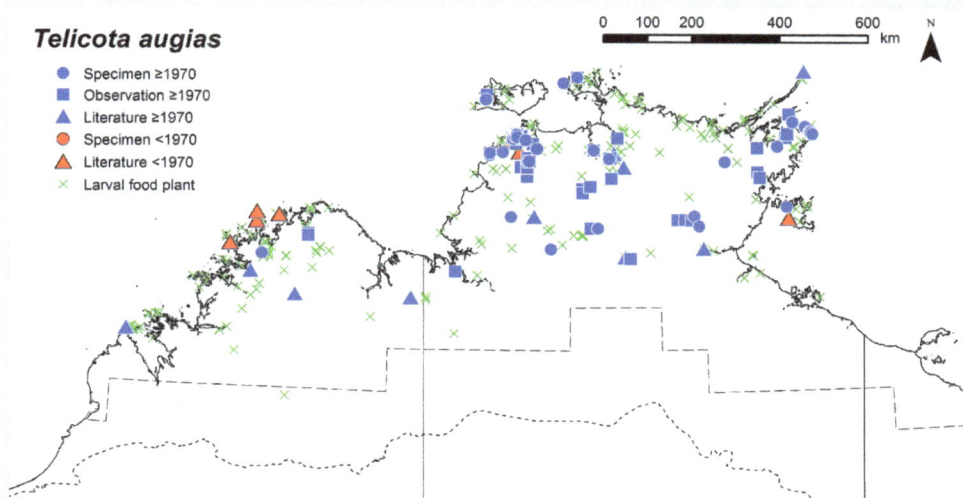

## Telicota augias

- ● Species record
- ▨ Geographic range
- -- Phytogeographical boundary
- ⋯ IBRA bioregional boundary

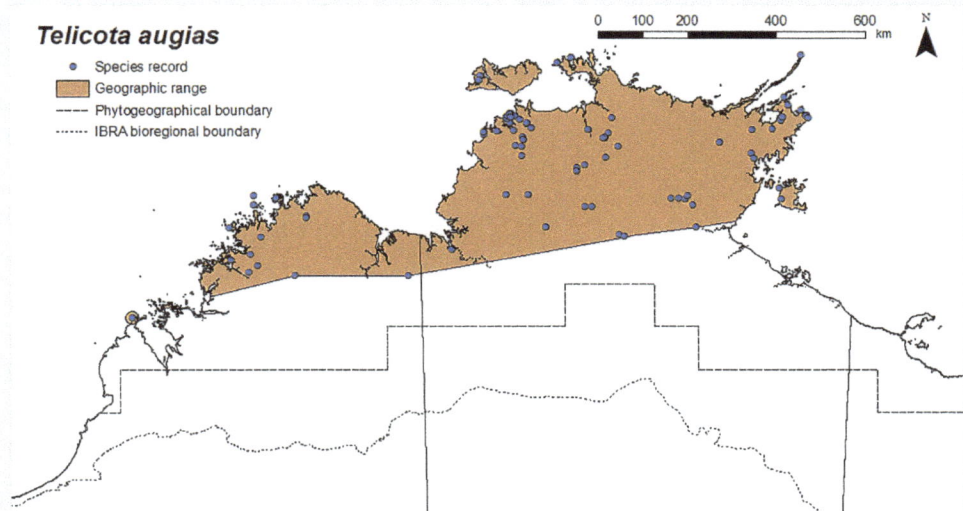

### Telicota augias (n = 100)

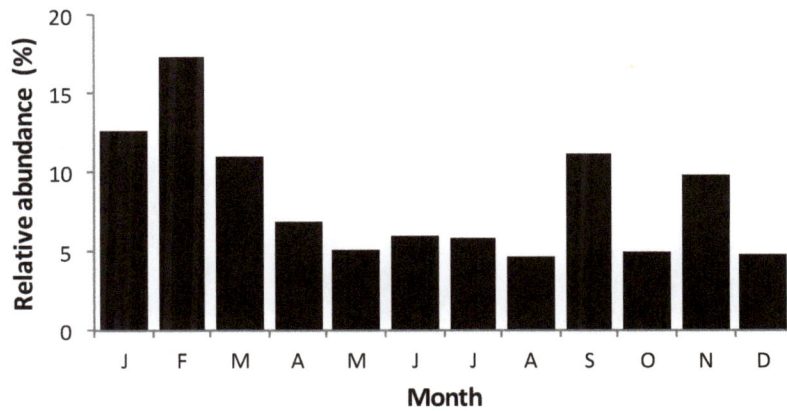

| Month | J | F | M | A | M | J | J | A | S | O | N | D |
|-------|---|---|---|---|---|---|---|---|---|---|---|---|
| Egg | | | | | | | | | | | | |
| Larva | | | | | | | | | | | | |
| Pupa | | | | | | | | | | | | |
| Adult | | | | | | | | | | | | |

# Orange Palm-dart
*Cephrenes augiades* (C. Felder, 1860)

Plate 53 Brisbane, Qld
Photo: M. F. Braby

## Distribution

This species is represented by an undescribed subspecies, which is endemic to the study region. Its presence in the Northern Territory was detected only as recently as 1991 (Braby 2000). It is restricted to the northern half of the Top End, where it occurs in the higher rainfall areas (> 1,200 mm mean annual rainfall). Its distribution extends from Darwin and Litchfield National Park (Wangi Falls) to Gove Peninsula (Koolatong River, Balma outstation and Rocky Bay). The geographic range of *C. augiades* is closely tied to the spatial distribution of its two native larval food plants. The food plants, however, have a slightly wider extent, occurring near the Joseph Bonaparte Gulf (e.g. Fitzmaurice River area). Further field surveys are therefore required to determine whether *C. augiades* also occurs in this area. Outside the study region, *C. augiades* occurs from Indonesia, through mainland New Guinea and adjacent islands and north-eastern and eastern Australia to New Britain and the Solomon Islands.

## Habitat

*Cephrenes augiades* breeds mainly in evergreen monsoon vine forest associated with springs and permanent streams where the native larval food plants grow as palms (Braby 2011a). In the Darwin area, it also occurs in suburban parks and gardens where the primary food plant (*Carpentaria acuminata*) and ornamental palms have been established.

## Larval food plants

*Carpentaria acuminata*, *Livistona benthamii* (Arecaceae). The main food plant is *C. acuminata* (Braby 2011a). Dunn (2009) listed *Livistona* sp. as the food plant near Cahills Crossing on the East Alligator River, and this record refers to *L. benthamii*. It may also utilise ornamental palms in suburban areas.

## Seasonality

Adults occur throughout the year, but they are more abundant during the wet season and early dry season (November–May). The immature stages have been recorded sporadically throughout the year; they may be particularly numerous in the late wet season. Presumably, the species breeds continuously throughout the year.

## Breeding status

This species is resident in the study region.

## Conservation status

LC. The putative subspecies *C. augiades* ssp. 'Top End' has a restricted range within which it occurs in several conservation reserves, including Litchfield National Park, Mary River National Park, Kakadu National Park, Laynhapuy IPA and several smaller reserves close to Darwin. Despite its restricted occurrence, there are no known threats facing the taxon.

## Cephrenes augiades

- ● Specimen ≥1970
- ■ Observation ≥1970
- ▲ Literature ≥1970
- ✕ Larval food plants

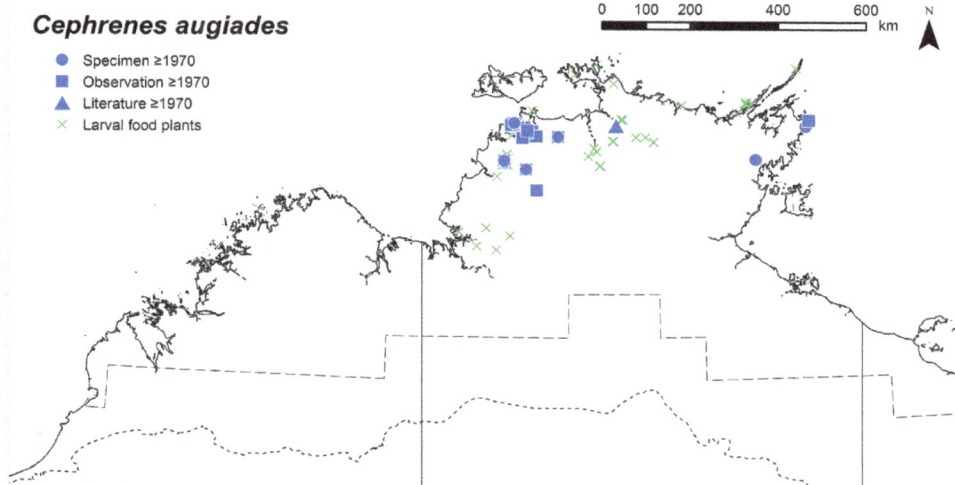

## Cephrenes augiades

- ● Species record
- ▨ Geographic range
- – – – Phytogeographical boundary
- ········ IBRA bioregional boundary

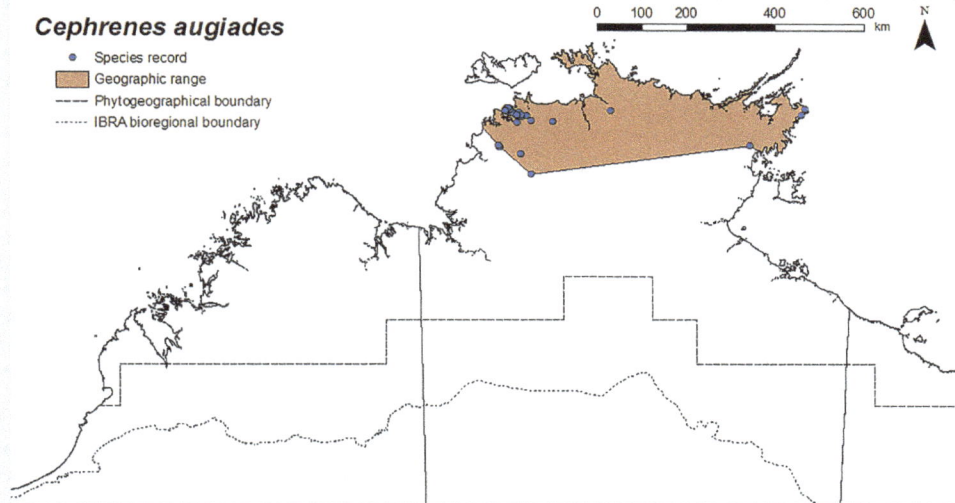

### Cephrenes augiades (n = 90)

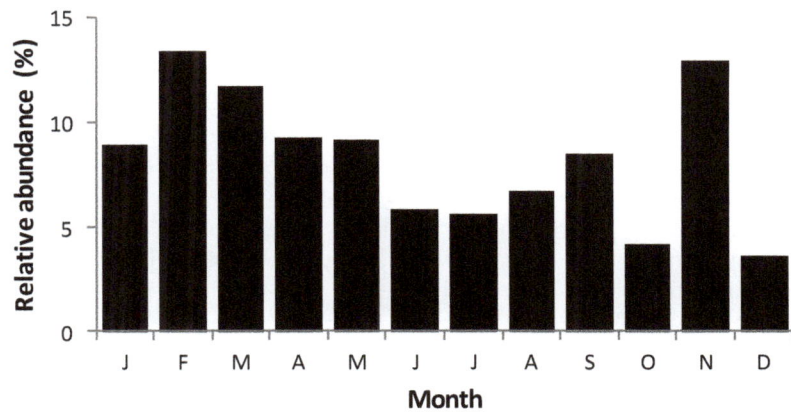

| Month | J | F | M | A | M | J | J | A | S | O | N | D |
|-------|---|---|---|---|---|---|---|---|---|---|---|---|
| Egg   |   |   |   |   |   |   |   |   |   |   |   |   |
| Larva |   |   |   |   |   |   |   |   |   |   |   |   |
| Pupa  |   |   |   |   |   |   |   |   |   |   |   |   |
| Adult |   |   |   |   |   |   |   |   |   |   |   |   |

# Yellow Palm-dart
## *Cephrenes trichopepla* (Lower, 1908)

Plate 54 Holmes Jungle, Darwin, NT
Photo: M. F. Braby

Plate 55 El Questro Wilderness Park, WA
Photo: M. F. Braby

## Distribution

This species occurs widely throughout the study region, extending further inland into the lower rainfall areas of the semi-arid zone (c. 600 mm mean annual rainfall) than *Cephrenes augiades*. Its geographic range corresponds well with the spatial distribution of its larval food plants, although records of *C. trichopepla* from the western Gulf Country are sparse. The species reaches its westernmost limit in the western Kimberley at Broome, WA (Dunn 1980; Common and Waterhouse 1981), which lies outside the range extent of its known native food plants. Further field surveys are required to determine the extent to which the species occurs in the Limmen Bight area and elsewhere in the Gulf Country. Outside the study region, *C. trichopepla* occurs naturally in Queensland and New South Wales.

## Habitat

*Cephrenes trichopepla* breeds mainly in savannah woodland where the larval food plants grow as palms (Braby 2011a, 2015e). In the drier areas, breeding populations are restricted to riparian evergreen monsoon vine forest or mixed woodland–monsoon forest along gorges with permanent water. It also occurs in suburban parks and gardens where ornamental palms have been established.

## Larval food plants

*Livistona benthamii, L. humilis, L. inermis, L. lorophylla, L. nasmophila, L. rigida, L. victoriae* (Arecaceae). The main food plants are *L. humilis* and *L. inermis* in the Top End, but it uses *L. lorophylla, L. nasmophila* and *L. victoriae* in the Kimberley (Braby 2011a, 2015e). The food plant is *L. rigida* at Mataranka, NT, and in the western Gulf Country at Boodjamulla/Lawn Hill National Park, Qld. The species also exploits several ornamental palms propagated in suburban parks and gardens (Dunn 2015b).

## Seasonality

Adults occur throughout the year, but they are more abundant during the late wet season and early dry season (March–May). The immature stages have been recorded in all months, except January, with no evidence of diapause in any of the life history stages, suggesting the species breeds continuously throughout the year.

## Breeding status

This species is resident in the study region.

## Conservation status

LC.

## Cephrenes trichopepla

0  100  200    400        600 km

- ● Specimen ≥1970
- ■ Observation ≥1970
- ▲ Literature ≥1970
- ● Specimen <1970
- ▲ Literature <1970
- ✕ Larval food plants

## Cephrenes trichopepla

0  100  200    400        600 km

- ● Species record
- ▬ Geographic range
- — — — Phytogeographical boundary
- ·········· IBRA bioregional boundary

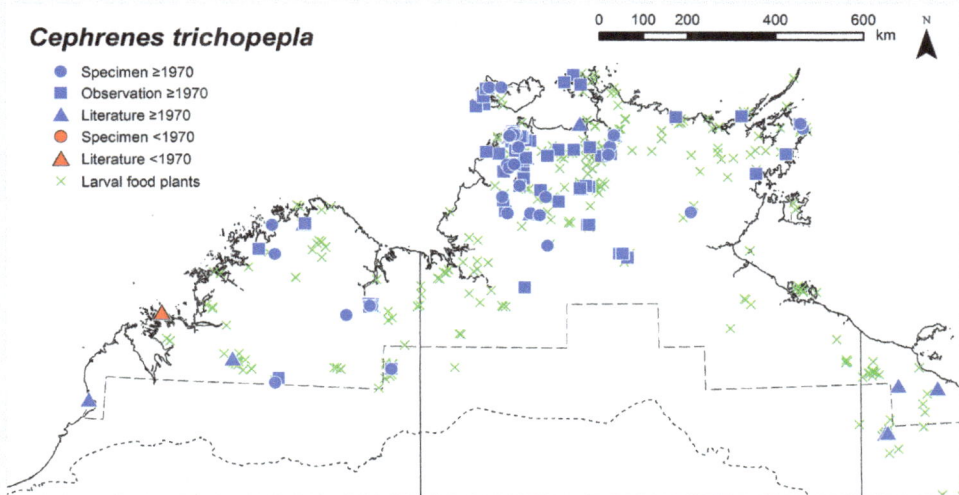

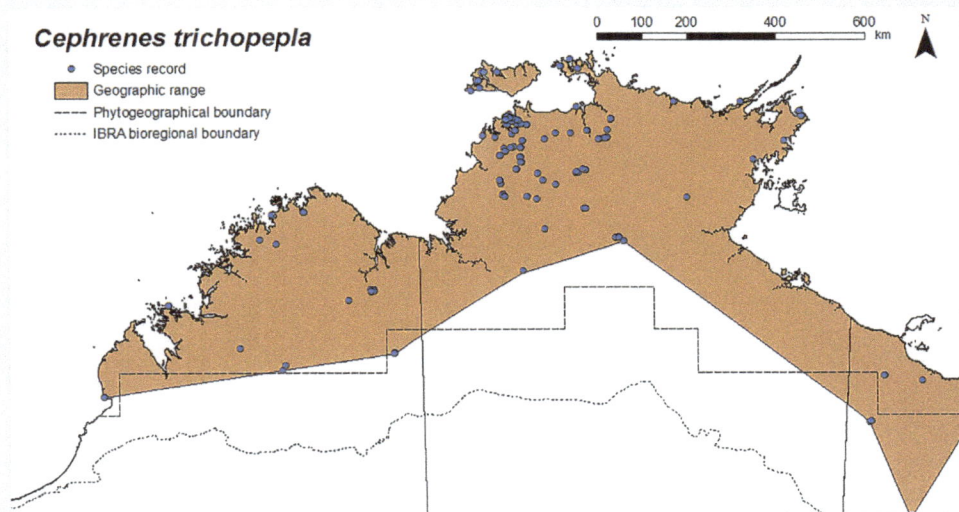

## Cephrenes trichopepla (n = 193)

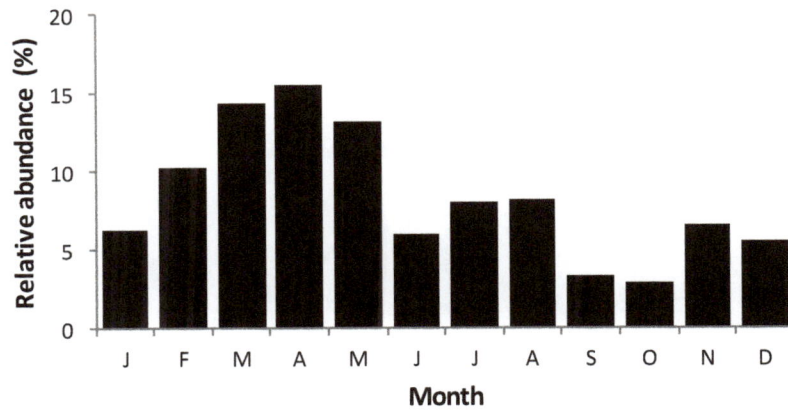

| Month | J | F | M | A | M | J | J | A | S | O | N | D |
|-------|---|---|---|---|---|---|---|---|---|---|---|---|
| Egg   |   |   |   |   |   |   |   |   |   |   |   |   |
| Larva |   |   |   |   |   |   |   |   |   |   |   |   |
| Pupa  |   |   |   |   |   |   |   |   |   |   |   |   |
| Adult |   |   |   |   |   |   |   |   |   |   |   |   |

# Whites and yellows

## (Pieridae)

# 7

# White Migrant
## *Catopsilia pyranthe* (Linnaeus, 1758)

Plate 56 Canberra, ACT
Photo: Richard Allen

Plate 57 Canberra, ACT
Photo: Stephanie Haygarth

## Distribution

This species is represented in the study region by the subspecies *C. pyranthe crokera* (W. S. Macleay, 1826). It occurs widely in the region, mostly in areas that receive less than 1,300 mm mean annual rainfall. It is particularly common in the semi-arid zone, but less common in the northern parts of the range (north of latitude 14°S). It is rare in the north-western coastal areas of the Top End (Dundee Beach–Darwin–Gunn Point), where it occurs very irregularly during migration/temporary range expansion (Braby 2014a). The putative native larval food plants (*Senna* spp.) occur widely in the semi-arid areas of northern Australia, and their combined spatial distribution corresponds well with the geographic range of *C. pyranthe*. Outside the study region, *C. pyranthe* occurs widely from India, southern China and South-East Asia to eastern Australia.

## Habitat

The breeding habitat of *C. pyranthe* has not been recorded in the study region. Adults have been collected mainly in savannah woodland and they undoubtedly breed in this habitat.

## Larval food plants

*Senna occidentalis* (Fabaceae: Caesalpinioideae); probably *Senna planitiicola* and *S. venusta*, which are two native food plants in Queensland (Braby 2000). The native food plants have not been recorded in the study region, but the immature stages have been reared on cultivated *S. occidentalis* at Berrimah Farm near Darwin, NT (M. Neal), and on an undetermined species at Tennant Creek, NT (C. Materne).

## Seasonality

Adults have been recorded sporadically during the year, but they are most abundant during March and April following good wet seasons of average or above average rainfall. The breeding phenology and seasonal history of the immature stages have not been recorded, but it is possible the species breeds seasonally. In the Darwin area and nearby locations—a region in which the species is normally absent—*C. pyranthe* is a rare seasonal visitor, with influxes likely to occur during March and April and less frequently from June to August and in December (Braby 2014a).

## Breeding status

This species is assumed to be resident in the study region, but populations appear to be temporary in the northernmost areas of the range.

## Conservation status

LC.

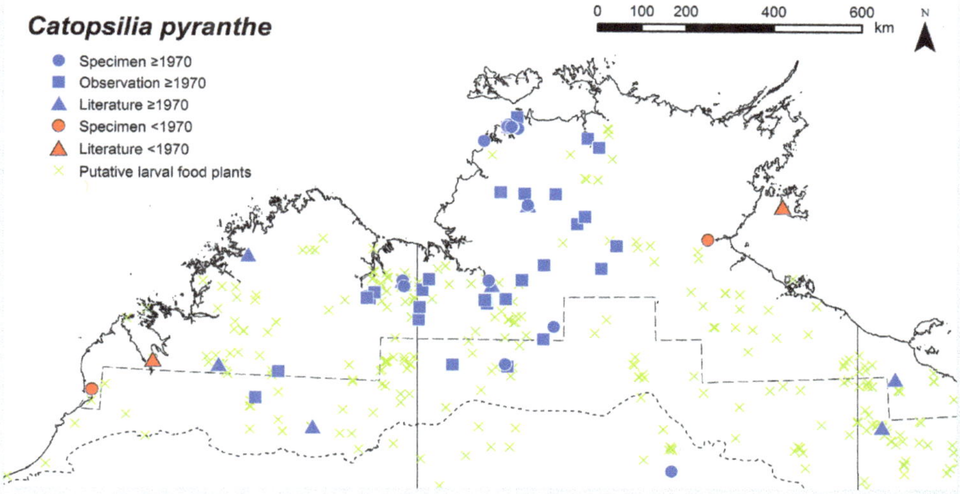

Catopsilia pyranthe
0 100 200 400 600
km
N
Specimen ≥1970
Observation ≥1970
Literature ≥1970
Specimen <1970
Literature <1970
Putative larval food plants

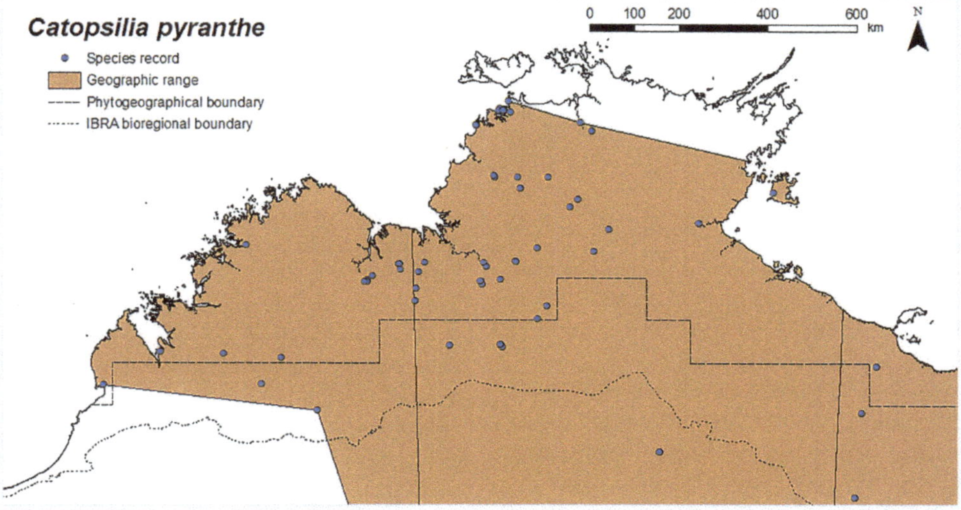

Catopsilia pyranthe
0 100 200 400 600
km
N
Species record
Geographic range
Phytogeographical boundary
IBRA bioregional boundary

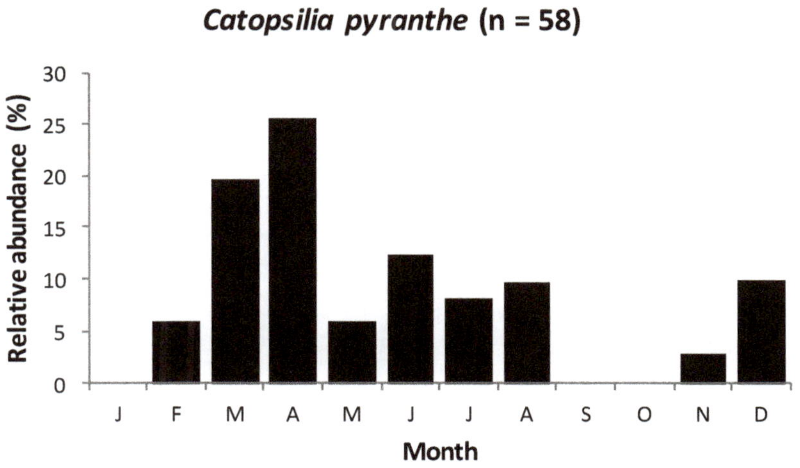

Catopsilia pyranthe (n = 58)
Relative abundance (%)
30
25
20
15
10
5
0
J F M A M J J A S O N D
Month

| Month | J | F | M | A | M | J | J | A | S | O | N | D |
|---|---|---|---|---|---|---|---|---|---|---|---|---|
| Egg | | | | | | | | | | | | |
| Larva | | | | | | | | | | | | |
| Pupa | | | | | | | | | | | | |
| Adult | | | | | | | | | | | | |

# Lemon Migrant
## *Catopsilia pomona* (Fabricius, 1775)

Plate 58 Herberton, Qld
Photo: Don Franklin

## Distribution

This species occurs very widely in the study region. It extends from the Kimberley, through the Top End to the western Gulf Country, and from moist coastal areas to drier inland areas of the semi-arid zone, as well as the arid zone of central Australia beyond the southern boundary of the study region. The putative native larval food plants (*Senna* spp.) occur in the semi-arid areas of northern Australia, and their spatial distribution corresponds well with the geographical range of *C. pomona*. Outside the study region, *C. pomona* occurs widely from India, southern China and South-East Asia, through mainland New Guinea and adjacent islands to Australia, where it occurs throughout the northern half of the continent.

## Habitat

The natural breeding habitat of *C. pomona* has not been recorded in the study region. The species frequently breeds in suburban parks and gardens where the introduced larval food plants are propagated as ornamental trees. Elsewhere, adults have been recorded in a wide variety of habitats in which they undoubtedly breed.

## Larval food plants

*Cassia fistula*, *C. siamea*, *Senna alata*, *S. occidentalis* (Fabaceae: Caesalpinioideae); probably *Senna magnifolia* and *S. venusta*, which are two food plants in Queensland (Braby 2000). The native food plants have not been recorded in the study region, but the immature stages are frequently found on introduced *C. fistula* and *C. siamea* in suburban areas.

## Seasonality

Adults occur throughout the year. They appear to show little seasonal variation in abundance, although they tend to be more numerous after the wet season (April and May) and least abundant during the cooler mid dry season (July), but they are also abundant during the wet season (December and January). Franklin (2011) found similar trends near Darwin, NT, based on quantitative studies conducted over 14 months during 2008–09. At Darwin, NT, the immature stages (eggs or larvae) have been recorded only during the warmer, more humid months (October–April), when the larval food plants produce flushes of new leaf growth. Presumably, the species breeds seasonally and females remain in reproductive diapause during the dry season (May–September), similar to populations in northern Queensland (Jones 1987). Although *C. pomona* a well-known migrant (Smithers 1983b), we have no records of migratory flights in the study region.

## Breeding status

This species is resident in the study region.

## Conservation status

LC.

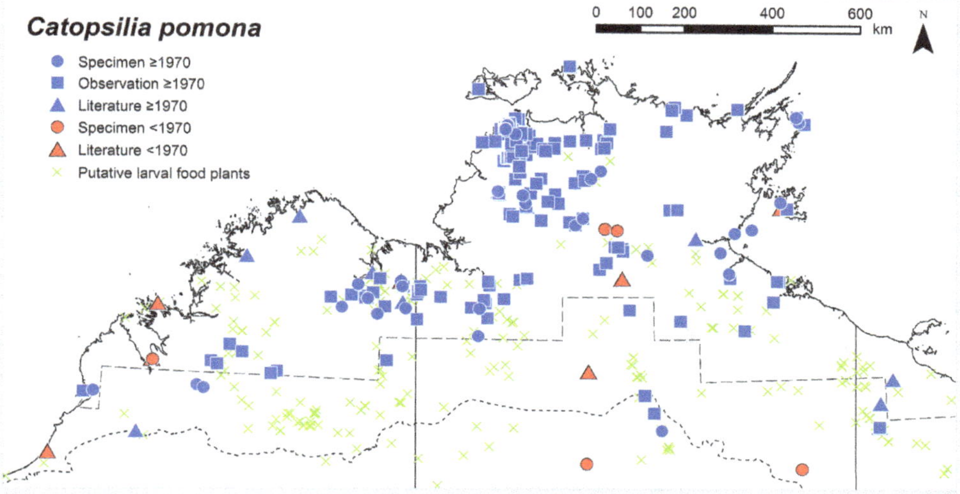

## Catopsilia pomona

Legend:
- ● Specimen ≥1970
- ■ Observation ≥1970
- ▲ Literature ≥1970
- ● Specimen <1970
- ▲ Literature <1970
- × Putative larval food plants

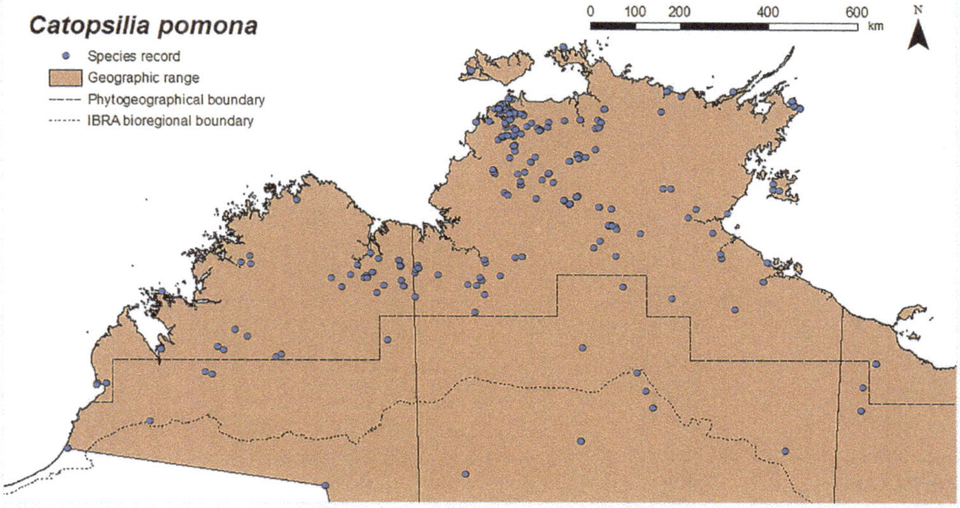

## Catopsilia pomona

Legend:
- • Species record
- ▨ Geographic range
- --- Phytogeographical boundary
- ····· IBRA bioregional boundary

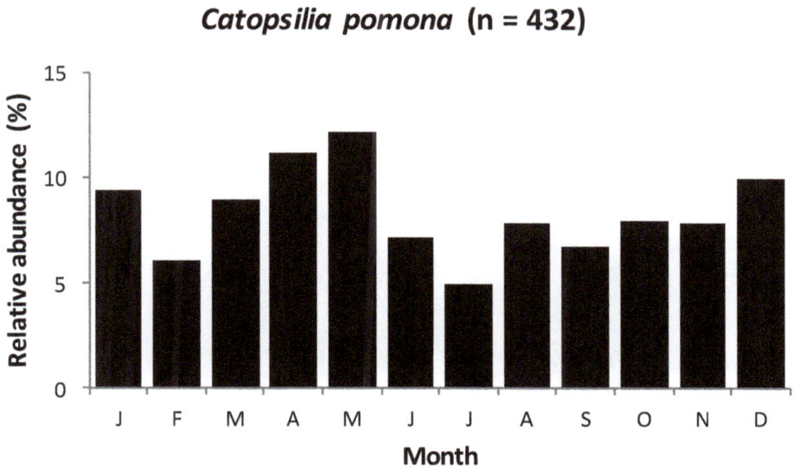

## Catopsilia pomona (n = 432)

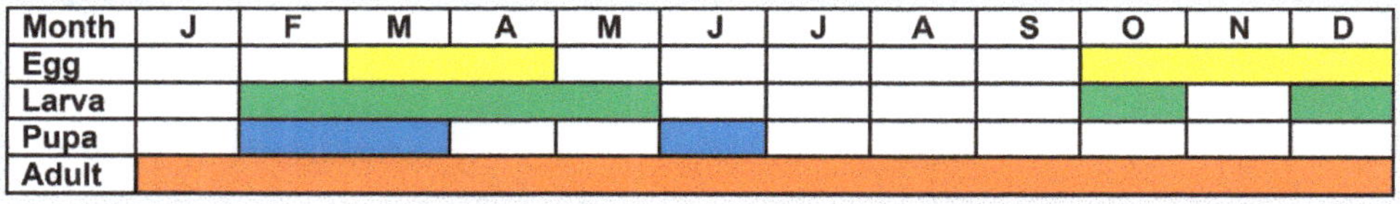

| Month | J | F | M | A | M | J | J | A | S | O | N | D |
|-------|---|---|---|---|---|---|---|---|---|---|---|---|
| Egg   |   |   | ▨ | ▨ |   |   |   |   |   | ▨ | ▨ | ▨ |
| Larva |   | ▨ | ▨ | ▨ | ▨ |   |   |   |   | ▨ |   | ▨ |
| Pupa  |   | ▨ |   |   | ▨ |   |   |   |   |   |   |   |
| Adult |   |   |   |   |   |   |   |   |   |   |   |   |

# Orange Migrant
## *Catopsilia scylla* (Linnaeus, 1763)

Plate 59 Wanguri, Darwin, NT
Photo: M. F. Braby

## Distribution

This species is represented in the study region by the subspecies *C. scylla etesia* (Hewitson, 1867). It occurs widely in the region, extending from moist coastal areas to drier inland areas of the semi-arid zone, as well as the arid zone of central Australia beyond the southern boundary of the study region. Its geographic range closely corresponds with the spatial distribution of its larval food plants. There are no records of *C. scylla* from Melville Island and Groote Eylandt, although the food plants occur on these islands; thus, further field surveys are required to determine whether *C. scylla* is present in these areas. Outside the study region, *C. scylla* occurs from South-East Asia, through mainland New Guinea and adjacent islands and central, north-eastern and eastern Australia to Fiji.

## Habitat

*Catopsilia scylla* breeds in a variety of habitats. In coastal areas, it breeds in monsoon vine thicket on laterite or sand dunes where the larval food plant (*Senna surattensis*) grows as a shrub, but in inland areas it breeds in woodland, often on sandstone, where alternative food plants grow as perennial understorey shrubs.

## Larval food plants

*Senna leptoclada*, *S. oligoclada*, *S. surattensis* (Fabaceae: Caesalpinioideae). *Senna surattensis* is the main food plant in coastal areas, but *S. oligoclada* and sometimes *S. leptoclada* are used in the more inland areas.

## Seasonality

Adults occur throughout the year, but they are most abundant from March to May and again in October, usually after rainfall. At Darwin, NT, breeding is seasonal, with the immature stages (eggs and larvae) recorded mostly around these two periods (i.e. March–May and September–November). Meyer et al. (2006) also noted that the species was seasonal in Darwin, with adults recorded only during the period March–May. It is not clear whether the species breeds at other times of the year, remains in reproductive diapause or the population disperses to breed elsewhere. There appear to be no previous reports of migration or movement patterns for *C. scylla* in Australia. However, on several occasions in the Top End and Northern Deserts, directional flights in which adults were flying rapidly in a constant direction (north, north-east or east) have been observed during April and May (Braby 2016b). Its regular seasonal appearance in Darwin also suggests the arrival of migratory populations.

## Breeding status

This species is assumed to be resident in the study region, but populations appear to be nomadic and are possibly temporary in many areas.

## Conservation status

LC.

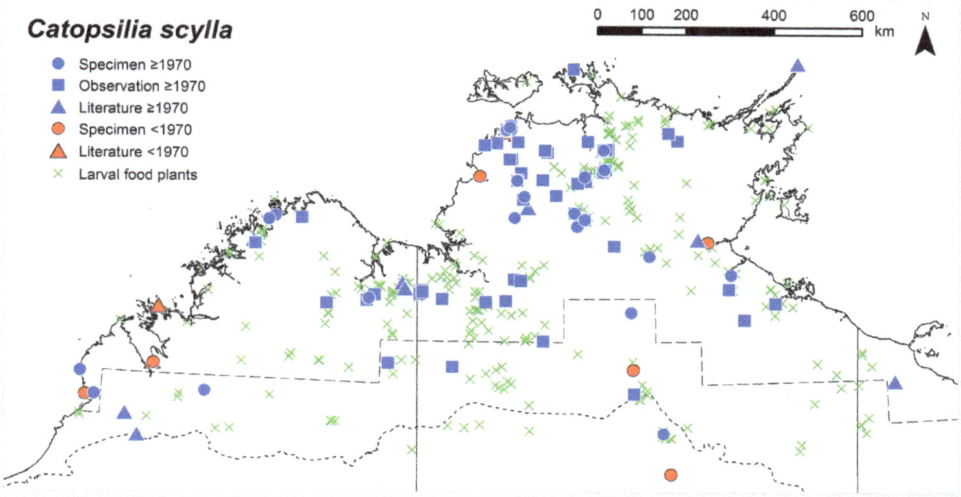

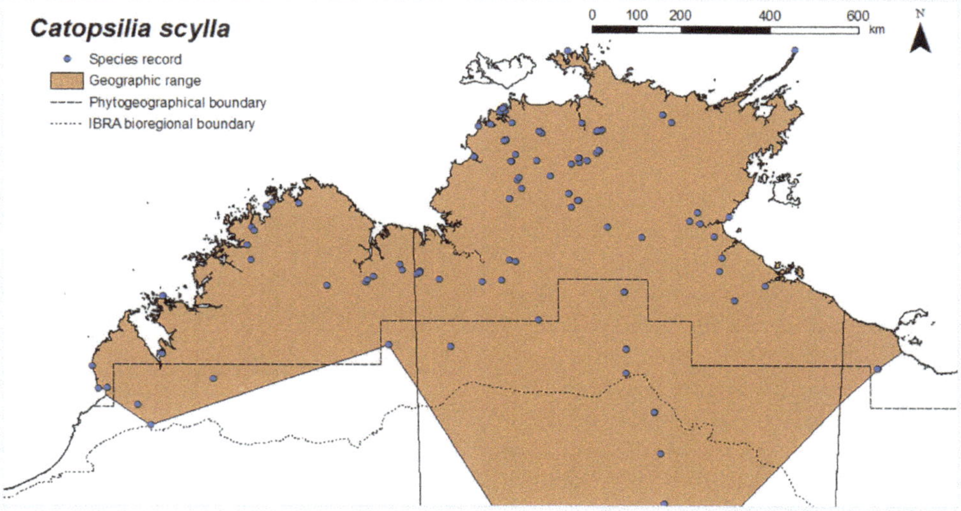

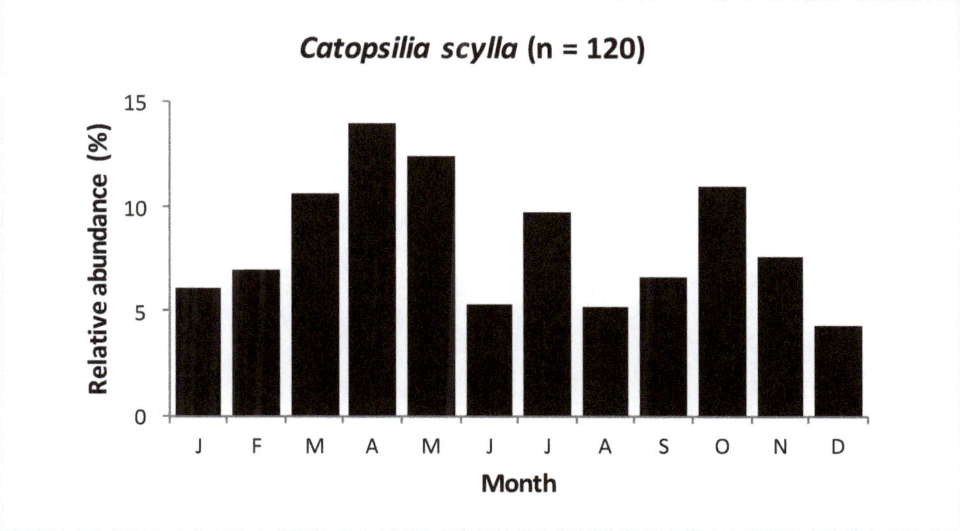

## Catopsilia scylla

- ● Specimen ≥1970
- ■ Observation ≥1970
- ▲ Literature ≥1970
- ● Specimen <1970
- ▲ Literature <1970
- ✕ Larval food plants

## Catopsilia scylla

- ● Species record
- ▨ Geographic range
- - - Phytogeographical boundary
- ⋯ IBRA bioregional boundary

## Catopsilia scylla (n = 120)

| Month | J | F | M | A | M | J | J | A | S | O | N | D |
|-------|---|---|---|---|---|---|---|---|---|---|---|---|
| Egg   |   |   | ▨ | ▨ | ▨ |   | ▨ |   | ▨ | ▨ | ▨ |   |
| Larva |   |   | ▨ | ▨ | ▨ |   |   |   | ▨ | ▨ | ▨ |   |
| Pupa  |   |   |   |   | ▨ |   |   |   | ▨ | ▨ | ▨ | ▨ |
| Adult | ▨ | ▨ | ▨ | ▨ | ▨ | ▨ | ▨ | ▨ | ▨ | ▨ | ▨ | ▨ |

# No-brand Grass-yellow
## *Eurema brigitta* (Stoll, [1780])

Plate 60 Big Mitchell Creek, Qld
Photo: Frank Pierce

## Distribution

This species is represented in the study region by the subspecies *E. brigitta australis* (Wallace, 1867). It is known from only a single worn female specimen collected from the Wessel Islands (Rimbija Island) in the north-eastern corner of the Top End by E. D. Edwards (Braby 2014a). The putative larval food plant (*Chamaecrista nomame*) occurs widely in the Kimberley and Top End. Outside the study region, *E. brigitta* occurs widely from Africa, India, southern China and South-East Asia, through mainland New Guinea and north-eastern Australia to New Caledonia and Fiji.

## Excluded data

Previous records for Darwin (Waterhouse and Lyell 1914; Meyer et al. 2006), Pine Creek (Angel 1951) and Daly River (Hutchinson 1978), NT, are considered to be erroneous (Braby 2014a). Waterhouse and Lyell's (1914) published record for 'Port Darwin' was based on F. P. Dodd's material, but examination of these historical specimens (and additional material from Groote Eylandt by N. B. Tindale) in the AM revealed that several catalogued under the name *E. brigitta* had been misidentified and were actually *Eurema laeta*.

## Habitat

The breeding habitat of *E. brigitta* has not been recorded in the study region.

## Larval food plants

Not recorded in the study region; possibly *Chamaecrista nomame* (Fabaceae: Caesalpinioideae), which is one of the food plants in eastern Australia (Braby 2016a).

## Seasonality

The seasonal abundance and breeding phenology of this species are not well understood. The single specimen was collected on 18 January 1977.

## Breeding status

*Eurema brigitta* does not appear to have become permanently established in the study region. It remains to be determined, however, whether it breeds temporarily on the Wessel Islands or the single individual was a vagrant that dispersed from Cape York Peninsula, Qld, across the Gulf of Carpentaria to north-eastern Arnhem Land.

## Conservation status

NA.

## Eurema brigitta

● Specimen ≥1970

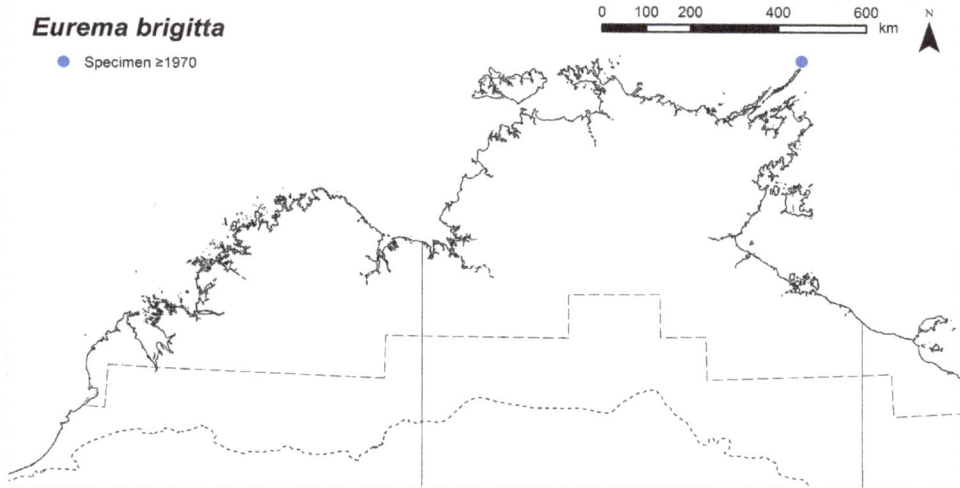

| Month | J | F | M | A | M | J | J | A | S | O | N | D |
|-------|---|---|---|---|---|---|---|---|---|---|---|---|
| Egg   |   |   |   |   |   |   |   |   |   |   |   |   |
| Larva |   |   |   |   |   |   |   |   |   |   |   |   |
| Pupa  |   |   |   |   |   |   |   |   |   |   |   |   |
| Adult |   |   |   |   |   |   |   |   |   |   |   |   |

Photo: Dhurputjpi Gove Penninsula, NT, M.F. Braby

# Lined Grass-yellow
## *Eurema laeta* (Boisduval, 1836)

Plate 61 Herbert Lagoon, NT
Photo: Deb Bisa

## Distribution

This species is represented in the study region by the subspecies *E. laeta sana* (Butler, 1877). It occurs widely in the Kimberley, Top End and western Gulf Country. It extends from moist coastal areas to drier inland areas of the semi-arid zone (c. 700 mm mean annual rainfall), reaching its southernmost limits in the Buccaneer Archipelago at Koolan Island in Yampi Sound (Koch and van Ingen 1969; McKenzie et al. 1995) and El Questro Wilderness Park (Pentecost River crossing) (Braby 2012b), WA; Judbarra/Gregory National Park (Limestone Gorge) (Braby and Archibald 2016) and Bessie Spring (Dunn and Dunn 1991), NT; and 60 km north-west of Doomadgee, Qld (Puccetti 1991). The putative larval food plants (*Chamaecrista* spp.) are widespread in northern Australia and their overall extent includes the geographic range of *E. laeta*. Outside the study region, *E. laeta* occurs widely from India, southern China, Japan and South-East Asia (although it is absent from the Malay Peninsula, Borneo and parts of Indonesia), through mainland New Guinea to north-eastern Australia.

## Habitat

The breeding habitat of *E. laeta* has not been recorded in the study region. Adults occur in a wide variety of habitats, especially savannah woodland, in which they undoubtedly breed. During the dry season they typically aggregate in the ground layer of moister habitats, such as riparian woodland, paperbark forest and monsoon vine forest along the banks of creeks and rivers or in semi-deciduous monsoon vine thicket on rocky outcrops and breakaways.

## Larval food plants

Not recorded in the study region; probably *Chamaecrista* spp. (Fabaceae: Caesalpinioideae), which are the food plants in eastern Queensland (Braby 2000, 2016a).

## Seasonality

Adults occur throughout the year, but they are more numerous during the dry season (June–September) after the peak of *Eurema herla*. Presumably, adults stop breeding during the dry season, remain in reproductive diapause and contract to moist refuges where they may be observed in relatively large numbers, similar to populations in northern Queensland (Jones and Rienks 1987). The breeding phenology and seasonal history of the immature stages have not been recorded, but it is likely several generations are completed during the wet season.

## Breeding status

This species is resident in the study region.

## Conservation status

LC.

## Eurema laeta

- ● Specimen ≥1970
- ■ Observation ≥1970
- ▲ Literature ≥1970
- ● Specimen <1970
- ▲ Literature <1970
- × Putative larval food plants

## Eurema laeta

- ● Species record
- ▧ Geographic range
- – – – Phytogeographical boundary
- ········ IBRA bioregional boundary

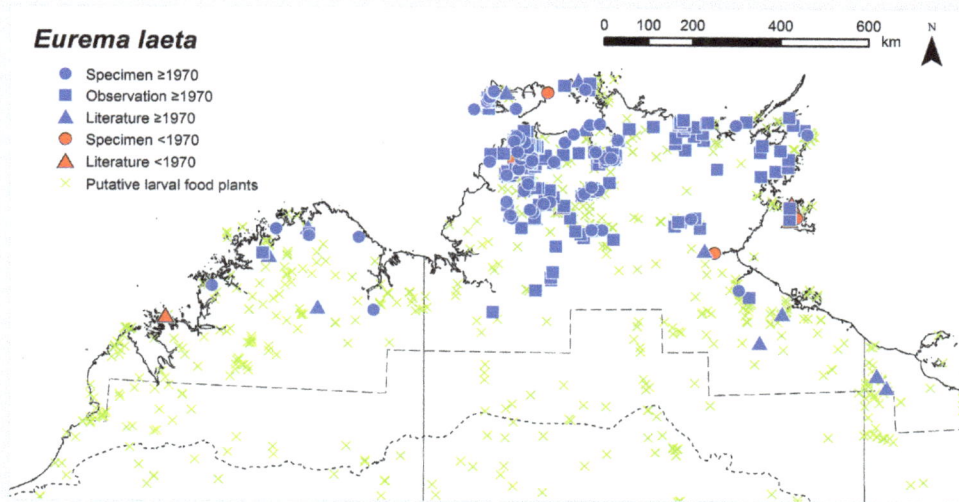

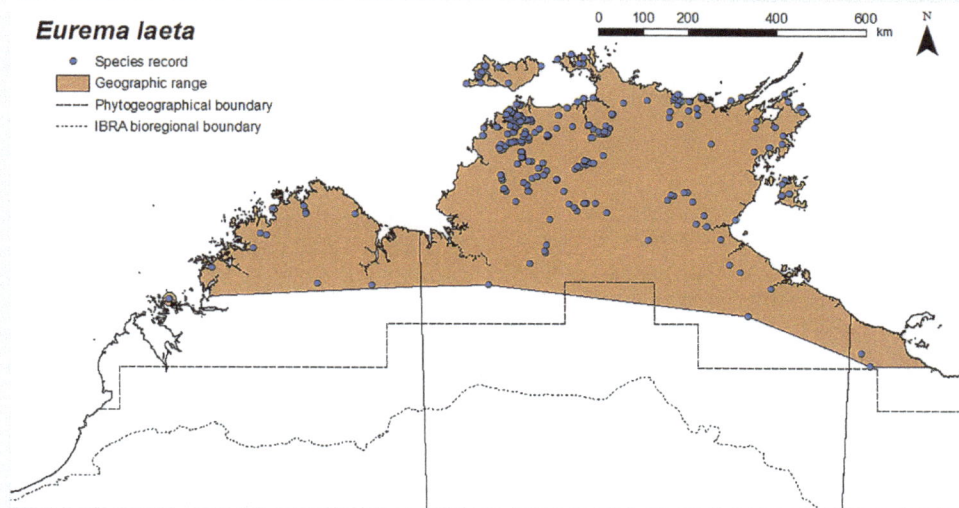

### Eurema laeta (n = 352)

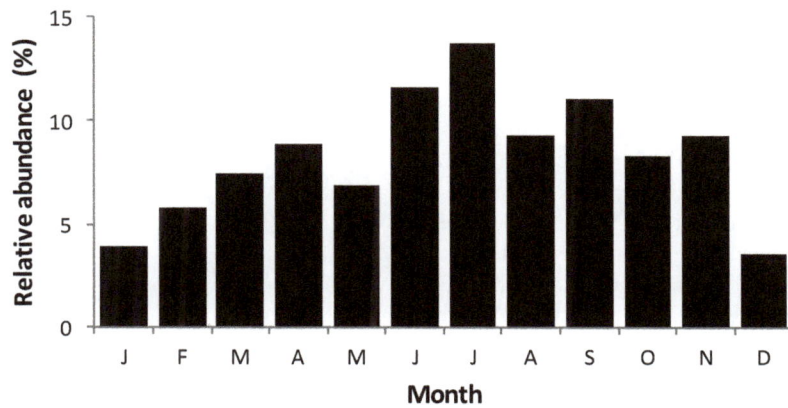

| Month | J | F | M | A | M | J | J | A | S | O | N | D |
|-------|---|---|---|---|---|---|---|---|---|---|---|---|
| Egg   |   |   |   |   |   |   |   |   |   |   |   |   |
| Larva |   |   |   |   |   |   |   |   |   |   |   |   |
| Pupa  |   |   |   |   |   |   |   |   |   |   |   |   |
| Adult |   |   |   |   |   |   |   |   |   |   |   |   |

# Macleay's Grass-yellow
## *Eurema herla* (W. S. Macleay, 1826)

Plate 62 Wongalara Wildlife Sanctuary, NT
Photo: M. F. Braby

## Distribution

This species has a very broad distribution, occurring throughout most of the study region. Its geographic range corresponds well with the spatial distribution of its larval food plants; however, *E. herla* extends further inland into semi-arid areas of lower rainfall (< 500 mm mean annual rainfall), including the Edgar Ranges, WA (Common 1981); 40 km south of Lajamanu, Rabbit Flat Roadhouse in the Tanami Desert (J. Archibald) and Tennant Creek, NT (Peters 1969); and Boodjamulla/Lawn Hill National Park, Qld (Daniels and Edwards 1998). It has also been recorded from Devils Marbles, NT (Dunn 1980), just outside the southern boundary of the study region. It is not known whether the species is permanently established in the semi-arid zone and breeds on alternative species of *Chamaecrista* or the populations are temporary. Most records from the southern edge of the range are in the dry season (April–August), which suggests *E. herla* makes temporary southern expansions after rainfall or with the onset of cooler, dry weather. However, a specimen has been collected at Tennant Creek (Mary Anne Dam) in February (J. Archibald). Outside the study region, *E. herla* occurs in north-eastern and eastern Australia.

## Habitat

*Eurema herla* breeds in savannah woodland, favouring open grassy areas where the larval food plants grow in the ground layer as annual herbs (Braby 2011a, 2015e). In the drier inland areas, it is usually associated with woodland on rocky outcrops and breakaways or along rocky seasonal gullies and creeks.

## Larval food plants

*Chamaecrista mimosoides*, *C. nigricans* (Fabaceae: Caesalpinioideae).

## Seasonality

Adults occur throughout the year, but they are more abundant during the late wet season and early dry season (March–June), before the peak of *Eurema laeta*. The breeding phenology and seasonal history of the immature stages are poorly known, but females have been observed laying eggs in March and May. Presumably, the life cycle strategy is similar to populations in northern Queensland in which breeding is limited to the wet season and early dry season and females remain in reproductive diapause during the mid to late dry season (June–October), when the food plants are not available (Jones and Rienks 1987).

## Breeding status

This species is resident in the study region.

## Conservation status

LC.

## Eurema herla

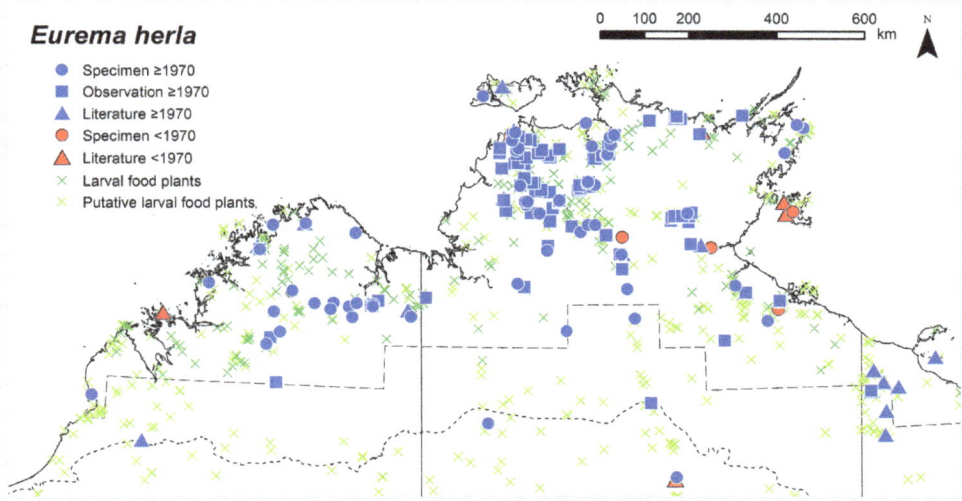

Legend:
- Specimen ≥1970
- Observation ≥1970
- Literature ≥1970
- Specimen <1970
- Literature <1970
- Larval food plants
- Putative larval food plants

## Eurema herla

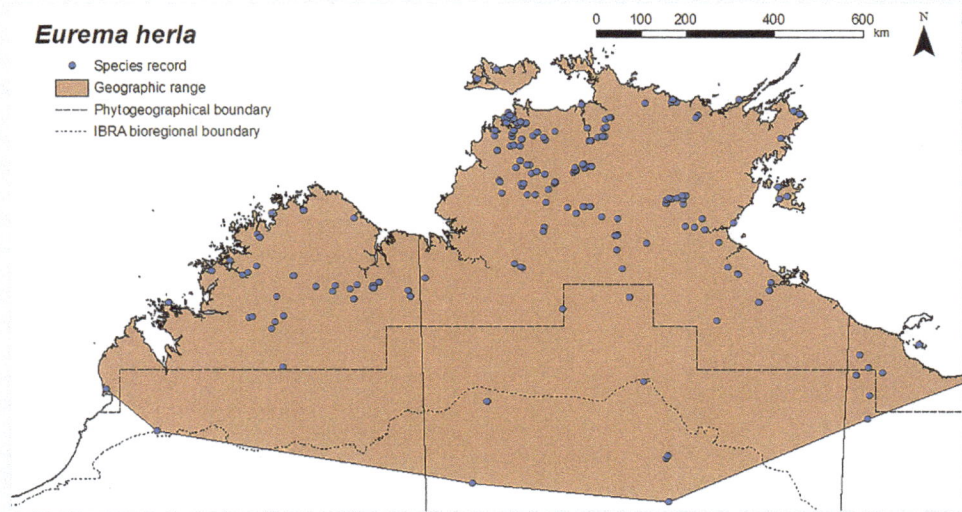

Legend:
- Species record
- Geographic range
- Phytogeographical boundary
- IBRA bioregional boundary

## Eurema herla (n = 217)

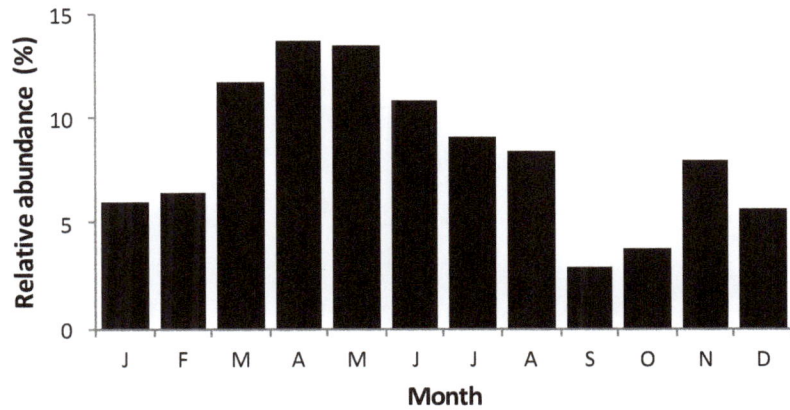

| Month | J | F | M | A | M | J | J | A | S | O | N | D |
|-------|---|---|---|---|---|---|---|---|---|---|---|---|
| Egg   |   |   |   |   |   |   |   |   |   |   |   |   |
| Larva |   |   |   |   |   |   |   |   |   |   |   |   |
| Pupa  |   |   |   |   |   |   |   |   |   |   |   |   |
| Adult |   |   |   |   |   |   |   |   |   |   |   |   |

# Small Grass-yellow

## *Eurema smilax* (Donovan, 1805)

Plate 63 Yenda, NSW
Photo: M. F. Braby

## Distribution

This species occurs throughout much of the study region. It also extends into the arid zone of central Australia beyond the southern boundary of the study region. It is more common in the semi-arid areas and less frequently recorded in the higher rainfall areas, such as the north-western corner of the Top End, where it reaches its northernmost limit at Darwin, NT (Waterhouse and Lyell 1914; Campbell 1947; Angel 1951; Le Souëf 1971; Meyer et al. 2006). Surprisingly, there are no records of the species from the Tiwi Islands, Cobourg Peninsula or northern and eastern Arnhem Land despite the presence of the putative larval food plants, which occur widely in northern Australia; *Neptunia gracilis* occurs in monsoon and semi-arid areas, *N. monosperma* mainly in semi-arid areas and *Senna artemisioides* in semi-arid and arid areas. Outside the study region, *E. smilax* occurs widely throughout most of the Australian continent, particularly in semi-arid and arid areas.

## Habitat

The breeding habitat of *E. smilax* has not been recorded in the study region. Adults occur in a wide variety of habitats, including savannah woodland, eucalypt low open woodland, grassland and *Acacia* shrubland, in which they no doubt breed.

## Larval food plants

Not recorded in the study region; probably *Neptunia gracilis*, *N. monosperma*, *Senna artemisioides* (Fabaceae: Mimosoideae, Caesalpinioideae), which are three food plants in Queensland (Braby 2000).

## Seasonality

Adults occur throughout the year, but they are more numerous during the dry season (May–August), although they are rarely observed in large numbers compared with other species of *Eurema*. The breeding phenology and seasonal history of the immature stages have not been recorded. Presumably, the species breeds throughout the year but populations are opportunistic and highly mobile.

## Breeding status

*Eurema smilax* is assumed to be resident in the study region, but populations appear to be nomadic and are possibly temporary in the northern areas of the range, particularly in the north of the Northern Territory.

## Conservation status

LC.

*Eurema smilax*

- Specimen ≥1970
- Observation ≥1970
- Literature ≥1970
- Specimen <1970
- Literature <1970
- Putative larval food plants

*Eurema smilax*

- Species record
- Geographic range
- Phytogeographical boundary
- IBRA bioregional boundary

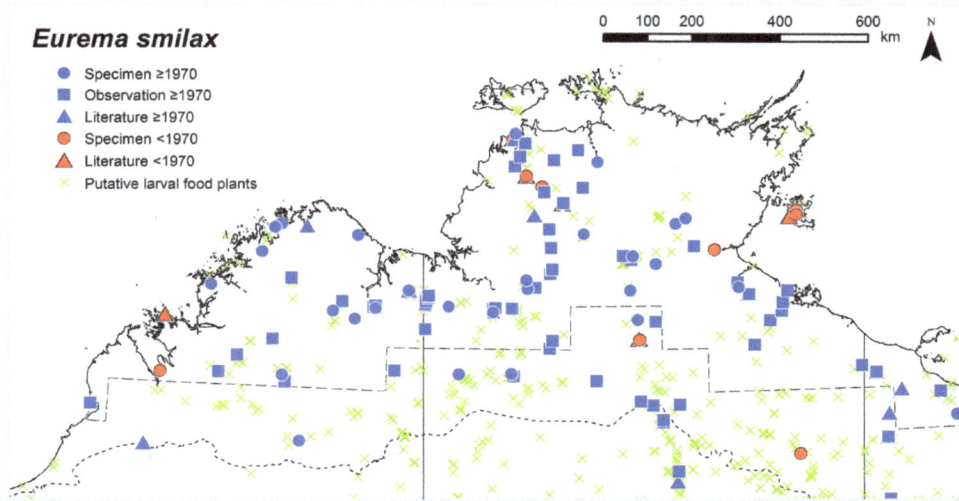

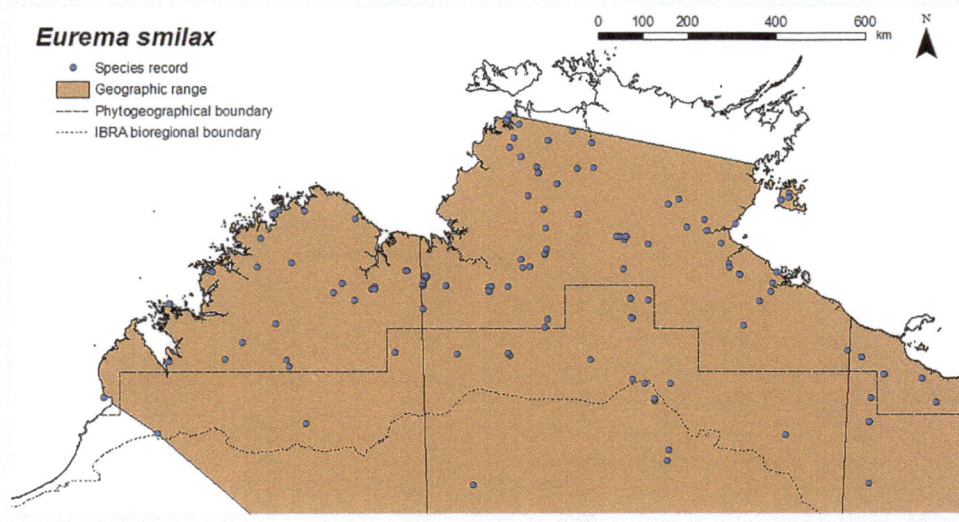

*Eurema smilax* (n = 137)

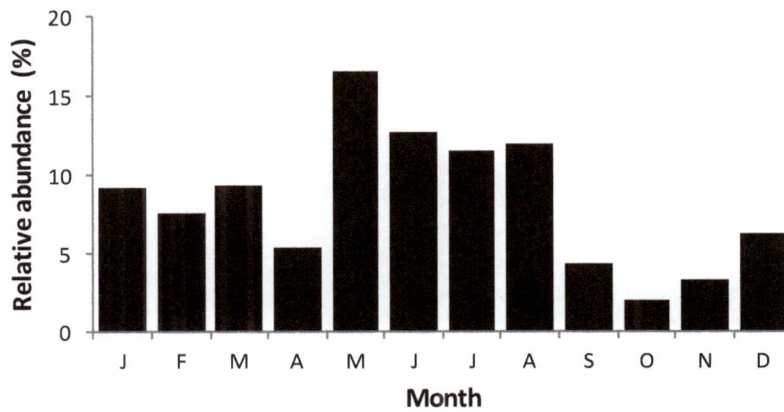

| Month | J | F | M | A | M | J | J | A | S | O | N | D |
|-------|---|---|---|---|---|---|---|---|---|---|---|---|
| Egg   |   |   |   |   |   |   |   |   |   |   |   |   |
| Larva |   |   |   |   |   |   |   |   |   |   |   |   |
| Pupa  |   |   |   |   |   |   |   |   |   |   |   |   |
| Adult |   |   |   |   |   |   |   |   |   |   |   |   |

# Scalloped Grass-yellow
*Eurema alitha* (C. & R. Felder, 1862)

Plate 64 Fish River Station, NT
Photo: M. F. Braby

## Distribution

This species is represented in the study region by the subspecies *E. alitha novaguineensis* Shriôzu & Yata, 1982. Although historical records go back as far as 1909, the species was previously confused with *Eurema hecabe*, such that *E. alitha* was not formally recognised in the fauna until quite recently (Braby 1997). It occurs in the western and northern Kimberley, where it extends from Augustus Water (J. E. and A. Koeyers) to Kalumburu (Grund and Hunt 2001), WA, and more widely in the Top End. It also extends into the Northern Deserts and western Gulf Country, where it occurs in the lower rainfall areas of the semi-arid zone (c. 700 mm mean annual rainfall). Its southernmost limits are Judbarra/Gregory National Park (Limestone Gorge) (Braby and Archibald 2016), Daly Waters (Braby 2000) and Cape Crawford (Braby 1997), NT. The larval food plant is considerably more widespread, occurring throughout much of the Kimberley, Northern Deserts and western Queensland in the Gulf Country. Further field surveys are therefore required to determine whether *E. alitha* is established in these areas. Outside the study region, *E. alitha* occurs from South-East Asia, through mainland New Guinea and adjacent islands to north-eastern and eastern Australia.

## Habitat

*Eurema alitha* breeds in semi-deciduous monsoon vine thicket and savannah woodland with a sparse cover of grasses where the larval food plant grows as a spreading ground cover or climber, usually over rock scree or on rocky hill-slopes (Braby 2011a).

## Larval food plants

*Galactia tenuiflora* (Fabaceae: Faboideae).

## Seasonality

Adults occur throughout the year, but, like *Eurema hecabe*, they are most abundant during the wet season, with a peak in abundance in April. They become less numerous as the dry season progresses and are scarce towards the end of the dry (September and October). The immature stages have been recorded during the wet season (January–April), when adults are most abundant. It is not known whether females enter reproductive diapause during the dry season or breed continuously, but reproductive activity declines as the dry season progresses.

## Breeding status

This species is resident in the study region.

## Conservation status

LC.

## Eurema alitha

- ● Specimen ≥1970
- ■ Observation ≥1970
- ▲ Literature ≥1970
- ▲ Literature <1970
- ✕ Larval food plant

## Eurema alitha

- ● Species record
- ▨ Geographic range
- – – Phytogeographical boundary
- ····· IBRA bioregional boundary

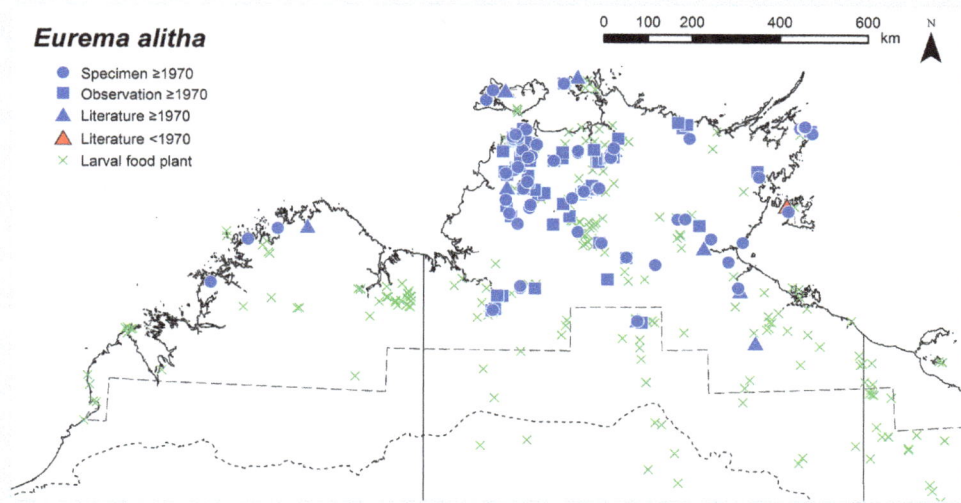

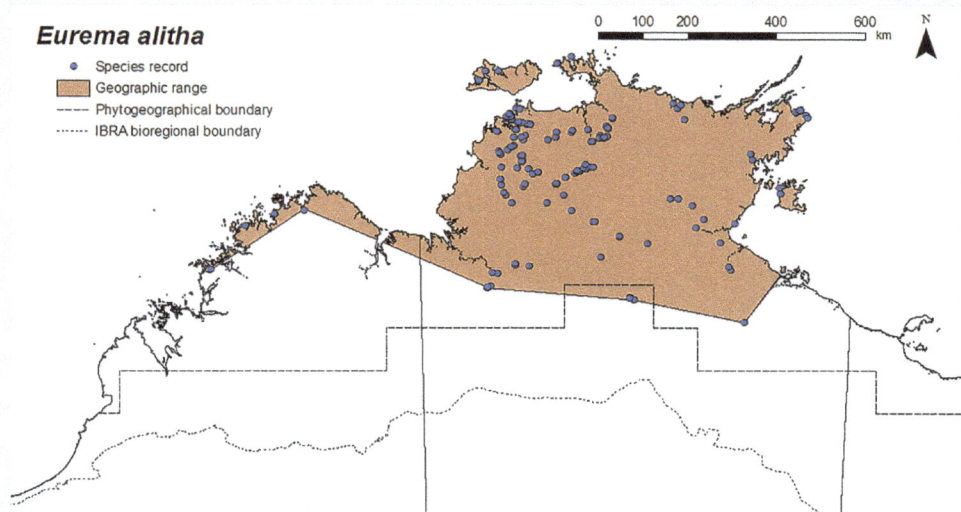

### Eurema alitha (n = 202)

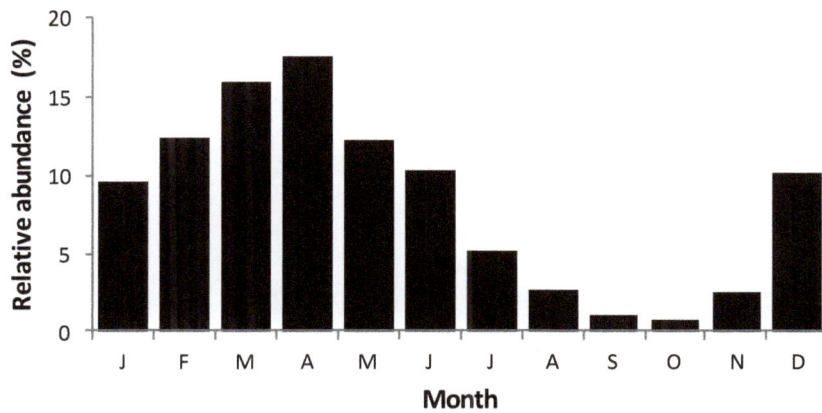

| Month | J | F | M | A | M | J | J | A | S | O | N | D |
|-------|---|---|---|---|---|---|---|---|---|---|---|---|
| Egg   |   |   |   |   |   |   |   |   |   |   |   |   |
| Larva |   |   |   |   |   |   |   |   |   |   |   |   |
| Pupa  |   |   |   |   |   |   |   |   |   |   |   |   |
| Adult |   |   |   |   |   |   |   |   |   |   |   |   |

# Large Grass-yellow
## Eurema hecabe (Linnaeus, 1758)

Plate 65 Wanguri, Darwin, NT
Photo: M. F. Braby

## Distribution

This species occurs throughout much of the study region. It extends from moist coastal areas to drier inland areas of the semi-arid zone (< 400 mm mean annual rainfall), where it has been recorded as far south as Kelly Creek to the south of Tennant Creek, NT (M. Malipatil and J. Hawkins). Its geographic range corresponds with the broad spatial distribution of its native larval food plants, indicating that *E. hecabe* has been well sampled in the region. Outside the study region, *E. hecabe* occurs widely from Africa, India, southern China, Japan and South-East Asia, through mainland New Guinea and adjacent islands and north-eastern and eastern Australia to the islands of the South Pacific.

## Habitat

*Eurema hecabe* occurs in a wide variety of habitats. It frequently breeds in both wet and dry monsoon forests, including evergreen monsoon vine forest associated with permanent freshwater springs or streams and semi-deciduous monsoon vine thicket associated with rock outcrops or coastal sand dunes and lateritic cliffs, where its native food plant (*Breynia cernua*) grows as a shrub. However, it also breeds in other habitats, including the edges of mangroves and savannah woodland where alternative food plants grow (particularly *Sesbania cannabina*).

## Larval food plants

*Senna surattensis*, *Sesbania cannabina* (Fabaceae: Caesalpinioideae, Faboideae), *Breynia cernua*, *Phyllanthus* sp. (Phyllanthaceae); also *Senna alata*, *S. obtusifolia*, *Leucaena leucocephala* (Fabaceae: Caesalpinioideae, Mimosoideae).

## Seasonality

Adults occur throughout the year, but, like *Eurema alitha*, they are most abundant during the wet season, with a peak in abundance in April. They may also be very abundant in May. They become less numerous as the dry season progresses and are scarce towards the end of the dry (August–October). The immature stages have been recorded during the wet season and early dry season (November–June). It is not known whether females enter reproductive diapause or breed continuously, but reproductive activity declines during the mid to late dry season (July–October). *Eurema hecabe* is a well-known migrant in eastern Australia (Smithers 1983b), but there few records of population movement in northern Australia. In April 2015, a large easterly flight was observed along the Stuart Highway between Noonamah and Adelaide River, NT (Braby 2016b).

## Breeding status

This species is resident in the study region.

## Conservation status

LC.

## Eurema hecabe

Legend:
- ● Specimen ≥1970
- ■ Observation ≥1970
- ▲ Literature ≥1970
- ● Specimen <1970
- ▲ Literature <1970
- ✕ Larval food plants

0  100  200      400      600 km

N

## Eurema hecabe

Legend:
- ● Species record
- ▬ Geographic range
- – – – Phytogeographical boundary
- ········ IBRA bioregional boundary

0  100  200      400      600 km

N

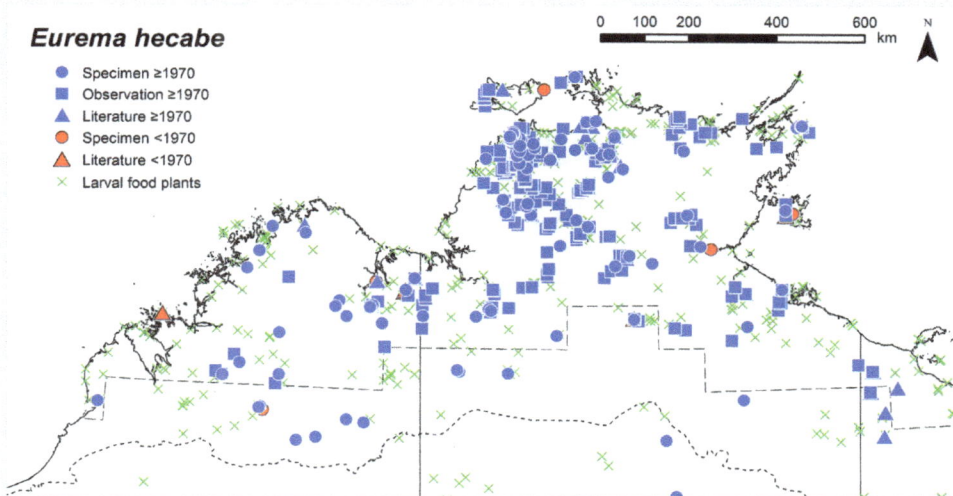

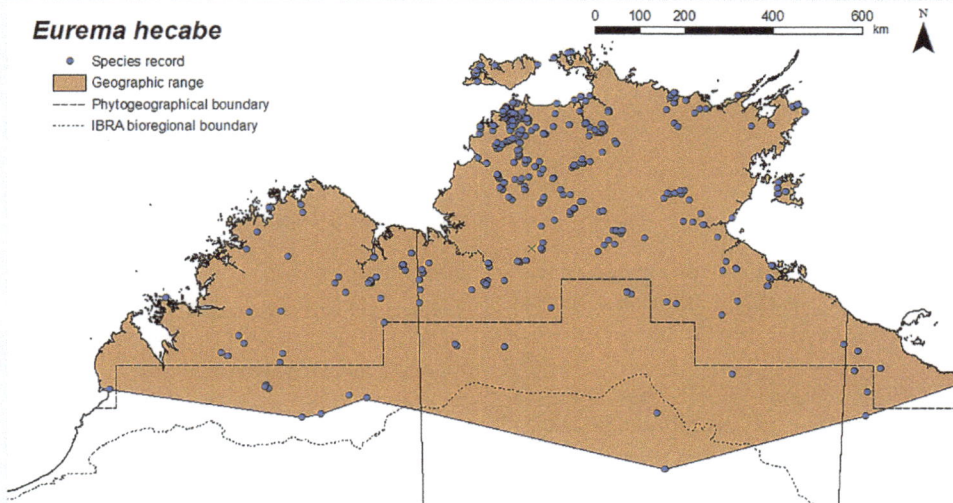

### Eurema hecabe (n = 621)

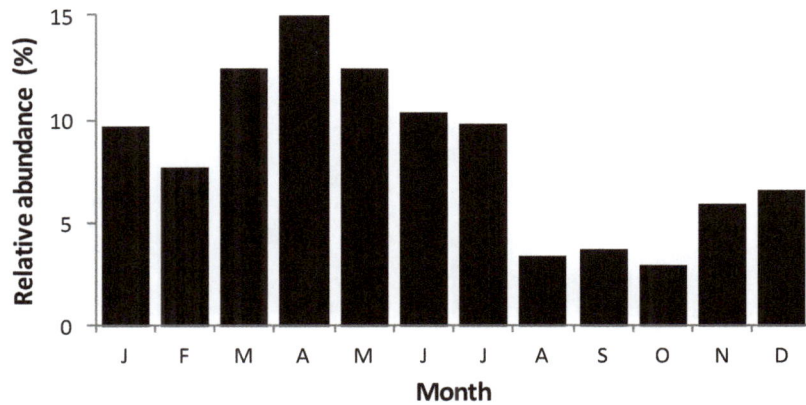

Relative abundance (%) vs Month

| Month | J | F | M | A | M | J | J | A | S | O | N | D |
|-------|---|---|---|---|---|---|---|---|---|---|---|---|
| Egg   |   |   |   |   |   |   |   |   |   |   |   |   |
| Larva |   |   |   |   |   |   |   |   |   |   |   |   |
| Pupa  |   |   |   |   |   |   |   |   |   |   |   |   |
| Adult |   |   |   |   |   |   |   |   |   |   |   |   |

# Black-spotted White

## *Leptosia nina* (Fabricius, 1793)

Plate 66 South of Kalumburu, WA
Photo: M. F. Braby

## Distribution

This species is restricted to the north-western Kimberley of the study region. *Leptosia nina* does not occur elsewhere in Australia and its presence in the region was detected only as recently as 1980 (Common and Waterhouse 1981; Naumann et al. 1991). It occurs only in the higher rainfall areas of the Kimberley (> 1,100 mm mean annual rainfall). Outside the study region, *L. nina* occurs throughout South-East Asia.

The subspecific status has not been determined, although Sands and New (2002) attributed the population to *L. nina comma* Fruhstorfer, 1902. However, Australian material appear to represent an undescribed subspecies because they differ from the two nearest subspecies in South-East Asia—*L. nina comma* from Timor to Tanimbar and *L. nina fumigata* Fruhstorfer, 1902 from Lombok to Flores—according to the diagnosis of Yata (1985).

## Excluded data

Dunn and Dunn (1991) listed the species from Sandfire Flat, south of Broome, WA, based on a possible sighting made in 1979, but this record is considered erroneous because of its location, habitat and lack of voucher material (Braby 2000).

## Habitat

*Leptosia nina* breeds in semi-deciduous monsoon vine thicket on steep basalt rock scree and breakaways, on bauxite with groundwater seepage and along seasonal rocky creeks where the larval food plant grows, preferring young vines growing very close to the ground. The dry monsoon forest in which it lives typically occurs in small patches protected from fire.

## Larval food plant

*Capparis sepiaria* (Capparaceae).

## Seasonality

The seasonal abundance and breeding phenology of this species are not well understood. Adults have been recorded from February to July and in November, but there are too few records (n = 14) to assess any seasonal changes in abundance. Presumably, the species breeds continuously throughout the year.

## Breeding status

This species is resident in the study region.

## Conservation status

LC. The putative subspecies *L. nina* ssp. 'Kimberley' is a short-range endemic (EOO = 5,930 sq km), with part of the population occurring in two conservation reserves: Mitchell River National Park and Uunguu IPA. Despite the restricted occurrence of the taxon, Sands and New (2002) considered it not to be threatened and therefore of no conservation significance.

## Leptosia nina

- ● Specimen ≥1970
- ▲ Literature ≥1970
- ✕ Larval food plant

## Leptosia nina

- ● Species record
- ▬ Geographic range
- — — Phytogeographical boundary
- ······ IBRA bioregional boundary

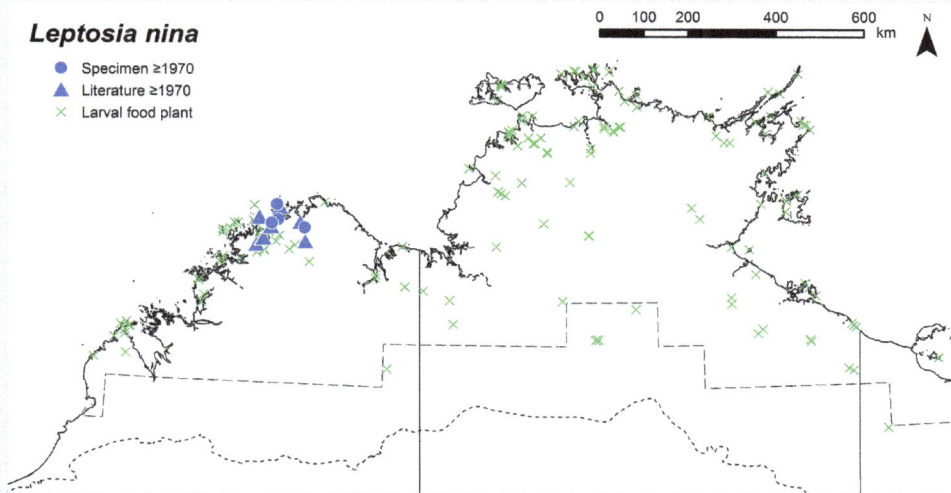

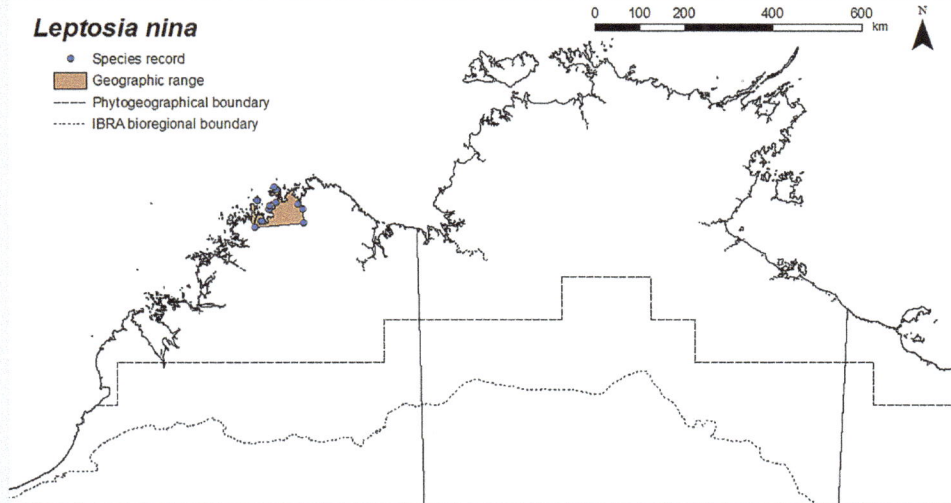

| Month | J | F | M | A | M | J | J | A | S | O | N | D |
|-------|---|---|---|---|---|---|---|---|---|---|---|---|
| Egg |  |  |  |  |  |  |  |  |  |  |  |  |
| Larva |  |  |  |  |  |  |  |  |  |  |  |  |
| Pupa |  |  |  |  |  |  |  |  |  |  |  |  |
| Adult |  |  |  |  |  |  |  |  |  |  |  |  |

# Narrow-winged Pearl-white
## *Elodina padusa* (Hewitson, 1853)

Plate 67 Boodjamulla (Lawn Hill) National Park, Qld
Photo: Don Franklin

Plate 68 Boodjamulla (Lawn Hill) National Park, Qld
Photo: Don Franklin

## Distribution

This species occurs widely in the study region. It occurs mainly in semi-arid areas (south of latitude 13ºS), reaching its northernmost limit at Kakadu National Park (Mary River Ranger Station), NT. In the semi-arid zone, it has been recorded as far south as Three Ways Roadhouse 25 km north of Tennant Creek, NT. It also extends into the arid zone of central Australia beyond the southern boundary of the study region. Its geographic range closely corresponds with the spatial distribution of its larval food plant, which is absent from the higher rainfall areas of the Top End (> 1,300 mm mean annual rainfall), where there are no records of *E. padusa*. Outside the study region, *E. padusa* occurs in the Pilbara of Western Australia and widely in central and eastern Australia.

## Habitat

*Elodina padusa* breeds in mixed lancewood–vine thicket associations where the larval food plant grows as a large scrambling vine often reaching into the canopy (Braby 2011a). Adults have also been collected in mixed eucalypt woodland–monsoon vine thicket on rocky outcrops and riparian woodland with rainforest elements along rocky seasonal creeks and gullies, and they may well breed in these moister habitats, which are embedded within the savannah landscape.

## Larval food plant

*Capparis lasiantha* (Capparaceae).

## Seasonality

Adults have been recorded during most months of the year, but they are apparently more abundant during the mid dry season (May–July). The breeding phenology and seasonal history of the immature stages have not been recorded, but presumably the species breeds throughout the year. Smithers (1983b) did not list this species as a migrant, but Grund and Hunt (2001) noted adults on the Mitchell Plateau in the northern Kimberley, WA, migrating in large numbers in a southerly direction during June and July.

## Breeding status

This species is resident in the study region.

## Conservation status

LC.

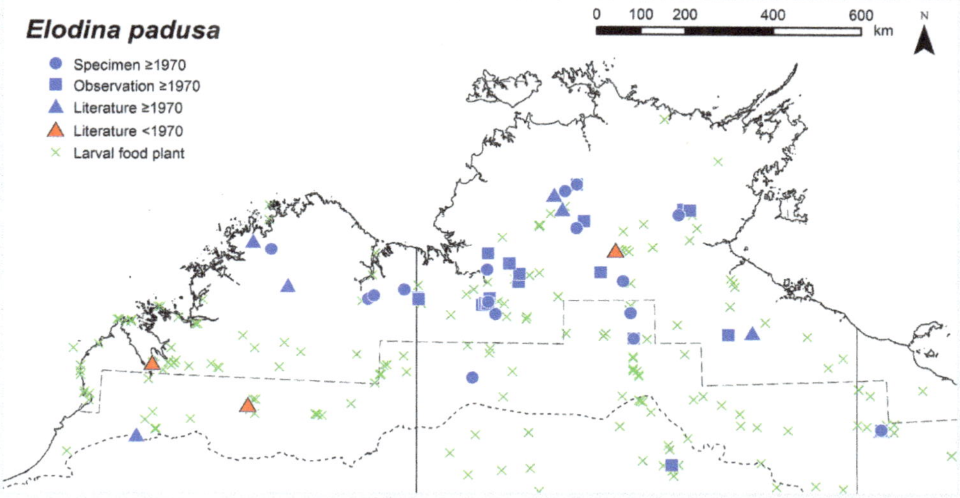

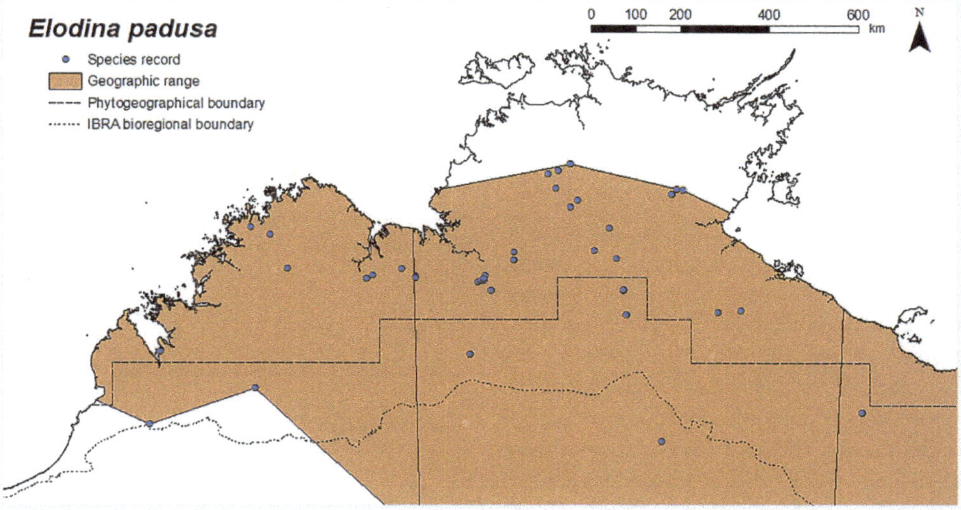

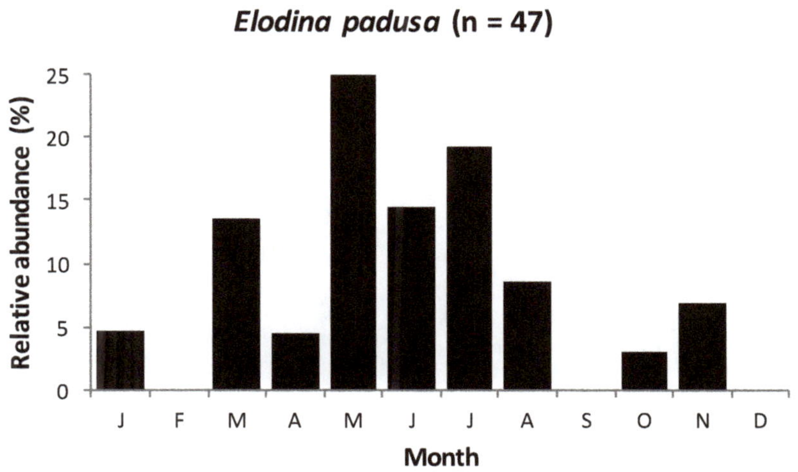

*Elodina padusa*

*Elodina padusa*

*Elodina padusa* (n = 47)

| Month | J | F | M | A | M | J | J | A | S | O | N | D |
|-------|---|---|---|---|---|---|---|---|---|---|---|---|
| Egg | | | | | | | | | | | | |
| Larva | | | | | | | | | | | | |
| Pupa | | | | | | | | | | | | |
| Adult | | | | | | | | | | | | |

# Small Pearl-white

*Elodina walkeri* Butler, 1898

Plate 69 Buffalo Creek, Leanyer Swamp, NT
Photo: M. F. Braby

## Distribution

This species occurs in the northern and eastern Kimberley, Top End and coastal areas of the western Gulf Country of the study region. It extends from moist coastal areas to drier inland areas of the semi-arid zone (c. 700 mm mean annual rainfall), reaching its southernmost limits at El Questro Wilderness Park (El Questro Gorge), WA; Pinkerton Range on the Victoria Highway, Judbarra/Gregory National Park (Limestone Gorge) and Killarney Station 60 km north of Top Springs, NT; and the Wellesley Islands (Bentinck Island), Qld (Fisher 1992). It also occurs in the Gulf of Carpentaria 5 km north of Karumba, Qld, just outside the boundary of the study region (Pierce 2017). The geographic range closely corresponds with the spatial distribution of its larval food plant. The food plant, however, extends slightly further south. Further field surveys are therefore required to determine whether *E. walkeri* is more widely distributed in semi-arid areas, particularly in coastal areas of the western Kimberley and near-coastal areas of the Gulf Country. Outside the study region, *E. walkeri* occurs in north-eastern Queensland.

We have provisionally included all records of *E. tongura* Tindale, 1923 with *E. walkeri* until the taxonomic status of *E. tongura* is clarified, primarily because the two species are indistinguishable on phenotypic characters.

## Habitat

*Elodina walkeri* breeds in semi-deciduous monsoon vine thicket in both coastal and inland areas, particularly on rocky outcrops, where the larval food plant grows as a scrambling vine.

## Larval food plant

*Capparis sepiaria* (Capparaceae).

## Seasonality

Adults occur throughout the year. They appear to show little seasonal variation in abundance, although they tend to be more numerous during the wet season (November–February)—possibly depending on the timing and amount of rainfall—and again in the mid dry season (July). Franklin (2011) found similar trends near Darwin, NT, based on quantitative studies conducted over 14 months during 2008–09. The immature stages have been recorded sporadically throughout the year, suggesting the species breeds continuously, with several generations completed annually.

## Breeding status

This species is resident in the study region.

## Conservation status

LC.

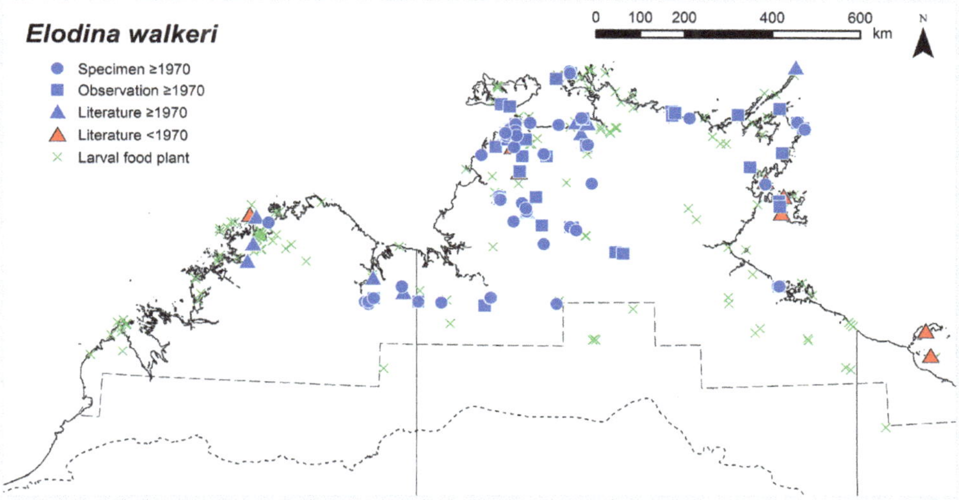

Elodina walkeri
0 100 200 400 600
km
N
Specimen ≥1970
Observation ≥1970
Literature ≥1970
Literature <1970
Larval food plant

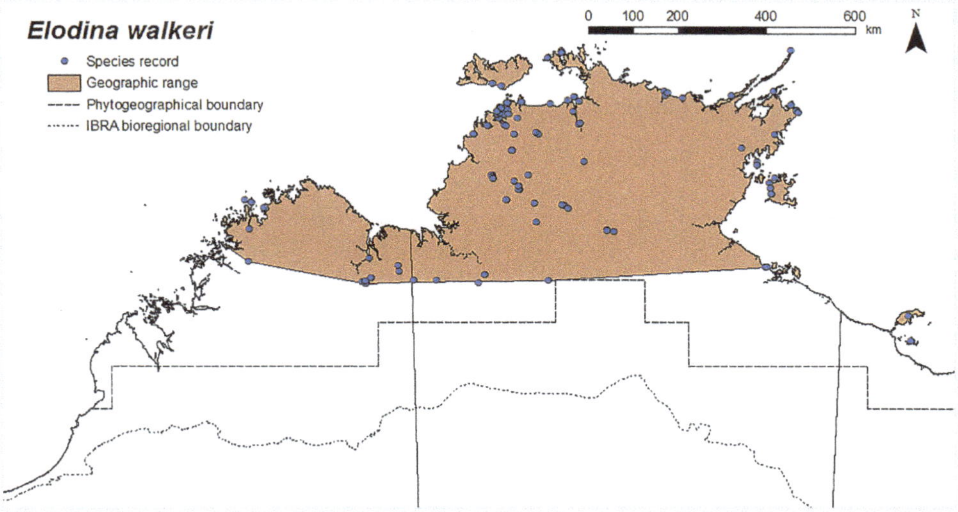

Elodina walkeri
0 100 200 400 600
km
N
Species record
Geographic range
Phytogeographical boundary
IBRA bioregional boundary

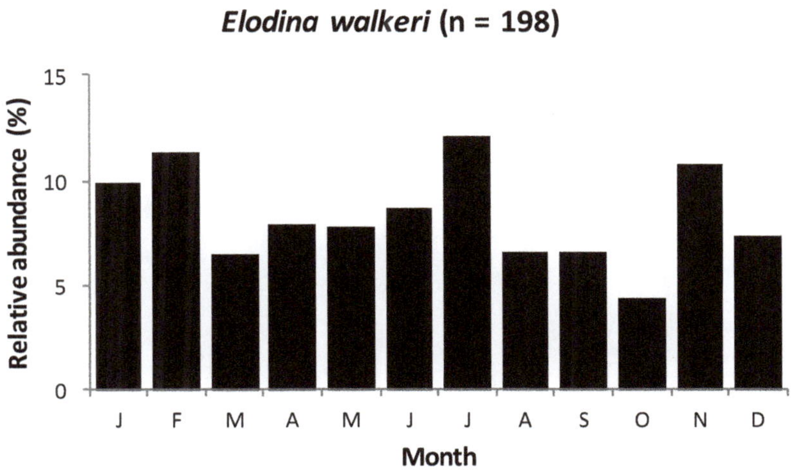

Elodina walkeri (n = 198)
Relative abundance (%)
15
10
5
0
J F M A M J J A S O N D
Month

| Month | J | F | M | A | M | J | J | A | S | O | N | D |
|---|---|---|---|---|---|---|---|---|---|---|---|---|
| Egg | ■ | ■ | ■ | | | | ■ | | | ■ | ■ | |
| Larva | ■ | | | ■ | | | ■ | | | ■ | | |
| Pupa | | | | ■ | | | | | | | | |
| Adult | ■ | ■ | ■ | ■ | ■ | ■ | ■ | ■ | ■ | ■ | ■ | ■ |

# White Albatross
## *Appias albina* (Boisduval, 1836)

Plate 70 Black Point, Cobourg Peninsula, NT
Photo: M. F. Braby

## Distribution

This species is represented by two subspecies: *A. albina albina* (Boisduval, 1836) and *A. albina infuscata* Fruhstorfer, 1910. Both subspecies occur in the north of the Northern Territory of the study region. *Appias albina albina* occurs along the northern coast of the Top End (Braby et al. 2010a), whereas *A. albina infuscata* is known only from Humpty Doo 30 km east–south-east of Darwin, NT, where a female was observed on a rural property in January 2010 (Braby et al. 2010b). Although the larval food plant is substantially more widespread, *A. albina* is restricted to coastal locations in the higher rainfall areas (> 1,400 mm mean annual rainfall). Outside the study region, *A. albina* occurs widely from India and South-East Asia, through mainland New Guinea to the Torres Strait Islands, Qld.

## Habitat

*Appias albina* breeds in coastal semi-deciduous monsoon vine thicket where the larval food plant grows as a tall shrub or small tree on laterite cliffs above the beach (Braby et al. 2010a).

## Larval food plant

*Drypetes deplanchei* (Putranjivaceae).

## Seasonality

Adults are seasonal, occurring only during the wet season (January–March). They are most abundant in January and February during or just after the monsoon rains, and in good wet seasons of average or above average rainfall they may be particularly abundant for a few weeks. Breeding is strictly seasonal, with the immature stages recorded only during the wet season (January–April), when the larval food plant produces flushes of new soft leaf growth. It is not clear how the species survives the dry season; it is suspected the pupae remain in diapause for many months, but pupae reared from larvae in captivity at the end of the flight season developed directly and produced adults in March and April. Alternatively, it is possible that populations in the Northern Territory are temporary, with adults migrating from South-East Asia followed by a return flight of the next generation at the end of the wet season, but this has not been confirmed.

## Breeding status

*Appias albina albina* is assumed to be resident in the study region, whereas *A. albina infuscata* is a rare vagrant from Indonesia and is non-resident.

## Conservation status

*Appias albina albina*: LC. *Appias albina infuscata*: NA. The subspecies *A. albina albina* has a narrow range in the study region (geographic range = 22,370 sq km) within which it occurs in several conservation reserves, including East Point Reserve, Casuarina Coastal Reserve and Garig Gunak Barlu National Park on Cobourg Peninsula, NT. Despite its restricted occurrence, there are no known threats facing the taxon.

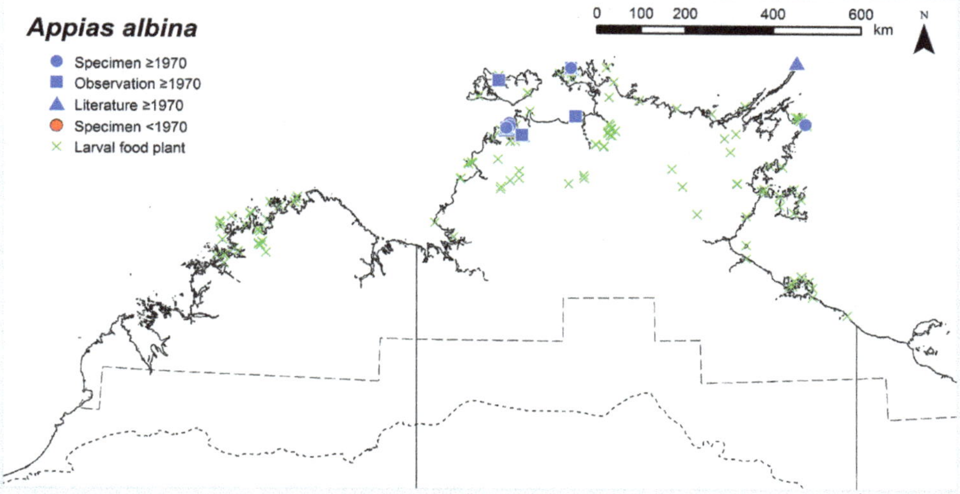

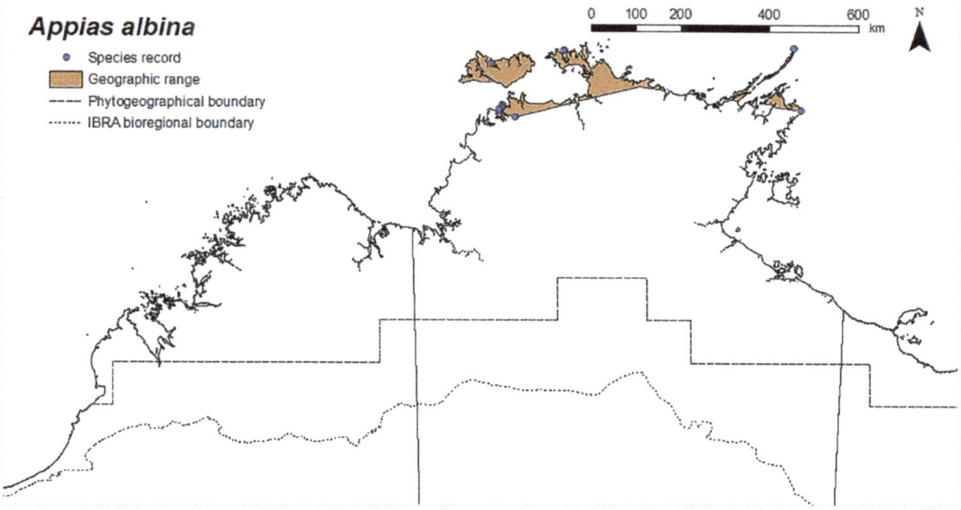

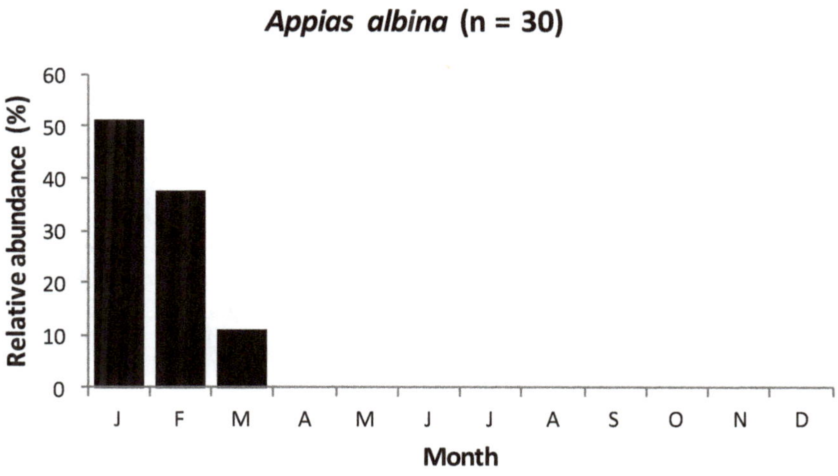

## Appias albina

- ● Specimen ≥1970
- ■ Observation ≥1970
- ▲ Literature ≥1970
- ● Specimen <1970
- ✕ Larval food plant

## Appias albina

- ● Species record
- Geographic range
- --- Phytogeographical boundary
- ····· IBRA bioregional boundary

### Appias albina (n = 30)

| Month | J | F | M | A | M | J | J | A | S | O | N | D |
|-------|---|---|---|---|---|---|---|---|---|---|---|---|
| Egg   |   |   |   |   |   |   |   |   |   |   |   |   |
| Larva |   |   |   |   |   |   |   |   |   |   |   |   |
| Pupa  |   |   |   |   |   |   |   |   |   |   |   |   |
| Adult |   |   |   |   |   |   |   |   |   |   |   |   |

# Yellow Albatross

## *Appias paulina* (Cramer, [1777])

Plate 71 East Point, Darwin, NT
Photo: M. F. Braby

Plate 72 Lakefield National Park, Cape York Peninsula, Qld
Photo: Don Franklin

## Distribution

This species is represented in the study region by the subspecies *A. paulina ega* (Boisduval, 1836). It has a disjunct distribution, occurring mainly in the higher rainfall areas (> 1,100 mm mean annual rainfall) of the north-western Kimberley and in the Top End. The geographic range corresponds well with the spatial distribution of its larval food plant. However, *A. paulina* has been recorded substantially further south in the Kimberley, at Cockatoo Island and Derby, WA (Warham 1957), and in the western Gulf Country at Doomadgee, Qld (Puccetti 1991). Presumably, these records, if reliable, represent vagrant individuals that have dispersed beyond the normal breeding range. The larval food plant also occurs in the Tiwi Islands and Sir Edward Pellew Group, NT; thus, further field surveys are required to determine whether *A. paulina* occurs in these areas. Outside the study region, *A. paulina* occurs widely from India and South-East Asia, through mainland New Guinea and adjacent islands and eastern Australia to Samoa.

## Excluded data

Bailey and Richards (1975) recorded this species from the Prince Regent Nature Reserve (Mt Trafalgar, Site W4), WA; however, examination of their material in the WADA indicated the specimen had been misidentified and was in fact a female *Cepora perimale*—a species they did not record.

## Habitat

*Appias paulina* breeds in semi-deciduous monsoon vine thicket in both coastal and inland areas and evergreen monsoon vine forest associated with permanent freshwater streams where the larval food plant grows as a tall shrub or small tree.

## Larval food plants

*Drypetes deplanchei* (Putranjivaceae). In addition, a female was observed laying several eggs on new leaf shoots of *Capparis sepiaria* (Capparaceae) at Black Point on Cobourg Peninsula, NT (Braby 2011a); however, the suitability of this vine as a food plant requires confirmation.

## Seasonality

Adults occur throughout the year, but they are most abundant during the wet season (December–March), with a peak in abundance in February. The immature stages have also been recorded during the wet season (November–April) and, like *Appias albina*, the larvae feed only on the new soft leaves and shoots of the food plant. It is not known whether the species breeds continuously throughout the year (and uses alternative food plants during the dry season), reproductive activity declines or females stop breeding as the dry season progresses or breeding is strictly seasonal and the immature stages (e.g. pupa) remain in diapause, with adults emerging spasmodically during the dry season, similar to that seen in *Graphium eurypylus*. Although adults are known to be migratory (Smithers 1983b), we have no records of population movements.

## Breeding status

This species is resident in the study region.

## Conservation status

LC.

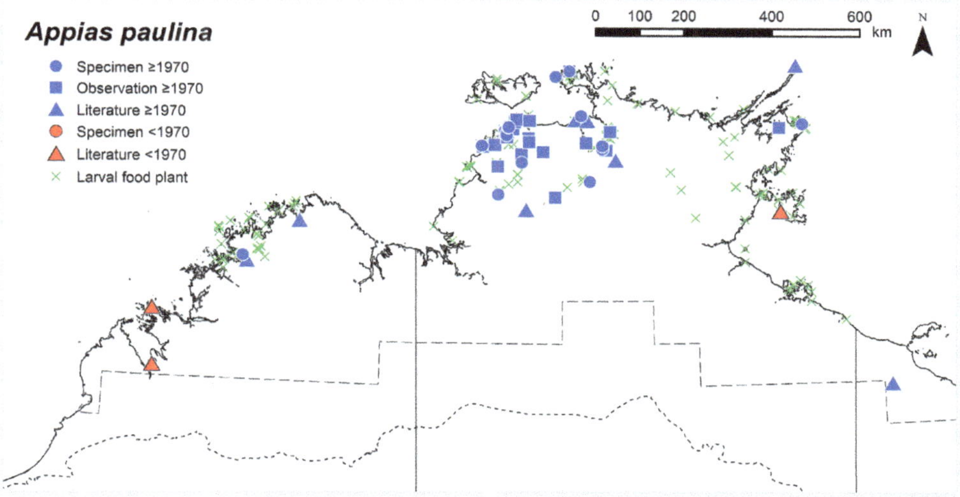

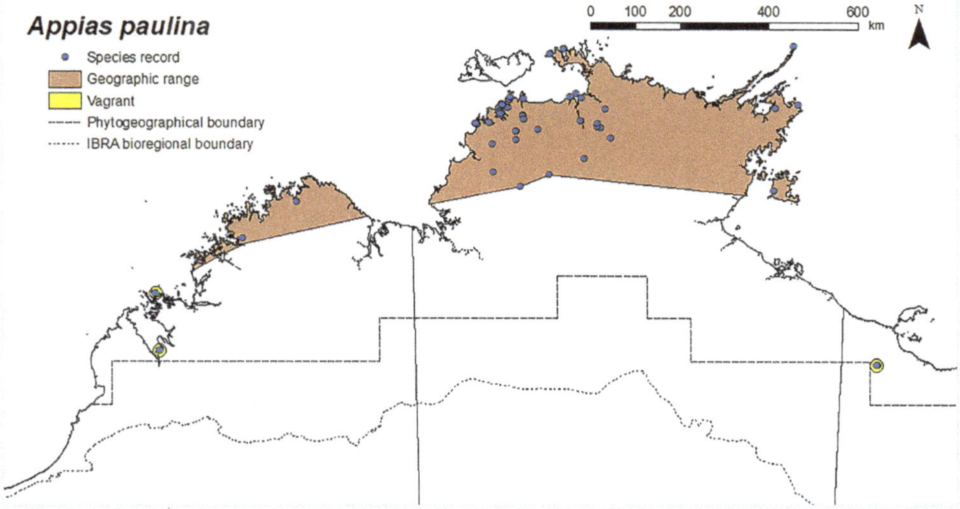

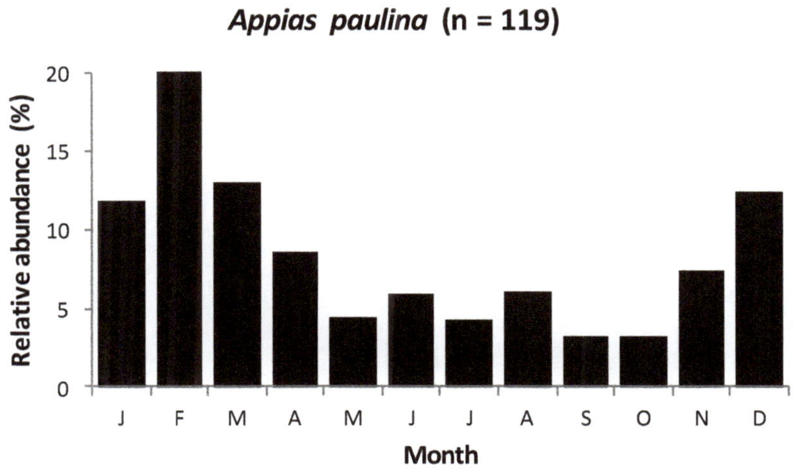

## Appias paulina

Appias paulina

- Specimen ≥1970
- Observation ≥1970
- Literature ≥1970
- Specimen <1970
- Literature <1970
- Larval food plant

Appias paulina

- Species record
- Geographic range
- Vagrant
- Phytogeographical boundary
- IBRA bioregional boundary

*Appias paulina* (n = 119)

| Month | J | F | M | A | M | J | J | A | S | O | N | D |
|-------|---|---|---|---|---|---|---|---|---|---|---|---|
| Egg   |   | ▓ | ▓ |   |   |   |   |   |   |   | ▓ |   |
| Larva | ▓ | ▓ | ▓ |   |   |   |   |   |   |   | ▓ | ▓ |
| Pupa  |   |   | ▓ | ▓ |   |   |   |   |   |   |   |   |
| Adult | ▓ | ▓ | ▓ | ▓ | ▓ | ▓ | ▓ | ▓ | ▓ | ▓ | ▓ | ▓ |

# Caper White
## *Belenois java* (Linnaeus, 1768)

Plate 73 Alice Springs, NT
Photo: M. F. Braby

## Distribution

This species is represented in the study region by the subspecies *B. java teutonia* (Fabricius, 1775). It occurs throughout the region, as well as the arid zone of central Australia beyond the southern boundary of the study region—its broad geographic range closely matching the spatial distribution of its larval food plants. Surprisingly, there are no records of *B. java* from the Tiwi Islands, NT, despite the presence of the food plants in this area. Outside the study region, *B. java* occurs from the Lesser Sunda Islands, through mainland New Guinea and adjacent islands to the islands of the South Pacific. It also occurs throughout the Australian continent.

## Habitat

*Belenois java* breeds in a variety of habitats, particularly savannah woodland and semi-deciduous monsoon vine thicket where the larval food plants grow as scrambling vines or small trees.

## Larval food plants

*Capparis jacobsii, C. lasiantha, C. sepiaria, C. umbonata* (Capparaceae).

## Seasonality

Adults occur throughout the year, but they are most abundant from the late dry season to the mid wet season (September–January). In the Darwin area, they are seasonal in appearance: adults usually arrive during migration (with adults generally flying in a north–north-westerly or north-westerly direction) during the 'build-up' in September or October; they breed for a few months, during which several overlapping generations are completed and they may be extremely abundant; and then disappear by January (Braby 2016b). During the remainder of the year (February–August), adults and immatures are infrequent and usually absent in Darwin, although on one occasion the immature stages (eggs and larvae) were recorded in March during an influx of adults. Further work is needed to determine whether the patterns of seasonal migration and breeding in the Darwin area are replicated elsewhere in the northern coastal areas of the Kimberley and Top End.

## Breeding status

This species is resident in the study region; however, in the northern parts of the range, the species is an immigrant and the populations are temporary.

## Conservation status

LC.

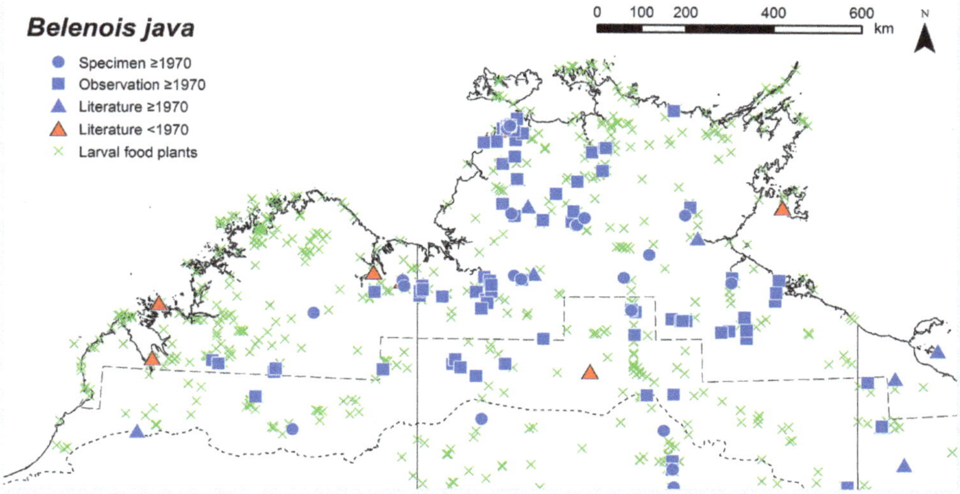

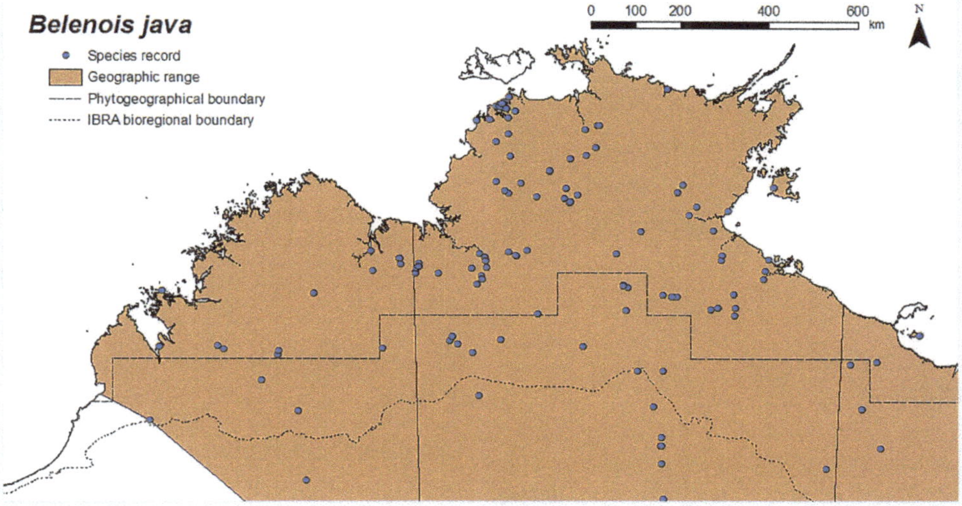

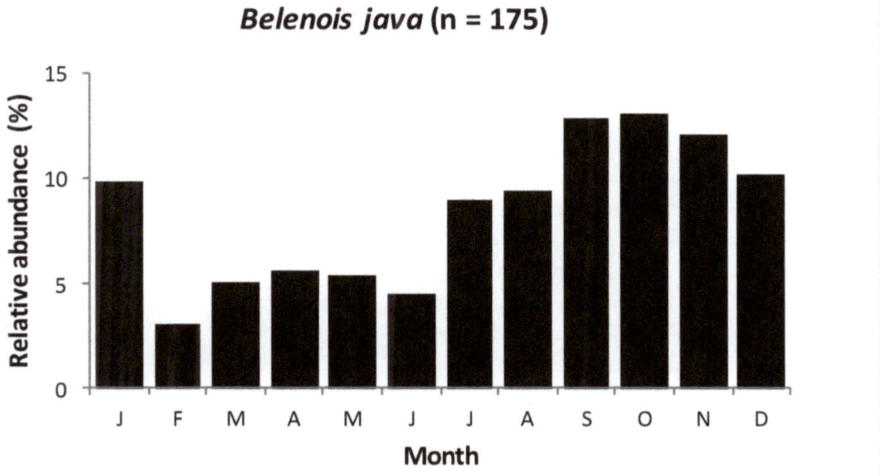

Belenois java
Specimen ≥1970
Observation ≥1970
Literature ≥1970
Literature <1970
Larval food plants
0 100 200 400 600 km
N

Belenois java
Species record
Geographic range
Phytogeographical boundary
IBRA bioregional boundary
0 100 200 400 600 km
N

Belenois java (n = 175)
Relative abundance (%)
0
5
10
15
J F M A M J J A S O N D
Month

| Month | J | F | M | A | M | J | J | A | S | O | N | D |
|---|---|---|---|---|---|---|---|---|---|---|---|---|
| Egg | ■ | | ■ | | | | | | | ■ | | |
| Larva | ■ | | ■ | | ■ | | | | | ■ | ■ | ■ |
| Pupa | ■ | | | | | | | ■ | | ■ | | ■ |
| Adult | ■ | ■ | ■ | ■ | ■ | ■ | ■ | ■ | ■ | ■ | ■ | ■ |

# Caper Gull
## *Cepora perimale* (Donovan, 1805)

Plate 74 Shoal Bay, NT
Photo: M. F. Braby

## Distribution

This species occurs widely in the study region. It occurs widely in the Kimberley, Top End and western Gulf Country, extending from moist coastal areas to drier inland areas of the semi-arid zone (c. 600 mm mean annual rainfall). It reaches its southernmost limits at Broome, WA (Williams et al. 2006); Montejini Station 43 km south–south-west of Top Springs, NT; and Boodjamulla/Lawn Hill National Park (Lawn Hill Gorge), Qld (Daniels and Edwards 1998; Franklin 2007). The geographic range closely corresponds with the spatial distribution of its larval food plants, except for one food plant (*Capparis umbonata*), which extends further inland into the arid zone of central Australia. Outside the study region, *C. perimale* occurs from Sulawesi and Lombok, through mainland New Guinea and adjacent islands and north-eastern and eastern Australia to Vanuatu, New Caledonia, Fiji and Norfolk Island.

The subspecific status has not been determined; it may be *C. perimale scyllara* (Macleay, 1826) from eastern Australia, but the adults are phenotypically distinct.

## Habitat

*Cepora perimale* breeds in a variety of habitats, particularly semi-deciduous monsoon vine thicket and savannah woodland where the larval food plants grow as strangling vines, shrubs or small trees. It also breeds along the edges of evergreen monsoon vine forest and in mixed mangrove–monsoon forest associations along the banks of perennial streams and rivers.

## Larval food plants

*Capparis jacobsii*, *C. sepiaria*, *C. umbonata* (Capparaceae).

## Seasonality

Adults occur throughout the year, but they are most abundant during the early dry season (April–June) and least abundant during the late dry season (August–October). The immature stages have been recorded during all months except August, indicating that it breeds continuously throughout the year, during which several generations are completed.

## Breeding status

This species is resident in the study region.

## Conservation status

LC.

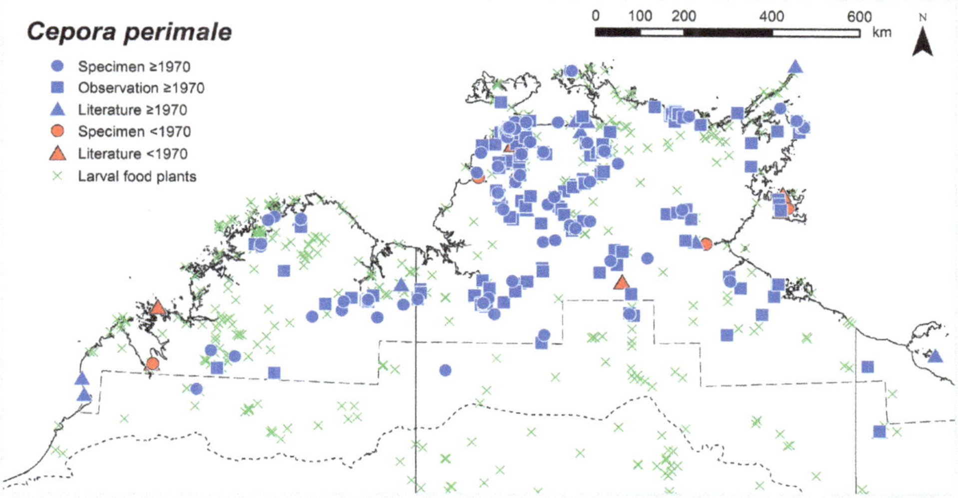

Cepora perimale
Specimen ≥1970
Observation ≥1970
Literature ≥1970
Specimen <1970
Literature <1970
Larval food plants
0 100 200 400 600
km
N

Cepora perimale
Species record
Geographic range
Phytogeographical boundary
IBRA bioregional boundary
0 100 200 400 600
km
N

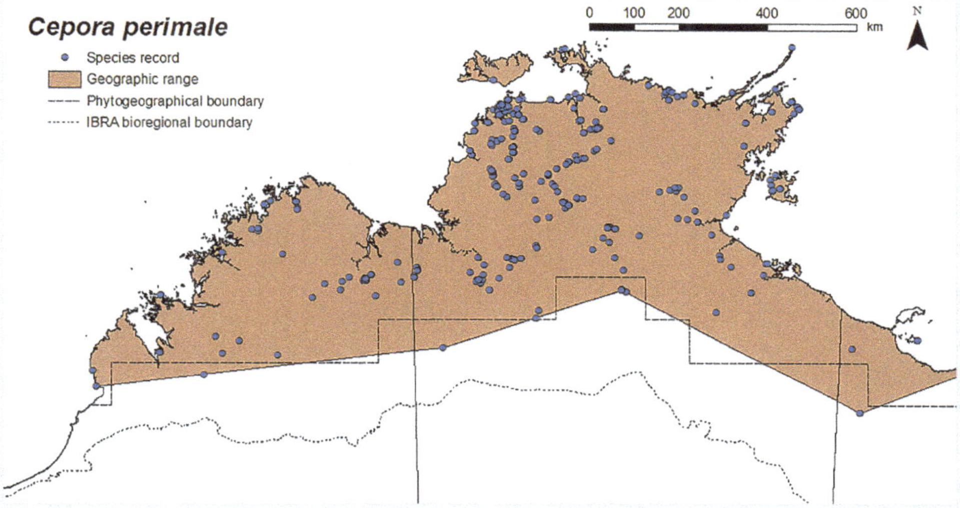

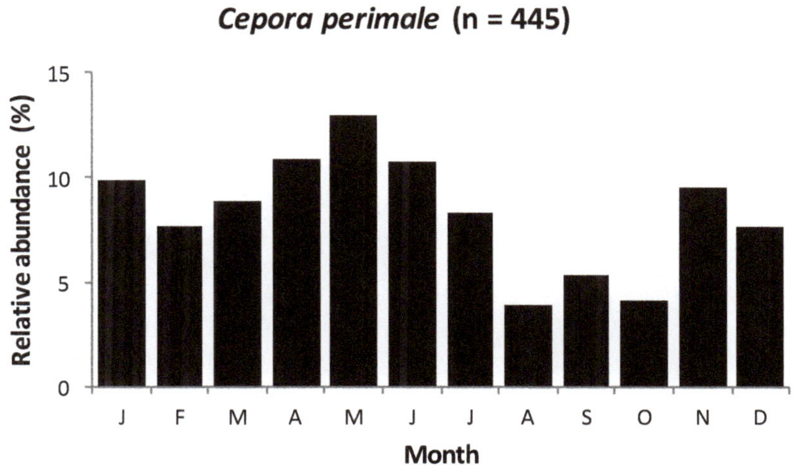

Cepora perimale (n = 445)
Relative abundance (%)
0
5
10
15
J F M A M J J A S O N D
Month

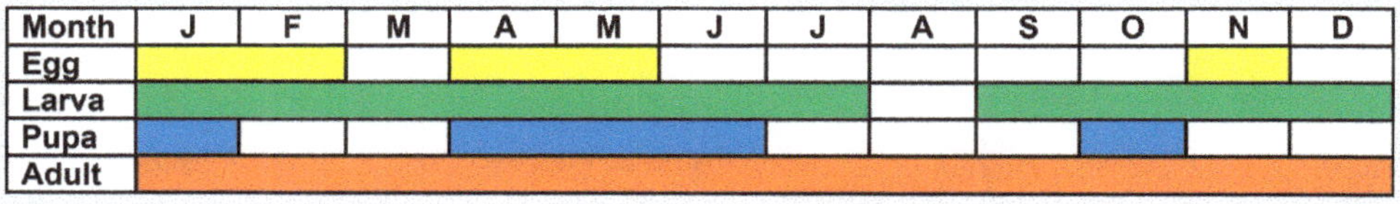

Month
J F M A M J J A S O N D
Egg
Larva
Pupa
Adult

# Mangrove Jezebel
## *Delias aestiva* Butler, 1897

Plate 75 Buffalo Creek, Leanyer Swamp, NT
Photo: M. F. Braby

## Distribution

This species is represented by the subspecies *D. aestiva aestiva* Butler, 1897, which is endemic to the study region. It occurs in the Top End, where it is restricted to coastal areas north of latitude 15°S. The geographic range corresponds well with the spatial distribution of its known and putative larval food plants (*Excoecaria* spp.). The food plants, however, are far more extensive, occurring in coastal areas throughout the Kimberley and western Gulf Country. Further field surveys are therefore required to determine whether *D. aestiva* also occurs in these areas. Inland records from Adelaide River and Daly River (Oolloo crossing) (Hutchinson 1978), NT, probably represent vagrants that dispersed upstream along rivers. Outside the study region, *D. aestiva* occurs in the Gulf Country of western Cape York Peninsula, Qld.

The subspecies *D. aestiva smithersi* (Daniels 2012) occurs in the Gulf of Carpentaria at Karumba, Qld, just outside the boundary of the study region (Daniels 2012; Braby 2014b); thus, further field surveys are required to determine whether it extends further west into the Northern Territory.

## Excluded data

Warham (1957) listed the species (under the name *D. mysis*) from Derby, Cockatoo Island and Wotjalum Mission (on the mainland to the south of the eastern end of Koolan Island) in Yampi Sound, WA. Braby (2012b), however, considered these records to be unreliable on the basis that no vouchered specimens were retained, the butterfly was likely to have been confused with *D. argenthona*—a common and

widespread species that occurs on Koolan Island (see McKenzie et al. 1995), but which was absent from Warham's list—and there have been no subsequent records from these areas for more than 50 years. More recently, Sands and New (2002) recorded the species in the Kimberley based on observations at Cape Leveque (in moist eucalypt woodland) and the Mitchell Plateau, WA; however, these locations/habitat do not accord with the biology of the species.

## Habitat

*Delias aestiva* breeds in the landward edge of mangroves in coastal estuarine areas where the larval food plant grows as a deciduous tree (Braby 2012c). Adults frequently disperse into adjacent habitats such as paperbark woodland, monsoon vine thicket and even suburban parks and gardens, but they do not breed in these habitats.

## Larval food plants

*Excoecaria ovalis* (Euphorbiaceae); probably *Excoecaria agallocha*.

## Seasonality

Adults have been recorded during most months of the year, but they are most abundant during the dry season (particularly July–September). In some years, they may also be very abundant during the late wet season and early dry season (March–May), possibly depending on the timing of the last monsoon rains. They are generally scarce and often absent during the early wet season (November and December or January), when the larval food plant is seasonally deciduous. The immature stages (eggs or larvae) have been recorded mainly in the dry season (April–October), during which several generations are completed, with larvae feeding only on the mature foliage. It is not known how the species survives the period of food shortfall during the pre-monsoon 'build-up', but it is likely it remains in pupal diapause (Braby 2012c).

## Breeding status

This species is resident in the study region.

## Conservation status

LC. The subspecies *D. aestiva aestiva* has a restricted range within which it occurs in several conservation reserves, including Casuarina Coastal Reserve, Djukbinj National Park, Kakadu National Park and Djelk and Laynhapuy IPAs.

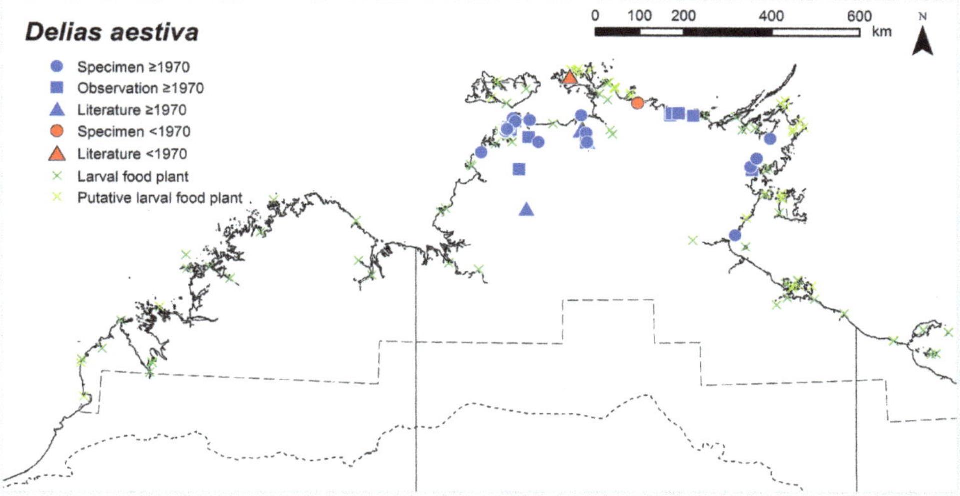

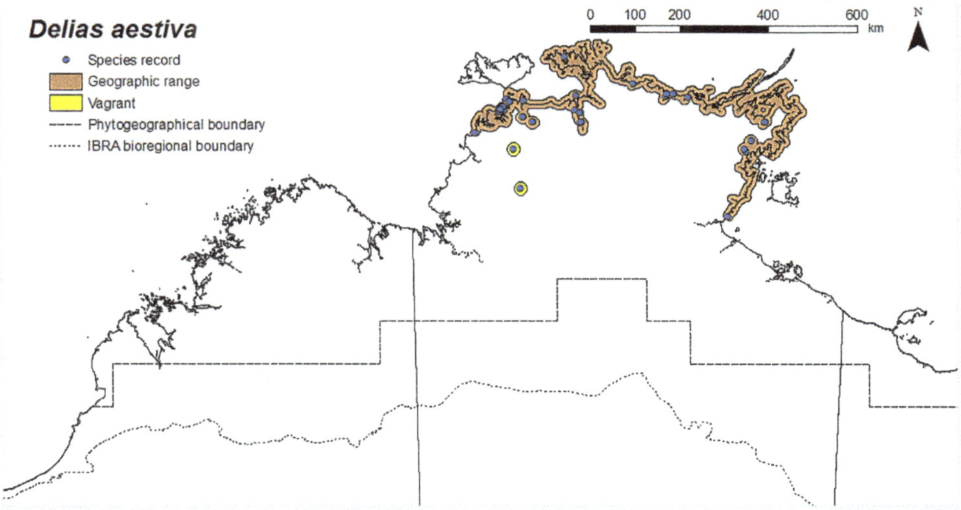

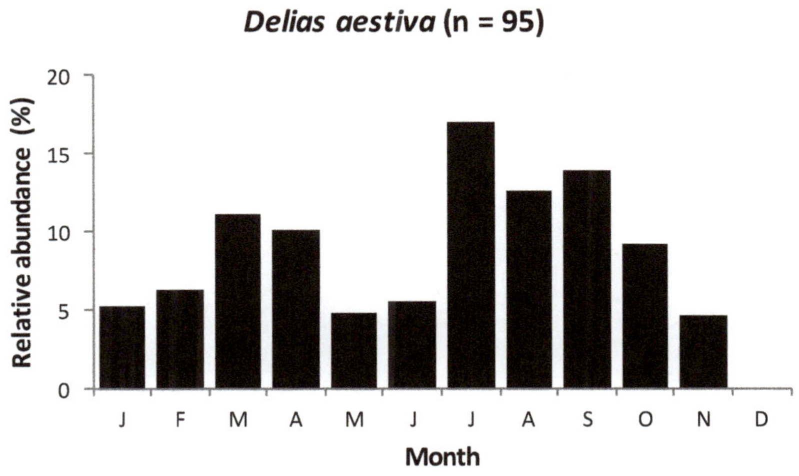

Delias aestiva
Specimen ≥1970
Observation ≥1970
Literature ≥1970
Specimen <1970
Literature <1970
Larval food plant
Putative larval food plant
0 100 200 400 600 km
N

Delias aestiva
Species record
Geographic range
Vagrant
Phytogeographical boundary
IBRA bioregional boundary
0 100 200 400 600 km
N

Delias aestiva (n = 95)
Relative abundance (%)
20
15
10
5
0
J F M A M J J A S O N D
Month

| Month | J | F | M | A | M | J | J | A | S | O | N | D |
|---|---|---|---|---|---|---|---|---|---|---|---|---|
| Egg | | | | | | | | | | | | |
| Larva | | | | | | | | | | | | |
| Pupa | | | | | | | | | | | | |
| Adult | | | | | | | | | | | | |

# Red-banded Jezebel
## *Delias mysis* (Fabricius, 1775)

Plate 76 Paluma, Qld
Photo: M. F. Braby

## Distribution

This species is represented in the study region by the subspecies *D. mysis mysis* (Fabricius, 1775). It is known only from a single historical male specimen from Groote Eylandt in the Gulf of Carpentaria (Talbot 1928–37). Braby (2012c) reexamined this specimen and concluded it was *D. mysis mysis*, a subspecies otherwise endemic to the Wet Tropics and nearby areas of north-eastern Queensland. The putative larval food plant (*Dendrophthoe glabrescens*) occurs widely in the Kimberley and Top End. Outside the study region, *D. mysis* occurs in the Aru Islands, south-eastern West Papua and north-eastern Australia.

## Habitat

The breeding habitat of *D. mysis* has not been recorded in the study region.

## Larval food plants

Not recorded in the study region; possibly *Dendrophthoe glabrescens* (Loranthaceae), which is one of the food plants in north-eastern Queensland (Braby 2000).

## Seasonality

The seasonal abundance and breeding phenology of this species are not well understood. The single specimen was collected in January 1926.

## Breeding status

*Delias mysis* does not appear to have become permanently established in the study region. It is likely the specimen from Groote Eylandt represents a vagrant that dispersed from north-eastern Queensland across the Gulf of Carpentaria; however, the possibility it may have been mislabelled, and therefore the locality erroneous, should not be discounted.

## Conservation status

NA.

*Delias mysis*

▲ Literature <1970

| Month | J | F | M | A | M | J | J | A | S | O | N | D |
|-------|---|---|---|---|---|---|---|---|---|---|---|---|
| Egg   |   |   |   |   |   |   |   |   |   |   |   |   |
| Larva |   |   |   |   |   |   |   |   |   |   |   |   |
| Pupa  |   |   |   |   |   |   |   |   |   |   |   |   |
| Adult |   |   |   |   |   |   |   |   |   |   |   |   |

# Scarlet Jezebel

## *Delias argenthona* (Fabricius, 1793)

Plate 77 Mt Burrell, NT
Photo: M. F. Braby

## Distribution

This species is represented by the subspecies *D. argenthona fragalactea* (Butler, 1869), which is endemic to the study region. It occurs throughout the Kimberley, Top End and western Gulf Country, extending from moist coastal areas to drier inland areas of the semi-arid zone (c. 700 mm mean annual rainfall). Its southernmost limits are Fitzroy Crossing, WA (Common and Waterhouse 1981), and Doomadgee, Qld (Puccetti 1991). Although the larval food plants are more widespread, particularly in the semi-arid areas, it is considered unlikely that *D. argenthona* is established in the hot, dry inland areas. Outside the study region, *D. argenthona* occurs in south-eastern New Guinea and eastern Australia.

The subspecies *D. argenthona argenthona* (Fabricius, 1793) occurs in the Gulf of Carpentaria at Walker Creek 36 km east of Karumba, Qld, just outside the eastern boundary of the study region (Braby 2015d). Further field surveys are therefore required to determine whether it extends further west or is continuous in extent with *D. argenthona fragalactea*.

## Habitat

*Delias argenthona* typically breeds in a variety of woodland and open woodland habitats where the larval food plants grow as mistletoes (parasitic shrubs) on various trees, including *Eucalyptus*, *Alstonia*, *Erythrophleum* and *Grevillea* (Braby 2011a, 2015e). It also breeds in suburban parks and gardens.

## Larval food plants

*Amyema miquelii*, *A. sanguinea*, *Decaisnina signata*, *Dendrophthoe glabrescens*, *Dendrophthoe odontocalyx* (Loranthaceae). *Decaisnina signata* is commonly used in suburban areas in Darwin, NT, where it frequently parasitises *Alstonia actinophylla* (Wade 1978; Anderson and Braby 2009).

## Seasonality

Adults occur throughout the year, but they are most abundant during the dry season (June–October), with a substantial peak in abundance in August. The immature stages have been recorded during most months, with larvae detected most frequently in August, indicating the species breeds throughout the year. Presumably, several overlapping generations are completed annually.

## Breeding status

This species is resident in the study region.

## Conservation status

LC.

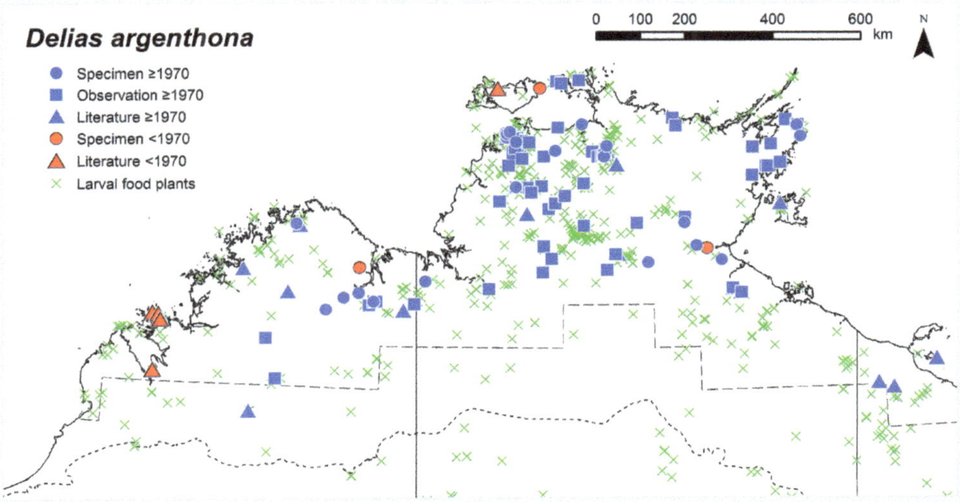

## Delias argenthona

- ● Specimen ≥1970
- ■ Observation ≥1970
- ▲ Literature ≥1970
- ● Specimen <1970
- ▲ Literature <1970
- ✕ Larval food plants

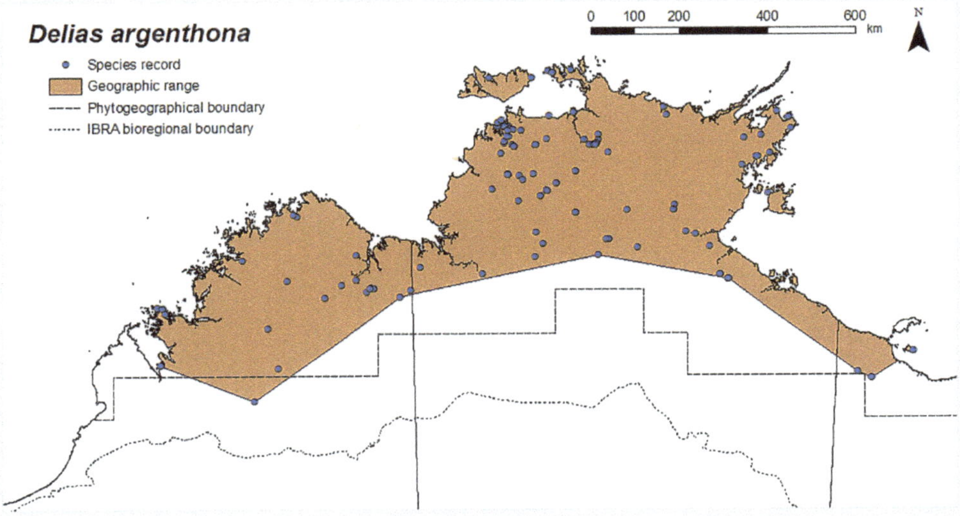

0   100   200        400        600
km

## Delias argenthona

- • Species record
- ▨ Geographic range
- – – – Phytogeographical boundary
- ....... IBRA bioregional boundary

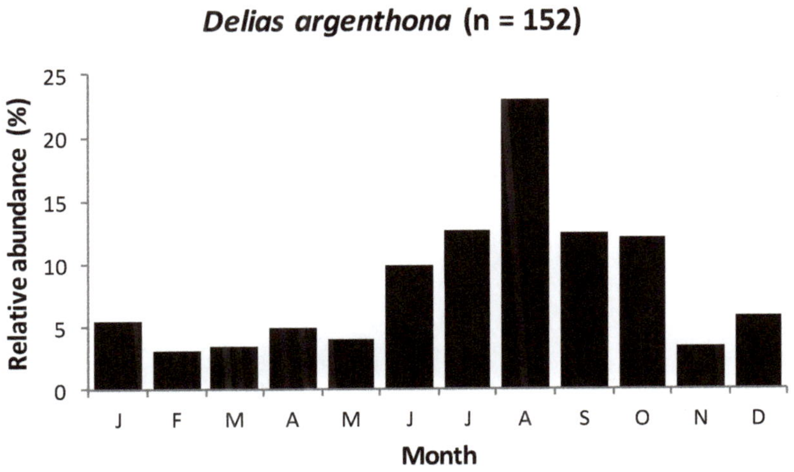

0   100   200        400        600
km

### Delias argenthona (n = 152)

| Month | J | F | M | A | M | J | J | A | S | O | N | D |
|-------|---|---|---|---|---|---|---|---|---|---|---|---|
| Egg   |   |   |   | ▢ |   |   | ▢ | ▢ |   |   | ▢ |   |
| Larva |   | ▢ | ▢ | ▢ |   | ▢ | ▢ | ▢ |   |   | ▢ |   |
| Pupa  |   |   |   |   |   |   | ▢ | ▢ | ▢ |   |   |   |
| Adult | ▢ | ▢ | ▢ | ▢ | ▢ | ▢ | ▢ | ▢ | ▢ | ▢ | ▢ | ▢ |

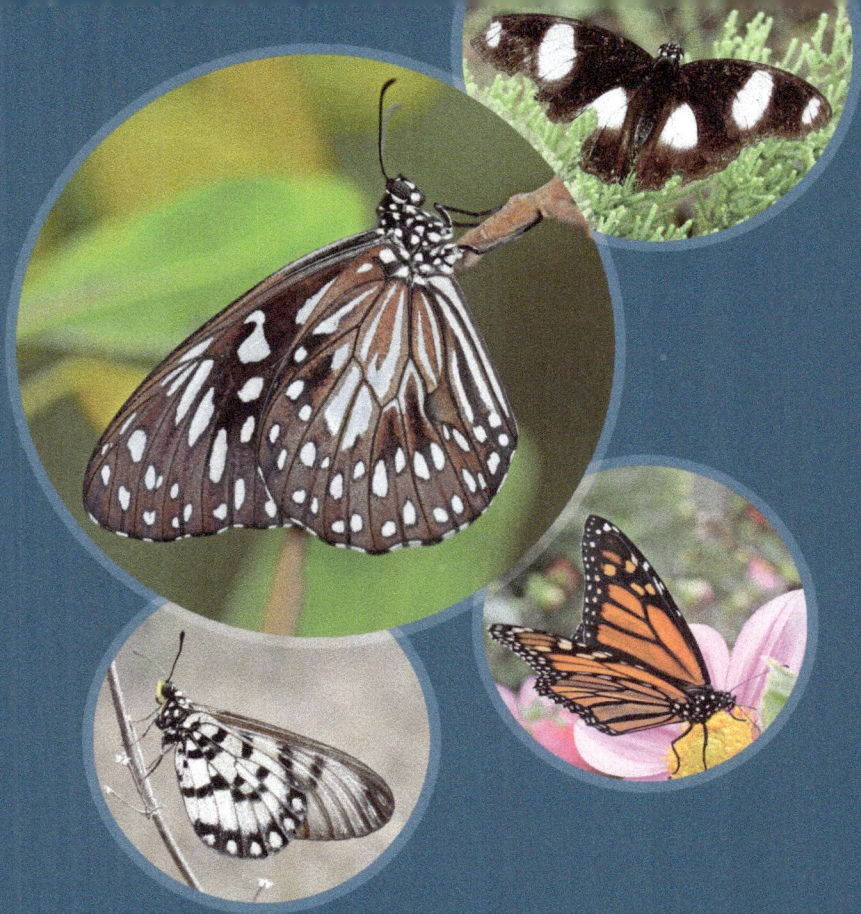

# Nymphs

(Nymphalidae)

8

# Purple Beak
## *Libythea geoffroyi* Godart, 1822

Plate 78 Emma Gorge, El Questro Wilderness Park, WA
Photo: M. F. Braby

Plate 79 Limestone Gorge, Judbarra/Gregory National Park, NT
Photo: M. F. Braby

## Distribution

This species is represented by the subspecies *L. geoffroyi genia* Waterhouse, 1938, which is endemic to the study region. It occurs widely but sporadically in the Kimberley and the western half of the Top End, as well as on Rimbija Island in the Wessel Islands, NT. It is locally common in the Kimberley, such as at Windjana Gorge (Williams et al. 2006) and Kalumburu (Yeates 1990), WA, but much rarer in the Northern Territory, where only a few specimens have been recorded from five locations: Judbarra/Gregory National Park (Limestone Gorge) (Braby 2014a), Fish River Station (Braby and Kessner 2012), Palmerston (Meyer et al. 2006), Darwin (Waterhouse and Lyell 1914; Waterhouse 1938) and Rimbija Island (Common and Waterhouse 1981). The geographical range of *L. geoffroyi* corresponds with the spatial distribution of its larval food plant. However, the food plant, although patchy in extent, is far more widely distributed, occurring in the western Gulf Country and coastal areas of the north of the Northern Territory. Further field surveys are therefore required to determine whether *L. geoffroyi* occurs in western Queensland and adjacent areas in the Northern Territory, and elsewhere in the northern coastal areas of the Top End. Outside the study region, *L. geoffroyi* occurs from South-East Asia, through mainland New Guinea and north-eastern Australia to the Bismarck Archipelago, the Solomon Islands, New Caledonia and the Loyalty Islands.

## Habitat

*Libythea geoffroyi* breeds in patches of semi-deciduous monsoon vine thicket on rocky outcrops or gorges composed of limestone, sandstone or basalt usually protected from fire where the larval food plant grows as a small tree in high density (Braby 2014a). Adults have also been collected in coastal areas of higher rainfall (> 1,100 mm mean annual rainfall) where a distinct larger-leafed variety with glabrous, ovate or oblong leaves of the food plant grows on lateritic cliffs and sand dunes; presumably, *L. geoffroyi* also breeds in these habitats.

## Larval food plant

*Celtis australiensis* (Cannabaceae).

## Seasonality

Adults are seasonal, occurring from January to August. They are most abundant during the late wet season, particularly in March, when they are usually in fresh condition (Williams et al. 2006; Meyer et al. 2013; G. Swann), after which their abundance appears to steadily decline as the dry season progresses. The breeding phenology and seasonal history of the immature stages are not known. Braby (2014a) found an empty pupal shell in good condition on the underside of a leaf of the larval food plant in early May, indicating recent adult emergence. In northern Queensland, the larvae feed only on new soft leaves following the first substantial rainfall event at the end of the dry season (Johnson and Valentine 1988, 1989). Thus, in northern

Australia it is likely the species breeds seasonally during the height of and/or just after the wet season, when *C. australiensis* produces new leaf growth in response to monsoon rains; however, it is not clear how the species survives the nongrowing period during the dry season, when the leaves are tough and nonedible to larvae. Presumably, only one or two generations are completed annually. Overseas, other species of *Libythea* remain in adult diapause during the nonbreeding period, but this behaviour has not been observed during the late dry season in northern Australia and thus it is not clear whether this strategy is adopted by *L. geoffroyi*. It is possible the eggs—which are tiny and concealed deep within the axils of leaf buds—remain in diapause during the long dry season.

## Breeding status

This species is resident in the study region.

## Conservation status

LC. The subspecies *L. geoffroyi genia* in the Northern Territory has a restricted range, with two locations occurring in conservation reserves: Fish River and Judbarra/Gregory National Park. Some areas within the overall range may be impacted by inappropriate fire regimes, particularly an increase in the frequency and severity of dry season burns. The habitat is highly susceptible to fire and the larval food plant is intolerant of, or sensitive to, fire.

Photo: Limestone Gorge, Judbarra/Gregory National Park, NT, M.F. Braby

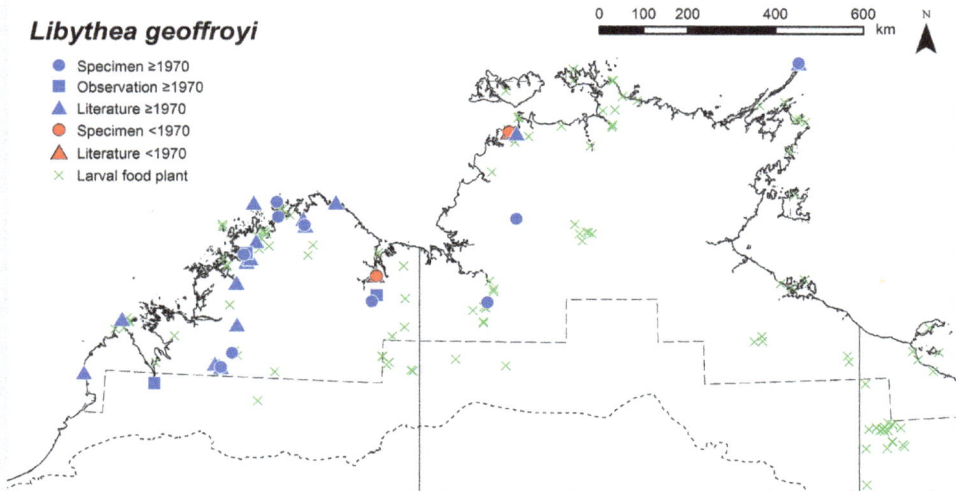

*Libythea geoffroyi*

| | Specimen ≥1970 |
| | Observation ≥1970 |
| | Literature ≥1970 |
| | Specimen <1970 |
| | Literature <1970 |
| | Larval food plant |

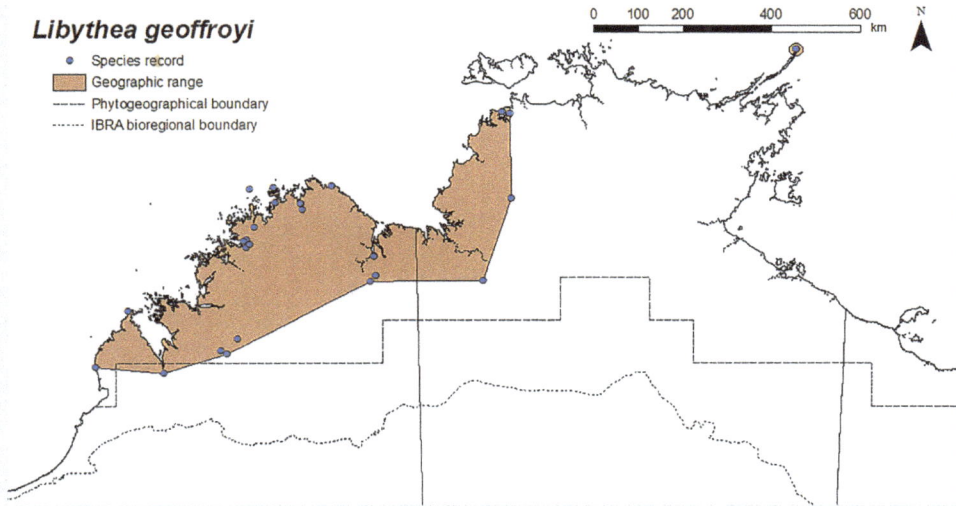

*Libythea geoffroyi*

| | Species record |
| | Geographic range |
| | Phytogeographical boundary |
| | IBRA bioregional boundary |

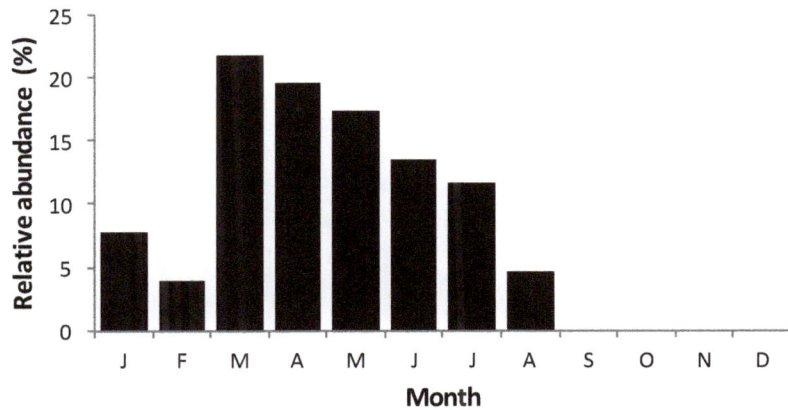

*Libythea geoffroyi* (n = 32)

| Month | J | F | M | A | M | J | J | A | S | O | N | D |
|-------|---|---|---|---|---|---|---|---|---|---|---|---|
| Egg | | | | | | | | | | | | |
| Larva | | | | | | | | | | | | |
| Pupa | | | | | | | | | | | | |
| Adult | | | | | | | | | | | | |

# Blue Tiger
## *Tirumala hamata* (W. S. Macleay, 1826)

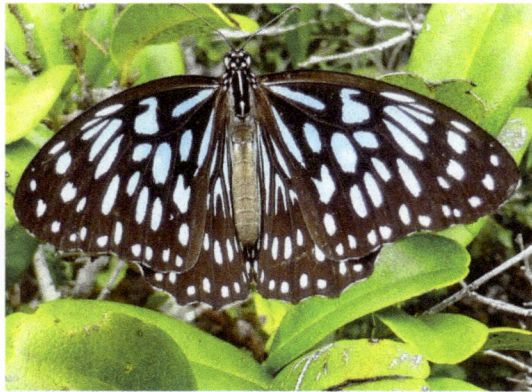

Plate 80 Daintree, Qld
Photo: Roger Farrow

Plate 81 Fogg Dam, NT
Photo: M. F. Braby

## Distribution

This species occurs in the Kimberley, Top End and western Gulf Country of the study region. It occurs mainly in the higher rainfall areas (> 900 mm mean annual rainfall), but it has been recorded from drier inland areas (700–800 mm), where it reaches its southernmost limits at Koolan Island in Yampi Sound (McKenzie et al. 1995), Kununurra (Koch 1957), WA; Limmen National Park (Butterfly Springs), NT (Franklin 2007); and Doomadgee, Qld (Puccetti 1991). Its geographic range corresponds well with the spatial distribution of its larval food plants, although the plants extend considerably further inland. Outside the study region, *T. hamata* occurs from South-East Asia, through mainland New Guinea and adjacent islands and eastern Australia to the Solomon Islands, Fiji, Tonga and Samoa.

## Habitat

*Tirumala hamata* breeds in both wet and dry monsoon forests, including evergreen monsoon vine forest and semi-deciduous monsoon vine thicket, where the larval food plants grow as vines (Forster and Martin 1990; Braby 2015e).

## Larval food plants

*Cynanchum carnosum, Marsdenia glandulifera, M. velutina, Secamone elliptica* (Apocynaceae). The main larval food plant appears to be *S. elliptica* (Forster and Martin 1990), but larvae have also been found and reared on *M. glandulifera* (Braby 2015e), *M. velutina* and *C. carnosum* (Forster and Martin 1990).

## Seasonality

Adults occur throughout the year and they do not appear to show any pronounced seasonal changes in abundance. They are generally observed in comparatively low numbers, do not migrate and do not form large overwintering aggregations, unlike the populations in north-eastern Queensland (see Monteith 1982; Scheermeyer 1993, 1999). The breeding phenology is not well understood for northern Australia, but the following observations suggest the breeding season is limited to a short period during the wet season. At Fogg Dam, NT, the immature stages were recorded during the 'build-up' (October and November) of high moisture and high temperatures, when the larval food plant (*Marsdenia glandulifera*) was producing new soft leaves on which the eggs were laid and larvae were feeding (Braby 2015e). Near Darwin, NT, Forster and Martin (1990: 131) noted that on *M. velutina* 'eggs are laid only on the fresh young leaves which are available only in the wet season'. During the dry season, adults have been recorded overwintering in relatively small numbers with *Euploea* spp. in moist refuges, mainly confined to areas of deep shade, in the Prince Regent Nature Reserve in the western Kimberley, WA (Bailey and Richards 1975). Presumably, the species stops breeding during the long dry season, similar to populations in northern Queensland (Scheermeyer 1993).

## Breeding status

This species is resident in the study region.

## Conservation status

LC.

## Tirumala hamata

Specimen ≥1970
Observation ≥1970
Literature ≥1970
Specimen <1970
Literature <1970
Larval food plants

## Tirumala hamata

Species record
Geographic range
Phytogeographical boundary
IBRA bioregional boundary

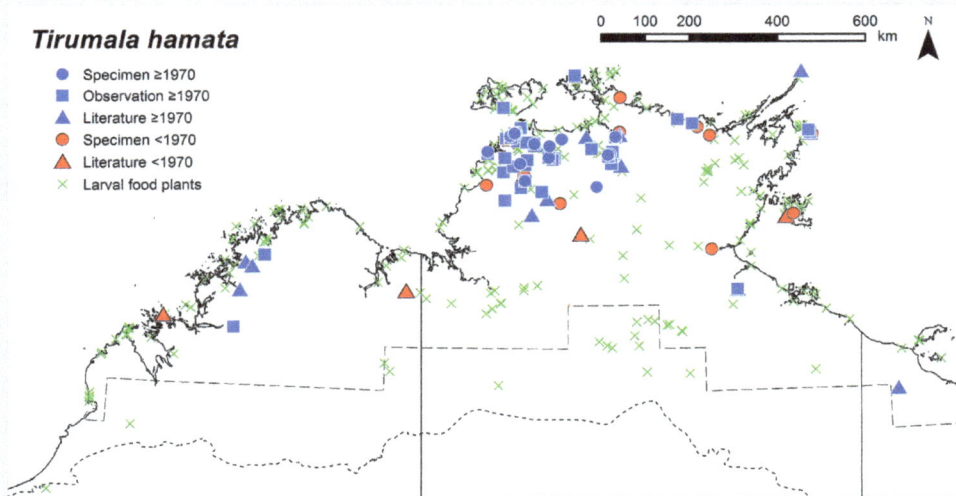

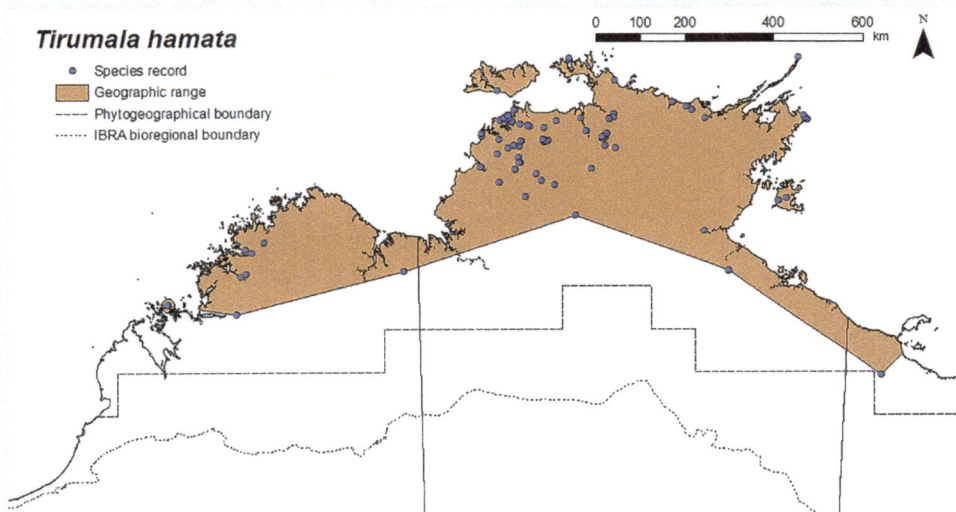

### Tirumala hamata (n = 109)

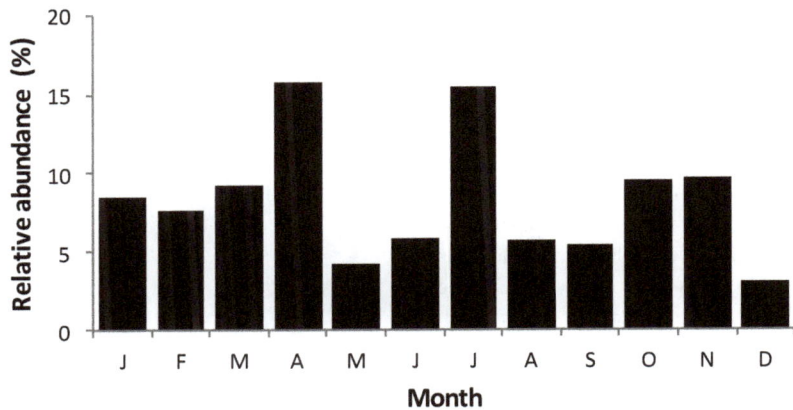

| Month | J | F | M | A | M | J | J | A | S | O | N | D |
|-------|---|---|---|---|---|---|---|---|---|---|---|---|
| Egg   |   |   |   |   |   |   |   |   |   | ■ |   |   |
| Larva |   |   |   |   |   |   |   |   |   | ■ | ■ |   |
| Pupa  |   |   |   |   |   |   |   |   |   | ■ | ■ |   |
| Adult | ■ | ■ | ■ | ■ | ■ | ■ | ■ | ■ | ■ | ■ | ■ | ■ |

# Plain Tiger
## *Danaus chrysippus* (Linnaeus, 1758)

Plate 82 Wanguri, Darwin, NT
Photo: M. F. Braby

Plate 83 Wanguri, Darwin, NT
Photo: M. F. Braby

## Distribution

This species is represented in the study region by the subspecies *D. chrysippus cratippus* (C. Felder, 1860). It is restricted to the northern coastal areas of the Top End, where it has been recorded from several sites on Cobourg Peninsula (Black Point, 2 km east–north-east of Black Point and Smith Point) (Braby 2014a; Braby et al. 2015) and in Darwin (Braby 2015c), NT. Outside the study region, *D. chrysippus* occurs widely in Africa, India and South-East Asia.

## Habitat

The breeding habitat of *D. chrysippus* has not been confirmed, but most adults have been collected in coastal paperbark swampland bordering freshwater lagoons where the putative larval food plant (*Cynanchum carnosum*) grows as a vine in high density (Braby 2014a). In addition, a freshly emerged female was collected from this habitat and an empty pupal shell (possibly this species) was found within a metre of the female; another female was observed flying close to the ground searching the putative food plant but did not oviposit. Also, males have been observed to establish mate-location behaviour by perching close to the ground on reeds and dead branches for short periods.

## Larval food plants

Not recorded in the study region; possibly *Cynanchum carnosum* (Apocynaceae) (Braby 2014a).

## Seasonality

The seasonal abundance and breeding phenology of this species are not well understood. Adults are seasonal, occurring only during the wet season (January–April). Most specimens have been collected in February and March, but there are too few records (n = 10) to assess seasonal changes in abundance.

## Breeding status

*Danaus chrysippus* does not appear to have become permanently established in the study region. It is most likely a vagrant or a rare immigrant from South-East Asia, occasionally colonising the region and breeding temporarily during favourable conditions (Braby et al. 2015).

## Conservation status

NA.

## *Danaus chrysippus*

● Specimen ≥1970
■ Observation ≥1970
▲ Literature ≥1970

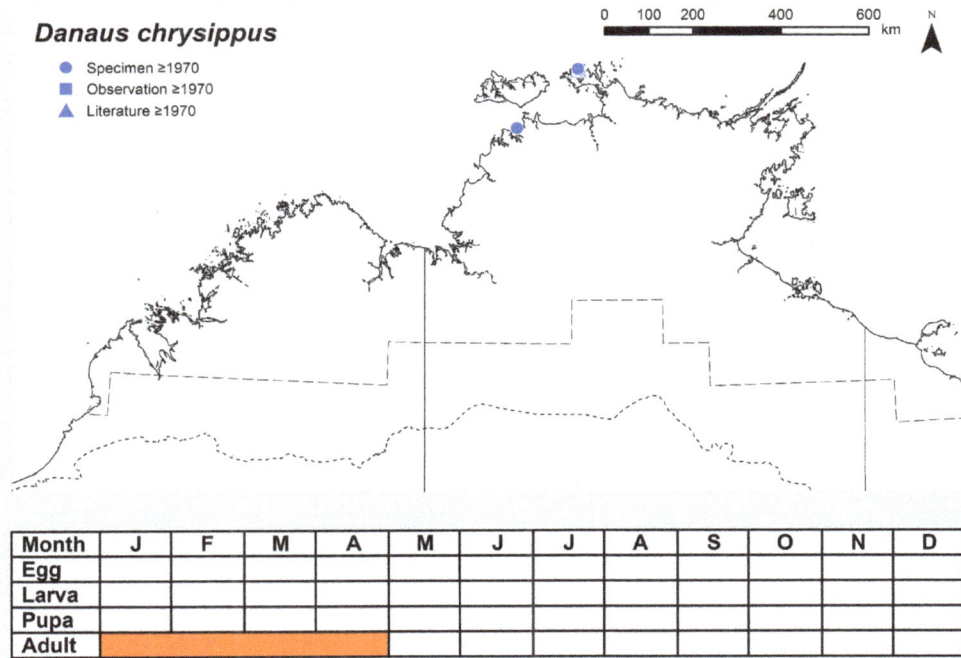

| Month | J | F | M | A | M | J | J | A | S | O | N | D |
|-------|---|---|---|---|---|---|---|---|---|---|---|---|
| Egg | | | | | | | | | | | | |
| Larva | | | | | | | | | | | | |
| Pupa | | | | | | | | | | | | |
| Adult | | | | | | | | | | | | |

Photo: Black Point, Cobourg Peninsula, NT, M.F. Braby

# Lesser Wanderer
## *Danaus petilia* (Stoll, 1790)

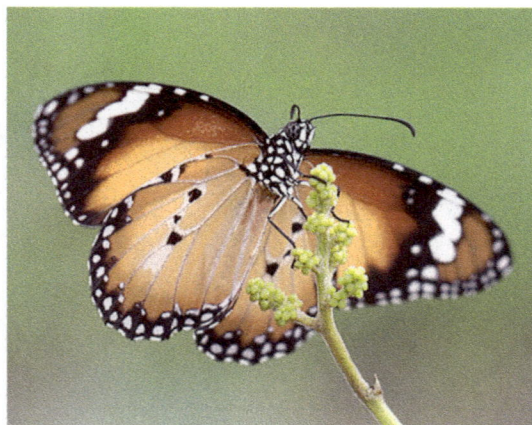

Plate 84 Holmes Jungle, Darwin, NT
Photo: Tissa Ratnayeke

Plate 85 Katherine, NT
Photo: M. F. Braby

## Distribution

This species occurs almost throughout the entire study region. It extends from moist coastal areas to drier inland areas of the semi-arid zone, as well as in the arid zone of central Australia beyond the southern boundary of the study region. The spatial distribution of the known native larval food plants is similarly widespread, although they are less well represented in the semi-arid and arid zones, indicating that several other (as yet unreported) food plants are probably used in these areas in addition to the introduced larval food plant (*Calotropis procera*). Outside the study region, *D. petilia* occurs in mainland New Guinea, throughout Australia and the islands of the South Pacific.

## Habitat

*Danaus petilia* occurs in a wide variety of habitats, but it breeds mainly in savannah woodland and paperbark woodland where the larval food plants grow as evergreen or seasonal perennial vines.

## Larval food plants

*Brachystelma glabriflorum, Cynanchum carnosum, C. christineae, C. leibianum, C. pedunculatum, Oxystelma esculentum, Tylophora flexuosa* (Apocynaceae); also *Calotropis procera* (Apocynaceae); possibly *Cynanchum floribundum, Marsdenia australis* and *Rhyncharrhena linearis* (Apocynaceae), which are food plants in the arid areas of South Australia and Queensland (Braby 2000).

## Seasonality

Adults occur throughout the year, but are generally more abundant towards the end of the wet season (April) and least abundant in the late dry season (September and October). The immature stages are found mainly during the wet season (November–April), when the larval food plants produce new shoots and buds (Forster 1991), but breeding may continue until at least the mid dry season (July). However, it is not clear whether breeding is seasonal and most females stop breeding during the late dry season or breeding continues but reproductive activity declines as the dry season progresses. Although it is a well-known migrant (Smithers 1983a), we have no records of migratory flights in the study region.

## Breeding status

This species is resident in the study region.

## Conservation status

LC.

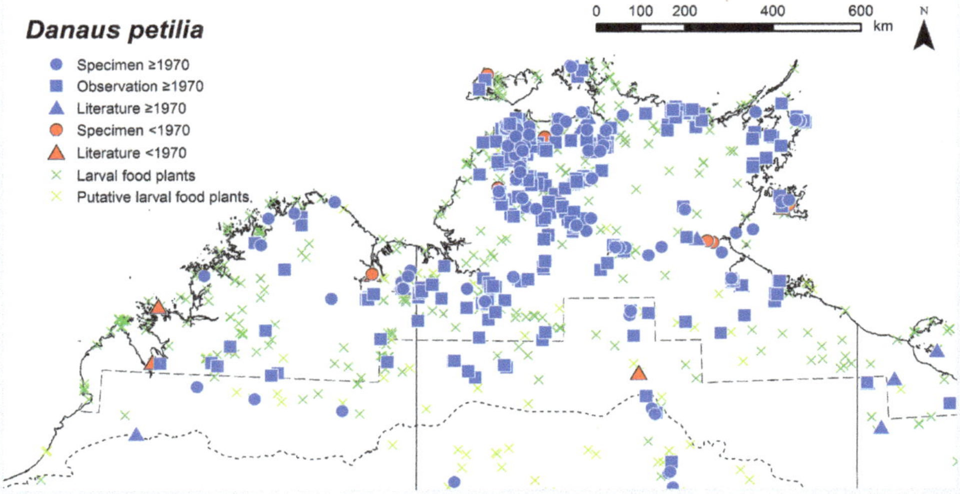

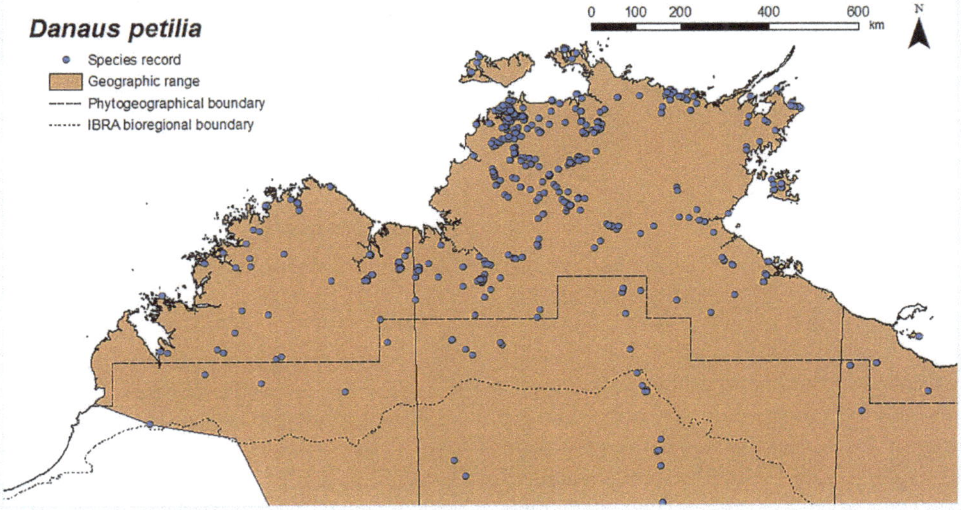

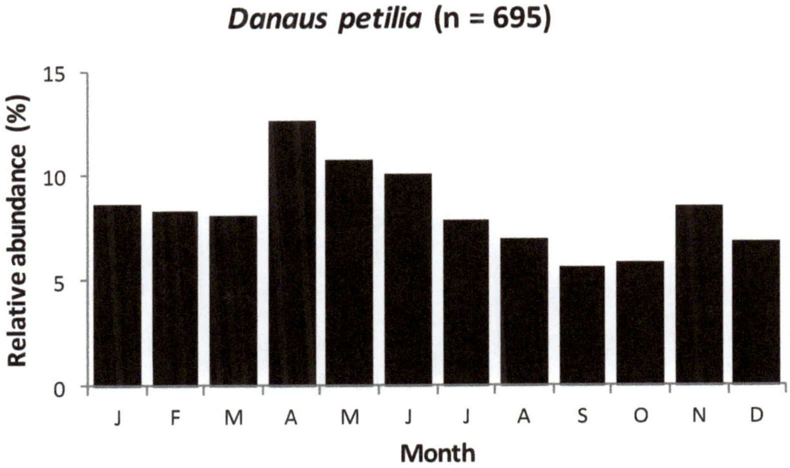

## Danaus petilia

- ● Specimen ≥1970
- ■ Observation ≥1970
- ▲ Literature ≥1970
- ● Specimen <1970
- ▲ Literature <1970
- ✕ Larval food plants
- ✕ Putative larval food plants.

## Danaus petilia

- • Species record
- ▨ Geographic range
- — Phytogeographical boundary
- ⋯ IBRA bioregional boundary

## Danaus petilia (n = 695)

| Month | J | F | M | A | M | J | J | A | S | O | N | D |
|-------|---|---|---|---|---|---|---|---|---|---|---|---|
| Egg | | | | | | | | | | | | |
| Larva | | | | | | | | | | | | |
| Pupa | | | | | | | | | | | | |
| Adult | | | | | | | | | | | | |

# Orange Tiger
## *Danaus genutia* (Cramer, [1779])

Plate 86 Elsey National Park, Mataranka, NT
Photo: M. F. Braby

Plate 87 Elsey National Park, Mataranka, NT
Photo: M. F. Braby

## Distribution

This species is represented by the subspecies *D. genutia alexis* (Waterhouse & Lyell, 1914), which is endemic to the study region. *Danaus genutia* does not occur elsewhere in Australia. It has a restricted and possibly disjunct distribution, occurring mainly in the eastern Kimberley and the north-western corner of the Top End. There is also a historical record from Derby in the western Kimberley, WA (Waterhouse and Lyell 1914), and an isolated population at Mataranka, NT, which is associated with the upper headwaters of the Roper River (McCubbin 1971; Le Souëf 1971). The geographic range broadly corresponds with the spatial distribution of the larval food plant, which likewise is restricted in extent; however, this plant is absent from Mataranka, indicating that an alternative (as yet unreported) food plant is used at this location. Outside the study region, *D. genutia* occurs widely in India, southern China and South-East Asia, although it is absent from Maluku.

## Habitat

*Danaus genutia* breeds mainly in upper estuarine swamps supporting dense stands of the reed *Typha* sp. on which the larval food plant grows as a vine (Meyer 1995). At Mataranka (Elsey National Park), adults occur in paperbark swampland along the edge of inland perennial freshwater streams and they no doubt breed in this habitat. Small numbers of adults have also been recorded in other habitats in the Top End, but these are thought to be individuals dispersing from nearby breeding areas.

## Larval food plant

*Oxystelma esculentum* (Apocynaceae).

## Seasonality

The seasonal abundance and breeding phenology of this species are not well understood. Adults have been recorded sporadically during the year; they have been noted to be abundant during the mid dry season (July) (McCubbin 1971) and late dry season (October) (Field 1990b), as well as in the wet season (December and April) (Meyer 1995). Our data suggest a peak in abundance in July. Presumably, adults occur throughout the year; the paucity of records during the wet season (January–March) is probably related to the fact that the breeding habitats are frequently inaccessible at that time of year. However, it remains to be determined whether they breed seasonally or continuously throughout the year.

## Breeding status

This species is resident in the study region.

## Conservation status

LC. Despite the wide geographical range of the subspecies *D. genutia alexis*, available data suggest it has a disjunct distribution with a restricted AOO. The larval food plant is currently listed as DD under the *TPWCA*. Further field data are needed on the ecological requirements of this species, particularly its critical habitat and larval food plants in noncoastal areas of range.

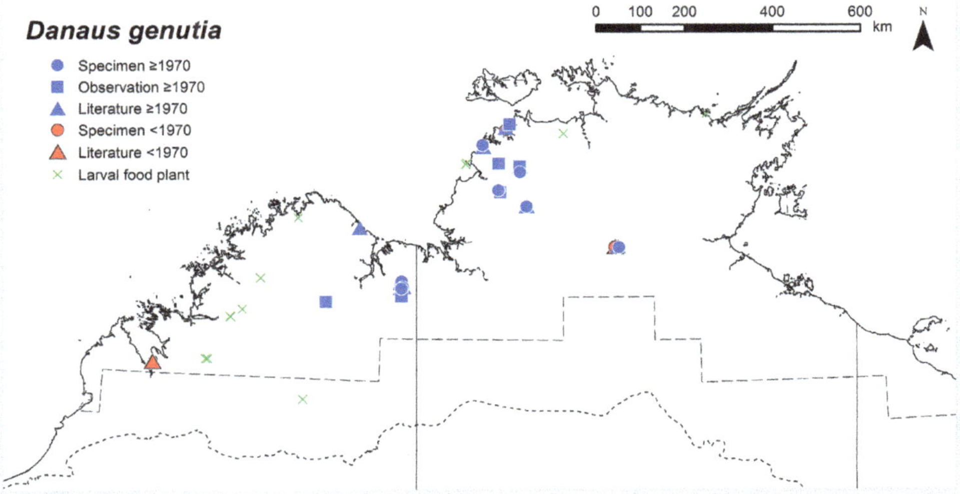

Danaus genutia
0 100 200 400 600 km
N
Specimen ≥1970
Observation ≥1970
Literature ≥1970
Specimen <1970
Literature <1970
Larval food plant

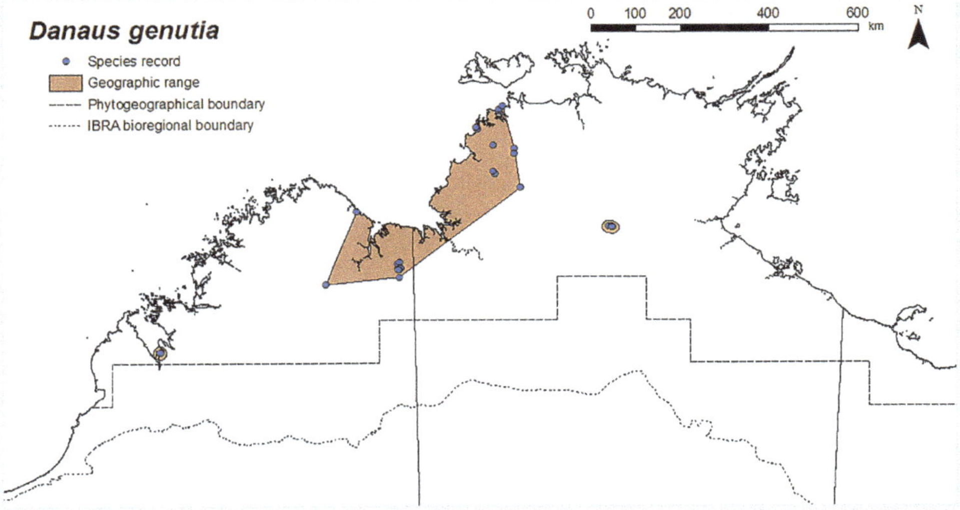

Danaus genutia
0 100 200 400 600 km
N
Species record
Geographic range
Phytogeographical boundary
IBRA bioregional boundary

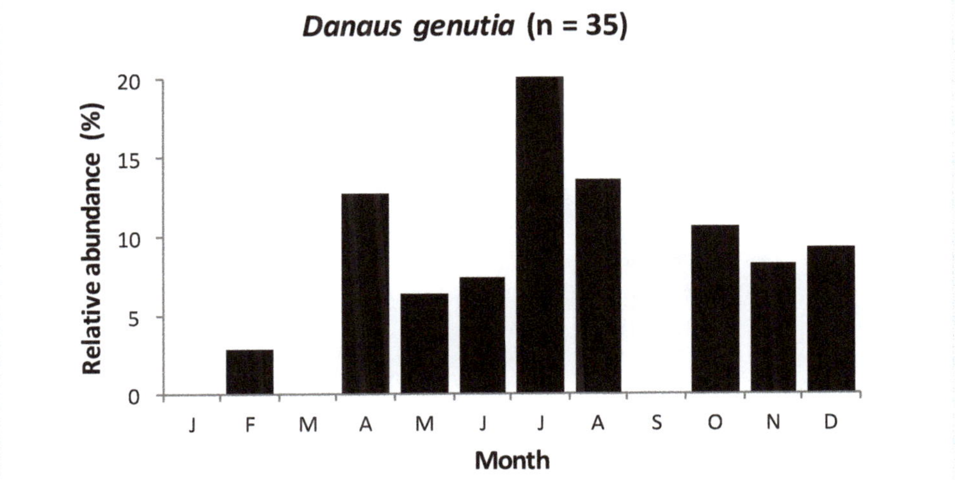

Danaus genutia (n = 35)
Relative abundance (%)
0
5
10
15
20
J F M A M J J A S O N D
Month

| Month | J | F | M | A | M | J | J | A | S | O | N | D |
|---|---|---|---|---|---|---|---|---|---|---|---|---|
| Egg | | | | | | | | | | | | |
| Larva | | | | | | | | | | | | |
| Pupa | | | | | | | | | | | | |
| Adult | | | | | | | | | | | | |

# Swamp Tiger
*Danaus affinis* (Fabricius, 1775)

Plate 88 Holmes Jungle, Darwin, NT
Photo: Tissa Ratnayeke

Plate 89 Buffalo Creek, Leanyer Swamp, NT
Photo: M. F. Braby

## Distribution

This species is represented in the study region by the subspecies *D. affinis affinis* (Fabricius, 1775). It occurs from the Kimberley, through the Top End to coastal and near-coastal areas in the western Gulf Country, generally in areas that receive more than 800 mm mean annual rainfall. The larval food plants are more widely distributed, particularly *Marsdenia viridiflora*, which extends further south into the semi-arid zone, but it is considered unlikely that *D. affinis* is established in these areas. Outside the study region, *D. affinis* occurs sporadically from South-East Asia, through mainland New Guinea and adjacent islands and eastern Australia to the Solomon Islands, Vanuatu and New Caledonia.

## Habitat

*Danaus affinis* breeds mainly in paperbark swampland and paperbark woodland adjacent to swamps in coastal areas, often associated with estuarine creeks and rivers, where its primary larval food plant (*Cynanchum carnosum*) grows as a vine. However, in the more inland areas, it breeds in eucalypt open forest where alternative food plants grow (Braby 2015e), as well as paperbark swamps adjacent to perennial freshwater rivers where *C. carnosum* grows.

## Larval food plants

*Cynanchum carnosum, Marsdenia viridiflora, Sarcolobus hullsii* (Apocynaceae). The main food plant is *C. carnosum*.

## Seasonality

Adults occur throughout the year and show little seasonal variation in abundance, although they are generally more abundant in the early dry season (April–June), when they may be exceedingly common. The immature stages, like *Danaus petilia*, have only been recorded during the wet season (November–April), during which several generations are completed. However, it is not clear whether breeding is seasonal and most females stop breeding and remain in reproductive diapause during the dry season or breeding continues but reproductive activity declines as the dry season progresses.

## Breeding status

This species is resident in the study region.

## Conservation status

LC.

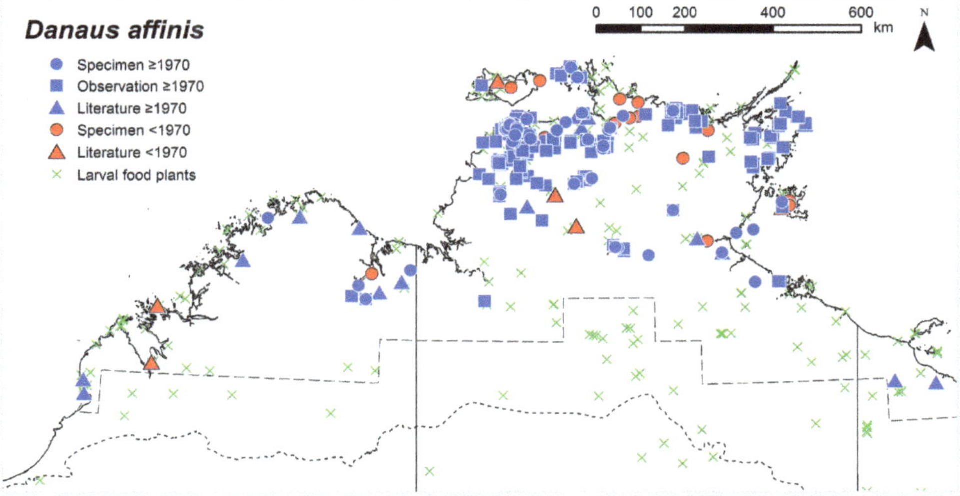

Danaus affinis
0 100 200 400 600 km
N
Specimen ≥1970
Observation ≥1970
Literature ≥1970
Specimen <1970
Literature <1970
Larval food plants

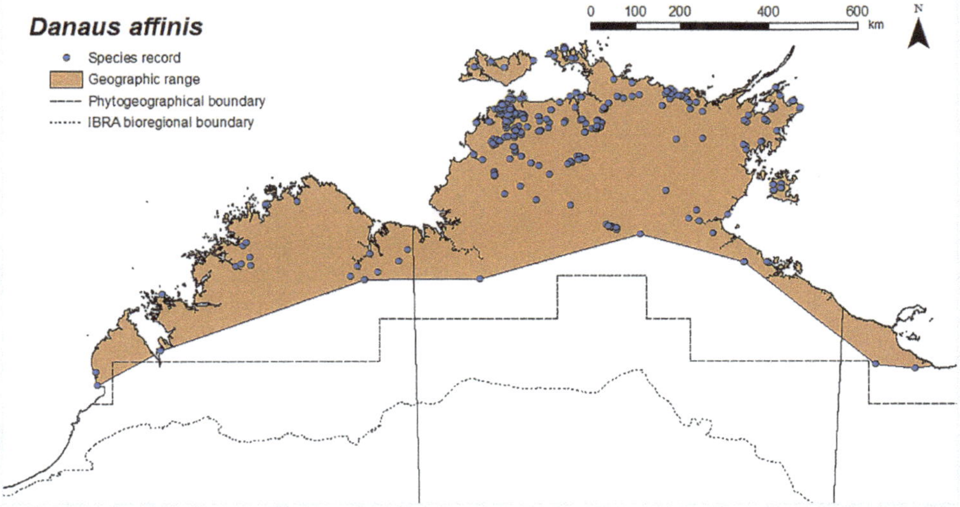

Danaus affinis
0 100 200 400 600 km
N
Species record
Geographic range
Phytogeographical boundary
IBRA bioregional boundary

***Danaus affinis*** **(n = 548)**

15
10
5
0
Relative abundance (%)
J F M A M J J A S O N D
Month

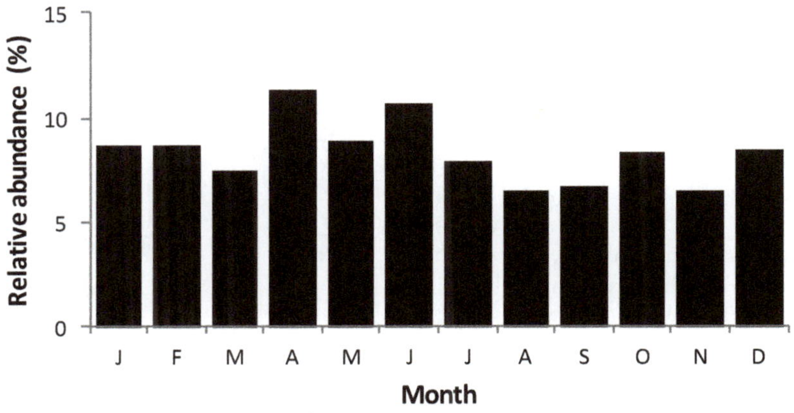

| Month | J | F | M | A | M | J | J | A | S | O | N | D |
|---|---|---|---|---|---|---|---|---|---|---|---|---|
| Egg | | | | | | | | | | | | |
| Larva | | | | | | | | | | | | |
| Pupa | | | | | | | | | | | | |
| Adult | | | | | | | | | | | | |

# Monarch

*Danaus plexippus* (Linnaeus, 1758)

Plate 90
Photo: Eleanor P. Williams

## Distribution

This species has been recorded sporadically from the study region, including Kununurra, WA (Dunn 1980); near Darwin, Kakadu National Park (South Alligator River) and 50 km south–south-east of Elliott, NT; and from several locations in the western Gulf Country (Braby 2014a). Most of the records for the Northern Territory and Queensland were made when a large-scale migration occurred during April–May 2013 (Braby 2014a). It has also been recorded 40 km north–north-east of Burke & Wills Roadhouse, Qld, on the eastern boundary of the study region (Dunn 2015a). More recently, a female specimen in worn condition flying in a northerly direction was netted at Humpty Doo, NT, on 26 November 2016 (K. McLachlan). Two earlier records from near Darwin (Holmes Jungle Conservation Park, 5 March 2013; Gunn Point, 30 November 2013) previously thought to be deliberate introductions (see Braby 2014a) are tentatively accepted as natural occurrences; the record for Holmes Jungle may have been part of the north-west migration observed two months later, in May 2013. The putative larval food plant (*Asclepias curassavica*) is cultivated in and near Darwin, NT. Outside the study region, *D. plexippus* occurs widely in eastern and southern Australia. It is native to the New World, but became established in Australia in the 1870s.

## Excluded data

A record from Darwin, NT (Bullocky Point, 2 September 2009), is an artificial introduction based on a wedding butterfly release (Braby 2014a) and is therefore excluded. An observation made at Kununurra, WA, by Dunn (1980) was considered doubtful by Braby (2012b) because at that time the species was not known elsewhere in the study region and it may have been confused with the similar looking *Danaus genutia*, which occurs commonly at the location. However, K. L. Dunn (pers. comm.) advises that the specimen was observed at close proximity, enabling it to be readily distinguished by its large size and characteristic wing pattern.

## Habitat

The breeding habitat of *D. plexippus* has not been recorded in the study region.

## Larval food plants

Not recorded in the study region; possibly *Asclepias curassavica* (Apocynaceae) and other introduced and naturalised species of milkweeds, which are the food plants in Queensland (Braby 2000).

## Seasonality

The seasonal abundance and breeding phenology of this species are not well understood. Adults have been recorded sporadically during the year, but there are too few records (n = 13) to assess any seasonal changes in abundance.

## Breeding status

*Danaus plexippus* does not appear to have become permanently established in the study region. It is most likely a vagrant or infrequent visitor, entering the region occasionally during migration, but it remains to be determined whether it colonises and breeds temporarily.

## Conservation status

NA.

## Danaus plexippus

- ● Specimen ≥1970
- ■ Observation ≥1970
- ▲ Literature ≥1970

| Month | J | F | M | A | M | J | J | A | S | O | N | D |
|-------|---|---|---|---|---|---|---|---|---|---|---|---|
| Egg   |   |   |   |   |   |   |   |   |   |   |   |   |
| Larva |   |   |   |   |   |   |   |   |   |   |   |   |
| Pupa  |   |   |   |   |   |   |   |   |   |   |   |   |
| Adult |   |   | ▒ | ▒ | ▒ |   |   | ▒ |   | ▒ | ▒ |   |

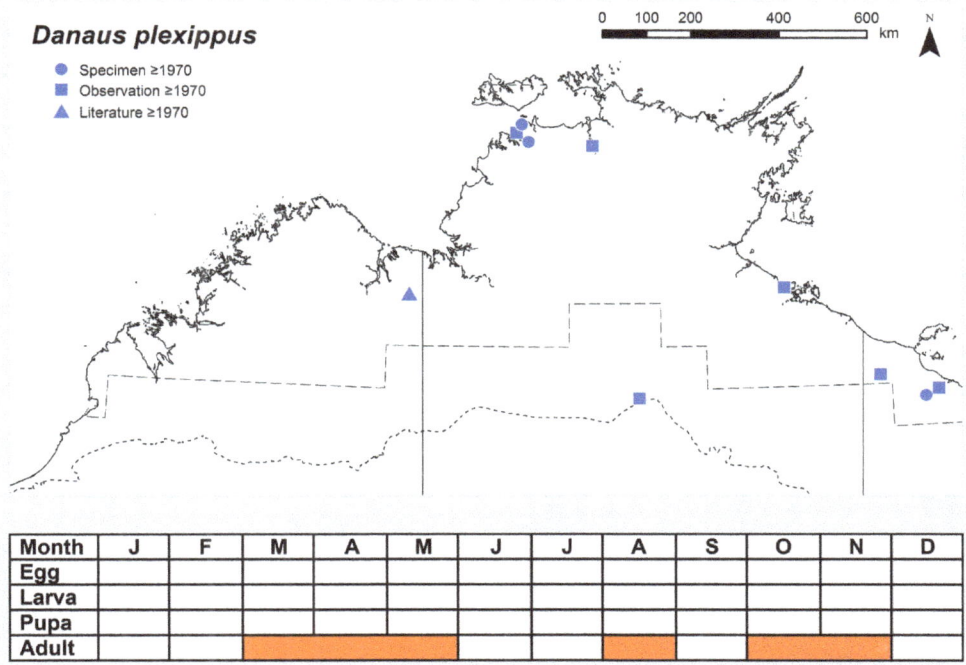

Photo: Nourlangie Rock, Kakadu National Park, NT, M.F. Braby

# Two-brand Crow

*Euploea sylvester* (Fabricius, 1793)

Plate 91 Maningrida, NT
Photo: Deb Bisa

## Distribution

This species is represented by the subspecies *E. sylvester pelor* Doubleday, 1847, which is endemic to the study region. It has a broad distribution, occurring in the Kimberley, Top End and western Gulf Country. It extends from moist coastal areas to lower rainfall areas of the semi-arid zone (c. 700 mm mean annual rainfall), reaching its southernmost limit at Bessie Spring, NT (Dunn and Dunn 1991). The geographic range corresponds well with the spatial distribution of its primary larval food plant (*Marsdenia geminata*), although the plant extends further inland. However, the record from Boodjamulla/Lawn Hill National Park (Murray Spring), Qld (Daniels and Edwards 1998), appears to lie outside the known range of the food plant, suggesting it may have been a vagrant that dispersed beyond the normal breeding range. Outside the study region, *E. sylvester* occurs widely from India, southern China and South-East Asia, through mainland New Guinea and adjacent islands and north-eastern Australia to Vanuatu and New Caledonia.

The subspecies *E. sylvester sylvester* (Fabricius, 1793) occurs in the Gulf of Carpentaria at Walker Creek 36 km east of Karumba, Qld, just outside the eastern boundary of the study region (Braby 2015d), and further surveys are required to determine whether it extends further west.

## Habitat

*Euploea sylvester* breeds in semi-deciduous monsoon vine thicket and riparian evergreen monsoon vine forest along the banks of rivers where the larval food plants grow as vines (Meyer 1997a).

## Larval food plants

*Marsdenia geminata*, *Parsonsia alboflavescens* (Apocynaceae). The main food plant is *M. geminata* (Meyer 1997a), but they occasionally use *P. alboflavescens* on Gove Peninsula, NT (Braby 2011a).

## Seasonality

Adults occur throughout the year. They appear to show little seasonal variation in abundance, although at Darwin, NT, Meyer (1997a) noted that adults were most numerous during the late wet season (late February and March). Breeding is seasonal, limited mainly to the wet season (December–January or April, depending on location), although on one occasion the immature stages were found in July on Gove Peninsula, where the wet season is protracted. During the dry season, females generally stop breeding and aggregate in small numbers in moist refuges, such as the deep shade within pockets of monsoon forest (Monteith 1982; Scheermeyer 1993).

## Breeding status

This species is resident in the study region.

## Conservation status

LC.

## Euploea sylvester

Specimen ≥1970
Observation ≥1970
Literature ≥1970
Specimen <1970
Literature <1970
Larval food plants

## Euploea sylvester

Species record
Geographic range
Vagrant
Phytogeographical boundary
IBRA bioregional boundary

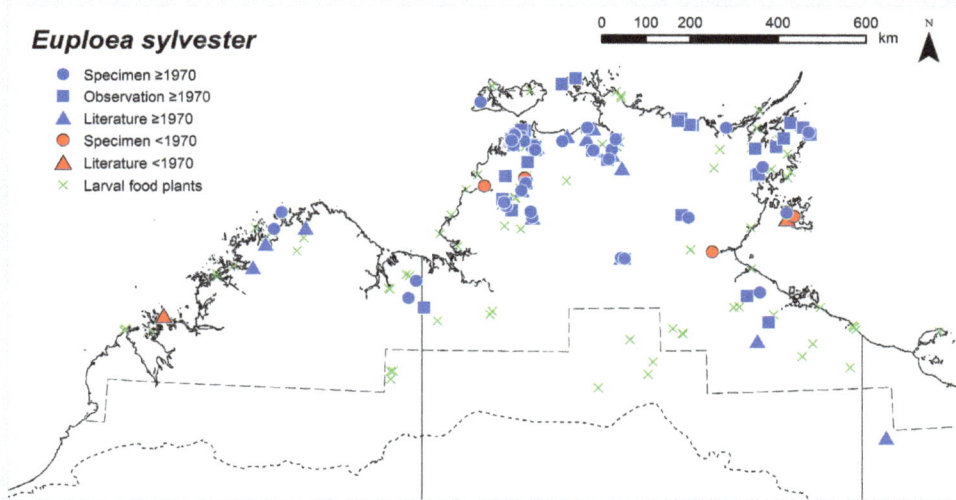

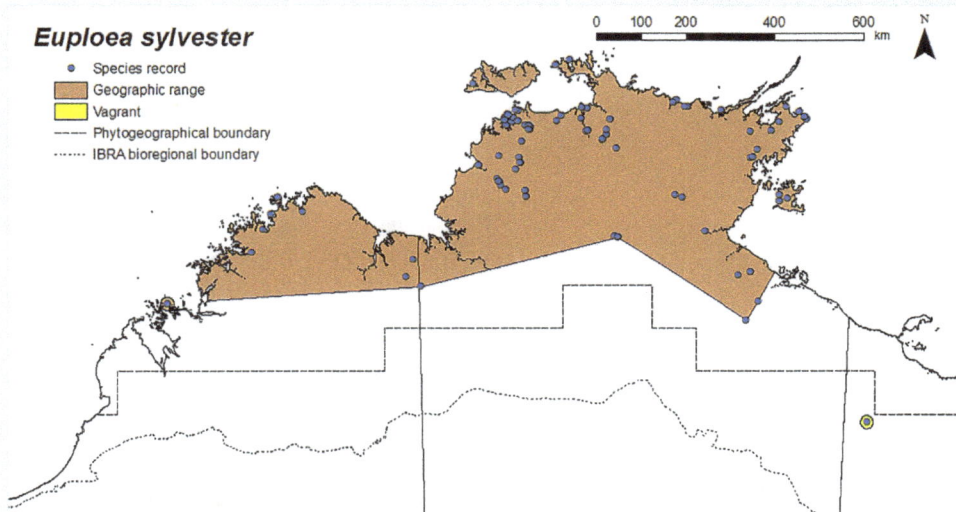

### Euploea sylvester (n = 124)

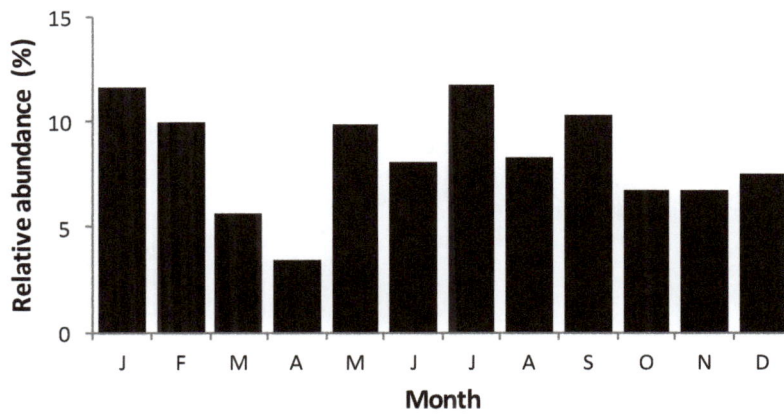

| Month | J | F | M | A | M | J | J | A | S | O | N | D |
|-------|---|---|---|---|---|---|---|---|---|---|---|---|
| Egg | | | | | | | | | | | | |
| Larva | | | | | | | | | | | | |
| Pupa | | | | | | | | | | | | |
| Adult | | | | | | | | | | | | |

# Small Brown Crow

*Euploea darchia* W. S. Macleay, 1826

Plate 92 Beatrice Hill, NT
Photo: M. F. Braby

## Distribution

This species is represented by the subspecies *E. darchia darchia* W. S. Macleay, 1826, which is endemic to the study region. It has a disjunct distribution, occurring in the northern Kimberley and Top End, where it is restricted to the higher rainfall areas (> 1,000 mm mean annual rainfall, but generally > 1,200 mm). Its geographic range closely corresponds with the spatial distribution of its larval food plant. The food plant, however, extends slightly further south, with occurrences in eastern Walcott Inlet in the western Kimberley, WA, and Groote Eylandt and Maria Island in Limmen Bight, NT. Further field surveys are therefore required to determine whether *E. darchia* occurs in these areas. Outside the study region, *E. darchia* occurs in Timor, Banda, the Kai Islands and north-eastern Australia.

## Habitat

*Euploea darchia* breeds in coastal semi-deciduous monsoon vine thicket and riparian evergreen monsoon vine forest along the banks of rivers where the larval food plant grows as a vine (Meyer 1996d).

## Larval food plant

*Trophis scandens* (Moraceae).

## Seasonality

Adults occur throughout the year, but their patterns of seasonal changes in abundance are not entirely clear. Near Darwin, NT, Meyer (1996d, and pers. comm.) noted that adults were freshly emerged and more abundant from February to May, and Franklin (2011) similarly found the species to be more abundant during the mid wet season and early dry season (January–May) based on quantitative studies conducted over 14 months during 2008–09. Our data suggest it is more abundant during the mid dry season and early wet season (June–November), but this may be due to biased sampling at that time of year when adults tend to aggregate in moist refuges, such as the deep shade in pockets of monsoon forest. Breeding, as for *Euploea sylvester*, is strictly seasonal and limited to the wet season (January–April).

## Breeding status

This species is resident in the study region.

## Conservation status

LC.

## Euploea darchia

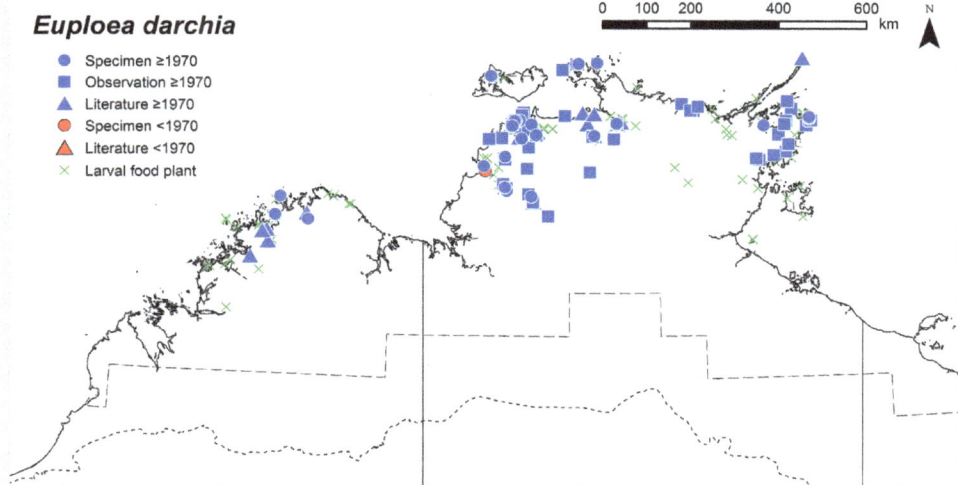

## Euploea darchia

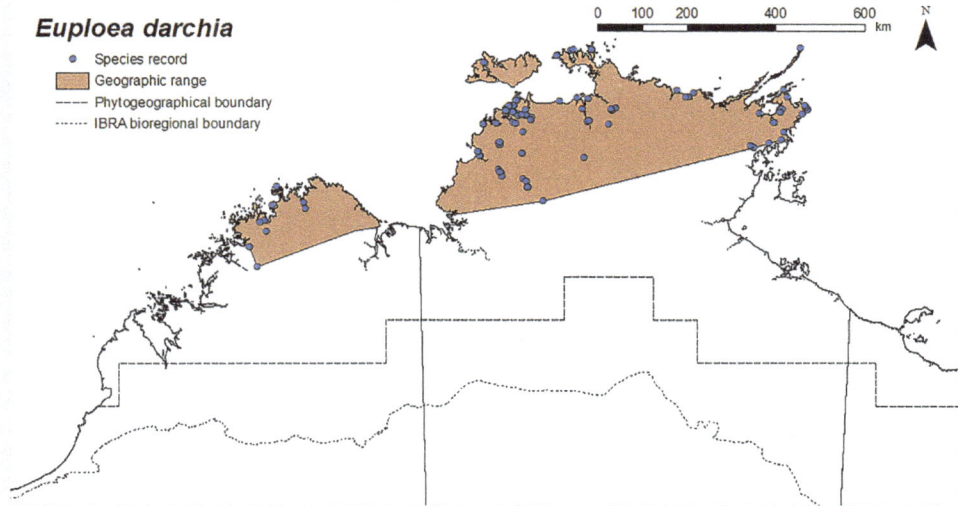

## Euploea darchia (n = 178)

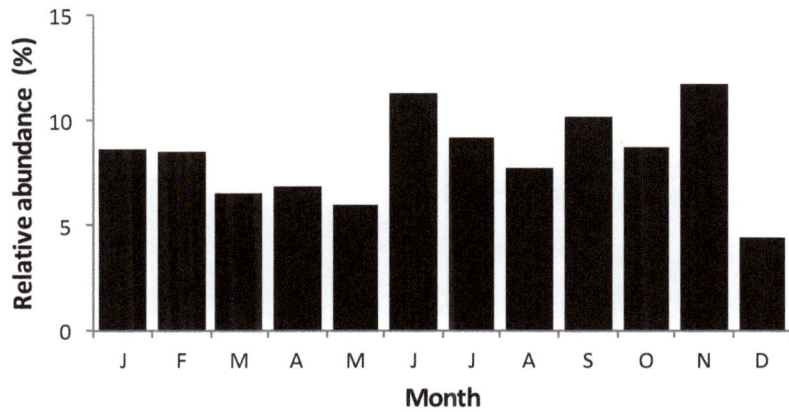

| Month | J | F | M | A | M | J | J | A | S | O | N | D |
|-------|---|---|---|---|---|---|---|---|---|---|---|---|
| Egg | | | | | | | | | | | | |
| Larva | | | | | | | | | | | | |
| Pupa | | | | | | | | | | | | |
| Adult | | | | | | | | | | | | |

# Common Crow
*Euploea corinna* (W. S. Macleay, 1826)

Plate 93 Palmerston, NT
Photo: Andrea Hope

## Distribution

This species has a very wide distribution, occurring throughout much of the study region. It extends from moist coastal areas to drier inland areas of the semi-arid zone, as well as the arid zone of central Australia beyond the southern boundary of the study region. It has been recorded as far south as Brunchilly Station 95 km north–north-east of Tennant Creek, NT. The broad geographic range corresponds well with the spatial distribution of its larval food plants. Outside the study region, *E. corinna* occurs in Cocos (Keeling) and Christmas islands, the Lesser Sunda Islands, central, north-eastern and eastern Australia and Lord Howe and Norfolk islands.

## Habitat

*Euploea corinna* breeds in a wide variety of habitats, including paperbark swampland, savannah woodland, open sandstone pavement and the edges of mixed riparian woodland and evergreen monsoon vine forest along creeks and riverbanks where the larval food plants usually grow as vines or sometimes as tall shrubs, as well as in suburban parks and gardens.

## Larval food plants

*Gymnanthera oblonga, Ichnocarpus frutescens, Marsdenia geminata, M. viridiflora, Parsonsia alboflavescens, Sarcolobus hullsii, Sarcostemma viminale, Secamone elliptica* (Apocynaceae), *Ficus virens* (Moraceae); also *\*Adenium obesum* (Apocynaceae). The larvae feed on many food plants, but they are most frequently found on *G. oblonga*.

## Seasonality

Adults occur throughout the year but show little seasonal variation in abundance, although they tend to be more abundant during the late wet season and early dry season (April–June). During the dry season (April–August), they aggregate in moist refuges—typically pockets of monsoon forest in sheltered gorges or as gallery forest along water courses, where they may be observed in immense numbers clustered on tree foliage, the trunks and roots of fig trees and sheltered rock faces, especially in the more inland areas (Le Souëf 1971; Bailey and Richards 1975; Monteith 1982). During the dry season aggregations, most females do not breed; however, the immature stages may be found throughout the year, indicating that some females continue to breed depending on local conditions—similar to that reported in Queensland (Scheermeyer 1993).

## Breeding status

This species is resident in the study region.

## Conservation status

LC.

## Euploea corinna

Specimen ≥1970
Observation ≥1970
Literature ≥1970
Specimen <1970
Literature <1970
Larval food plants

0   100   200      400         600
                                    km

## Euploea corinna

Species record
Geographic range
Phytogeographical boundary
IBRA bioregional boundary

0   100   200      400         600
                                    km

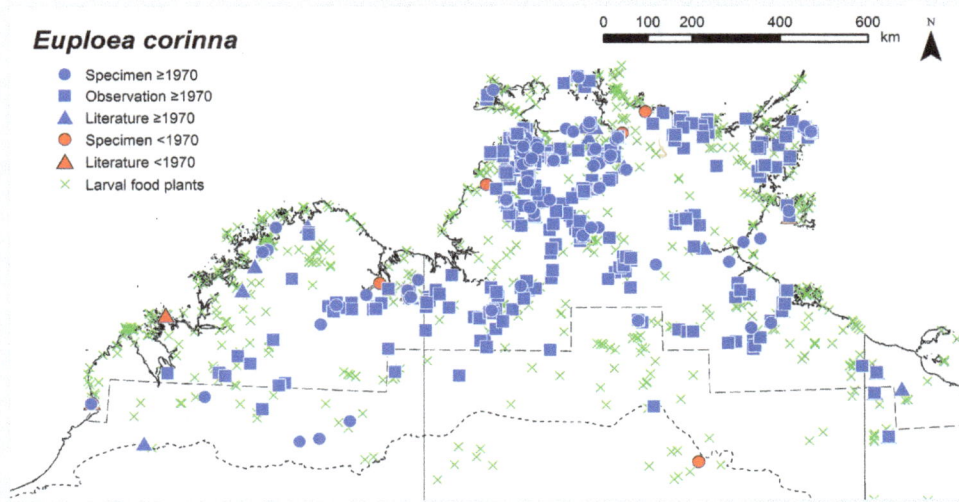

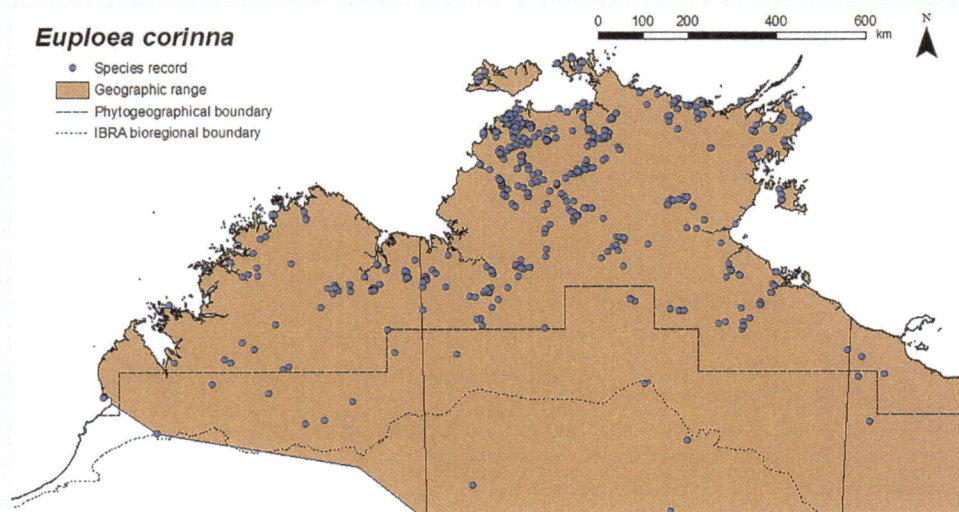

### Euploea corinna (n = 958)

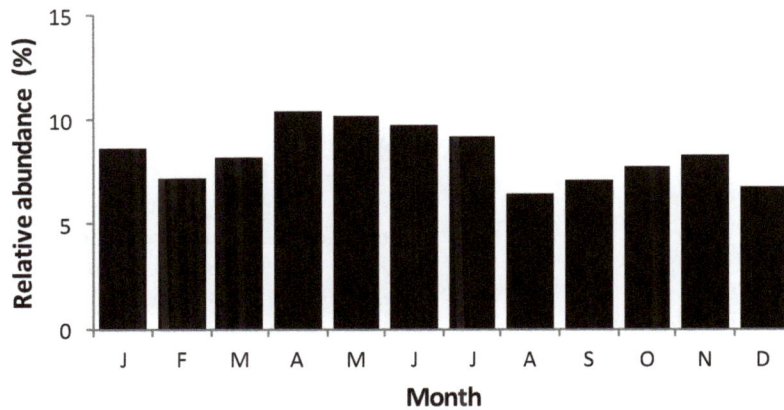

| Month | J | F | M | A | M | J | J | A | S | O | N | D |
|-------|---|---|---|---|---|---|---|---|---|---|---|---|
| Egg   |   |   |   |   |   |   |   |   |   |   |   |   |
| Larva |   |   |   |   |   |   |   |   |   |   |   |   |
| Pupa  |   |   |   |   |   |   |   |   |   |   |   |   |
| Adult |   |   |   |   |   |   |   |   |   |   |   |   |

# No-brand Crow, Gove Crow
## *Euploea alcathoe* (Godart, [1819])

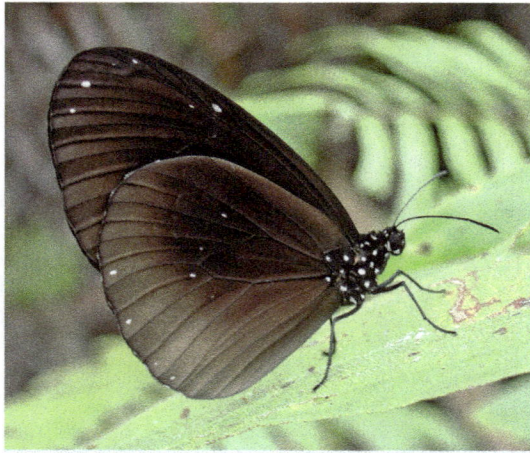

Plate 94 Gapuwiyak, Gove Peninsula, NT
Photo: M. F. Braby

## Distribution

This species is represented by the subspecies *E. alcathoe enastri* Fenner, 1991, which is endemic to the study region. It occurs in the Top End, its presence in the region detected only as recently as 1988 (Fenner 1991). It is restricted to Gove Peninsula, NT, where it occurs in lowland coastal or near-coastal areas. Its geographic range closely corresponds with the spatial distribution of its primary larval food plant (*Parsonsia alboflavescens*). This plant extends slightly further west to the Goyder River in Arnhem Land, but field surveys carried out at this location and nearby areas (Koolatong River) did not detect *E. alcathoe* (Braby 2010a). Outside the study region, *E. alcathoe* occurs from Maluku, the Kai and Aru islands, through mainland New Guinea to the D'Entrecasteaux Islands and the Torres Strait Islands, Qld (Lambkin et al. 2017).

## Habitat

*Euploea alcathoe* breeds in mixed paperbark tall open forest with rainforest elements in the understorey, usually in juxtaposition with evergreen monsoon vine forest, and in the edges of evergreen monsoon vine forest where the larval food plants grow as vines. Both breeding habitats are associated with perennial groundwater seepages or springs, usually along drainage lines or floodplains (Braby 2009, 2010a). The critical breeding habitats are subject to natural disturbance by both fire and flood, and occasionally cyclonic events.

## Larval food plants

*Gymnanthera oblonga*, *Marsdenia glandulifera*, *Parsonsia alboflavescens* (Apocynaceae). The main food plant is *P. alboflavescens* (Braby 2009), but occasionally other species are used, including *G. oblonga*, *M. glandulifera* (Braby 2009) and possibly *Tylophora benthamii* (Fenner 1991).

## Seasonality

Adults have been recorded during most months of the year, but their patterns of seasonal changes in abundance are not well understood. Adults have been recorded more frequently during the dry season (June–October), but this is due to a sampling bias because very little fieldwork has been conducted during the wet season, when the breeding habitats are largely inaccessible. Similarly, the immature stages have been recorded mainly during the dry season (June–October), as well as in the late wet season (March and April). Presumably, the species breeds continuously throughout the year and several generations are completed annually (Braby 2009).

## Breeding status

This species is resident in the study region.

## Conservation status

Near Threatened (NT). The subspecies *E. alcathoe enastri* is a short-range endemic (EOO = 9,100 sq km), with much of its range occurring within the Dhimurru and Laynhapuy IPAs. Braby (2010a) concluded there was no evidence of decline, but it was conservation-dependant in that management was required to mitigate threats from habitat modification through inappropriate fire regimes and disturbance by feral animals (buffalo and pig), while maintaining some level of natural disturbance. Moreover, the larval food plant is currently listed as NT under the *TPWCA*. Without management, the taxon may qualify for a threatened category in the near future because the population of *E. alcathoe enastri* is likely to be reduced based on a projected decline in the AOO and/or quality of its habitat (criterion A3c). Monitoring of the extent and/or quality of the critical habitat and occupancy of the butterfly is required for this species.

## Euploea alcathoe

- ● Specimen ≥1970
- ■ Observation ≥1970
- ▲ Literature ≥1970
- ✕ Larval food plants

## Euploea alcathoe

- ● Species record
- ▨ Geographic range
- --- Phytogeographical boundary
- ⋯ IBRA bioregional boundary

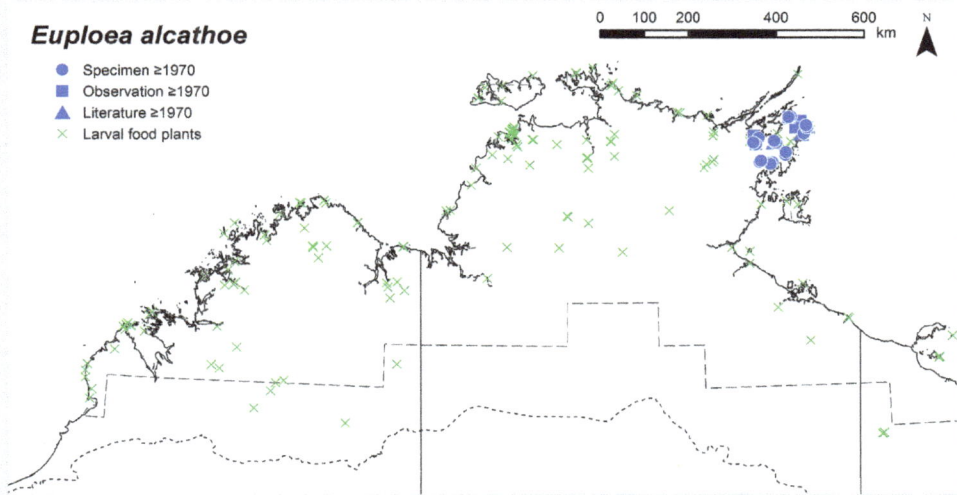

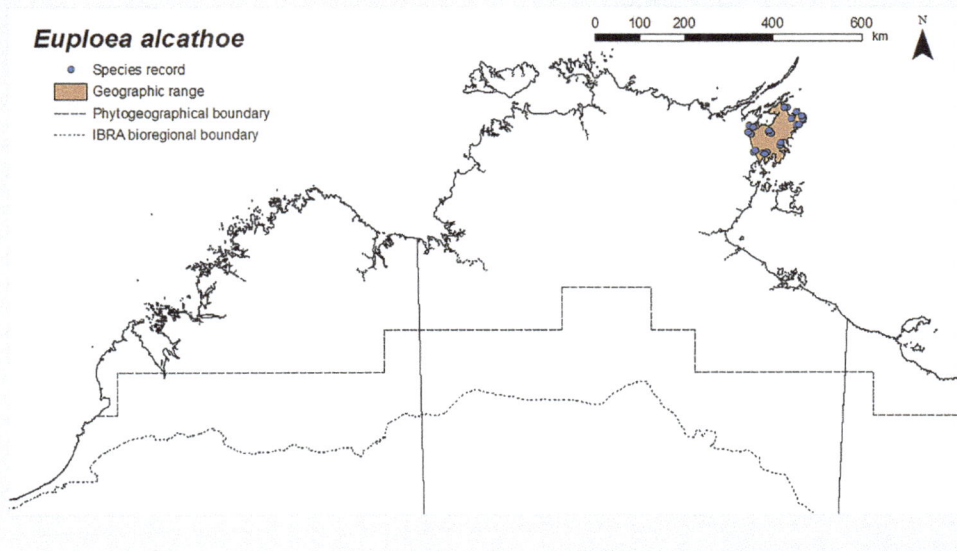

### Euploea alcathoe (n = 42)

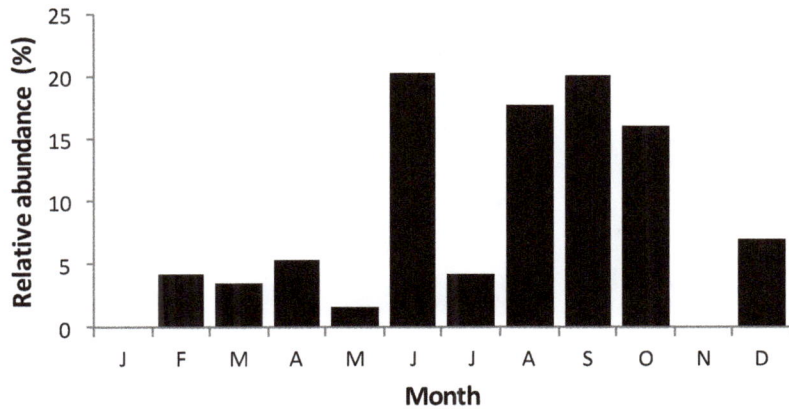

| Month | J | F | M | A | M | J | J | A | S | O | N | D |
|-------|---|---|---|---|---|---|---|---|---|---|---|---|
| Egg   |   |   | ▨ | ▨ |   | ▨ | ▨ | ▨ | ▨ | ▨ |   |   |
| Larva |   |   | ▨ | ▨ |   | ▨ | ▨ | ▨ | ▨ | ▨ |   |   |
| Pupa  |   |   |   | ▨ |   | ▨ | ▨ | ▨ | ▨ |   |   |   |
| Adult |   | ▨ | ▨ | ▨ | ▨ | ▨ | ▨ | ▨ | ▨ | ▨ |   | ▨ |

# Climena Crow
## *Euploea climena* (Stoll, [1782])

Plate 95 Christmas Island, WA
Photo: Frank Pierce

## Distribution

This species is represented in the study region by the subspecies *E. climena macleari* Butler, 1887. It is known only from two historical males from Derby, WA, in the western Kimberley (Waterhouse and Lyell 1914). Outside the study region, *E. climena* occurs in Christmas Island and Indonesia (from Sumatra to Ceram and the Kai Islands).

## Habitat

The breeding habitat of *E. climena* has not been recorded in the study region.

## Larval food plants

Not recorded in the study region. The larval food plant on Christmas Island is *Hoya aldrichii* (Apocynaceae), a vine that is absent from north-western Australia (Wilson and Johnson 2017).

## Seasonality

The seasonal abundance and breeding phenology of this species are not well understood. The two specimens were apparently collected in November.

## Breeding status

*Euploea climena* does not appear to have become permanently established in the study region. Presumably, it is a rare vagrant from Christmas Island.

## Conservation status

NA.

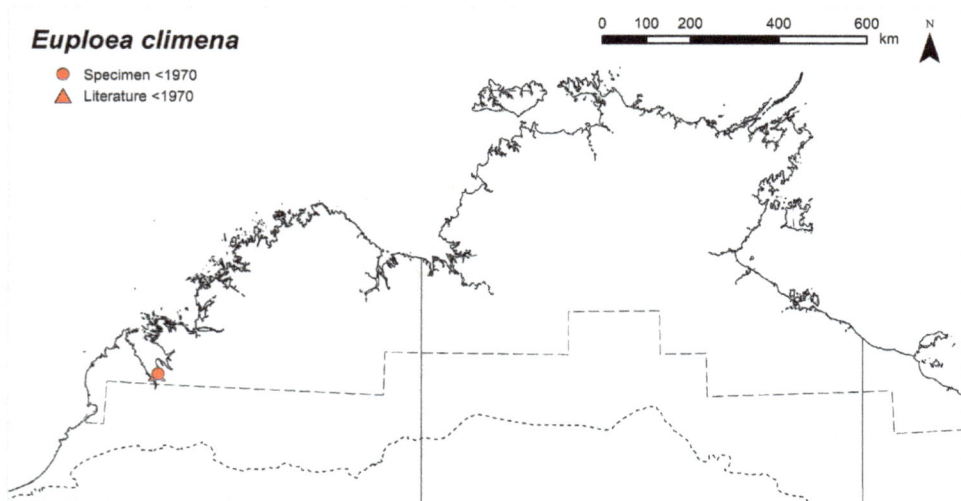

*Euploea climena*

● Specimen <1970
▲ Literature <1970

| Month | J | F | M | A | M | J | J | A | S | O | N | D |
|-------|---|---|---|---|---|---|---|---|---|---|---|---|
| Egg   |   |   |   |   |   |   |   |   |   |   |   |   |
| Larva |   |   |   |   |   |   |   |   |   |   |   |   |
| Pupa  |   |   |   |   |   |   |   |   |   |   |   |   |
| Adult |   |   |   |   |   |   |   |   |   |   | ▮ |   |

# Orange Lacewing
## *Cethosia penthesilea* (Cramer, 1777)

Plate 96 Fish River Station, NT
Photo: M. F. Braby

## Distribution

This species is represented in the study region by the subspecies *C. penthesilea paksha* Fruhstorfer, 1905. *Cethosia penthesilea* does not occur elsewhere in Australia. It is restricted to the Top End, where it occurs in the higher rainfall areas (> 800 mm mean annual rainfall, but mostly > 1,000 mm). Its southernmost limits are Katherine (Charles Darwin University) and Ngukurr (Normand 2009), NT, which are slightly drier areas. The geographic range closely follows the spatial distribution of its larval food plant in the Northern Territory. Although the food plant extends to the Kimberley, *C. penthesilea* is absent from this area. Outside the study region, *C. penthesilea* occurs in Java and the Lesser Sunda Islands, including Timor and its nearby islands.

## Habitat

*Cethosia penthesilea* breeds in semi-deciduous monsoon vine thicket and the edges of evergreen monsoon vine forest in both coastal and inland areas where the larval food plant grows as a large, woody perennial deciduous vine on a variety of rocky substrates (Hall 1981).

## Larval food plant

*Adenia heterophylla* (Passifloraceae).

## Seasonality

Adults occur throughout the year, but they are most abundant from the late wet season to the mid dry season (April–August), with a peak in abundance in July. However, in some years or locations they may be very numerous during the mid wet season (January and February). Their numbers decline rapidly as the dry season progresses and very few adults are present during the late dry season and early wet season. Hall (1981) also noted that adults occur throughout the year, but are usually more common during the first half of the dry season (April–July). The immature stages have been recorded sporadically from the early wet season to the mid dry season (November–June). The larvae feed and develop rapidly on the new soft leaf growth and several generations are completed annually, but it is not clear how *C. penthesilea* survives the late dry season (August–October), when the foliage of the food plant is frequently absent. Presumably, the species remains in pupal diapause during the late dry season.

## Breeding status

This species is resident in the study region.

## Conservation status

LC.

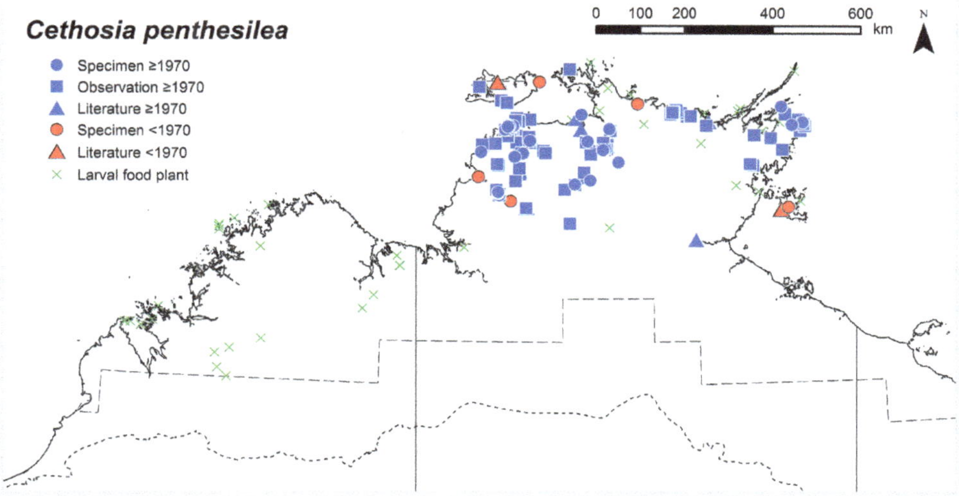

Cethosia penthesilea
0 100 200 400 600 km
N
Specimen ≥1970
Observation ≥1970
Literature ≥1970
Specimen <1970
Literature <1970
Larval food plant

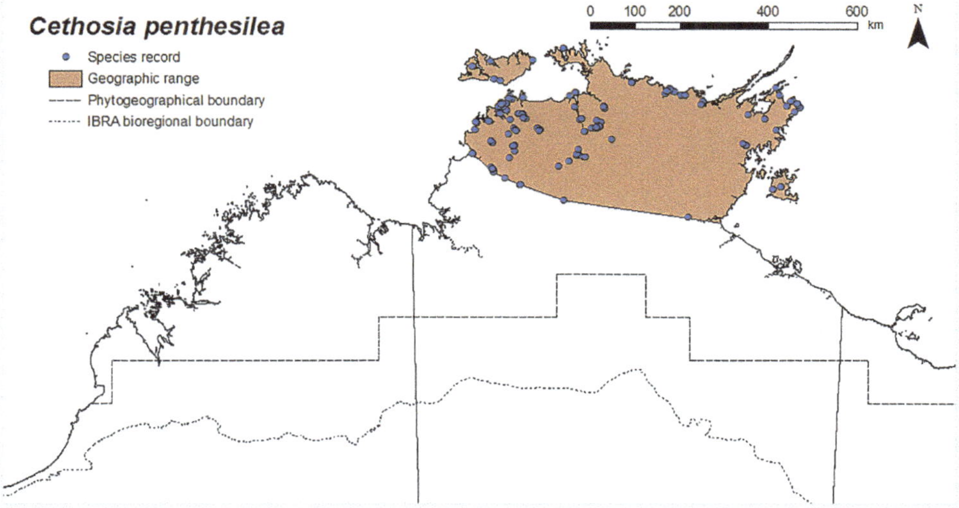

Cethosia penthesilea
0 100 200 400 600 km
N
Species record
Geographic range
Phytogeographical boundary
IBRA bioregional boundary

*Cethosia penthesilea* (n = 223)

20
15
10
5
0
Relative abundance (%)
J F M A M J J A S O N D
Month

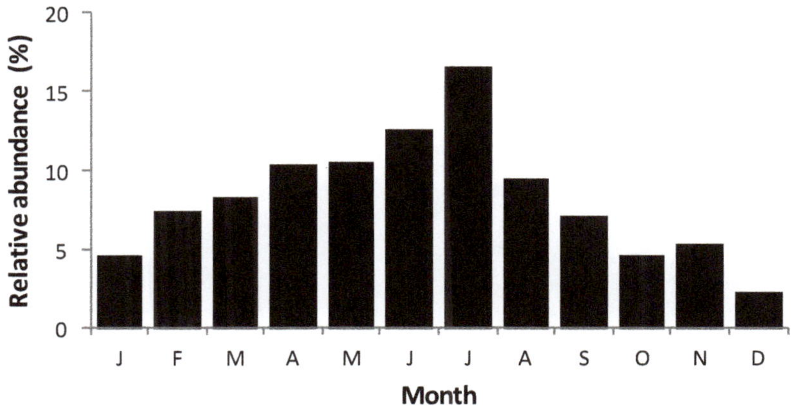

| Month | J | F | M | A | M | J | J | A | S | O | N | D |
|---|---|---|---|---|---|---|---|---|---|---|---|---|
| Egg | | ■ | | ■ | | ■ | | | | | ■ | |
| Larva | | ■ | ■ | | ■ | ■ | | | | | | |
| Pupa | | ■ | ■ | | | | | | | | | |
| Adult | ■ | ■ | ■ | ■ | ■ | ■ | ■ | ■ | ■ | ■ | ■ | ■ |

# Glasswing
## *Acraea andromacha* (Fabricius, 1775)

Plate 97 Wongalara Wildlife Sanctuary, NT
Photo: M. F. Braby

## Distribution

This species is represented in the study region by the subspecies *A. andromacha andromacha* (Fabricius, 1775). It occurs very widely throughout the region, extending from moist coastal areas to drier inland areas of the semi-arid zone (< 500 mm mean annual rainfall), where it has been recorded as far south as the Edgar Ranges (Common 1981) and Halls Creek (Koch 1957), WA; Three Ways Roadhouse, NT (Dunn and Dunn 1991); and Boodjamulla/Lawn Hill National Park, Qld (Daniels and Edwards 1998; Franklin 2007; Dunn 2015a). It also extends into the arid zone of central Australia beyond the southern boundary of the study region. The geographic range corresponds well with the spatial distribution of its larval food plants. Outside the study region, *A. andromacha* occurs from the Lesser Sunda Islands, through mainland New Guinea and the northern half of Australia to the Solomon Islands, New Caledonia, Fiji and Samoa.

## Habitat

*Acraea andromacha* breeds in a variety of habitats, including savannah woodland and disturbed open woodland where the larval food plants (*Hybanthus* spp.) grow as seasonal perennial herbs, and coastal semi-deciduous monsoon vine thicket and riparian evergreen monsoon vine forest along perennial streams within gorges where the alternative food plant (*Adenia heterophylla*) grows as a large climbing vine. Males also congregate on hilltops to locate females, but they do not breed in these habitats.

## Larval food plants

*Adenia heterophylla* (Passifloraceae), *Hybanthus aurantiacus*, *H. enneaspermus* (Violaceae); also *\*Passiflora foetida* (Passifloraceae).

## Seasonality

Adults occur throughout the year, but they are most abundant during the mid dry season (May–July). Their numbers diminish as the dry season progresses, and they are very scarce at the start of the wet season. The immature stages (eggs or larvae) have been recorded sporadically from the early wet season to the mid dry season (November–July), and are not known to undergo diapause. Several generations are completed annually, but it is not clear how the species survives the late dry season (August–October), when the foliage of the larval food plants is frequently absent or reduced in quality. Presumably, populations contract to moister areas, such as along the margins of riverbanks and perennial creeks, where breeding may continue.

## Breeding status

This species is resident in the study region.

## Conservation status

LC.

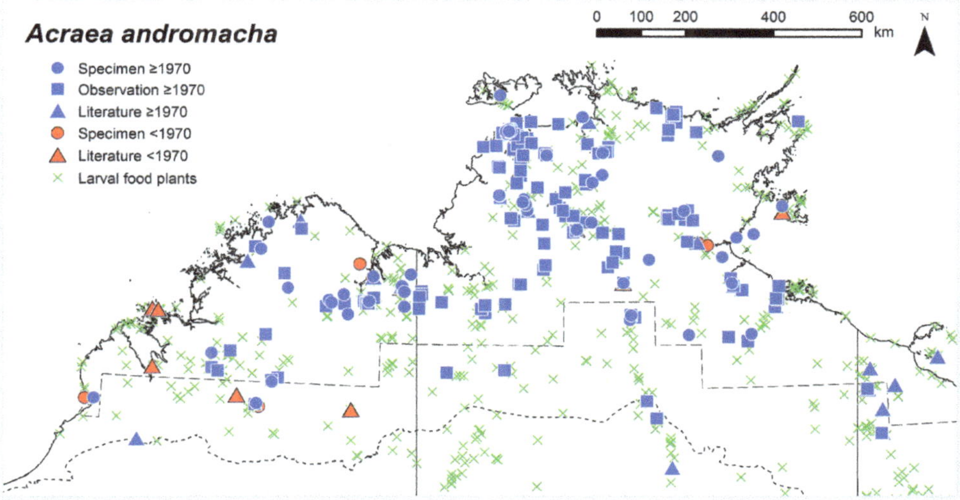

## Acraea andromacha

- ● Specimen ≥1970
- ■ Observation ≥1970
- ▲ Literature ≥1970
- ● Specimen <1970
- ▲ Literature <1970
- × Larval food plants

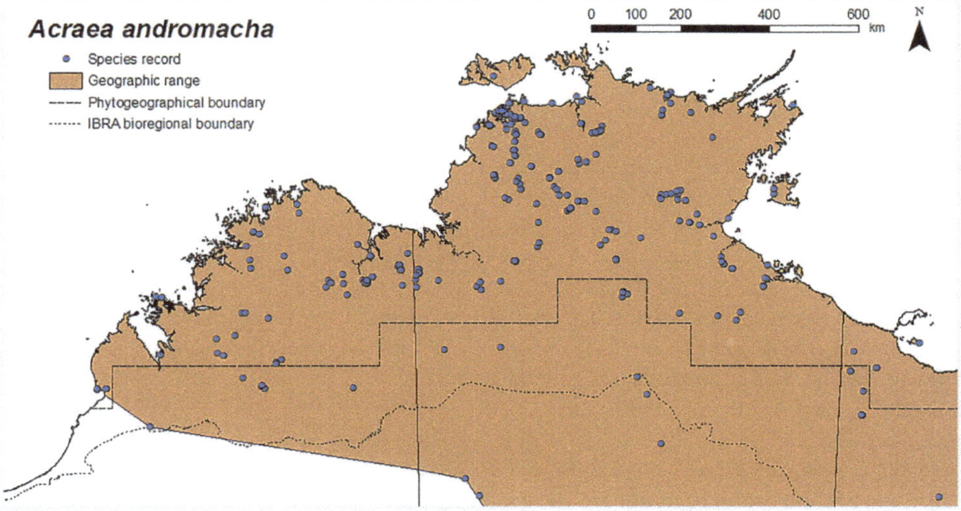

## Acraea andromacha

- ● Species record
- ▨ Geographic range
- — Phytogeographical boundary
- ⋯ IBRA bioregional boundary

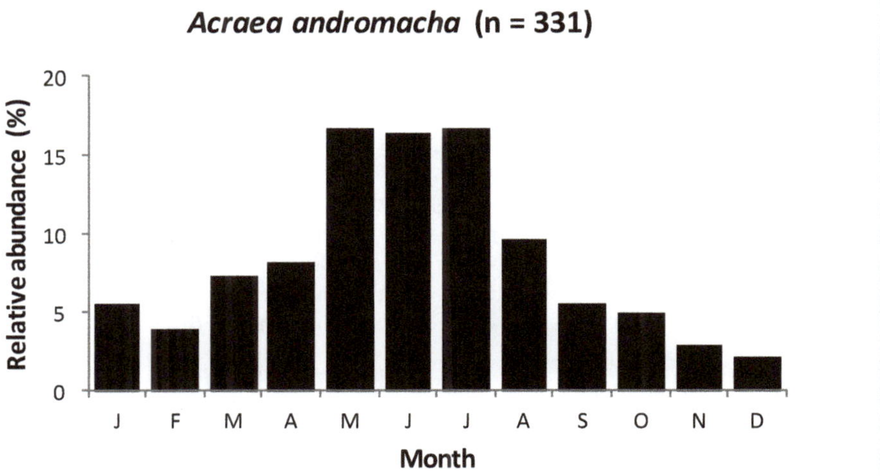

### Acraea andromacha (n = 331)

| Month | J | F | M | A | M | J | J | A | S | O | N | D |
|-------|---|---|---|---|---|---|---|---|---|---|---|---|
| Egg | | | | | | | ▨ | | | | ▨ | |
| Larva | ▨ | ▨ | ▨ | ▨ | ▨ | | ▨ | | | | | |
| Pupa | | | | | ▨ | | | | | | | |
| Adult | ▨ | ▨ | ▨ | ▨ | ▨ | ▨ | ▨ | ▨ | ▨ | ▨ | ▨ | ▨ |

# Tawny Coster
## *Acraea terpsicore* (Linnaeus, 1758)

Plate 98 Wagait Beach, Cox Peninsula, NT
Photo: M. F. Braby

Plate 99 Wagait Beach, Cox Peninsula, NT
Photo: M. F. Braby

## Distribution

This species occurs widely in the Kimberley and Top End of the study region and extends to the Northern Deserts. The geographic range of *A. terpsicore* has been expanding rapidly since it was first detected near Darwin, NT, in April 2012 (Braby et al. 2014a, 2014b). It has been recorded as far south as Broome, WA (G. Swann); Elliott (R. P. Weir) and Camfield Station at the northern edge of the Tanami Desert (J. Archibald), NT; and has recently been detected in the Gulf Country of western Cape York Peninsula (Kowanyama district), Qld (Wilson 2016), just outside the eastern boundary of the study region. Spatial modelling predicts that its range will eventually occupy the entire monsoon tropics of northern Australia, from moist coastal areas to drier inland areas of the semi-arid zone (c. 500 mm mean annual rainfall), which also coincides with the spatial distribution of its primary larval food plant (Braby et al. 2014a). Outside the study region, *A. terpsicore* occurs naturally in India and Sri Lanka, but it has now colonised Indochina, the Malay Peninsula, Borneo and the Greater and Lesser Sunda islands.

## Habitat

*Acraea terpsicore* breeds mainly in savannah woodland and grassland, favouring modified or disturbed open areas, such as suburban roadsides, where the native larval food plant grows as a seasonal perennial herb (Braby et al. 2014b). Males also congregate on hilltops to locate females, but they do not breed in these habitats.

## Larval food plants

*Hybanthus enneaspermus* (Violaceae); also *\*Passiflora foetida* (Passifloraceae). The main food plant is *H. enneaspermus*, but occasionally the larvae feed on introduced *P. foetida* (Braby et al. 2014b). In captivity, the larvae readily feed and develop successfully on *Adenia heterophylla*, but to date they are not known to use this plant in the field.

## Seasonality

Adults occur during most months of year, but they are most abundant during the wet season and early dry season (January–May). They become scarce as the dry season progresses (June–November) and are apparently absent in October. The immature stages have been recorded sporadically from the early wet season to the mid dry season (December–August), and are not known to undergo diapause. Several generations are completed annually, but it is not clear how the species survives the late dry season (September–November) when the foliage of both the native and the introduced larval food plants is frequently absent or reduced in quality. Presumably, populations contract to moister areas, such as along the margins of swamps and riverbanks, where breeding may continue.

## Breeding status

Although *A. terpsicore* is resident in the study region, it has recently colonised the area (within the past six years) and its range is currently expanding outside the region in northern Queensland (Wilson 2016; Field 2017; Franklin et al. 2017).

## Conservation status

NA.

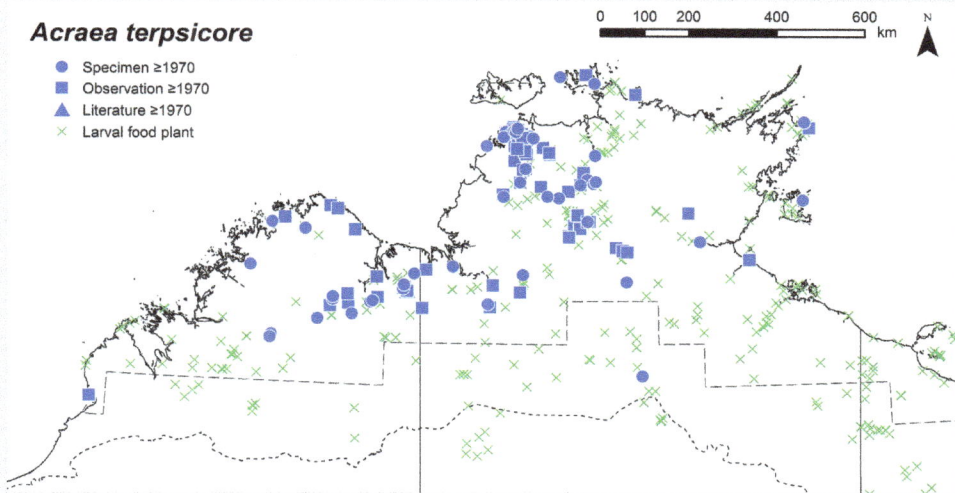

*Acraea terpsicore*

- ● Specimen ≥1970
- ■ Observation ≥1970
- ▲ Literature ≥1970
- × Larval food plant

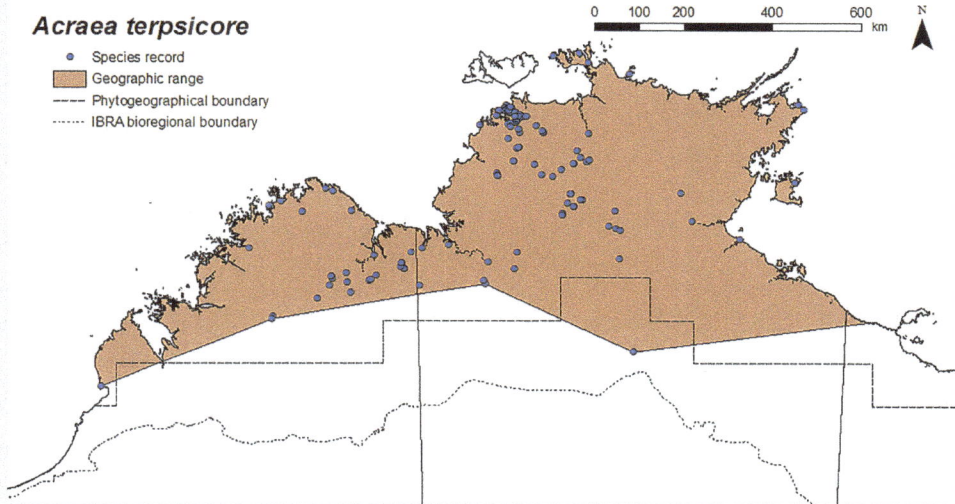

*Acraea terpsicore*

- ● Species record
- ▨ Geographic range
- — — Phytogeographical boundary
- ·········· IBRA bioregional boundary

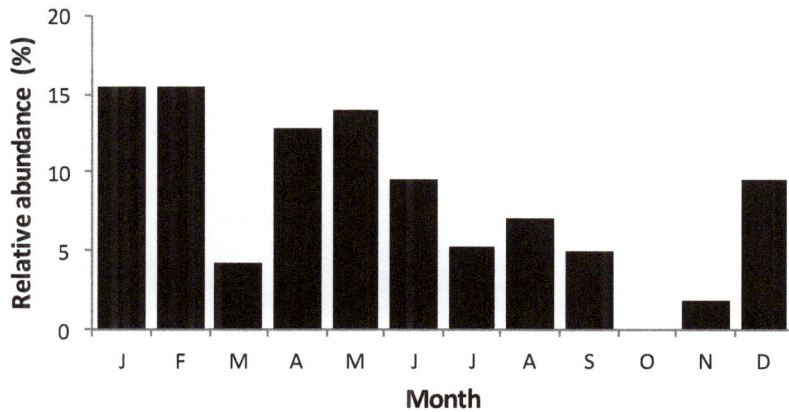

*Acraea terpsicore* (n = 165)

| Month | J | F | M | A | M | J | J | A | S | O | N | D |
|-------|---|---|---|---|---|---|---|---|---|---|---|---|
| Egg   |   |   |   |   |   |   |   |   |   |   |   |   |
| Larva |   |   |   |   |   |   |   |   |   |   |   |   |
| Pupa  |   |   |   |   |   |   |   |   |   |   |   |   |
| Adult |   |   |   |   |   |   |   |   |   |   |   |   |

# Spotted Rustic
## *Phalanta phalantha* (Drury, [1773])

Plate 100 Manton Dam, NT
Photo: Deb Bisa

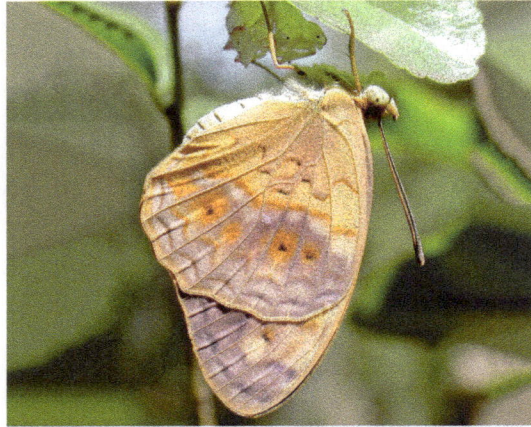

Plate 101 Fish River Station, NT
Photo: M. F. Braby

## Distribution

This species is represented by the subspecies *P. phalantha araca* (Waterhouse & Lyell, 1914), which is endemic to the study region. *Phalanta phalantha* does not occur elsewhere in Australia. It occurs in the north of the Northern Territory, where it is restricted to the higher rainfall areas (> 900 mm mean annual rainfall) of the north-western corner of the Top End. It reaches its southern limit at Flora River Nature Park and its easternmost limit at Kakadu National Park (Mt Brockman) (Kikkawa and Monteith 1980), NT. The geographical range broadly corresponds with the spatial distribution of its native larval food plant. The food plant, however, extends further east along the northern coast to Gove Peninsula; thus, further field surveys are required to determine whether *P. phalantha* occurs in north-eastern Arnhem Land. The isolated record from further south, at Daly Waters, NT, based on two specimens collected in January 1973 (A. Allwood and T. Weir), either constitutes vagrant specimens outside the normal breeding range or indicates that a temporary population became established on one of the introduced larval food plants cultivated in suburban gardens. Outside the study region, *P. phalantha* occurs widely in Africa, Madagascar, India, Japan and South-East Asia.

## Habitat

*Phalanta phalantha* breeds mainly in semi-deciduous monsoon vine thicket where the native larval food plant grows as a semi-deciduous shrub. It also breeds in riparian woodland with rainforest elements in the understorey and the edges of evergreen monsoon vine forest along the banks of perennial streams, as well as in suburban parks and gardens.

## Larval food plants

*Flacourtia territorialis* (Salicaceae); also *\*Flacourtia inermis*, *\*F. rukam* (Salicaceae).

## Seasonality

Adults occur throughout the year, but they are most abundant during the wet season and early dry season (November–July), with an apparent peak in abundance in April. Hutchinson (1978) recalls observing large numbers of adults—in excess of 100—during May on the Daly River, NT. They are very scarce and often absent during the late dry season (August–October) and in some years they may be absent during the 'build-up' (November and December). The immature stages have been recorded sporadically from the early wet season to the early dry season (November–June). The larvae feed and develop rapidly on the new soft leaf growth and several generations are completed annually, but it is not clear how the species survives the late dry season (August–October) when the foliage of the native larval food plant is frequently absent or reduced in quality. Presumably, the species remains in pupal diapause during the late dry season.

## Breeding status

This species is resident in the study region.

## Conservation status

LC. Although the subspecies *P. phalantha araca* has a restricted range, there are no known threats facing the taxon.

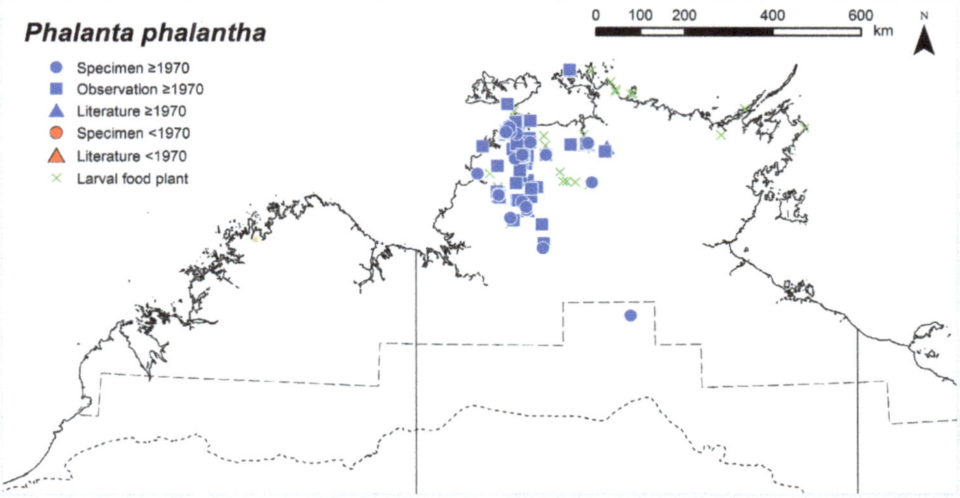

## Phalanta phalantha

- ● Specimen ≥1970
- ■ Observation ≥1970
- ▲ Literature ≥1970
- ● Specimen <1970
- ▲ Literature <1970
- ✕ Larval food plant

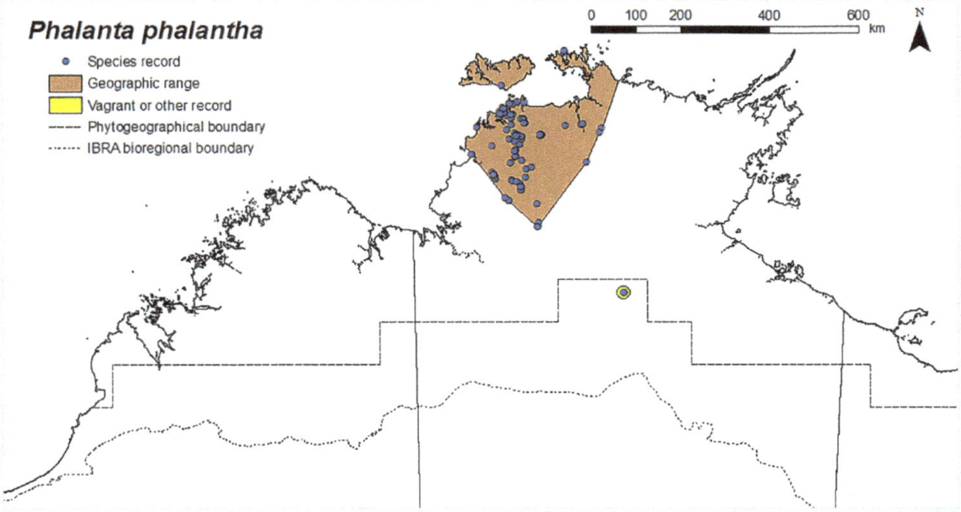

0  100  200     400     600 km  N

## Phalanta phalantha

- • Species record
- ▨ Geographic range
- ▨ Vagrant or other record
- — — — Phytogeographical boundary
- ········· IBRA bioregional boundary

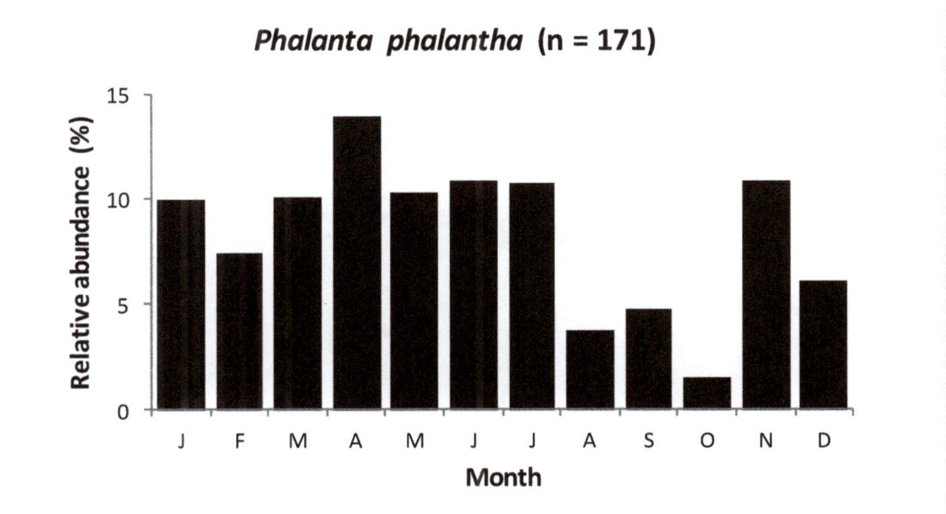

0  100  200     400     600 km  N

### Phalanta phalantha (n = 171)

| Month | J | F | M | A | M | J | J | A | S | O | N | D |
|-------|---|---|---|---|---|---|---|---|---|---|---|---|
| Egg   |   |   |   |   |   |   |   |   |   |   |   |   |
| Larva |   |   |   |   |   |   |   |   |   |   |   |   |
| Pupa  |   |   |   |   |   |   |   |   |   |   |   |   |
| Adult |   |   |   |   |   |   |   |   |   |   |   |   |

# Australian Painted Lady
## *Vanessa kershawi* (McCoy, 1868)

Plate 102 Ocean Grove, Vic
Photo: M. F. Braby

## Distribution

This species is known only from the western Gulf Country of the study region, where it has been recorded at Doomadgee (Puccetti 1991) and Boodjamulla/Lawn Hill National Park (Daniels and Edwards 1998), Qld. Outside the study region, *V. kershawi* occurs in the Cocos (Keeling) Islands, throughout southern Australia, Lord Howe and Norfolk islands and New Zealand.

## Habitat

The breeding habitat of *V. kershawi* has not been recorded in the study region.

## Larval food plants

Not recorded in the study region.

## Seasonality

The seasonal abundance and breeding phenology of this species are not well understood. Adults have been recorded in November and May (Puccetti 1991; Daniels and Edwards 1998). The species is a well-known migrant (Smithers 1985).

## Breeding status

*Vanessa kershawi* does not appear to have become permanently established in the study region. Presumably, it is a rare vagrant or migrant from south-eastern Australia.

## Conservation status

NA.

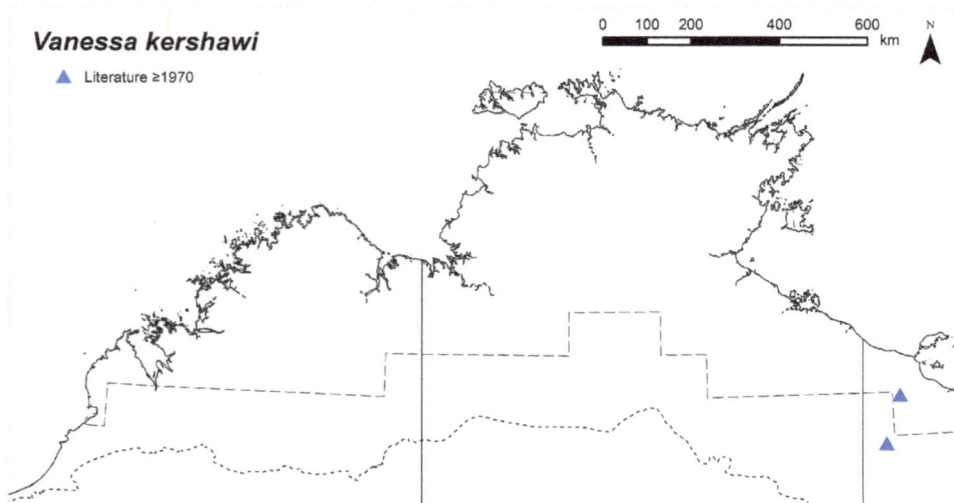

*Vanessa kershawi*

▲ Literature ≥1970

| Month | J | F | M | A | M | J | J | A | S | O | N | D |
|-------|---|---|---|---|---|---|---|---|---|---|---|---|
| Egg   |   |   |   |   |   |   |   |   |   |   |   |   |
| Larva |   |   |   |   |   |   |   |   |   |   |   |   |
| Pupa  |   |   |   |   |   |   |   |   |   |   |   |   |
| Adult |   |   |   |   | ▓ |   |   |   |   |   | ▓ |   |

# Yellow Admiral

## *Vanessa itea* (Fabricius, 1775)

Plate 103 Brindabella Range, ACT
Photo: M. F. Braby

## Distribution

This species is known only from the western Kimberley of the study region, where a single specimen was collected in the Buccaneer Archipelago at Koolan Island in Yampi Sound, WA, in December 1964 (McKenzie et al. 1995). Outside the study region, *V. itea* occurs mainly in southern Australia, Lord Howe and Norfolk islands and New Zealand.

## Habitat

The breeding habitat of *V. itea* has not been recorded in the study region.

## Larval food plants

Not recorded in the study region.

## Seasonality

The seasonal abundance and breeding phenology of this species are not well understood. Only a single adult has been recorded, in December (McKenzie et al. 1995). The species is known to migrate in south-eastern Australia (Smithers 1985), and in south-western Western Australia there is much evidence to suggest it is highly mobile and not permanently established (R. J. Powell, pers. comm.). For instance, on the islands off south-western Australia, *V. itea* is an immigrant in that it breeds temporarily during late autumn and winter when the native annual herbaceous larval food plants (*Parietaria* spp.) germinate and grow (Powell 1993; Williams and Powell 1998, 2006).

## Breeding status

*Vanessa itea* does not appear to have become permanently established in the study region. Presumably, it is a rare vagrant or migrant from south-western Australia or the arid zone of central Australia.

## Conservation status

NA.

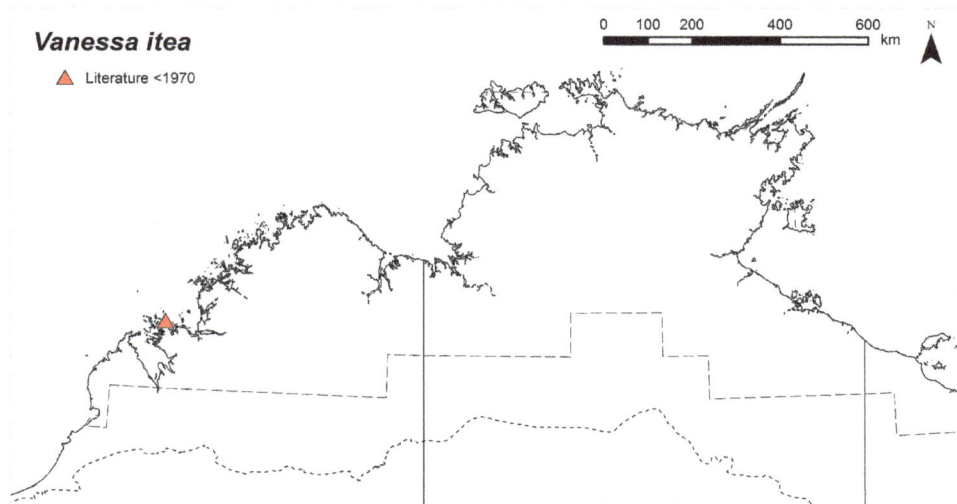

*Vanessa itea*

▲ Literature <1970

| Month | J | F | M | A | M | J | J | A | S | O | N | D |
|-------|---|---|---|---|---|---|---|---|---|---|---|---|
| Egg   |   |   |   |   |   |   |   |   |   |   |   |   |
| Larva |   |   |   |   |   |   |   |   |   |   |   |   |
| Pupa  |   |   |   |   |   |   |   |   |   |   |   |   |
| Adult |   |   |   |   |   |   |   |   |   |   |   | ▩ |

# Blue Argus
## *Junonia orithya* (Linnaeus, 1758)

Plate 104 Kingfisher Camp, north of Lawn Hill, Qld
Photo: Don Franklin

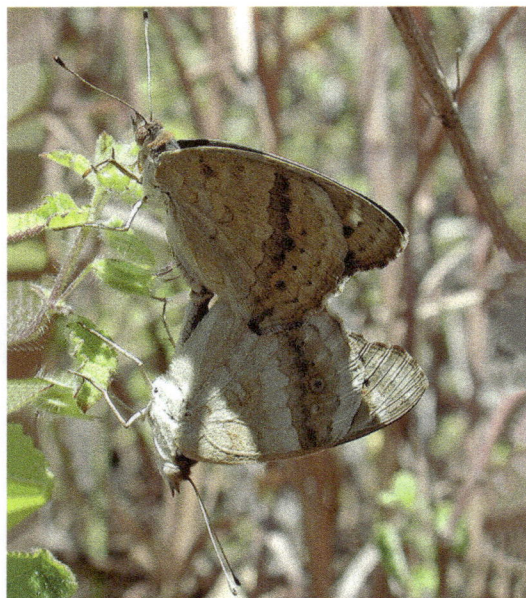

Plate 105 Mt Burrell, NT
Photo: M. F. Braby

## Distribution

This species is represented in the study region by the subspecies *J. orithya albicincta* Butler, 1875. It occurs widely throughout much of the region, from moist coastal areas to drier inland areas of the semi-arid zone (< 500 mm mean annual rainfall), where it has been recorded as far south as 24 km south-west of Halls Creek, WA; Powell Creek Outstation near Renner Springs, NT; and Thorntonia, north-east of Camooweal, Qld (Franklin 2007). It has also been recorded from McLaren Creek, NT, just outside the southern boundary of the study region. The geographic range falls within the spatial distribution of the larval food plants, although the food plants extend much further inland. Outside the study region, *J. orithya* occurs widely from Africa, India, southern China and South-East Asia, through mainland New Guinea and adjacent islands to north-eastern and eastern Australia.

## Habitat

*Junonia orithya* breeds mainly in savannah woodland and open woodland where the larval food plants (*Buchnera* spp.) grow as annual herbs (Braby 2011a, 2015e), but it may also breed along the edges of monsoon forest where alternative food plants (*Thunbergia arnhemica*, *Pseuderanthemum variabile*) grow.

## Larval food plants

*Pseuderanthemum variabile*, *Thunbergia arnhemica* (Acanthaceae), *Buchnera asperata*, *B. gracilis*, *B. linearis* (Orobanchaceae); also *\*Asystasia gangetica* (Acanthaceae).

## Seasonality

Adults occur throughout the year, but, like *Junonia villida*, they are most abundant during the late wet season and early dry season (April and May). Their numbers steadily diminish as the dry season progresses. The immature stages have been recorded sporadically during the wet season and early dry season (November–May). It is not clear whether the species continues to breed during the dry season, but a female has been observed in late August ovipositing on the food plant growing in moist sand near permanent water at Kakadu National Park (Maguk Plunge Pool). Presumably, most females stop breeding and remain in reproductive diapause during the late dry season, similar to populations in northern Queensland (Jones 1987).

## Breeding status

This species is resident in the study region.

## Conservation status

LC.

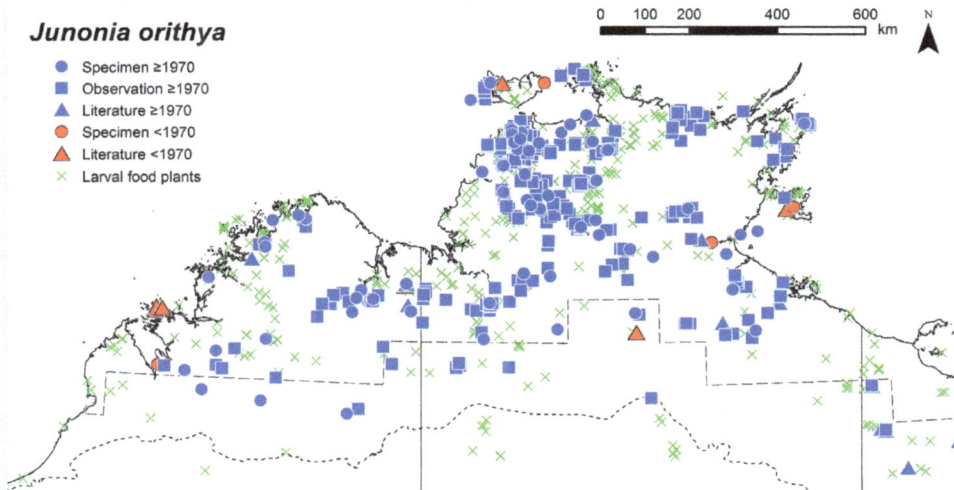

*Junonia orithya*

Specimen ≥1970
Observation ≥1970
Literature ≥1970
Specimen <1970
Literature <1970
Larval food plants

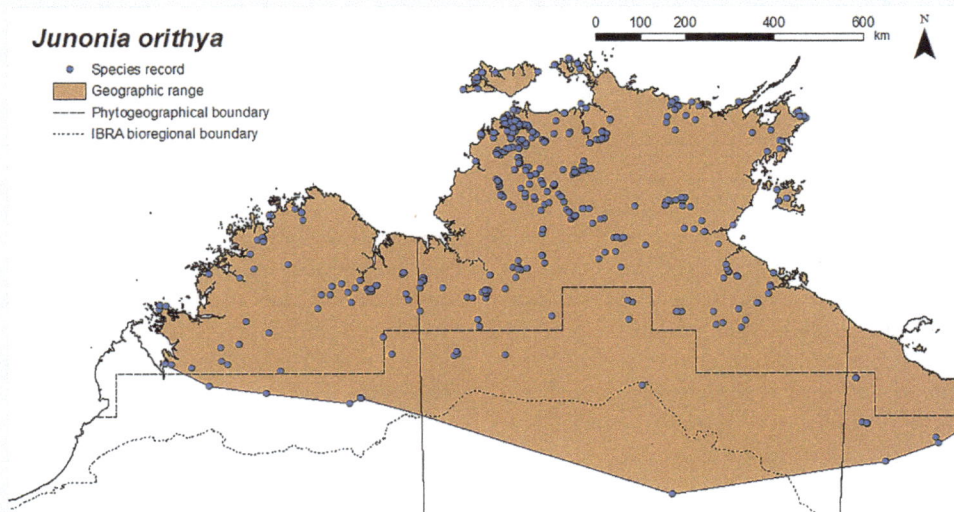

*Junonia orithya*

Species record
Geographic range
Phytogeographical boundary
IBRA bioregional boundary

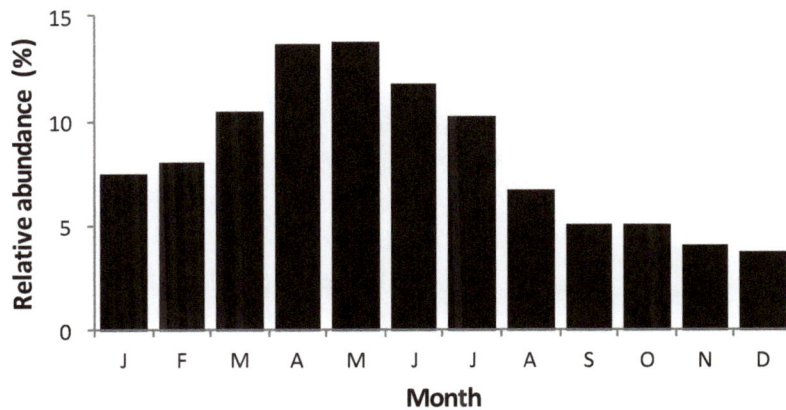

*Junonia orithya* (n = 625)

| Month | J | F | M | A | M | J | J | A | S | O | N | D |
|-------|---|---|---|---|---|---|---|---|---|---|---|---|
| Egg | | | | | | | | ▨ | | | ▨ | |
| Larva | | ▨ | ▨ | ▨ | ▨ | | | | | | ▨ | |
| Pupa | | | | | ▨ | | | | | | | ▨ |
| Adult | ▨ | ▨ | ▨ | ▨ | ▨ | ▨ | ▨ | ▨ | ▨ | ▨ | ▨ | ▨ |

# Meadow Argus
## *Junonia villida* (Fabricius, 1787)

Plate 106 Kakadu National Park, NT
Photo: Don Franklin

Plate 107 Berry Springs, NT
Photo: M. F. Braby

## Distribution

This species is represented in the study region by the subspecies *J. villida villida* (Fabricius, 1787). It occurs throughout the entire study region, extending well into the semi-arid zone, as well as into arid areas of central Australia south of the study region. Outside the study region, *J. villida* occurs in Christmas and Cocos (Keeling) islands, mainland New Guinea and adjacent islands, throughout Australia, Lord Howe and Norfolk islands, New Zealand and the islands of the South Pacific.

## Habitat

*Junonia villida* occurs in a variety of habitats, but breeds mainly in savannah woodland and open woodland.

## Larval food plants

Not recorded in the study region; probably *Hygrophila angustifolia* (Acanthaceae) and *Evolvulus alsinoides* (Convolvulaceae), as well as a wide variety of introduced herbaceous species, which are the food plants in eastern Australia (Braby 2000).

## Seasonality

Adults occur throughout the year, but, like *Junonia orithya*, they are most abundant during the late wet season and early dry season (March–May). Their numbers steadily diminish as the dry season progresses, and they are scarce or even absent in some locations during the late dry season (August and September). Franklin (2011) observed similar trends near Darwin, NT, based on quantitative studies conducted over 14 months during 2008–09, although adults peaked slightly earlier in the wet season. Surprisingly, we have few data on the breeding phenology and seasonal history of the immature stages of this common and widespread species. Larvae have been recorded in January and May, but the identity of the plants was not determined due to the tendency of final instar larvae to wander from their food plants. Presumably, the species breeds during the wet season and early dry season and most females stop breeding and remain in reproductive diapause during the late dry season, similar to populations in northern Queensland (Jones 1987).

## Breeding status

This species is resident in the study region.

## Conservation status

LC.

## Junonia villida

- Specimen ≥1970
- Observation ≥1970
- Literature ≥1970
- Specimen <1970
- Literature <1970
- Putative larval food plants

0  100  200      400      600 km

## Junonia villida

- Species record
- Geographic range
- Phytogeographical boundary
- IBRA bioregional boundary

0  100  200      400      600 km

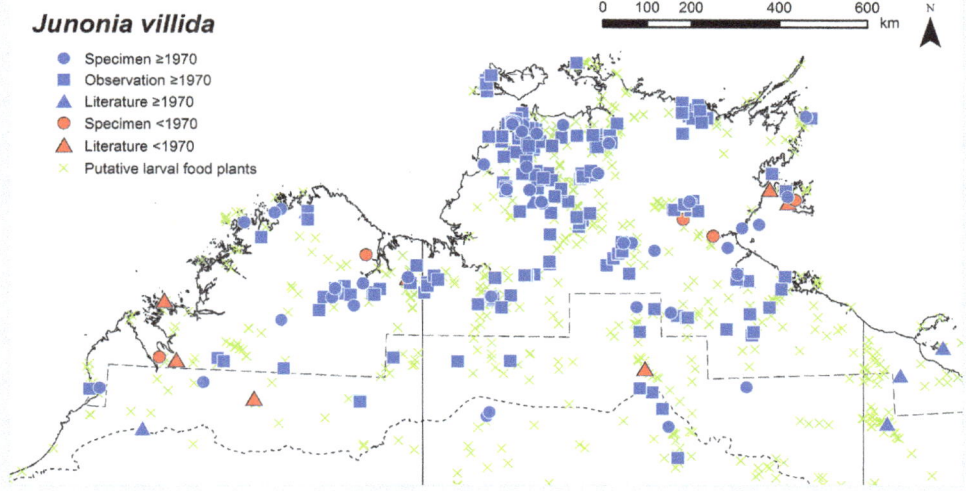

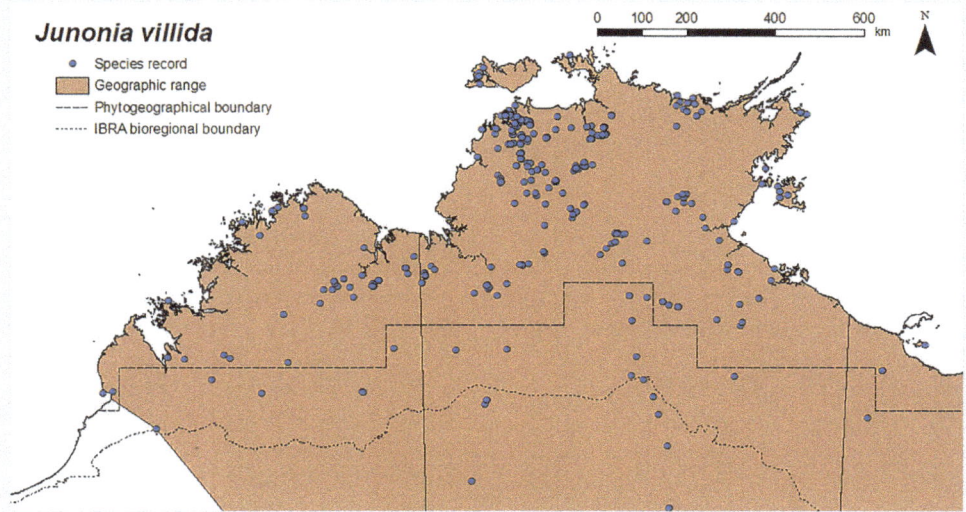

### Junonia villida (n = 500)

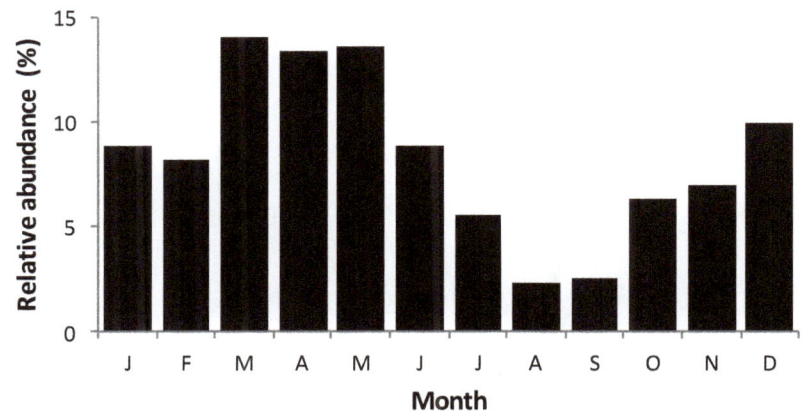

| Month | J | F | M | A | M | J | J | A | S | O | N | D |
|-------|---|---|---|---|---|---|---|---|---|---|---|---|
| Egg   |   |   |   |   |   |   |   |   |   |   |   |   |
| Larva |   |   |   |   |   |   |   |   |   |   |   |   |
| Pupa  |   |   |   |   |   |   |   |   |   |   |   |   |
| Adult |   |   |   |   |   |   |   |   |   |   |   |   |

# Chocolate Argus
*Junonia hedonia* (Linnaeus, 1764)

Plate 108 Wanguri, Darwin, NT
Photo: M. F. Braby

Plate 109 Cardwell, Queensland
Photo: M. F. Braby

## Distribution

This species is represented in the study region by the subspecies *J. hedonia zelima* (Fabricius, 1775). It occurs in the Kimberley, Top End and western Gulf Country, extending from moist coastal areas to drier inland areas of the semi-arid zone (c. 700 mm mean annual rainfall), where it has been recorded as far south as Adcock Gorge, WA (Franklin et al. 2005); and Wollogorang Station (Settlement Creek), NT (Franklin et al. 2005). Its geographic range corresponds well with the spatial distribution of its larval food plant, although the food plant extends further inland into lower rainfall areas. Its occurrence is more sporadic in the Kimberley and the food plant appears to be underreported in the northern Kimberley. Outside the study region, *J. hedonia* occurs from the Philippines and Indonesia, through mainland New Guinea and adjacent islands and north-eastern and eastern Australia to the Bismarck Archipelago and the Solomon Islands.

## Habitat

*Junonia hedonia* breeds mainly in paperbark swampland in floodplains and mixed paperbark–pandanus swamps adjacent to evergreen monsoon vine forest where the larval food plant grows as an annual herb, sometimes in shallow water (Braby 2011a). It also breeds in riparian paperbark woodland.

## Larval food plant

*Hygrophila angustifolia* (Acanthaceae).

## Seasonality

Adults occur throughout the year, but they are most abundant during the late wet season and first half of the dry season (April–July), with a peak in abundance in June. Their numbers steadily diminish as the dry season progresses, and they are scarce during the wet season. We have very few data on the breeding phenology and seasonal history of the immature stages. The immature stages (eggs and larvae) have been recorded towards the end of the wet season and early dry season (March and May), but undoubtedly they occur at other times of the year. The larval food plant germinates during the wet season and is available only during the late wet season and early dry season and then dies off in the mid to late dry season. Presumably, breeding is seasonal and most females stop breeding and remain in reproductive diapause during the late dry season and possibly early wet season. Although not previously known to migrate, migrations in the Top End have been observed on several occasions at the end of the wet season (Braby 2016b). During migration, adults fly rapidly 2–3 m above the ground from late morning to early afternoon in a northerly direction between mid April and early May. In 2012, a series of observations made over 13 days that involved large numbers of adults indicated that migration lasted for approximately two weeks.

## Breeding status

This species is resident in the study region.

## Conservation status

LC.

## Junonia hedonia

Legend:
- ● Specimen ≥1970
- ■ Observation ≥1970
- ▲ Literature ≥1970
- ● Specimen <1970
- ▲ Literature <1970
- ✕ Larval food plant

Scale: 0 100 200 400 600 km

## Junonia hedonia

Legend:
- ● Species record
- ▨ Geographic range
- — Phytogeographical boundary
- ⋯ IBRA bioregional boundary

Scale: 0 100 200 400 600 km

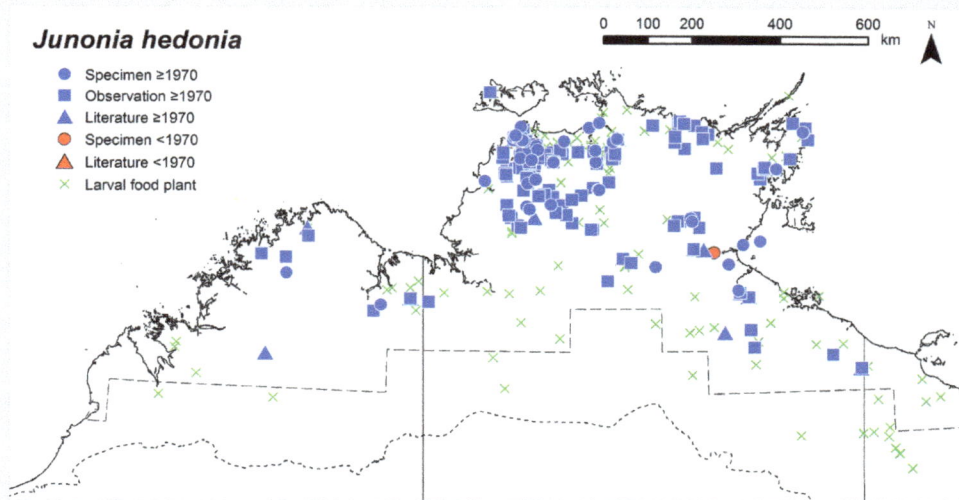

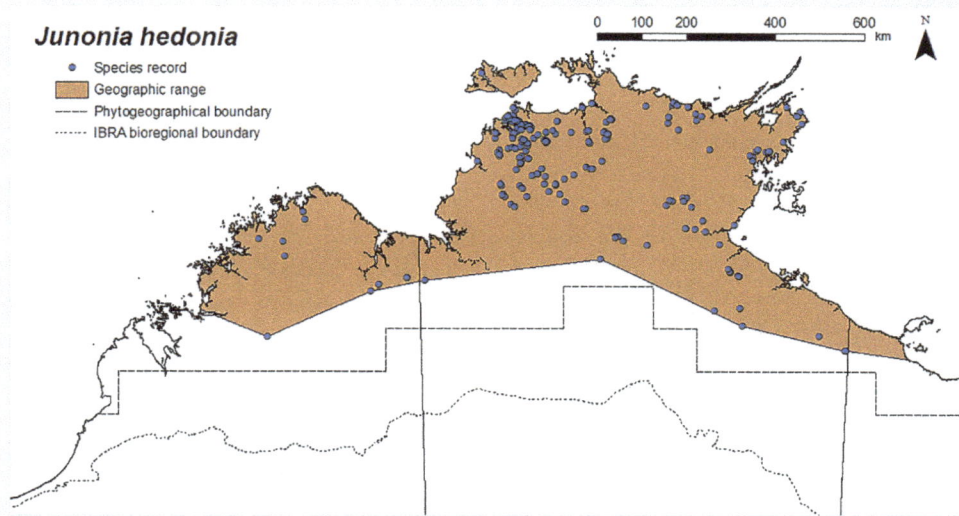

### Junonia hedonia (n = 286)

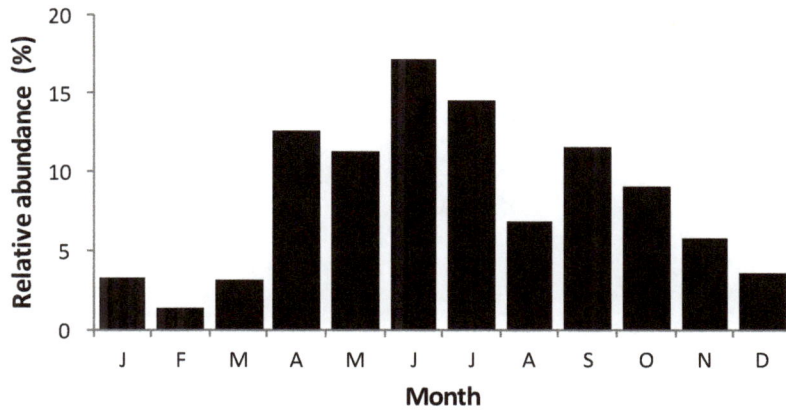

Bar chart: Relative abundance (%) by Month (J F M A M J J A S O N D)

| Month | J | F | M | A | M | J | J | A | S | O | N | D |
|-------|---|---|---|---|---|---|---|---|---|---|---|---|
| Egg | | | ▨ | | ▨ | | | | | | | |
| Larva | | | ▨ | | ▨ | | | | | | | |
| Pupa | | | | | | | | | | | | |
| Adult | ▬ | ▬ | ▬ | ▬ | ▬ | ▬ | ▬ | ▬ | ▬ | ▬ | ▬ | ▬ |

# Northern Argus
## *Junonia erigone* (Cramer, [1775])

Plate 110 Rimbija Island, Wessell Islands, NT
Photo: M. F. Braby

## Distribution

This species is known only from coastal areas of the Northern Territory of the study region. It is known only from the Wessel Islands (Rimbija Island), based on a female collected in January 1977 (Edwards 1977), and from Cobourg Peninsula (Black Point), NT, based on a male observed in February 2007 (Braby 2014a). Outside the study region, *J. erigone* occurs in Indonesia, Timor, mainland New Guinea, Fergusson Island and the Trobriand Islands.

The subspecific identity has not been determined, but it may be *J. erigone walkeri* (Butler, 1901) from Timor, although Sands and New (2002) suggested that Australian material may represent an undescribed subspecies.

## Habitat

The breeding habitat of *J. erigone* has not been recorded in the study region. Both adults were recorded in open areas adjacent to coastal semi-deciduous monsoon vine thicket.

## Larval food plants

Not recorded in the study region.

## Seasonality

The seasonal abundance and breeding phenology of this species are not well understood. The two adults were recorded during the wet season (January and February).

## Breeding status

*Junonia erigone* does not appear to have become permanently established in the study region. Presumably, it is a rare vagrant from the Lesser Sunda Islands.

## Conservation status

NA.

*Junonia erigone*

■ Observation ≥1970
▲ Literature ≥1970

| Month | J | F | M | A | M | J | J | A | S | O | N | D |
|---|---|---|---|---|---|---|---|---|---|---|---|---|
| Egg | | | | | | | | | | | | |
| Larva | | | | | | | | | | | | |
| Pupa | | | | | | | | | | | | |
| Adult | | | | | | | | | | | | |

# Lurcher
## *Yoma sabina* (Cramer, [1780])

Plate 111 Macassan Beach, Gove Peninsula, NT
Photo: M. F. Braby

## Distribution

This species is represented in the study region by the subspecies *Y. sabina parva* (Butler, 1876), although Lambkin and Kendall (2016) recently proposed that *Y. sabina parva* is a synonym of *Y. sabina sabina* (Cramer, [1780]). It was first recorded in the region by Common and Waterhouse (1981), who reported it from Groote Eylandt; however, the earliest record we are aware of is the series of specimens in the NTM collected in April 1976 from Nhulunbuy, NT, by A. J. Dartnall. *Yoma sabina* occurs in the Top End, where it is restricted to north-eastern Arnhem Land. It has been recorded from Elcho Island (Gäwa), Gove Peninsula and Groote Eylandt, NT. Its geographic range closely corresponds with the spatial distribution of its larval food plant, which is largely restricted to Gove Peninsula. Although the food plant also occurs further west at Kakadu National Park, *Y. sabina* is not established in this area. However, the food plant is absent from Elcho Island and Groote Eylandt, indicating that either alternative (as yet unreported) food plants are used at these locations or the known food plant is underreported. Outside the study region, *Y. sabina* occurs from South-East Asia, through mainland New Guinea and adjacent islands and north-eastern Australia to the Bismarck Archipelago and the Solomon Islands.

## Excluded data

A historical record from Darwin, NT, by Waterhouse and Lyell (1914) is considered to be erroneous (Braby 2014a).

## Habitat

*Yoma sabina* breeds in eucalypt woodland adjacent to mixed paperbark swampland–monsoon forest or the edges of evergreen monsoon vine forest where the larval food plant grows as a seasonal perennial herb in open areas (G. Martin, pers. comm.).

## Larval food plant

*Dipteracanthus bracteatus* (Acanthaceae).

## Seasonality

The seasonal abundance and breeding phenology of this species are not well understood. Adults have been recorded during most months of the year, but there are too few records (n = 23) to assess any seasonal changes in abundance. Breeding appears to be limited to the wet season, when the food plant is seasonally available (G. Martin). During the dry season, adults retract to moist refuges, such as evergreen monsoon vine forest or mixed paperbark tall swampland with rainforest elements in the understorey, where they remain settled in deep shade within the dry dead leaves of tall clumps of *Pandanus spiralis*. Presumably, females stop breeding and remain in reproductive diapause during the long dry season.

## Breeding status

This species is resident in the study region.

## Conservation status

LC. The species *Y. sabina* has a narrow range in the study region (geographic range = 14,900 sq km), within which it occurs in several conservation reserves, including the Dhimurru, Laynhapuy and Anindilyakwa IPAs. Despite its restricted occurrence, there are no major threats facing the taxon. However, the larval food plant is currently listed as NT under the *TPWCA* because of its limited AOO and impacts from mining activities.

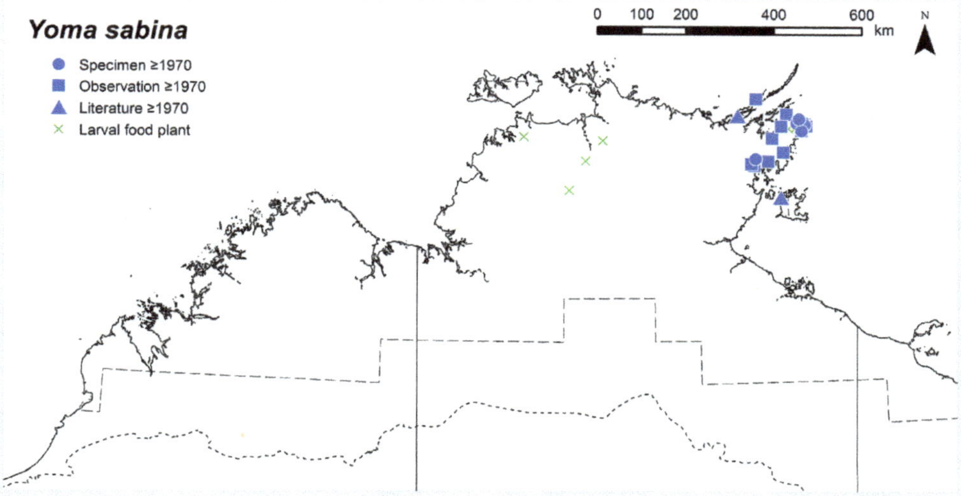

Yoma sabina
0 100 200 400 600
km
N
Specimen ≥1970
Observation ≥1970
Literature ≥1970
Larval food plant

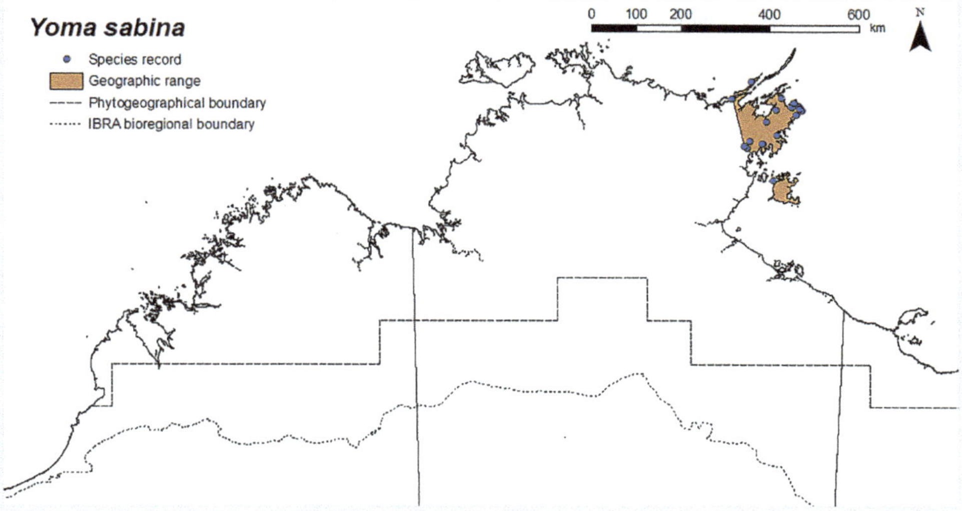

Yoma sabina
0 100 200 400 600
km
N
Species record
Geographic range
Phytogeographical boundary
IBRA bioregional boundary

| Month | J | F | M | A | M | J | J | A | S | O | N | D |
|---|---|---|---|---|---|---|---|---|---|---|---|---|
| Egg | | | | | | | | | | | | |
| Larva | | | | | | | | | | | | |
| Pupa | | | | | | | | | | | | |
| Adult | ■ | | ■ | ■ | ■ | ■ | ■ | ■ | ■ | ■ | | |

# Blue-banded Eggfly

## *Hypolimnas alimena* (Linnaeus, 1758)

Plate 112 Mosquito Creek, Gove Peninsula, NT
Photo: M. F. Braby

Plate 113 Mt Burrell, NT
Photo: M. F. Braby

## Distribution

This species is represented by the subspecies *H. alimena darwinensis* Waterhouse & Lyell, 1914, which is endemic to the study region. It is restricted to the Top End, where it generally occurs in the higher rainfall areas (> 1,000 mm mean annual rainfall), although it has been recorded in drier areas at Wongalara Wildlife Sanctuary (Braby 2012a) and Ngukurr (Normand 2009), NT. Its geographic range corresponds well with the spatial distribution of its native larval food plant. The food plant, however, extends slightly further south-west to the Port Keats area. Outside the study region, *H. alimena* occurs from Maluku, including the Kai and Aru islands, through mainland New Guinea and adjacent islands and north-eastern and eastern Australia to the Bismarck Archipelago and the Solomon Islands.

## Habitat

*Hypolimnas alimena* breeds in the ecotone between riparian evergreen monsoon vine forest and savannah woodland where the native and introduced larval food plants grow as herbs in the ground layer (G. Martin, pers. comm.). Males also congregate on hilltops in open woodland to locate females during the wet season, but they do not breed in these habitats.

## Larval food plants

*Pseuderanthemum variabile* (Acanthaceae); also *Asystasia gangetica* (Acanthaceae); possibly *Brunoniella* spp. (Acanthaceae).

## Seasonality

Adults occur throughout the year. They show little seasonal variation in abundance, although they appear to be more numerous during the 'build-up' (October and November) and again towards the end of the wet season (April). We have no records of the immature stages, but breeding is apparently limited to the wet season (November–April) (G. Martin, pers. comm.). During the dry season, adults retract to moist refuges, such as riparian evergreen monsoon vine forest or mixed paperbark forest with rainforest elements along perennial creeks or swamps, where they remain settled in deep shade. Presumably, females stop breeding and remain in reproductive diapause during the long dry season.

## Breeding status

This species is resident in the study region.

## Conservation status

LC.

## *Hypolimnas alimena*

- ● Specimen ≥1970
- ■ Observation ≥1970
- ▲ Literature ≥1970
- ● Specimen <1970
- ▲ Literature <1970
- ✕ Larval food plant
- ✕ Putative larval food plants

## *Hypolimnas alimena*

- ● Species record
- ▨ Geographic range
- —·— Phytogeographical boundary
- ········· IBRA bioregional boundary

## *Hypolimnas alimena* (n = 153)

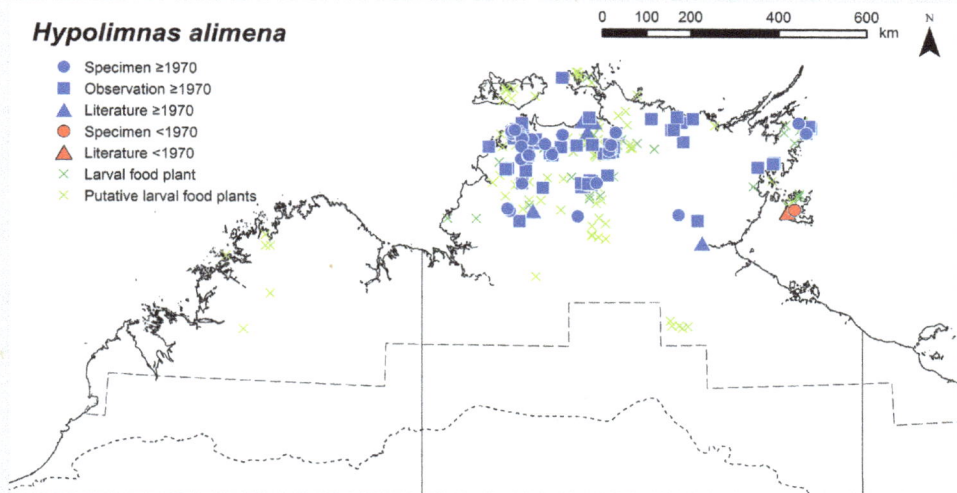

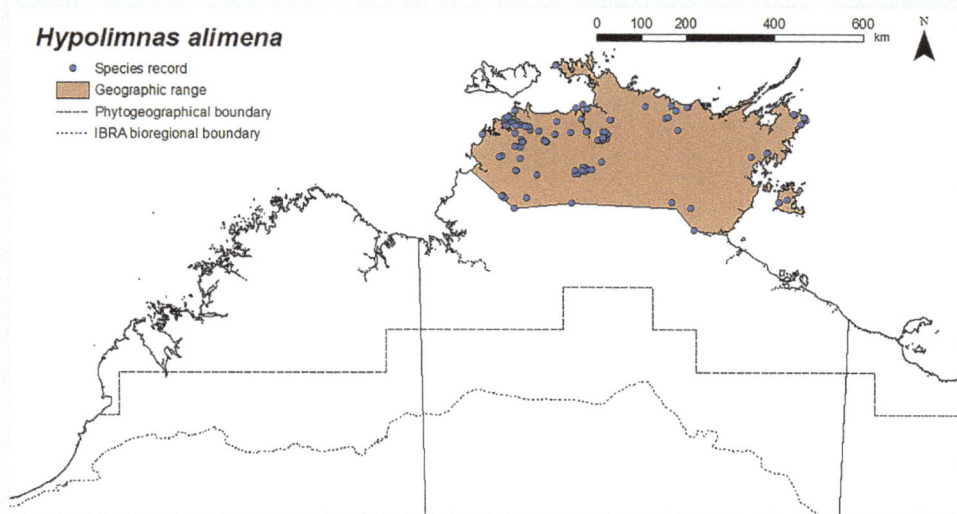

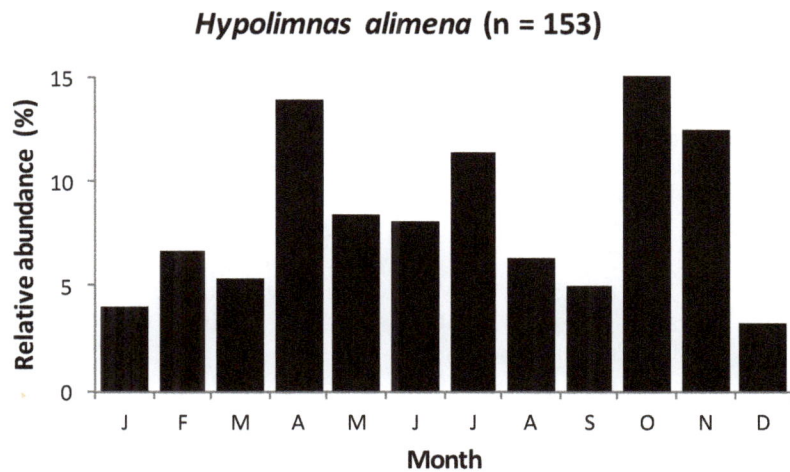

| Month | J | F | M | A | M | J | J | A | S | O | N | D |
|-------|---|---|---|---|---|---|---|---|---|---|---|---|
| Egg | | | | | | | | | | | | |
| Larva | | | | | | | | | | | | |
| Pupa | | | | | | | | | | | | |
| Adult | | | | | | | | | | | | |

# Varied Eggfly
## *Hypolimnas bolina* (Linnaeus, 1758)

Plate 114 Wongalara Wildlife Sanctuary, NT
Photo: M. F. Braby

Plate 115 Wanguri, Darwin, NT
Photo: M. F. Braby

## Distribution

This species is represented in the study region by the subspecies *H. bolina nerina* (Fabricius, 1775). It occurs very widely throughout the region, extending from moist coastal areas to drier inland areas of the semi-arid zone (< 400 mm mean annual rainfall), where it has been recorded as far south as Tennant Creek (Mary Ann Dam), NT (J. Archibald). It has also been recorded on Ashmore Reef in the Timor Sea, WA (D. C. Binns). The geographic range broadly corresponds with the spatial distribution of the putative native larval food plant (*Alternanthera denticulata*), as well as with the known food plants, all of which are introduced. Outside the study region, *H. bolina* occurs widely from Madagascar, India, Japan and South-East Asia, through mainland New Guinea and adjacent islands to the islands of the South Pacific. It also occurs throughout much of the Australian continent.

## Habitat

*Hypolimnas bolina* breeds in open woodland and open disturbed areas, including suburban parks and gardens, where the introduced larval food plants grow as annual or seasonal perennial herbs, often in moist shaded areas beneath trees.

## Larval food plants

*Asystasia gangetica* (Acanthaceae), *Synedrella nodiflora*, *Tridax procumbens* (Asteraceae), *Sida acuta* (Malvaceae); probably *Alternanthera denticulata* (Amaranthaceae), which is a food plant in Queensland (Braby 2000). The native food plants have not been recorded in the study region.

## Seasonality

Adults occur throughout the year, but they are most abundant during the wet season (November–April), with a peak in abundance in March and April, and least abundant in the late dry season (August–October). During the dry season, adults retreat to moist refuges such as riparian forest or woodland along gullies and creeks, where they overwinter. The immature stages have been recorded sporadically only during the warmer months of high humidity (October–May). Presumably, females stop breeding and remain in reproductive diapause during the dry season (May–September), similar to populations in northern Queensland (Jones 1987; Kemp 2001).

## Breeding status

This species is resident in the study region.

## Conservation status

LC.

## Hypolimnas bolina

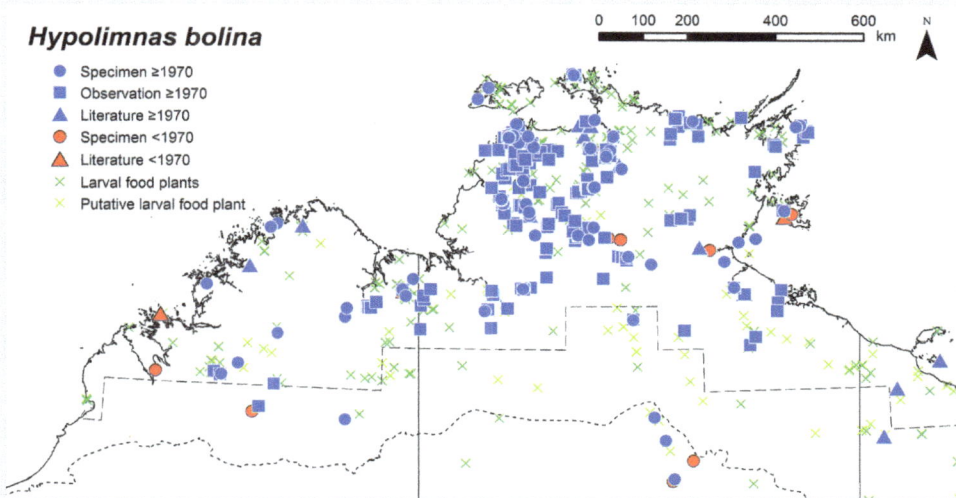

Legend:
- ● Specimen ≥1970
- ■ Observation ≥1970
- ▲ Literature ≥1970
- ● Specimen <1970
- ▲ Literature <1970
- × Larval food plants
- × Putative larval food plant

## Hypolimnas bolina

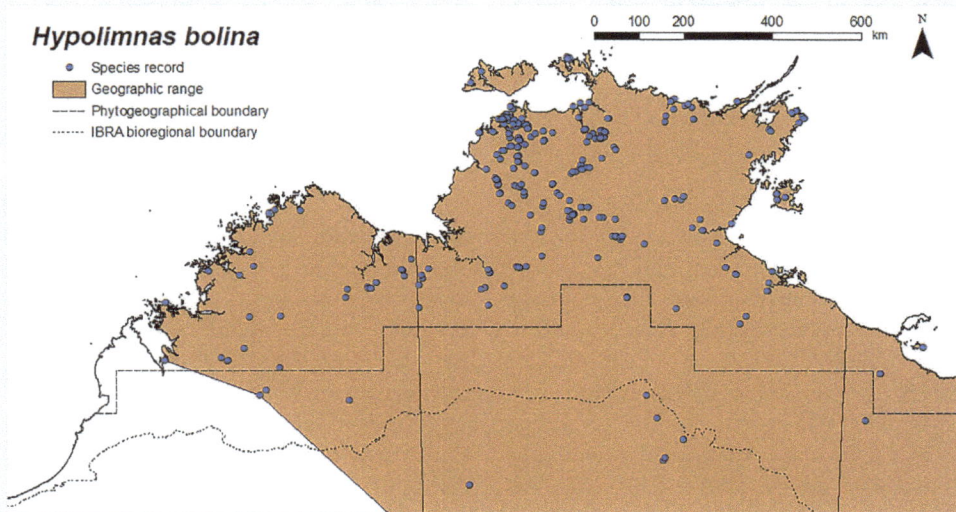

Legend:
- • Species record
- ▨ Geographic range
- — — Phytogeographical boundary
- ········ IBRA bioregional boundary

## Hypolimnas bolina (n = 478)

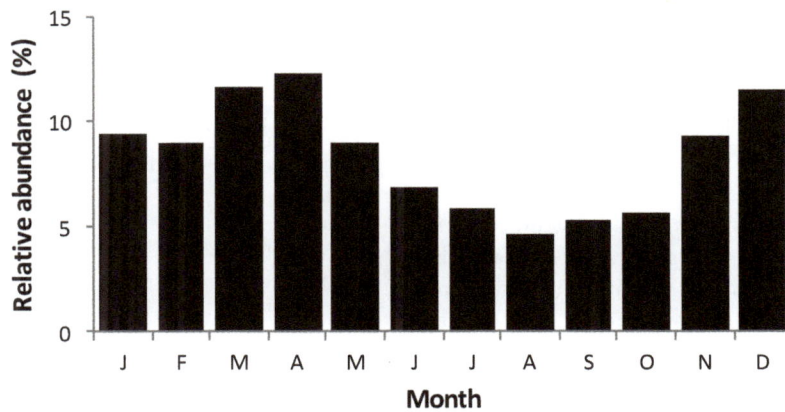

| Month | J | F | M | A | M | J | J | A | S | O | N | D |
|-------|---|---|---|---|---|---|---|---|---|---|---|---|
| Egg   |   |   | ▨ | ▨ |   |   |   |   |   | ▨ |   |   |
| Larva |   |   | ▨ | ▨ |   |   |   |   |   |   | ▨ |   |
| Pupa  |   |   |   | ▨ | ▨ |   |   |   |   |   | ▨ | ▨ |
| Adult | ▨ | ▨ | ▨ | ▨ | ▨ | ▨ | ▨ | ▨ | ▨ | ▨ | ▨ | ▨ |

# Danaid Eggfly
## *Hypolimnas misippus* (Linnaeus, 1764)

Plate 116 Mt Burrell, NT
Photo: M. F. Braby

Plate 117 Fish River Station, NT
Photo: M. F. Braby

## Distribution

This species occurs in the Kimberley and Top End of the study region. It extends from moist coastal areas to drier inland areas, generally above 800 mm mean annual rainfall. It occurs sporadically from the Buccaneer Archipelago at Koolan Island in Yampi Sound, WA (Koch and van Ingen 1969; McKenzie et al. 1995), to the Roper River area, NT, near the Gulf of Carpentaria. The larval food plant is considerably more widespread, extending to the semi-arid zone as well as across the Top End. Further field surveys are thus required to determine whether *H. misippus* occurs in the eastern half of the Top End, particularly north-eastern Arnhem Land. Outside the study region, *H. misippus* occurs widely from Africa, Madagascar, India and South-East Asia, through mainland New Guinea and eastern Australia to the Solomon Islands and Norfolk Island, as well as in coastal regions of north-western South America (K. Willmott, pers. comm.).

## Habitat

*Hypolimnas misippus* breeds in open disturbed areas where the larval food plant grows as an annual herb. Males also congregate on prominent hilltops to locate females during the wet season, but they do not breed in these habitats.

## Larval food plant

*Portulaca oleracea* (Portulacaceae).

## Seasonality

Adults occur throughout the year, but they are observed more frequently during the latter half of the wet season (February–April), although they are never encountered in large numbers or high densities. Very few adults have been recorded during the dry season (June–October). The breeding phenology is not well understood. The immature stages (larvae or pupae) have been recorded during the late wet season (March and April), when adults are most abundant, and freshly emerged adults have also been collected around this time (March–May). However, it is not clear how the species survives the dry season when foliage of the larval food plant is absent. Presumably, the species breeds seasonally during the wet season and females then stop breeding and remain in reproductive diapause during the dry season.

## Breeding status

This species is assumed to be resident in the study region.

## Conservation status

LC.

## Hypolimnas misippus

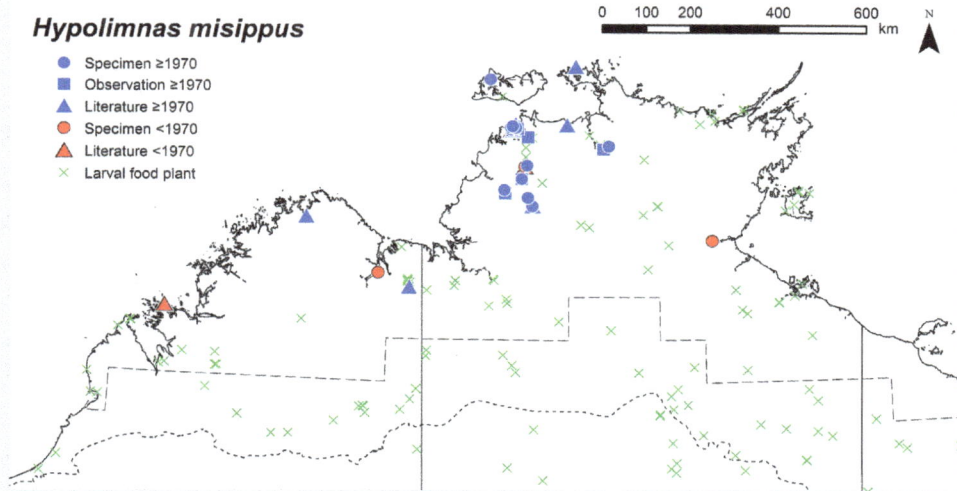

Legend:
- Specimen ≥1970
- Observation ≥1970
- Literature ≥1970
- Specimen <1970
- Literature <1970
- Larval food plant

## Hypolimnas misippus

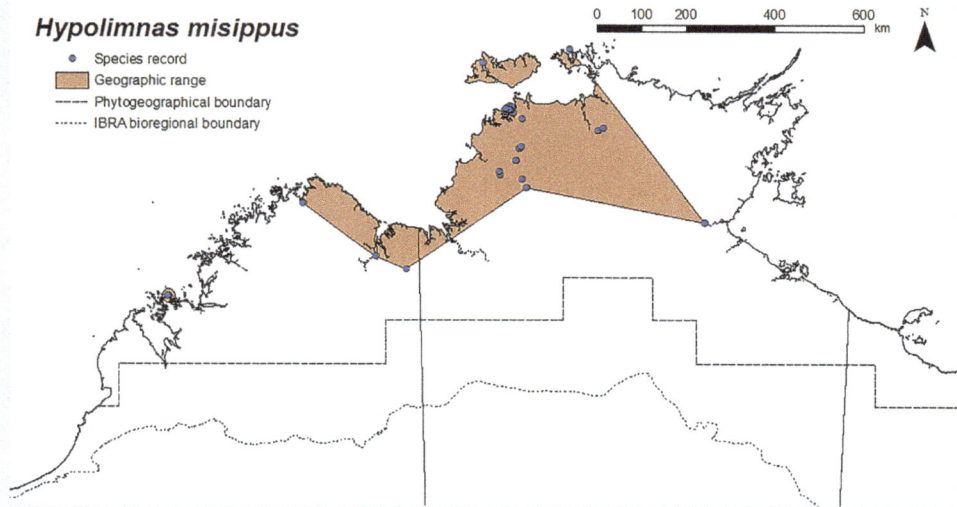

Legend:
- Species record
- Geographic range
- Phytogeographical boundary
- IBRA bioregional boundary

## Hypolimnas misippus (n = 50)

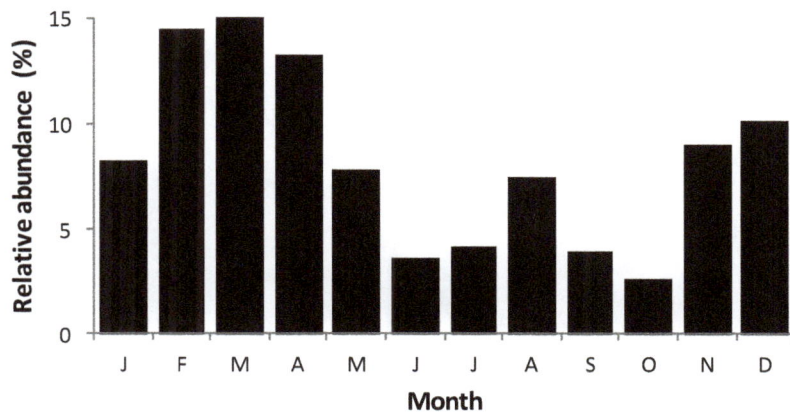

| Month | J | F | M | A | M | J | J | A | S | O | N | D |
|-------|---|---|---|---|---|---|---|---|---|---|---|---|
| Egg   |   |   |   |   |   |   |   |   |   |   |   |   |
| Larva |   |   |   |   |   |   |   |   |   |   |   |   |
| Pupa  |   |   |   |   |   |   |   |   |   |   |   |   |
| Adult |   |   |   |   |   |   |   |   |   |   |   |   |

# Crow Eggfly
## *Hypolimnas anomala* (Wallace, 1869)

Plate 118 Christmas Island, WA
Photo: Frank Pierce

Plate 119 Christmas Island, WA
Photo: Frank Pierce

## Distribution

This species is represented in the study region by the subspecies *H. anomala albula* (Wallace, 1869). It is known from only two specimens, both males, from Darwin, NT: one collected in March 1909 by F. P. Dodd (Waterhouse and Lyell 1914; Gibb 1977; Common 1978) and the other in February 1987 by C. G. Miller (Braby 2000). Outside the study region, *H. anomala* occurs in South-East Asia and north-eastern Australia, where it is probably not established.

## Habitat

The breeding habitat of *H. anomala* has not been recorded in the study region. One of the specimens was collected in semi-deciduous monsoon vine thicket at East Point Reserve, NT (Meyer et al. 2006).

## Larval food plants

Not recorded in the study region.

## Seasonality

The seasonal abundance and breeding phenology of this species are not well understood. The two adults were recorded during the wet season (February and March).

## Breeding status

*Hypolimnas anomala* does not appear to have become permanently established in the study region. Presumably, it is a rare vagrant from the Lesser Sunda Islands.

## Conservation status

NA.

## Hypolimnas anomala

▲ Literature ≥1970
▲ Literature <1970

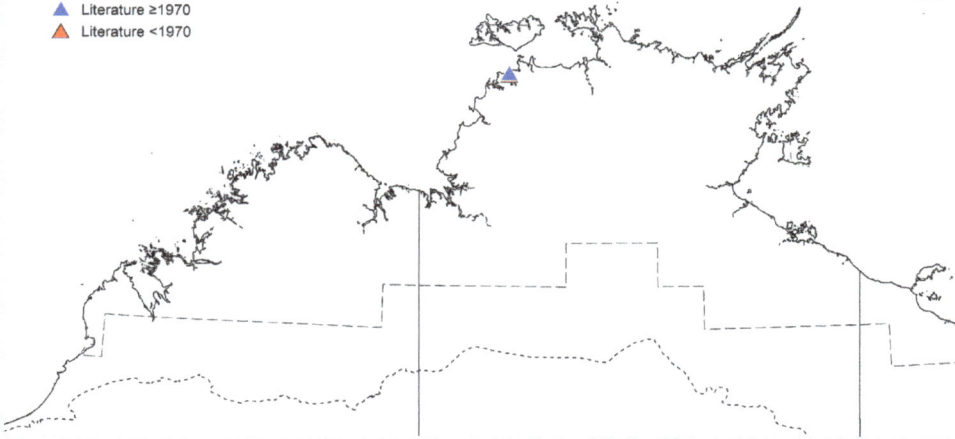

| Month | J | F | M | A | M | J | J | A | S | O | N | D |
|-------|---|---|---|---|---|---|---|---|---|---|---|---|
| Egg   |   |   |   |   |   |   |   |   |   |   |   |   |
| Larva |   |   |   |   |   |   |   |   |   |   |   |   |
| Pupa  |   |   |   |   |   |   |   |   |   |   |   |   |
| Adult |   |   |   |   |   |   |   |   |   |   |   |   |

Photo: Florence Creek, Litchfield National Park, NT, M.F. Braby

# Tailed Emperor

*Charaxes sempronius* (Fabricius, 1793)

Plate 120 Fish River Station, NT
Photo: M. F. Braby

## Distribution

This species is represented in the study region by the subspecies *C. sempronius sempronius* (Fabricius, 1793). It occurs widely in the Kimberley, Top End and western Gulf Country, extending from moist coastal areas to drier inland areas of the semi-arid zone (c. 400 mm mean annual rainfall). It has been recorded as far south as 16 km east–north-east of Anna Plains Station and Broome, WA (Warham 1957); 7 km north–north-east of Helen Springs Homestead near Renner Springs, NT; and Boodjamulla/Lawn Hill National Park (Constance Range), Qld (Dunn 2015a). The geographic range corresponds well with the spatial distribution of the known larval food plants. Outside the study region, *C. sempronius* occurs throughout north-eastern, eastern and south-eastern Australia, and Lord Howe Island.

## Habitat

*Charaxes sempronius* breeds in a wide variety of habitats, including monsoon forest and eucalypt woodland where the larval food plants grow as trees. It also occurs in suburban parks and gardens. Males readily congregate on hilltops to locate females, but the species does not breed in these habitats.

## Larval food plants

*Celtis australiensis* (Cannabaceae), *Acacia hemsleyi*, *Peltophorum pterocarpum* (Fabaceae), *Vitex acuminata* (Lamiaceae); also *\*Mimosa pigra* (Fabaceae). This highly polyphagous species (see Braby 2000) undoubtedly uses a much wider range of plants than current records from the study region indicate.

## Seasonality

Adults occur throughout the year, but they are observed more frequently during the late dry season, with a pronounced peak in abundance in September, although they are never encountered in large numbers or high densities. The breeding phenology is not well understood. The immature stages have been recorded sporadically during the dry season (April–October). Presumably, the species also breeds in the wet season and several generations are completed annually.

## Breeding status

This species is resident in the study region.

## Conservation status

LC.

## Charaxes sempronius

- ● Specimen ≥1970
- ■ Observation ≥1970
- ▲ Literature ≥1970
- ● Specimen <1970
- ▲ Literature <1970
- × Larval food plants

0  100  200  400  600 km  N

## Charaxes sempronius

- ● Species record
- ▨ Geographic range
- – – – Phytogeographical boundary
- ⋯⋯ IBRA bioregional boundary

0  100  200  400  600 km  N

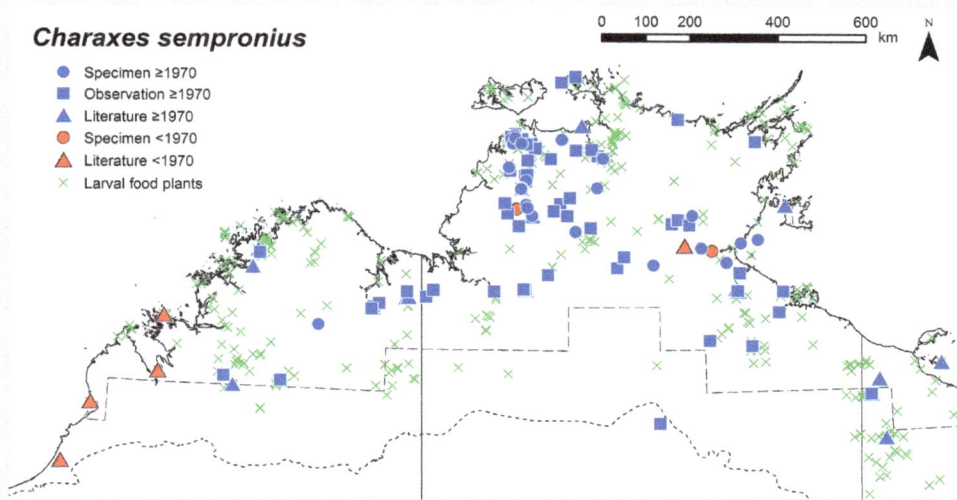

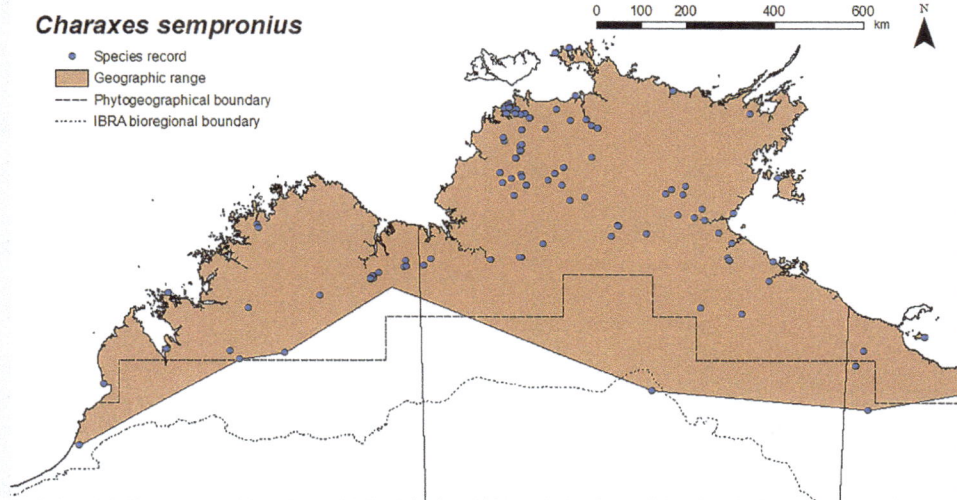

### Charaxes sempronius (n = 166)

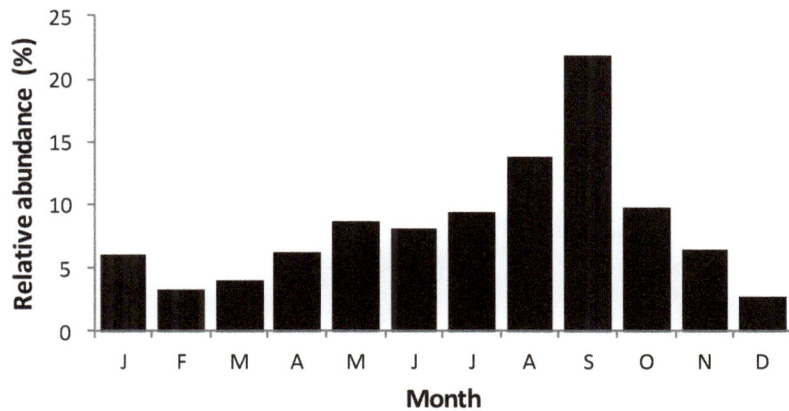

| Month | J | F | M | A | M | J | J | A | S | O | N | D |
|-------|---|---|---|---|---|---|---|---|---|---|---|---|
| Egg   |   |   |   |   |   |   |   | ▣ |   |   |   |   |
| Larva |   |   |   | ▣ |   | ▣ |   |   | ▣ |   |   |   |
| Pupa  |   |   | ▣ |   |   |   |   | ▣ |   |   |   |   |
| Adult | ▣ | ▣ | ▣ | ▣ | ▣ | ▣ | ▣ | ▣ | ▣ | ▣ | ▣ | ▣ |

# Evening Brown
## *Melanitis leda* (Linnaeus, 1758)

Plate 121 Connors Range, St Lawrence, Qld
Photo: M. F. Braby

## Distribution

This species is represented in the study region by the subspecies *M. leda bankia* (Fabricius, 1775). It occurs widely in the Kimberley, Top End and the western Gulf Country, extending from moist coastal areas to drier inland areas of the semi-arid zone (< 700 mm mean annual rainfall). It has been recorded as far inland as 4 km north of Argyle Diamond Mine, WA (G. Swann); Newcastle Creek near Newcastle Waters, NT (Dunn and Dunn 1991); and Gregory River Crossing, Gregory Downs, Qld (Dunn 2015a). The geographic range corresponds well with the spatial distribution of its known native larval food plants, indicating that *M. leda* has been well sampled in the region. Outside the study region, *M. leda* occurs widely from Africa, Madagascar, Mauritius, India and South-East Asia, through mainland New Guinea and adjacent islands and north-eastern and eastern Australia to the Solomon Islands, New Caledonia, Fiji and Tahiti.

## Habitat

*Melanitis leda* occurs in a variety of habitats, but it usually breeds in riparian areas along the banks of rivers, creeks and swamps that support evergreen monsoon vine forest, mixed eucalypt open forest and paperbark swampland or open forest with rainforest elements where the larval food plants grow as perennial or annual grasses (Braby 2011a, 2015e); it usually favours the ecotone or edges of these habitats.

## Larval food plants

*Imperata cylindrica*, *Ischaemum australe* (Poaceae); also *Cenchrus pedicellatus*, *Cynodon radiatus*, *Megathyrsus maximus*, *Oryza sativa*, *Sorghum vulgare* (Poaceae).

## Seasonality

Adults occur throughout the year, but they are most abundant after the monsoon rains and during the cooler part of the dry season (April–July). Breeding is strictly seasonal, with the immature stages occurring only during the wet season and early dry season (November–May), when their larval food plants, which consist of a suite of native and introduced grasses, are available and several generations are completed. Presumably, females stop breeding and remain in reproductive diapause and aggregate in moist refuges during the mid to late dry season (July–October), similar to populations in northern Queensland (Jones 1987).

## Breeding status

This species is resident in the study region.

## Conservation status

LC.

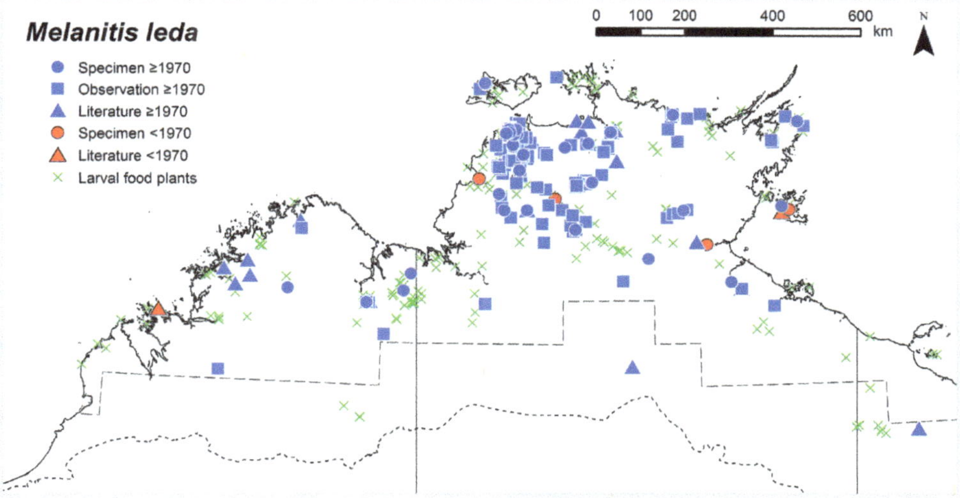
Melanitis leda
0 100 200 400 600 km
N
Specimen ≥1970
Observation ≥1970
Literature ≥1970
Specimen <1970
Literature <1970
Larval food plants

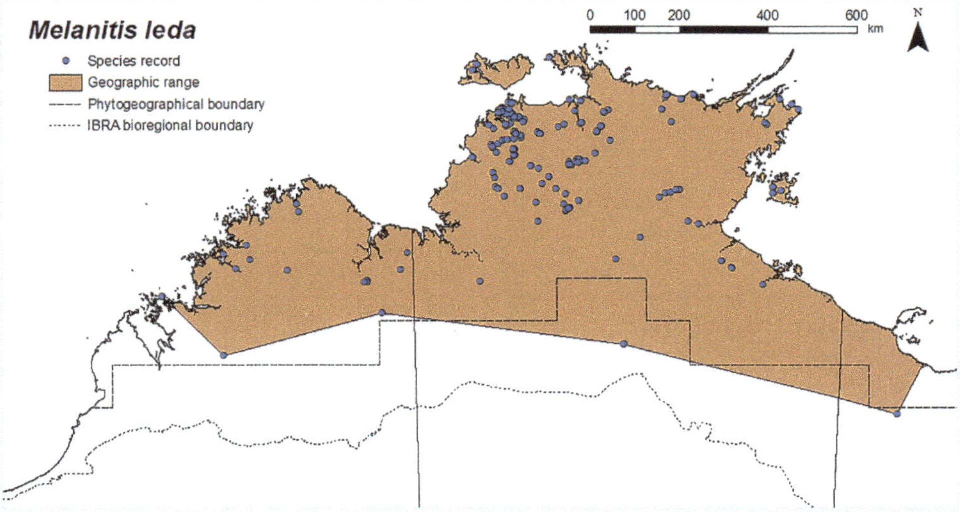
Melanitis leda
0 100 200 400 600 km
N
Species record
Geographic range
Phytogeographical boundary
IBRA bioregional boundary

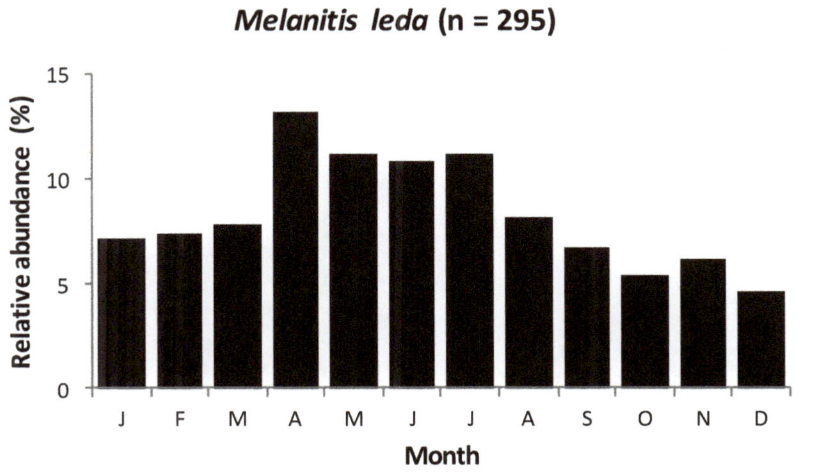
Melanitis leda (n = 295)
Relative abundance (%)
15
10
5
0
J F M A M J J A S O N D
Month

| Month | J | F | M | A | M | J | J | A | S | O | N | D |
|---|---|---|---|---|---|---|---|---|---|---|---|---|
| Egg | | | | | | | | | | | | |
| Larva | | | | | | | | | | | | |
| Pupa | | | | | | | | | | | | |
| Adult | | | | | | | | | | | | |

# Dingy Bush-brown

## *Mycalesis perseus* (Fabricius, 1775)

Plate 122 Cardwell, Qld
Photo: M. F. Braby

## Distribution

This species is represented in the study region by the subspecies *M. perseus perseus* (Fabricius, 1775). It is distributed throughout the Top End and more sporadically in the western Gulf Country. It occurs mainly in the higher rainfall areas (> 800 mm mean annual rainfall); however, it has been recorded in lower rainfall areas of the semi-arid zone, including 20 km south of Mataranka (L. and B. R. Reid) and Bessie Spring (Braby 2000), NT; and Doomadgee, Qld (Puccetti 1991), as well as in the Gulf of Carpentaria at Walker Creek 36 km east of Karumba, Qld, just outside the eastern boundary of the study region (Dunn 2015a). Its geographic range falls within the spatial distribution of its putative larval food plants (*Dichanthium sericeum*, *Heteropogon* spp., *Themeda triandra*), which occur widely in northern Australia. Outside the study region, *M. perseus* occurs widely from India and South-East Asia, through mainland New Guinea and adjacent islands and north-eastern Australia to Vanuatu.

## Habitat

*Mycalesis perseus* occurs in a variety of woodland habitats where the putative larval food plants grow as seasonal perennial grasses.

## Larval food plants

Not recorded in the study region; probably *Dichanthium sericeum*, *Heteropogon* spp., *Themeda triandra* (Poaceae), as well as other native and introduced grasses, which are the food plants in north-eastern Queensland (Braby 2000).

## Seasonality

Adults occur throughout the year, but they are most abundant after the monsoon rains and during the cooler part of the dry season (April–August), with a pronounced peak in abundance in June. We have very limited information on the breeding phenology and incidence of the immature stages of this common species. Presumably, the seasonal history is similar to populations in northern Queensland in which breeding is strictly seasonal and limited to the wet season and early dry season, when their putative larval food plants are available and several generations are completed, and females stop breeding, remain in reproductive diapause and aggregate in moist refuges during the mid to late dry season (May/June–November/December), depending on the timing of rainfall (Braby 1995a, 1995b).

## Breeding status

This species is resident in the study region.

## Conservation status

LC.

## Mycalesis perseus

Legend:
- ● Specimen ≥1970
- ■ Observation ≥1970
- ▲ Literature ≥1970
- ● Specimen <1970
- ▲ Literature <1970
- × Putative larval food plants

0 100 200 400 600 km

N

## Mycalesis perseus

Legend:
- ● Species record
- ▨ Geographic range
- — — Phytogeographical boundary
- ······ IBRA bioregional boundary

0 100 200 400 600 km

N

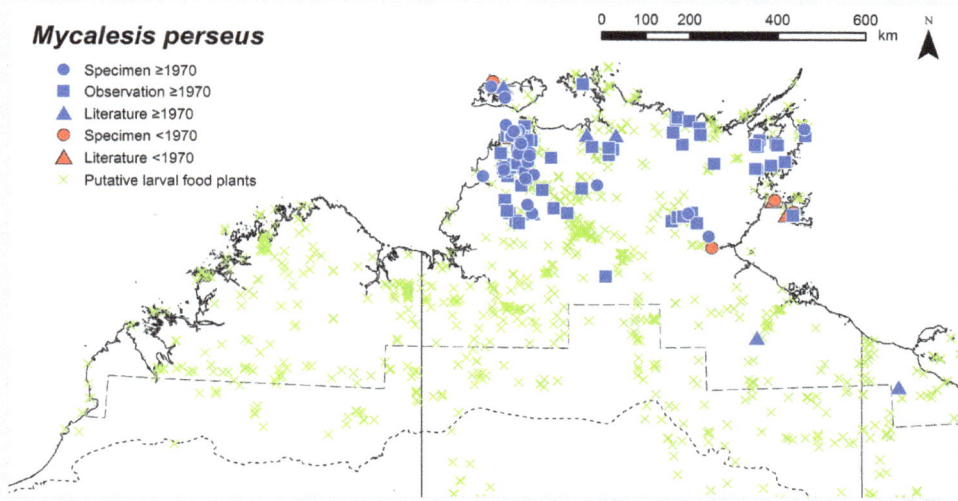

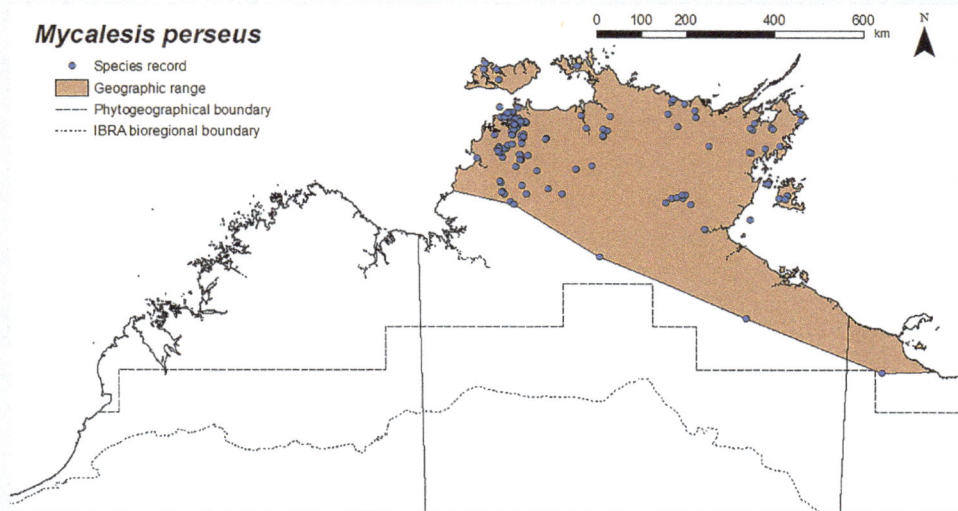

### Mycalesis perseus (n = 196)

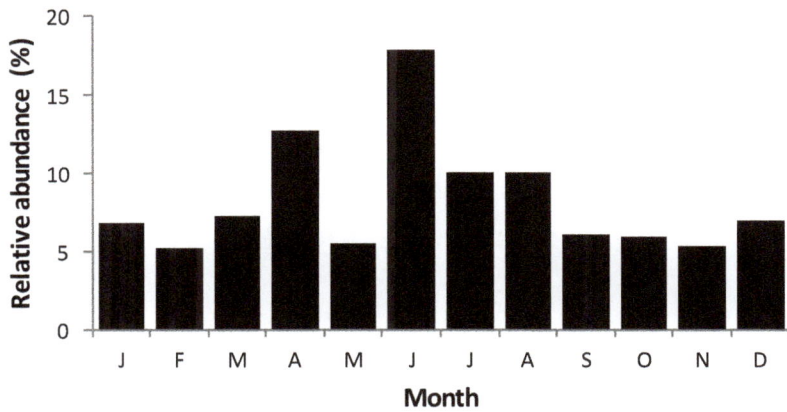

| Month | J | F | M | A | M | J | J | A | S | O | N | D |
|-------|---|---|---|---|---|---|---|---|---|---|---|---|
| Egg   |   |   |   | ▧ |   |   |   |   |   |   |   |   |
| Larva |   |   |   |   |   |   |   |   |   |   |   |   |
| Pupa  |   |   |   |   |   |   |   |   |   |   |   |   |
| Adult |   |   |   |   |   |   |   |   |   |   |   |   |

# Cedar Bush-brown
## *Mydosama sirius* (Fabricius, 1775)

Plate 123 Baralminar River near Gapuwiyak, NT
Photo: M. F. Braby

## Distribution

This species is represented in the study region by the subspecies *M. sirius sirius* (Fabricius, 1775). It is restricted to the Top End, where it occurs in the higher rainfall areas (> 800 mm mean annual rainfall). It reaches its southernmost extent at Bradshaw Field Training Area (Fitzmaurice River catchment) (Archibald and Braby 2017), Mataranka (Dunn and Dunn 1991) and Costello Outstation 25 km north-east of Ngukurr (S. Normand), NT. The geographic range falls within the spatial distribution of the known and putative larval food plants. Although the food plants are much more widespread, searches in the Kimberley and western Gulf Country have not detected *M. sirius* in these areas. Outside the study region, *M. sirius* occurs from Maluku, through mainland New Guinea and adjacent islands to north-eastern Australia.

## Habitat

*Mydosama sirius* occurs mainly in paperbark swampland where the putative larval food plant (*Ischaemum australe*) grows as a perennial grass, but it may also breed along the edges of riparian evergreen monsoon vine forest, where alternative grasses are used (Braby 2015e). It also occurs in the edges of floodplain wetlands, where it undoubtedly breeds.

## Larval food plants

*Imperata cylindrica* (Poaceae); probably *Ischaemum australe* (Poaceae), which is a food plant in north-eastern Queensland (Braby 2000).

## Seasonality

Adults occur throughout the year, but they are most abundant after the monsoon rains and during the cooler part of the dry season (April–September). We have very limited information on the breeding phenology and incidence of the immature stages. Presumably, the seasonal history is similar to populations in northern Queensland in which breeding occurs throughout the year, but reproductive activity declines as the dry season progresses, and adults are least abundant during the wet season (Braby 1995a, 1995b).

## Breeding status

This species is resident in the study region.

## Conservation status

LC.

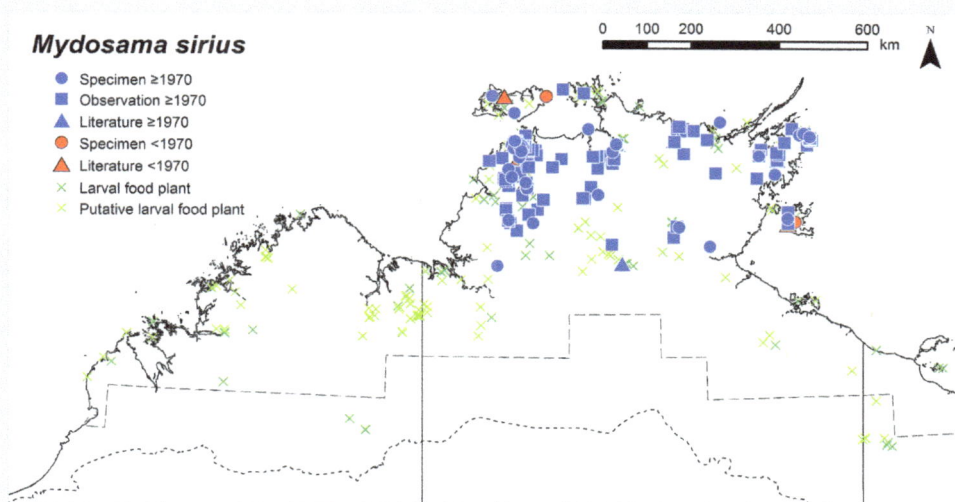

*Mydosama sirius*

| | 0 | 100 | 200 | | 400 | | 600 |
| km |

Specimen ≥1970
Observation ≥1970
Literature ≥1970
Specimen <1970
Literature <1970
Larval food plant
Putative larval food plant

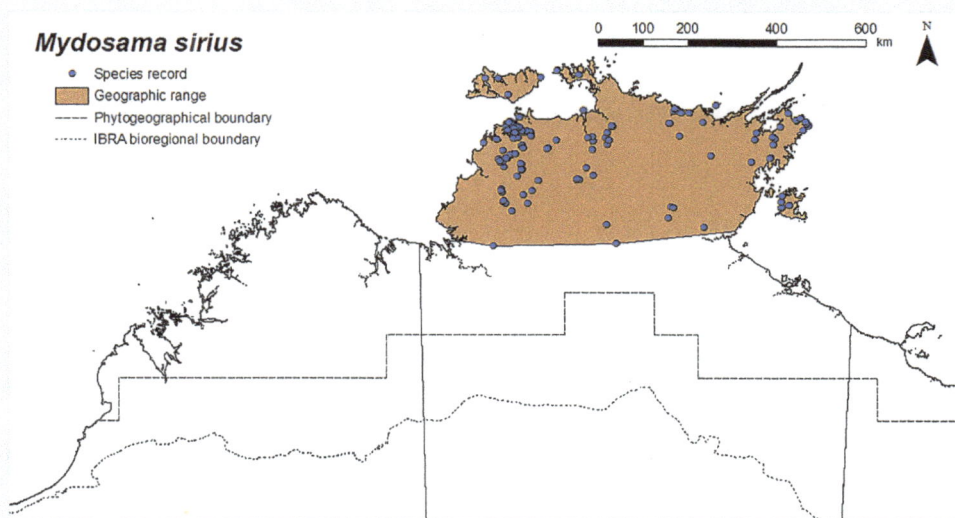

*Mydosama sirius*

Species record
Geographic range
Phytogeographical boundary
IBRA bioregional boundary

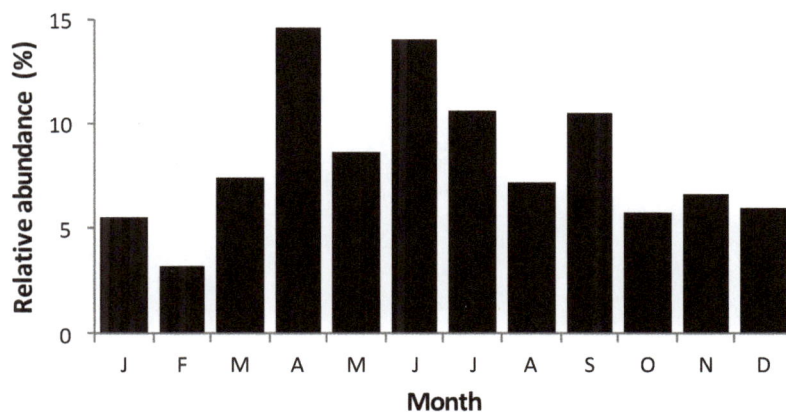

*Mydosama sirius* (n = 225)

| Month | J | F | M | A | M | J | J | A | S | O | N | D |
|-------|---|---|---|---|---|---|---|---|---|---|---|---|
| Egg | | | | | | | | | | | | |
| Larva | | | | ▮ | | | | | | | | |
| Pupa | | | | | | | | | | | | |
| Adult | | | | | | | | | | | | |

# Orange Ringlet
## *Hypocysta adiante* (Hübner, 1831)

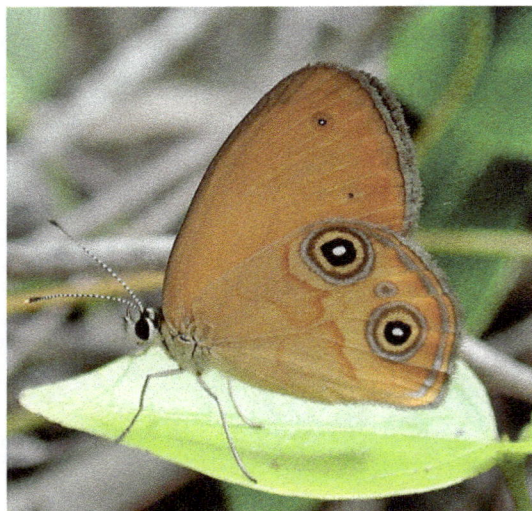

Plate 124 Nhulunbuy, NT
Photo: M. F. Braby

## Distribution

This species is represented by the subspecies *H. adiante antirius* Butler, 1868, which is endemic to the study region. It occurs throughout the Kimberley and Top End, extending from moist coastal areas to drier inland areas of the semi-arid zone (c. 500 mm mean annual rainfall), where it has been recorded as far south as Palm Springs 30 km south-east of Halls Creek, WA (Williams et al. 2006); and Daguragu Aboriginal Land Trust (Giles Creek), NT. Its broad geographical range closely corresponds with the spatial distribution of its known native larval food plants. Although the food plants extend further south-east into the western Gulf Country, recent surveys of this area have not detected *H. adiante* further south than Borroloola, NT. Outside the study region, *H. adiante* occurs in north-eastern and eastern Australia.

## Habitat

*Hypocysta adiante* breeds in a wide range of habitats, including savannah woodland, riparian woodland along seasonal and perennial creeks or riverbanks and the edges of wet and dry monsoon forests, where the larval food plants grow as perennial grasses (Braby 2011a, 2015e). It also breeds in suburban parks and gardens, where it uses a range of introduced grasses.

## Larval food plants

*Aristida macroclada, Arundinella nepalensis, Digitaria gibbosa, Ischaemum australe, I. tropicum* (Poaceae); also *\*Axonopus compressus, \*Chloris* sp., *\*Cynodon dactylon, \*Sporobolus* sp., *\*Urochloa mosambicensis* (Poaceae).

## Seasonality

Adults occur throughout the year. They appear to show little seasonal variation in abundance, although they tend to be more numerous after the monsoon rains and during the cooler part of the dry season (March–August). The immature stages have been recorded during a similar period (February–August). Presumably, the seasonal history is similar to populations in northern Queensland in which breeding occurs throughout the year, but reproductive activity may decline as the dry season progresses (Braby 1995a).

## Breeding status

This species is resident in the study region.

## Conservation status

LC.

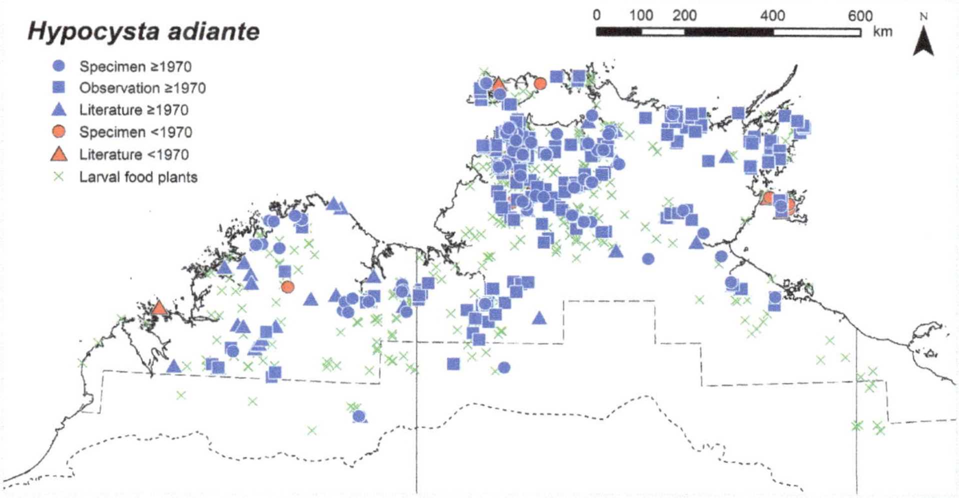

## Hypocysta adiante

**Legend (top map):**
- ● Specimen ≥1970
- ■ Observation ≥1970
- ▲ Literature ≥1970
- ● Specimen <1970
- ▲ Literature <1970
- × Larval food plants

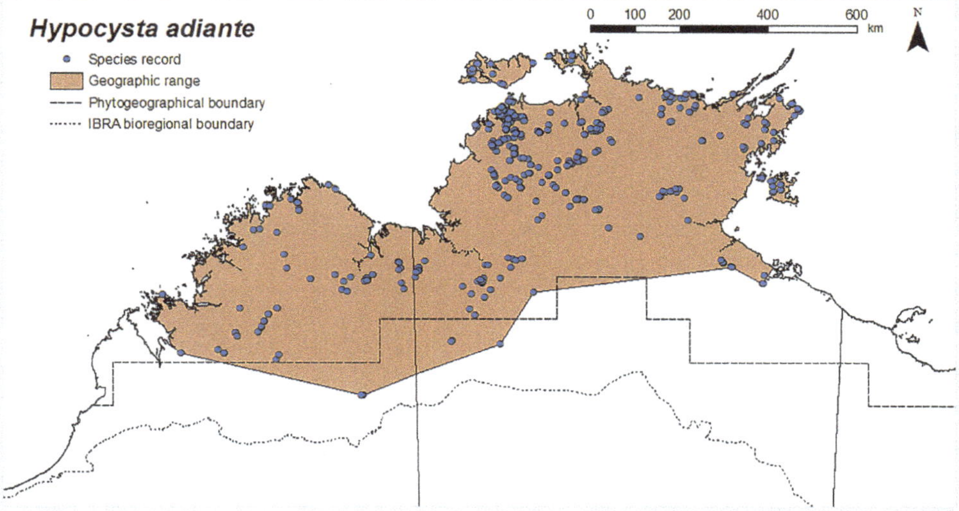

Scale: 0 100 200 400 600 km  N

## Hypocysta adiante

**Legend (middle map):**
- ● Species record
- ▬ Geographic range
- — — Phytogeographical boundary
- ········ IBRA bioregional boundary

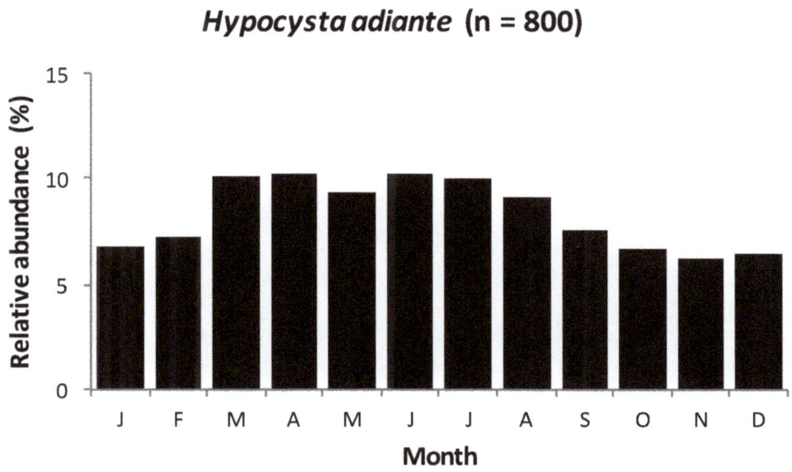

Scale: 0 100 200 400 600 km  N

## Hypocysta adiante (n = 800)

Relative abundance (%) vs Month (J F M A M J J A S O N D)

| Month | J | F | M | A | M | J | J | A | S | O | N | D |
|-------|---|---|---|---|---|---|---|---|---|---|---|---|
| Egg   |   |  yellow  |   |   |   |  yellow  |   |   |   |   |   |   |
| Larva |   | green |   | green |   |   | green |   |   |   |   |   |
| Pupa  |   |   |   | blue |   |   |   |   |   |   |   |   |
| Adult | orange | | | | | | | | | | | |

# Dusky Knight
## *Ypthima arctous* (Fabricius, 1775)

Plate 125 Murray Falls, Tully, Qld
Photo: M. F. Braby

Plate 126 Murray Falls, Tully, Qld
Photo: M. F. Braby

## Distribution

This species is represented in the study region by the subspecies *Y. arctous arctous* (Fabricius, 1775). It occurs widely in the Kimberley, Top End and western Gulf Country, extending from moist coastal areas to drier inland areas of the semi-arid zone (c. 700 mm mean annual rainfall). It has been recorded as far south as Mornington Wildlife Sanctuary (Annie Creek), WA; Bessie Spring, NT (Dunn and Dunn 1991); and Lagoon Creek crossing 17 km north-west of Hells Gate Roadhouse, Qld. The putative native larval food plants (*Imperata cylindrica* and *Themeda triandra*) occur widely in northern Australia, including both coastal and semi-arid areas, and their overall spatial distribution includes the geographical range of *Y. arctous*. Outside the study region, *Y. arctous* occurs in mainland New Guinea and adjacent islands and north-eastern and eastern Australia.

## Habitat

The natural breeding habitat of *Y. arctous* has not been recorded in the study region. Adults occur in a wide variety of habitats, but they are usually associated with woodland, mixed swampland and the edges of monsoon forest or, in the drier inland areas, riparian woodland along creeks and gullies, in which they no doubt breed.

## Larval food plants

\**Axonopus compressus* (Poaceae); probably *Imperata cylindrica* and *Themeda triandra* (Poaceae), which are two larval food plants in Queensland (Braby 2000). The native food plants have not been recorded in the study region, but the larvae probably feed on a range of grasses.

## Seasonality

Adults occur throughout the year, but they are most abundant after the monsoon rains and during the cooler part of the dry season (April–August). They appear to become scarce towards the end of the dry season during the 'build-up' (September–November). We have very limited information on the breeding phenology and incidence of the immature stages. Presumably, the seasonal history is similar to populations in northern Queensland in which breeding occurs throughout the year (Braby 1995a).

## Breeding status

This species is resident in the study region.

## Conservation status

LC.

*Ypthima arctous*

- ● Specimen ≥1970
- ■ Observation ≥1970
- ▲ Literature ≥1970
- ● Specimen <1970
- ▲ Literature <1970
- ✕ Putative larval food plants

*Ypthima arctous*

- ● Species record
- Geographic range
- – – – Phytogeographical boundary
- ········· IBRA bioregional boundary

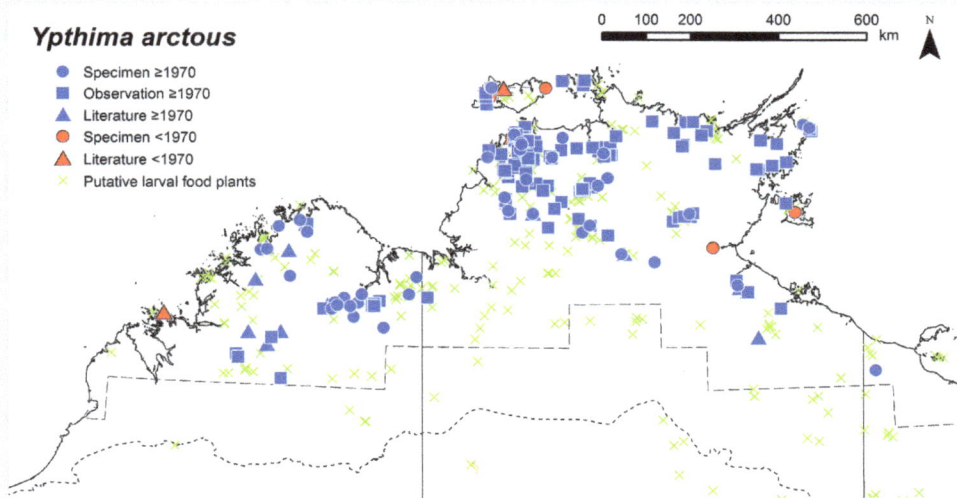

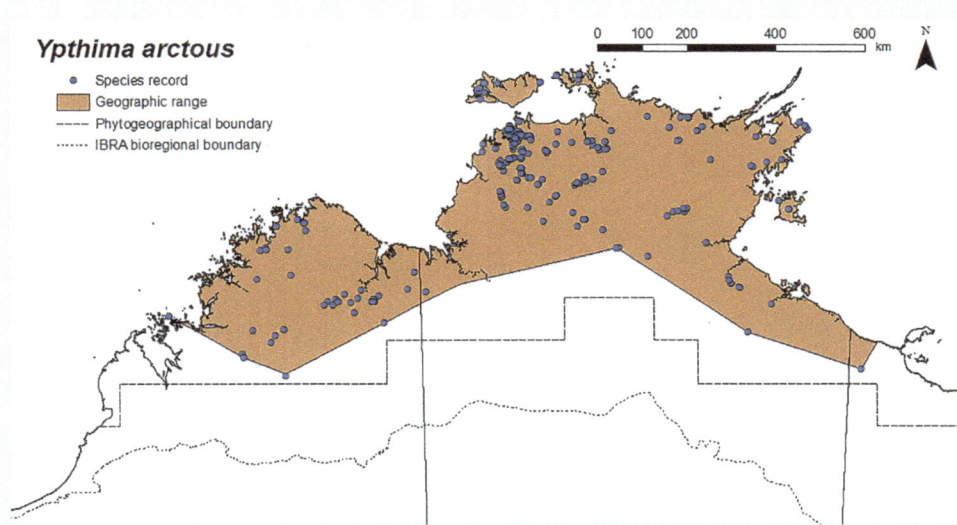

*Ypthima arctous* (n = 304)

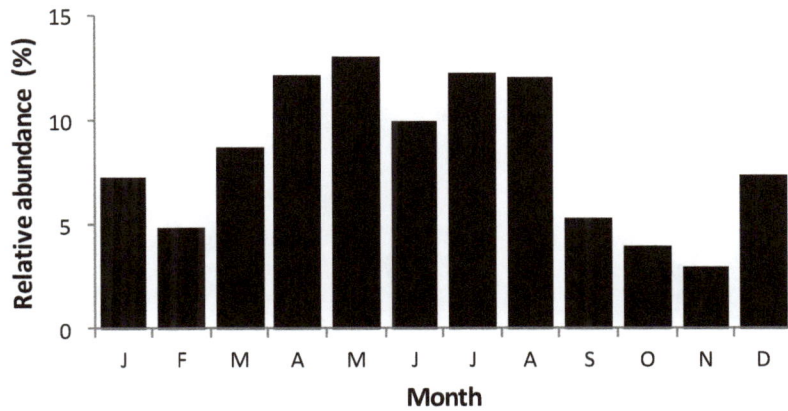

| Month | J | F | M | A | M | J | J | A | S | O | N | D |
|-------|---|---|---|---|---|---|---|---|---|---|---|---|
| Egg | | | | | | | | | | | | |
| Larva | | | | | | | | | | | | |
| Pupa | | | | | | | | | | | | |
| Adult | | | | | | | | | | | | |

# Blues

(Lycaenidae)

# 9

# Moth Butterfly
## *Liphyra brassolis* Westwood, [1864]

Plate 127 Holmes Jungle, Darwin, NT
Photo: Tissa Ratnayeke

Plate 128 Humpty Doo, NT
Photo: Alison Worsnop

## Distribution

This species is represented in the study region by the subspecies *L. brassolis major* Rothschild, 1898. It has a disjunct distribution, occurring mainly in the higher rainfall areas (> 1,000 mm mean annual rainfall) of the Kimberley and the Top End. It has been recorded as far inland as the King Leopold Ranges at Mount Elizabeth Station 6 km north–north-east of Joint Hill, WA (G. Swann), and Nitmiluk National Park (Katherine Gorge), and further east at Groote Eylandt (Yedikba) (Tindale 1923), NT. It has also been recorded at Walker Creek 36 km east of Karumba, Qld, just outside the eastern boundary of the study region (Braby 2015d). The geographic range closely corresponds with the spatial distribution of its associated ant. The ant, however, also occurs in the eastern Kimberley; thus, further field surveys are required to determine whether *L. brassolis* occurs in the intervening areas of its known range in the Joseph Bonaparte Gulf region, particularly the area between the Ord and Victoria rivers. Outside the study region, *L. brassolis* occurs widely from Sikkim in northern India and South-East Asia, through mainland New Guinea and adjacent islands and north-eastern Australia to the Solomon Islands.

## Habitat

*Liphyra brassolis* breeds in a wide range of habitats, wherever arboreal nests of the associated ant are established in the canopy of tall shrubs and trees. The ants are sensitive to fire and thus are more prevalent in wet and dry monsoon forests and riparian forest because these habitats are less prone to fire (Andersen et al. 2007a).

## Attendant ant/larval food

*Oecophylla smaragdina* (Formicidae: Formicinae). The larvae are predacious on the immature stages of the Green Tree Ant (*O. smaragdina*) (Eastwood and Fraser 1999; Braby 2000).

## Seasonality

Adults occur throughout the year. They appear to show little seasonal variation in abundance, although they have been recorded more frequently in June and November, but this may be an artefact of small sample size. We have very few data on the occurrence of the immature stages and the breeding phenology is not well understood. Presumably, the species breeds continuously throughout the year.

## Breeding status

This species is resident in the study region.

## Conservation status

LC.

## Liphyra brassolis

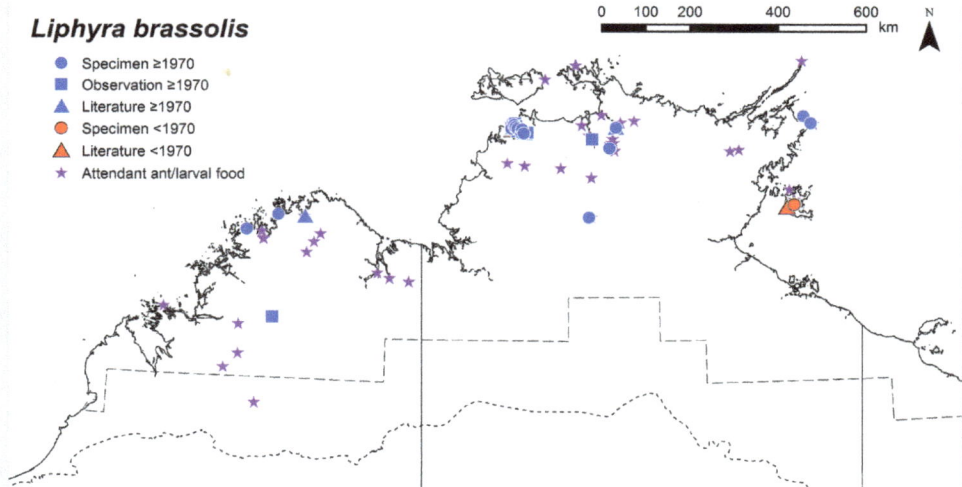

Legend:
- Specimen ≥1970
- Observation ≥1970
- Literature ≥1970
- Specimen <1970
- Literature <1970
- Attendant ant/larval food

## Liphyra brassolis

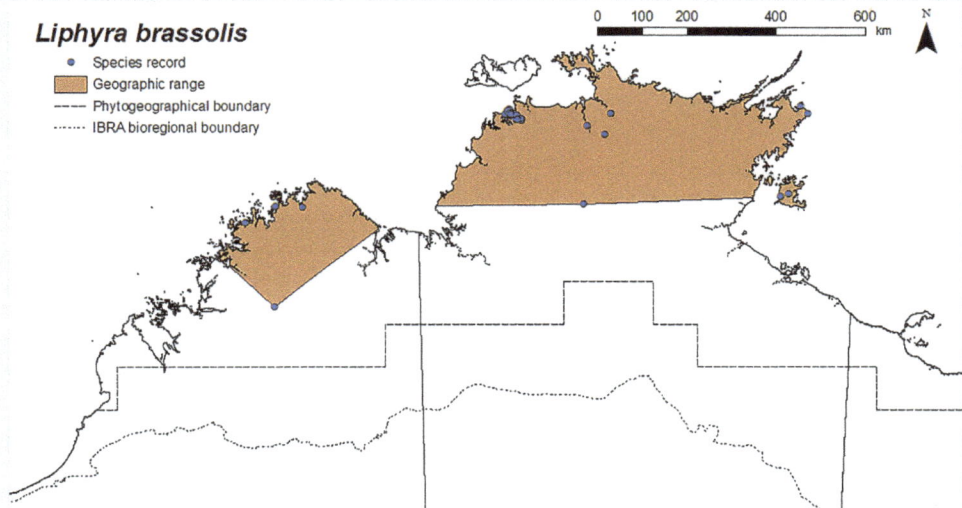

Legend:
- Species record
- Geographic range
- Phytogeographical boundary
- IBRA bioregional boundary

## Liphyra brassolis (n = 31)

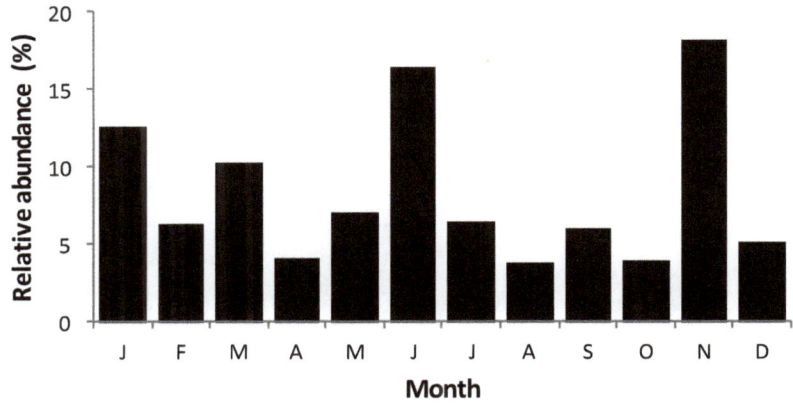

| Month | J | F | M | A | M | J | J | A | S | O | N | D |
|-------|---|---|---|---|---|---|---|---|---|---|---|---|
| Egg |  |  | ▉ |  |  |  |  |  |  |  |  |  |
| Larva |  |  |  |  |  |  |  |  |  |  |  |  |
| Pupa |  |  |  |  |  |  |  |  |  |  |  |  |
| Adult | ▉ | ▉ | ▉ | ▉ | ▉ | ▉ | ▉ | ▉ | ▉ | ▉ | ▉ | ▉ |

# Small Ant-blue
## *Acrodipsas myrmecophila* (Waterhouse & Lyell, 1913)

Plate 129 Canberra, ACT
Photo: M. F. Braby

## Distribution

This species is known from the Top End of the study region, where it has been recorded only at Mt Burrell on Tipperary Station, NT (Dunn and Dunn 1991; Braby 2000). Its presence in the region was detected only as recently as 1981. The putative attendant ants (*Papyrius* spp.) are very widespread in the region, occurring in the Kimberley, Top End and western Gulf Country, from moist coastal areas to drier inland areas of the semi-arid zone (> 500 mm mean annual rainfall). Further field surveys are thus required to determine whether *A. myrmecophila* is equally widespread. Outside the study region, *A. myrmecophila* occurs sporadically from northern Queensland, through NSW and ACT to Victoria.

## Habitat

The breeding habitat of *A. myrmecophila* has not been recorded in the study region. Adults have been collected in the canopy of trees in low open woodland on a hilltop. Presumably, the species breeds in savannah woodland where colonies of the putative associated ant are established.

## Attendant ant/larval food

Not recorded in the study region; probably *Papyrius* spp. (Formicidae: Dolichoderinae). In eastern Australia, the larvae are predacious on the immature stages of *Papyrius* ants (Eastwood and Fraser 1999; Braby 2000).

## Seasonality

The seasonal abundance and breeding phenology of this species are not well understood. Adults have been collected only during the dry season (July and October) (Dunn and Dunn 1991; Braby 2000), but there are too few records (n = 2) to assess any seasonal changes in abundance.

## Breeding status

This species is resident in the study region.

## Conservation status

DD. The species *A. myrmecophila* has an extremely limited range in the study region (AOO is likely to be < 2,000 sq km, with spatial buffering of records providing a first approximation of 700 sq km). It is currently known from only one site, which lies entirely within a pastoral property that lacks adequate conservation management due to threats from the invasion and spread of grassy weeds (particularly gamba grass, *Andropogon gayanus*), inappropriate fire regimes, particularly an increase in fire frequency, and extensive clearing for cattle grazing of savannah woodland surrounding the hilltop (i.e. potential breeding habitat). Moreover, the species appears to be rare in that very few specimens have been collected, especially during the past two decades. However, the geographic range is incomplete due to the difficulties of detecting and sampling the taxon; breeding colonies are very localised and adults are small, highly seasonal and frequently fly in the canopy at the summit of steep hills. The spatial distribution of the putative attendant ant—based on records in the TERC and the occurrence of the Fiery Jewel (*Hypochrysops ignitus*), which is associated with the same ant— suggest *A. myrmecophila* may be very widespread. Thus, the taxon may qualify as LC rather than Near Threatened (NT) once adequate data are available. Sands and New (2002) also concluded that DD was the most appropriate Red List category for this species in the Northern Territory, in contrast with its threatened status elsewhere in Australia. Nevertheless, targeted field surveys (e.g. on hilltops) to clarify the extent of its distribution should be a priority for this species.

## *Acrodipsas myrmecophila*

- ● Specimen ≥1970
- ▲ Literature ≥1970
- ★ Attendant ant

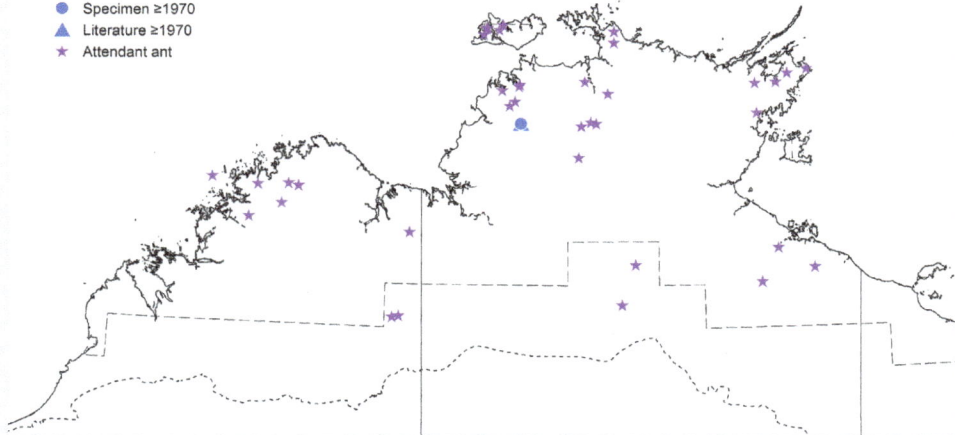

| Month | J | F | M | A | M | J | J | A | S | O | N | D |
|-------|---|---|---|---|---|---|---|---|---|---|---|---|
| Egg   |   |   |   |   |   |   |   |   |   |   |   |   |
| Larva |   |   |   |   |   |   |   |   |   |   |   |   |
| Pupa  |   |   |   |   |   |   |   |   |   |   |   |   |
| Adult |   |   |   |   |   |   |   |   |   |   |   |   |

Photo: Litchfield National Park, NT, M.F. Braby

# Northern Ant-blue
*Acrodipsas decima* Miller & Lane, 2004

Plate 130 Mt Burrell, NT
Photo: M. F. Braby

Plate 131 Mt Burrell, NT
Photo: M. F. Braby

## Distribution

This species is endemic to the study region. It has been recorded only at Mt Burrell on Tipperary Station, NT. Its presence in the Top End was detected only as recently as 1991 (Miller and Lane 2004). Further field surveys are required to determine whether *A. decima* occurs elsewhere in the Top End.

## Habitat

The breeding habitat of *A. decima* has not been recorded in the study region. Adults have been collected in the canopy of trees in low open woodland at the summit of a hill (Miller and Lane 2004). Presumably, the species breeds in savannah woodland and plains surrounding steep hills and mesas.

## Attendant ant/larval food

Not recorded.

## Seasonality

The seasonal abundance and breeding phenology of this species are not well understood. Adults have been recorded sporadically during the wet season and early dry season (November–May), with most specimens collected in April and May (Miller and Lane 2004), but there are too few records (n = 14) to assess any seasonal changes in abundance.

## Breeding status

This species is resident in the study region.

## Conservation status

DD. The species *A. decima* appears to be a short-range endemic (AOO is likely to be < 2,000 sq km, with spatial buffering of records providing a first approximation of 700 sq km). It is currently known from only one site, which lies entirely within a pastoral property that lacks adequate conservation management due to threats from the invasion and spread of grassy weeds (particularly gamba grass, *Andropogon gayanus*), inappropriate fire regimes, particularly an increase in fire frequency, and extensive clearing for cattle grazing of savannah woodland surrounding the hilltop (i.e. potential breeding habitat). Moreover, it is rarely encountered and few specimens have been collected during the past decade. Thus, the taxon may qualify as Near Threatened (NT) once adequate data confirm its restricted AOO and limited number of locations. Currently, the geographic range is incomplete due to the difficulties of detecting and sampling the taxon: adults are small, seasonal and appear to fly in the canopy at the summit of steep hills. Targeted field surveys (e.g. on hilltops) to clarify the extent of its distribution and determine its critical habitat should be a high priority for this species.

## Acrodipsas decima

- ● Specimen ≥1970
- ▲ Literature ≥1970

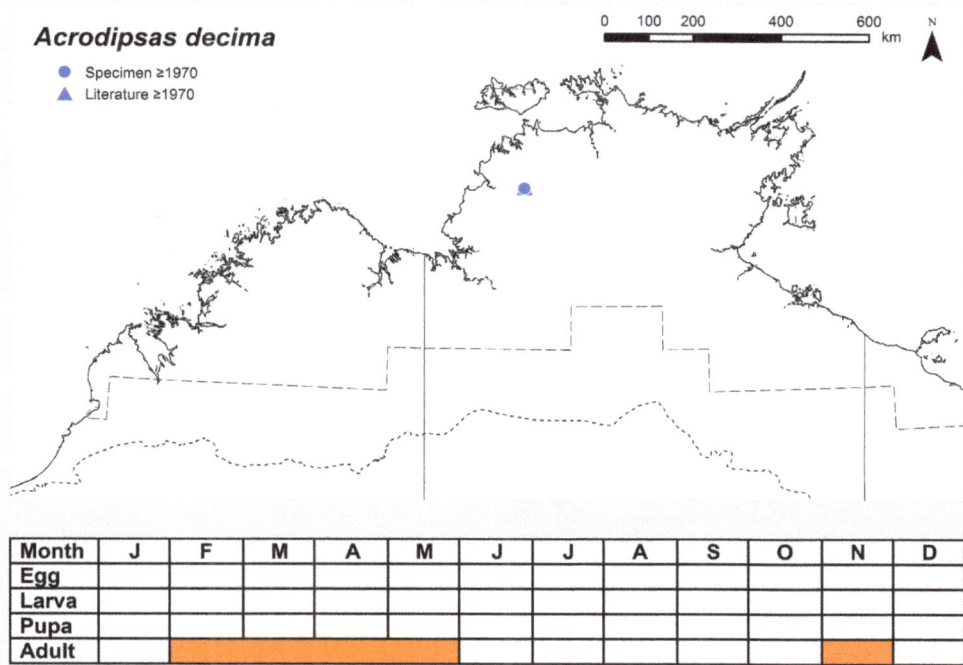

| Month | J | F | M | A | M | J | J | A | S | O | N | D |
|-------|---|---|---|---|---|---|---|---|---|---|---|---|
| Egg | | | | | | | | | | | | |
| Larva | | | | | | | | | | | | |
| Pupa | | | | | | | | | | | | |
| Adult | | | | | | | | | | | | |

Photo: Mt Burrell, NT, M.F. Braby

# Copper Jewel
## *Hypochrysops apelles* (Fabricius, 1775)

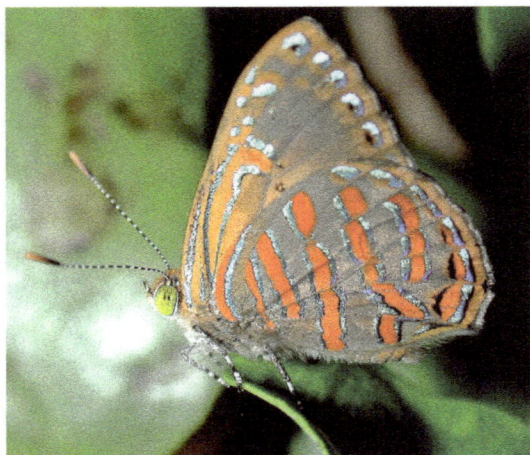

Plate 132 Buffalo Creek, Leanyer Swamp, NT
Photo: M. F. Braby

## Distribution

This species is represented by an undescribed subspecies, which is endemic to the study region. It occurs in the north of the Northern Territory, where it is restricted to coastal areas of the Top End, its presence in the region first reported by Common and Waterhouse (1981) based on a male specimen collected from Casuarina Beach near Darwin in 1973 by E. D. Edwards. More recently, it has been recorded further east near Maningrida, NT (Bisa 2013). The larval food plants occur very widely in coastal areas of the study region, suggesting the known geographic range of *H. apelles* may be incomplete and undersampled, possibly due to the difficulties of accessing the butterfly's habitat. Further field surveys are thus required to determine whether *H. apelles* occurs elsewhere, particularly in coastal areas of the Kimberley, north-eastern Arnhem Land and the western Gulf Country. Outside the study region, *H. apelles* occurs in the Aru Islands, mainland New Guinea and adjacent islands and north-eastern and eastern Australia.

## Habitat

*Hypochrysops apelles* breeds mainly in the landward edge of mangroves where its main larval food plant (*Lumnitzera racemosa*) grows as a small tree in relatively high densities and colonies of the attendant ant are established. Breeding colonies are very localised within these habitats.

## Larval food plants

*Lumnitzera racemosa* (Combretaceae), *Ceriops australis* (Rhizophoraceae). The main food plant is *L. racemosa* (Meyer 1996a; Eastwood et al. 2008), but occasionally *C. australis* is used during the early dry season when adult population numbers are higher (Braby 2011a).

## Attendant ant

*Crematogaster* sp. (Formicidae: Myrmicinae). The larvae and pupae are constantly attended by numerous small black ants in an obligatory myrmecophilous association (Eastwood and Fraser 1999; Braby 2000).

## Seasonality

Adults have been recorded during most months of the year, but there are too few records (n = 19) to assess any seasonal changes in abundance. In general, adults and the immature stages (larvae) appear to be more numerous during the late wet season and early dry season (April and May) and are scarce in the late dry season. The immature stages have been recorded sporadically during the dry season (April–September). Presumably, the species breeds continuously throughout the year and several generations are completed annually.

## Breeding status

This species is resident in the study region.

## Conservation status

Near Threatened (NT). Available data suggest the putative subspecies *H. apelles* ssp. 'Top End' is a short-range endemic (EOO = 4,320 sq km) and occurs in at least two conservation reserves: Casuarina Coastal Reserve and Djelk IPA. However, it may qualify for a threatened category in the near future because much of the intervening mangrove habitat between these two locations is not protected from coastal development impacts and therefore the butterfly populations may be at risk (Sands and New 2002), particularly in the Darwin area, where it is known only from a limited number of sites (Casuarina Coastal Reserve, Buffalo Creek, Channel Island and Palmerston). The population of *Hypochrysops apelles* ssp. 'Top End' is likely to be reduced in future based on a projected decline in the AOO and/or number of locations (criterion D2). Further field surveys to clarify its distribution and monitor its occupancy are required for this species.

## Hypochrysops apelles

- ● Specimen ≥1970
- ■ Observation ≥1970
- ▲ Literature ≥1970
- ✕ Larval food plants

## Hypochrysops apelles

- ● Species record
- ▨ Geographic range
- — — Phytogeographical boundary
- ......... IBRA bioregional boundary

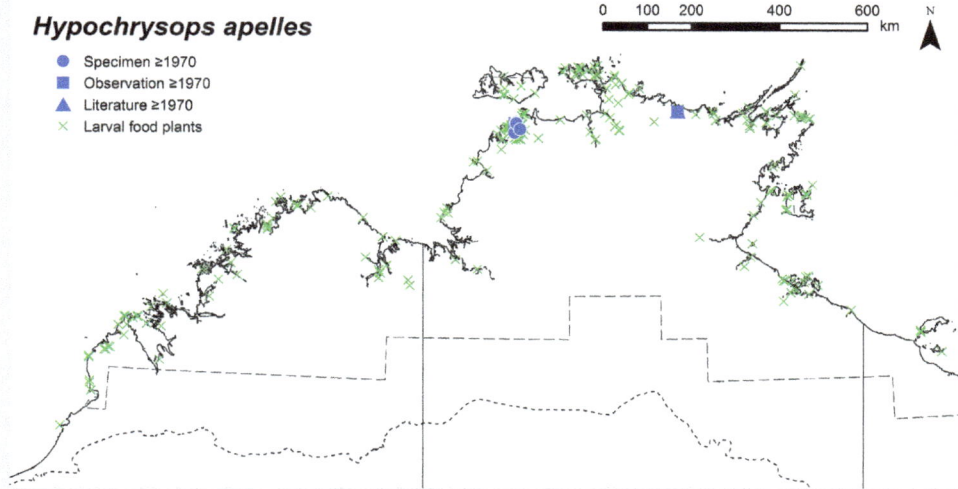

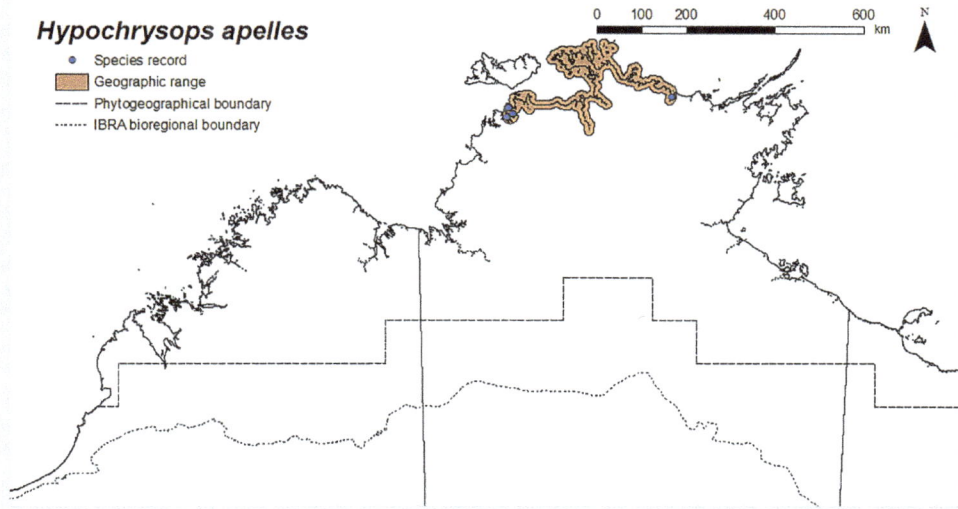

| Month | J | F | M | A | M | J | J | A | S | O | N | D |
|-------|---|---|---|---|---|---|---|---|---|---|---|---|
| Egg   |   |   |   |   | ▨ |   |   |   |   |   |   |   |
| Larva |   |   |   | ▨ | ▨ | ▨ | ▨ |   | ▨ | ▨ |   |   |
| Pupa  |   |   |   | ▨ | ▨ | ▨ |   |   | ▨ |   |   |   |
| Adult |   | ▨ | ▨ | ▨ | ▨ | ▨ | ▨ |   | ▨ | ▨ | ▨ |   |

# Fiery Jewel
## *Hypochrysops ignitus* (Leach, 1814)

Plate 133 Berry Springs, NT
Photo: M. F. Braby

## Distribution

This species is represented by the subspecies *H. ignitus erythrina* (Waterhouse & Lyell, 1909), which is endemic to the study region. It has a wide but sporadic distribution, extending from the high rainfall areas of the Tiwi Islands (> 1,800 mm mean annual rainfall) to drier inland areas of the semi-arid zone (< 500 mm). It occurs mainly in the northern and eastern Kimberley and in the Top End, where it has been recorded as far inland as 98 km south-west of Katherine, NT (Braby 2000). However, it has also been recorded from remote areas further south, including Broome (Williams et al. 1992; Johnson and Valentine 2004) and the Edgar Ranges (Common 1981), WA, in the southern Kimberley; and from Doomadgee (Puccetti 1991) and Bentinck Island in the South Wellesley Islands, Qld (Daniels 2005), in the western Gulf Country. These populations have not been placed to subspecies, but are likely to be *H. ignitus erythrina*, based on geographic grounds (the type locality is Darwin) (Waterhouse and Lyell 1909; Sands 1986). The geographical range of *H. ignitus* broadly corresponds with the spatial distribution of its associated ant. Although the larval food plants are more widely distributed, the ant occurs mainly in areas with more than 500 mm mean annual rainfall. Further field surveys are required in the Kimberley, Northern Deserts and the Limmen Bight area in the Gulf of Carpentaria to determine whether *H. ignitus* occurs in the intervening areas between its known geographic range. Outside the study region, *H. ignitus* occurs in southern New Guinea and throughout the coastal and near-coastal areas of north-eastern, eastern, south-eastern and south-western Australia.

## Habitat

*Hypochrysops ignitus* breeds mainly in savannah woodland on a variety of substrates, but often loamy sand or sand, but also in the ecotone between savannah woodland and monsoon forest where the larval food plants grow as shrubs or saplings regenerating after fire and where colonies of the attendant ant are established. Breeding colonies of the butterfly are very localised within these habitats.

## Larval food plants

*Maranthes corymbosa* (Chrysobalanaceae), *Glochidion apodogynum* (Phyllanthaceae), *Acacia leptocarpa*, *A. tumida* (Fabaceae), *Clerodendrum floribundum* (Lamiaceae), *Planchonia careya* (Lecythidaceae), *Brachychiton paradoxus* (Malvaceae), *Alphitonia excelsa* (Rhamnaceae), *Gardenia megasperma* (Rubiaceae), *Smilax australis* (Smilacaceae). The larvae feed on a wide range of plants, but they are frequently found on *Planchonia careya*. The record of *Brachychiton paradoxus* by Meyer (1996a) may well be *B. megaphyllus*, because these two species have been confused in the past (Franklin and Bate 2013).

## Attendant ants

*Papyrius* spp. (*nitidus* species group) (Formicidae: Dolichoderinae). The larvae and pupae are constantly attended by numerous Coconut Ants (*Papyrius* spp.) in an obligatory myrmecophilous association (Eastwood and Fraser 1999; Braby 2000).

## Seasonality

Adults occur during most months of the year. They appear to show little seasonal variation in abundance, although they tend to be more numerous during the wet season (December–March). The immature stages have been recorded during all months except October, indicating that the species breeds throughout the year, during which several generations are completed.

## Breeding status

This species is resident in the study region.

## Conservation status

LC.

 

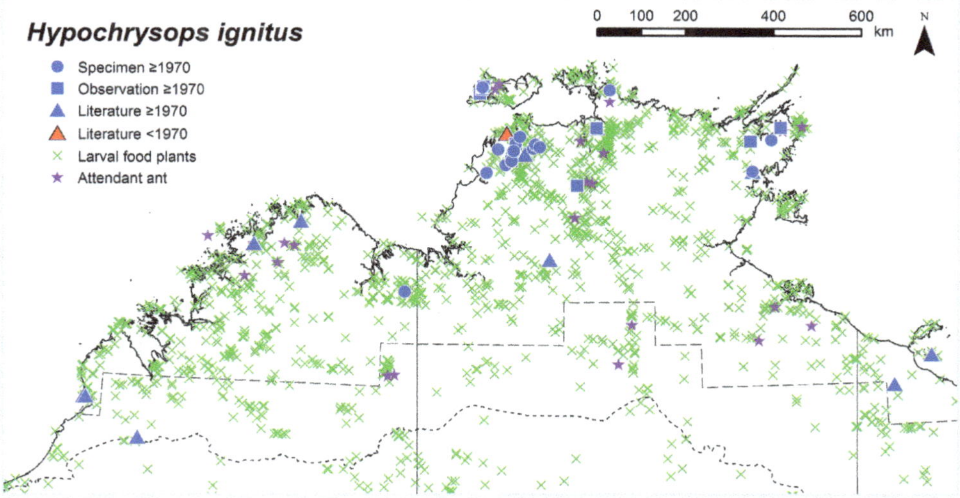
Hypochrysops ignitus
0 100 200 400 600
km
N
Specimen ≥1970
Observation ≥1970
Literature ≥1970
Literature <1970
Larval food plants
Attendant ant

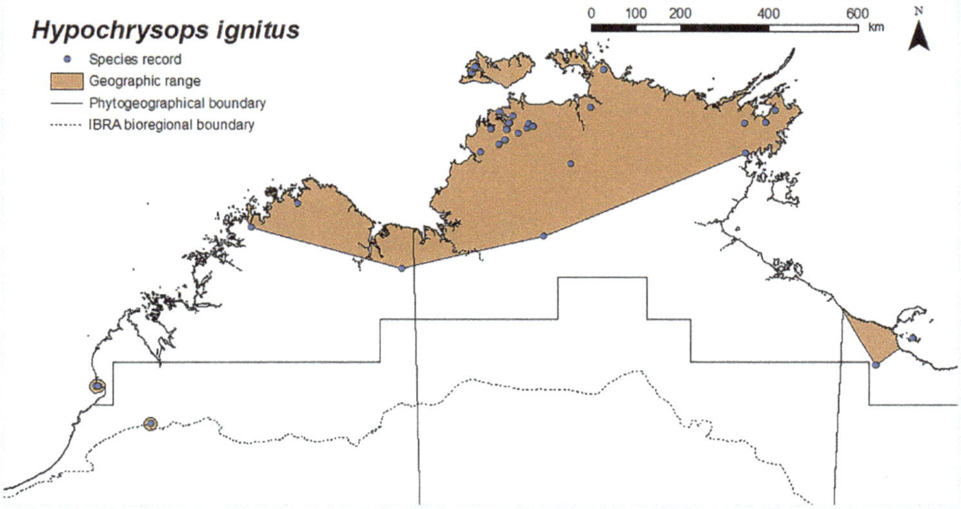
Hypochrysops ignitus
0 100 200 400 600
km
N
Species record
Geographic range
Phytogeographical boundary
IBRA bioregional boundary

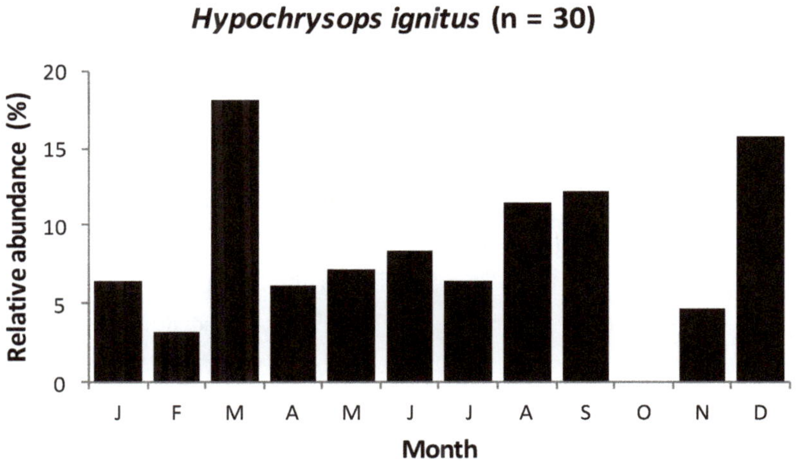
Hypochrysops ignitus (n = 30)
Relative abundance (%)
20
15
10
5
0
J F M A M J J A S O N D
Month

| Month | J | F | M | A | M | J | J | A | S | O | N | D |
|---|---|---|---|---|---|---|---|---|---|---|---|---|
| Egg | | | | | | | | | | | | |
| Larva | | | | | | | | | | | | |
| Pupa | | | | | | | | | | | | |
| Adult | | | | | | | | | | | | |

# Purple Oak-blue
## *Arhopala eupolis* (Miskin, 1890)

Plate 134 Mary River, NT
Photo: Don Franklin

## Distribution

This species is represented by the subspecies *A. eupolis asopus* Waterhouse & Lyell, 1914, which is endemic to the study region. It occurs in the Kimberley and throughout the Top End, extending from moist coastal areas to the semi-arid zone (c. 700 mm mean annual rainfall). Its southernmost limits include Windjana Gorge National Park, WA (Williams et al. 2006); and Keep River National Park (Jinumum Gorge), Beward Lagoon 20 km west of Hodgson Downs, and Limmen River Fishing Camp (S. Normand), NT. The geographical range of *A. eupolis* broadly corresponds with the spatial distribution of its associated ant. Although the larval food plants are more widely distributed, particularly in the Northern Deserts and western Gulf Country, the attendant ant is absent from these areas and is much more restricted in extent (Lokkers 1986). Outside the study region, *A. eupolis* occurs from the Kai and Aru islands, through mainland New Guinea and north-eastern Australia to the Louisiade Archipelago.

## Habitat

*Arhopala eupolis* occurs in a variety of habitats, but breeds mainly in savannah woodland where the larval food plants grow as understorey trees or as saplings regenerating after fire and where arboreal nests of the attendant ant are established (Braby 2011a, 2015e). It also breeds in mixed woodland–monsoon forest associations and along the edges of riparian evergreen monsoon vine forest, particularly in drier inland areas.

## Larval food plants

*Buchanania obovata* (Anacardiaceae), *Maranthes corymbosa* (Chrysobalanaceae), *Terminalia carpentariae*, *T. ferdinandiana* (Combretaceae), *Corymbia bella*, *Corymbia disjuncta*, *Eucalyptus miniata* (Myrtaceae), *Cupaniopsis anacardioides* (Sapindaceae); also *\*Syzygium aqueum* (Myrtaceae).

## Attendant ant

*Oecophylla smaragdina* (Formicidae: Formicinae). The larvae and pupae are constantly attended by numerous Green Tree Ants (*O. smaragdina*) in an obligatory myrmecophilous association (Eastwood and Fraser 1999; Braby 2000).

## Seasonality

Adults occur throughout the year, but they are generally more abundant during the wet season (November–April). The immature stages have also been recorded over a similar period, as well as in the dry season (August). The larvae feed on the new soft leaf growth and the life cycle is completed relatively quickly, with no evidence of diapause in any of the life history stages. Presumably, the species breeds continuously throughout the year and several generations are completed annually by switching its food plants on a seasonal basis according to the availability of new foliage.

## Breeding status

This species is resident in the study region.

## Conservation status

LC.

## Arhopala eupolis

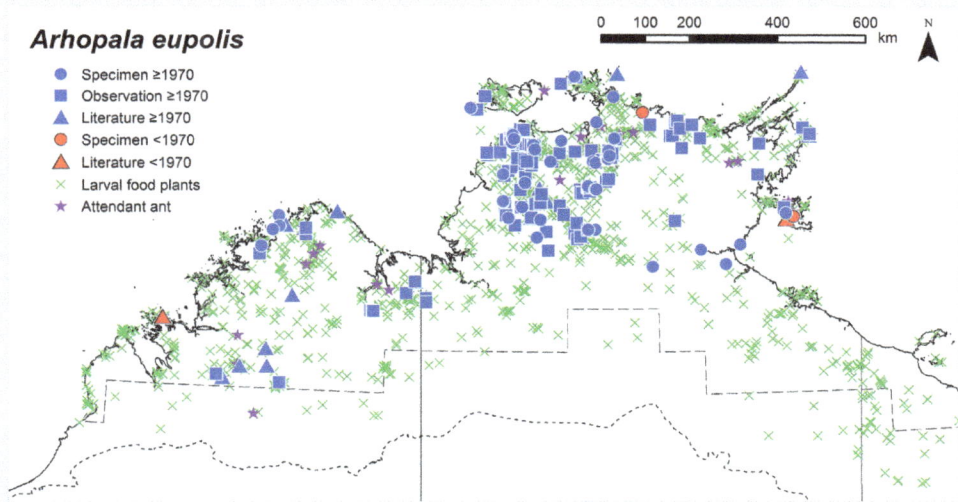

Legend:
- ● Specimen ≥1970
- ■ Observation ≥1970
- ▲ Literature ≥1970
- ● Specimen <1970
- ▲ Literature <1970
- ✕ Larval food plants
- ★ Attendant ant

## Arhopala eupolis

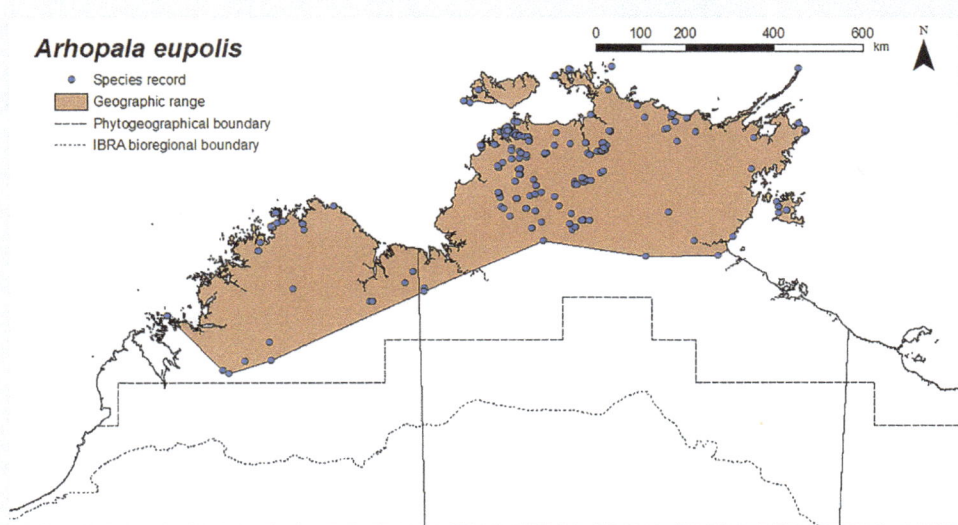

Legend:
- ● Species record
- ▨ Geographic range
- – – – Phytogeographical boundary
- ········· IBRA bioregional boundary

## Arhopala eupolis (n = 331)

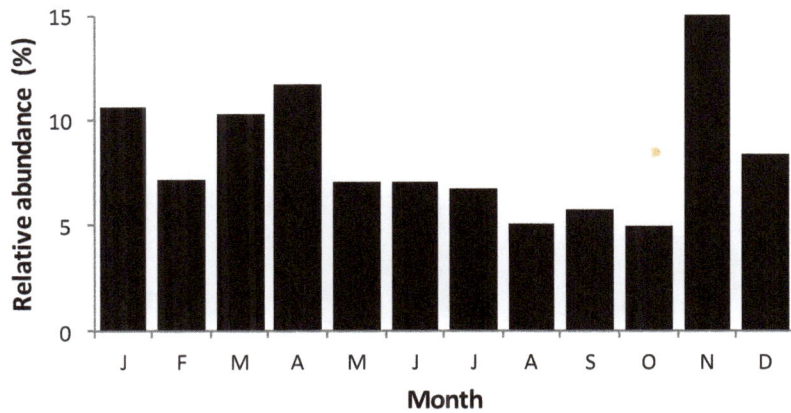

| Month | J | F | M | A | M | J | J | A | S | O | N | D |
|-------|---|---|---|---|---|---|---|---|---|---|---|---|
| Egg   |   |   |   | ▣ |   |   |   |   |   |   | ▣ |   |
| Larva |   | ▣ | ▣ |   |   |   |   | ▣ |   |   | ▣ | ▣ |
| Pupa  |   |   | ▣ | ▣ |   |   |   | ▣ |   |   | ▣ | ▣ |
| Adult | ▣ | ▣ | ▣ | ▣ | ▣ | ▣ | ▣ | ▣ | ▣ | ▣ | ▣ | ▣ |

# Shining Oak-blue
## *Arhopala micale* Blanchard, [1848]

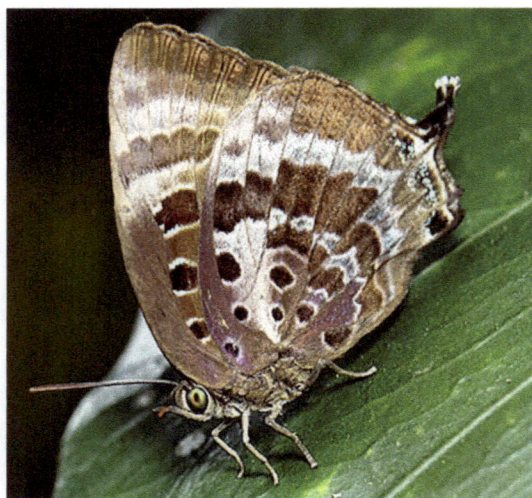

Plate 135 Holmes Jungle, Darwin, NT
Photo: Tissa Ratnayeke

Plate 136 Kakadu National Park, NT
Photo: M. F. Braby

## Distribution

This species is represented by an undescribed subspecies, which is endemic to the study region. It is restricted to the Top End, where it occurs mainly in the higher rainfall areas (> 800 mm mean annual rainfall). It has been recorded as far south as Bradshaw Field Training Area (Angalarri River catchment) (Archibald and Braby 2017), Beward Lagoon 20 km west of Hodgson Downs, and Limmen River Fishing Camp (S. Normand), NT. The geographical range of *A. micale* closely corresponds with the spatial distribution of its associated ant within the Top End. Although the larval food plants and attendant ant also occur widely in the Kimberley, searches have not detected *A. micale* in this area. Similarly, although the larval food plants are more widely distributed in the Northern Deserts and western Gulf Country, the ant is absent from these areas and is much more restricted in extent (Lokkers 1986). Outside the study region, *A. micale* occurs from Maluku, including the Kai and Aru islands, through mainland New Guinea and adjacent islands and north-eastern Australia to the Louisiade Archipelago.

## Habitat

*Arhopala micale* breeds mainly in coastal monsoon vine thicket and evergreen monsoon vine forest associated with seepages or permanent freshwater streams where the larval food plants grow as understorey trees and where arboreal nests of the attendant ant are established.

## Larval food plants

*Calophyllum inophyllum* (Clusiaceae), *Brachychiton diversifolius*, *Sterculia quadrifida* (Malvaceae), *Cupaniopsis anacardioides* (Sapindaceae).

## Attendant ant

*Oecophylla smaragdina* (Formicidae: Formicinae). The larvae and pupae are constantly attended by numerous Green Tree Ants (*O. smaragdina*) in an obligatory myrmecophilous association (Eastwood and Fraser 1999; Braby 2000).

## Seasonality

Adults occur throughout the year, but they are generally more abundant during the dry season (April–July). The immature stages have been recorded more sporadically during the year. The larvae feed on the new soft leaf growth and the life cycle is completed relatively quickly, with no evidence of diapause in any of the life history stages. Presumably, the species breeds continuously throughout the year and several generations are completed annually by switching its food plants on a seasonal basis according to the availability of new foliage.

## Breeding status

This species is resident in the study region.

## Conservation status

LC.

## Arhopala micale

- ● Specimen ≥1970
- ■ Observation ≥1970
- ▲ Literature ≥1970
- ● Specimen <1970
- ▲ Literature <1970
- ✕ Larval food plants
- ★ Attendant ant

0  100  200    400    600
km                        N

## Arhopala micale

- ● Species record
- ▨ Geographic range
- — — Phytogeographical boundary
- ······· IBRA bioregional boundary

0  100  200    400    600
km                        N

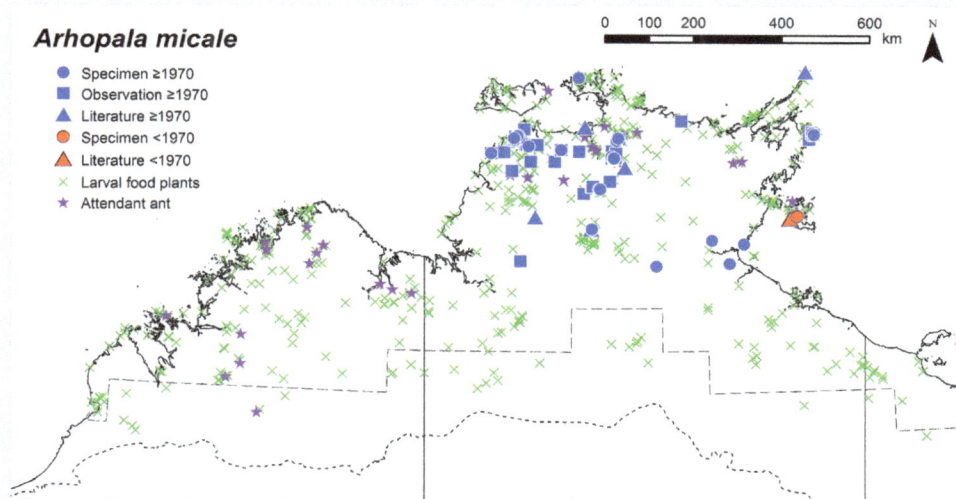

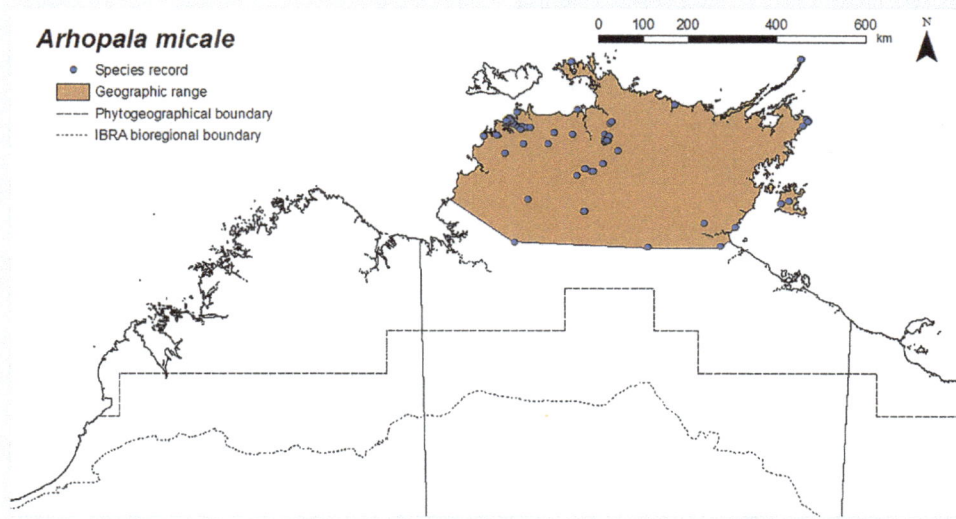

## Arhopala micale (n = 99)

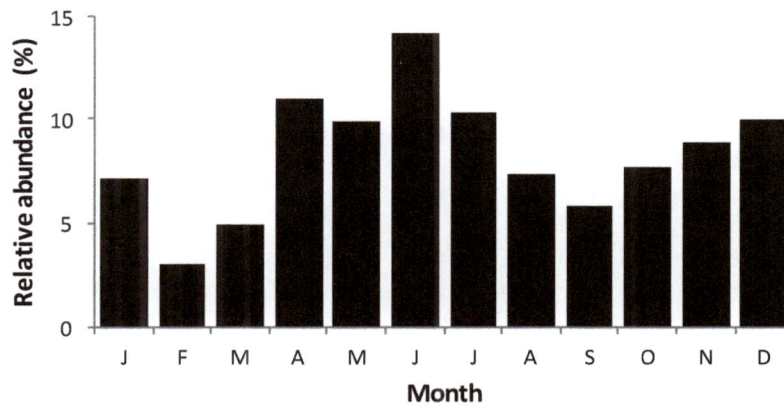

| Month | J | F | M | A | M | J | J | A | S | O | N | D |
|-------|---|---|---|---|---|---|---|---|---|---|---|---|
| Egg   | ■ |   |   | ■ |   |   |   |   |   |   |   | ■ |
| Larva | ■ |   |   | ■ |   |   |   |   |   |   |   | ■ |
| Pupa  |   |   |   |   |   |   | ■ |   |   | ■ |   | ■ |
| Adult | ■ | ■ | ■ | ■ | ■ | ■ | ■ | ■ | ■ | ■ | ■ | ■ |

# Silky Azure

## *Ogyris oroetes* (Hewitson, 1862)

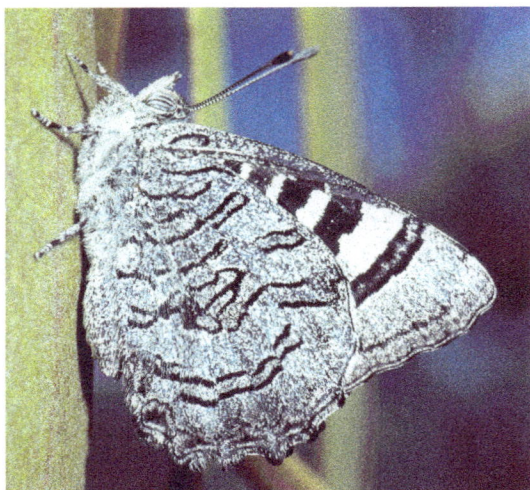

Plate 137 Calperum Station, SA
Photo: M. F. Braby

## Distribution

This species is represented in the study region by the subspecies *O. oroetes oroetes* (Hewitson, 1862). It has a wide but sporadic distribution in the Kimberley, Top End and western Gulf Country, where it occurs mainly in the lower rainfall areas (700–1,400 mm mean annual rainfall). It has been recorded mainly from inland areas in the eastern Kimberley and the western half of the Top End, but it has also been recorded as far west as Derby, WA (Waterhouse and Lyell 1914), and as far east as Borroloola, NT. The larval food plant is considerably more widespread than the known geographic range of *O. oroetes*, particularly in the Kimberley, Northern Deserts and western Gulf Country. Further field surveys are therefore required to determine whether *O. oroetes* occurs in the intervening areas between its known ranges. Outside the study region, *O. oroetes* occurs widely in eastern, central and southern Australia.

## Habitat

*Ogyris oroetes* breeds in savannah woodland where the larval food plant grows as a mistletoe (parasitic shrub) on eucalypts, particularly *Corymbia* spp. (Braby 2011a). Males also briefly visit hilltops to locate females, but they do not breed in these habitats.

## Larval food plant

*Amyema bifurcata* (Loranthaceae).

## Attendant ant

Not recorded in the study region. In the arid zone of the Northern Territory outside the study region, larvae and pupae are occasionally attended by *Crematogaster* sp. (*longiceps* group) ants (Eastwood et al. 2008; Braby 2011a). Elsewhere, they are attended by a few ants from several genera in a facultative myrmecophilous association (Eastwood and Fraser 1999; Braby 2000).

## Seasonality

The seasonal abundance and breeding phenology of this species are not well understood. Adults have been recorded during most months of the year, but we have too few records (n = 15) to assess any seasonal changes in abundance. Preliminary observations suggest that adults, like *Ogyris amaryllis*, are more numerous in the second half of the year during the mid to late dry season and 'build-up' (July–December), but are scarce during the wet season and early dry season (January–June). We have few data on the incidence of the immature stages, with pupae recorded sporadically from February to July. Presumably, the species breeds continuously throughout the year and several generations are completed annually.

## Breeding status

This species is resident in the study region.

## Conservation status

LC. Despite its wide, sporadic range, *O. oroetes* appears to be poorly represented in conservation reserves, and is currently known only from Keep River National Park.

 

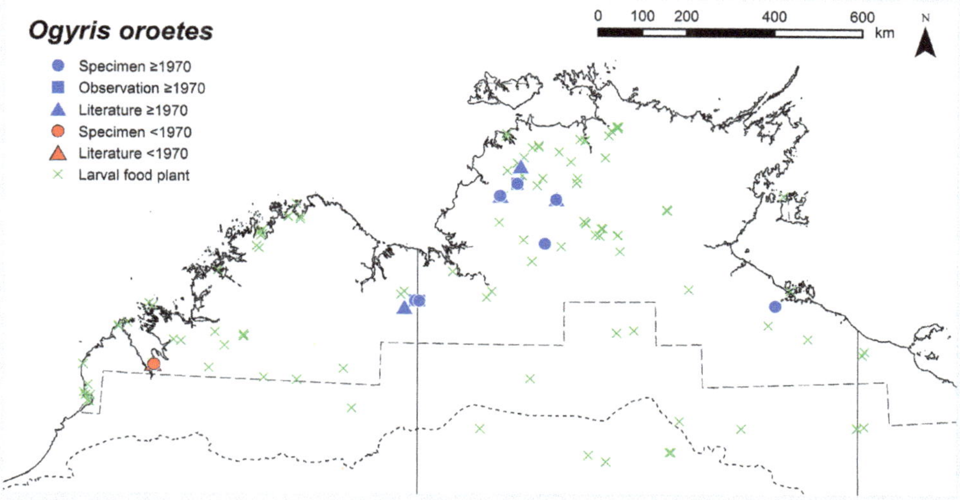

Ogyris oroetes
0 100 200 400 600
km
N
Specimen ≥1970
Observation ≥1970
Literature ≥1970
Specimen <1970
Literature <1970
Larval food plant

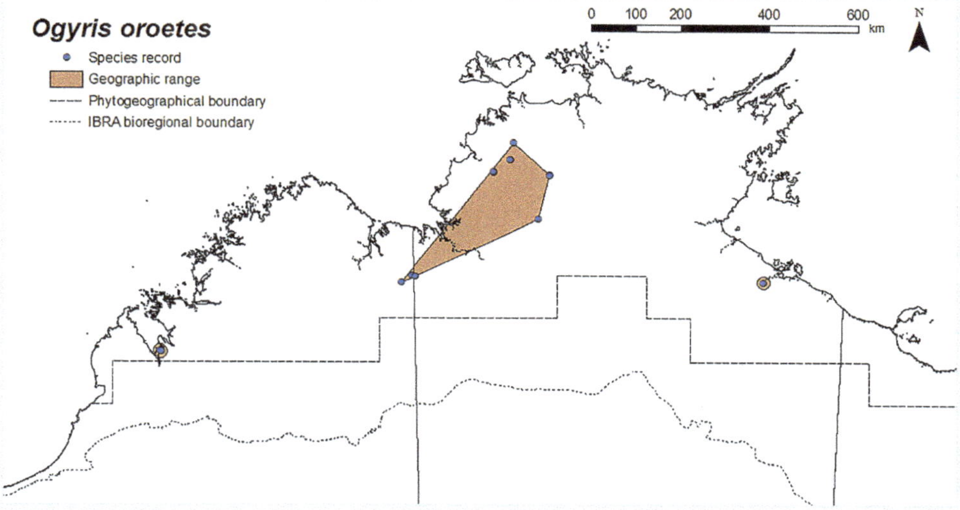

Ogyris oroetes
0 100 200 400 600
km
N
Species record
Geographic range
Phytogeographical boundary
IBRA bioregional boundary

| Month | J | F | M | A | M | J | J | A | S | O | N | D |
|---|---|---|---|---|---|---|---|---|---|---|---|---|
| Egg | | | | | | | | | | | | |
| Larva | | | | | | | | | | | | |
| Pupa | | ■ | | | ■ | | ■ | | | | | |
| Adult | | ■ | | ■ | ■ | ■ | ■ | ■ | ■ | ■ | ■ | ■ |

# Bright Purple Azure
## *Ogyris barnardi* (Miskin, 1890)

Plate 138 Burra Range, Qld
Photo: M. F. Braby

## Distribution

This species is represented in the study region by the subspecies *O. barnardi barnardi* (Miskin, 1890). It is restricted to western Queensland in the Gulf Country, where it occurs in the semi-arid zone (400 mm mean annual rainfall), its presence in the region detected only as recently as 2011 (Dunn 2013). The larval food plant occurs slightly further west in the Northern Territory; however, targeted searches at several sites supporting suitable habitat (in September 2015) failed to detect *O. barnardi*. Outside the study region, *O. barnardi* occurs in the inland areas of Queensland, northern NSW and central southern South Australia.

## Habitat

*Ogyris barnardi* breeds in *Acacia* low open woodland where the larval food plant grows as a mistletoe (parasitic shrub) on *Acacia georginae*.

## Larval food plant

*Amyema quandang* (Loranthaceae).

## Attendant ant

Not recorded in the study region. In Queensland, the larvae and pupae are attended by *Crematogaster* sp. ants in a facultative myrmecophilous association (Eastwood and Fraser 1999; Braby 2000).

## Seasonality

The seasonal abundance and breeding phenology of this species are not well understood. Adults have been recorded only during the warmer 'spring' months of the late dry season (September–November), but we have too few records (n = 9) to assess any seasonal changes in abundance.

## Breeding status

This species is resident in the study region.

## Conservation status

DD. The subspecies *O. barnardi barnardi* has a short range in the study region (EOO = 9,090 sq km). All known sites occur on roadsides and pastoral lands with inadequate protection or management, and are potentially threatened from habitat loss as a result of land clearing for cattle production and/or degradation from roadworks. That is, the population may be reduced in future based on a projected decline in the AOO and/or quality of its habitat. Thus, the taxon may qualify as NT once adequate data are available. Monitoring the extent and/or quality of the critical habitat and identification of the nature and extent of key threatening processes are required for this species.

## Ogyris barnardi

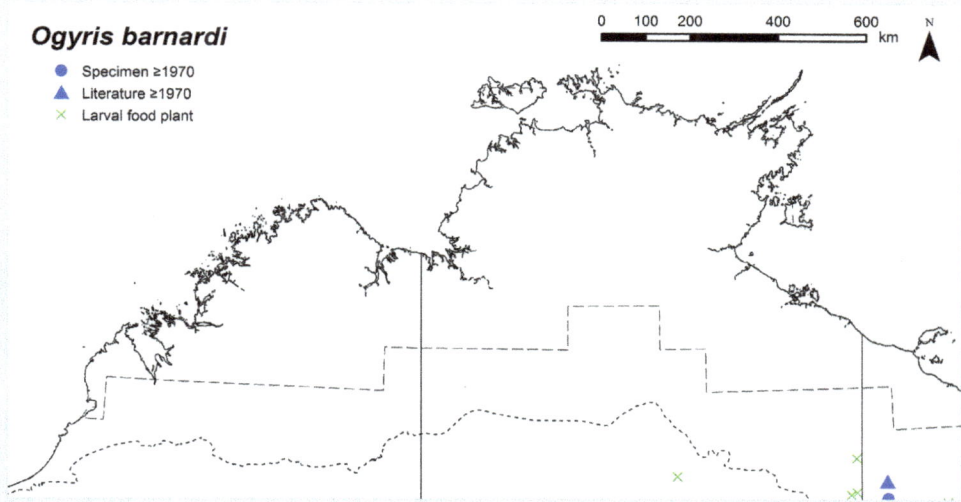

- ● Specimen ≥1970
- ▲ Literature ≥1970
- ✕ Larval food plant

0   100   200      400      600
km

N

## Ogyris barnardi

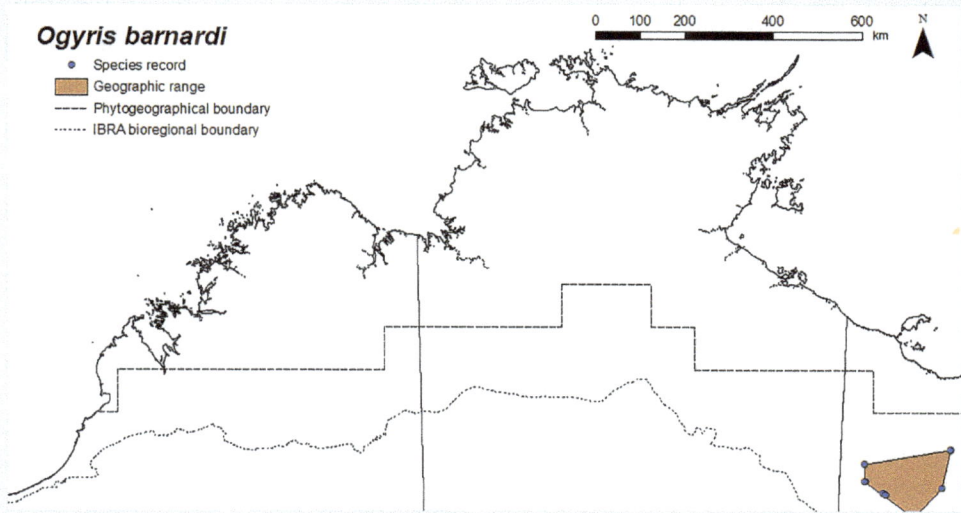

- • Species record
- ▨ Geographic range
- — — Phytogeographical boundary
- ······ IBRA bioregional boundary

0   100   200      400      600
km

N

| Month | J | F | M | A | M | J | J | A | S | O | N | D |
|-------|---|---|---|---|---|---|---|---|---|---|---|---|
| Egg   |   |   |   |   |   |   |   |   |   |   |   |   |
| Larva |   |   |   |   |   |   |   |   |   |   |   |   |
| Pupa  |   |   |   |   |   |   |   |   |   |   |   |   |
| Adult |   |   |   |   |   |   |   |   |   |   |   |   |

# Satin Azure
## *Ogyris amaryllis* (Hewitson, 1862)

Plate 139 South-west of Charters Towers, Qld
Photo: M. F. Braby

Plate 140 South-west of Charters Towers, Qld
Photo: M. F. Braby

## Distribution

This species is represented in the study region by the variable subspecies *O. amaryllis meridionalis* (Bethune-Baker, 1905). It has a very wide distribution, extending from moist coastal areas to drier inland areas of the semi-arid zone (< 400 mm mean annual rainfall), as well as the arid zone of central Australia beyond the southern boundary of the study region. Its geographic range broadly corresponds with the spatial distribution of its larval food plants. The food plants, however, are wider in extent, occurring also in the northern Kimberley, on the Tiwi Islands and Cobourg Peninsula, NT; thus, further field surveys are required to determine whether *O. amaryllis* occurs in these areas. Outside the study region, *O. amaryllis* occurs throughout most of the Australian continent.

## Habitat

*Ogyris amaryllis* breeds in a variety of habitats, including coastal mangroves, savannah woodland, riparian woodland, eucalypt woodland with a hummock grass understorey and *Acacia* low open woodland where the larval food plants grow as mistletoes (parasitic shrubs) on various trees.

## Larval food plants

*Amyema benthamii*, *A. quandang*, *A. sanguinea*, *A. thalassia*, *Diplatia grandibractea* (Loranthaceae).

## Attendant ant

*Crematogaster* sp. (Formicidae: Myrmicinae). The larvae and pupae are occasionally attended by a few ants in a facultative myrmecophilous association (Meyer 1996a; Paton 2013; Braby 2015e).

## Seasonality

Adults occur throughout the year, but, like *Ogyris oroetes*, they are more abundant during the second half of the year, particularly during the late dry season (August and September). The immature stages have also been recorded frequently during August and September. Presumably, the species breeds continuously throughout the year and several generations are completed annually.

## Breeding status

This species is resident in the study region.

## Conservation status

LC.

## Ogyris amaryllis

Specimen ≥1970
Observation ≥1970
Literature ≥1970
Specimen <1970
Literature <1970
Larval food plants

## Ogyris amaryllis

Species record
Geographic range
Phytogeographical boundary
IBRA bioregional boundary

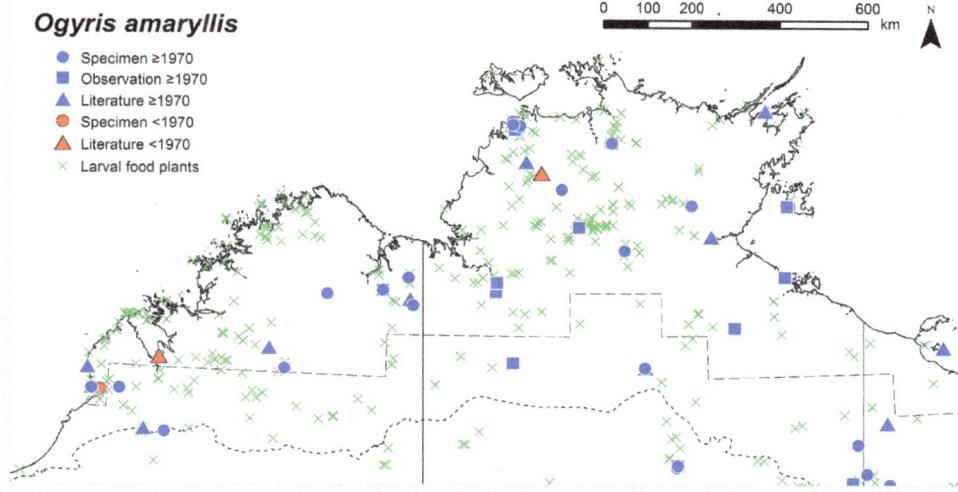

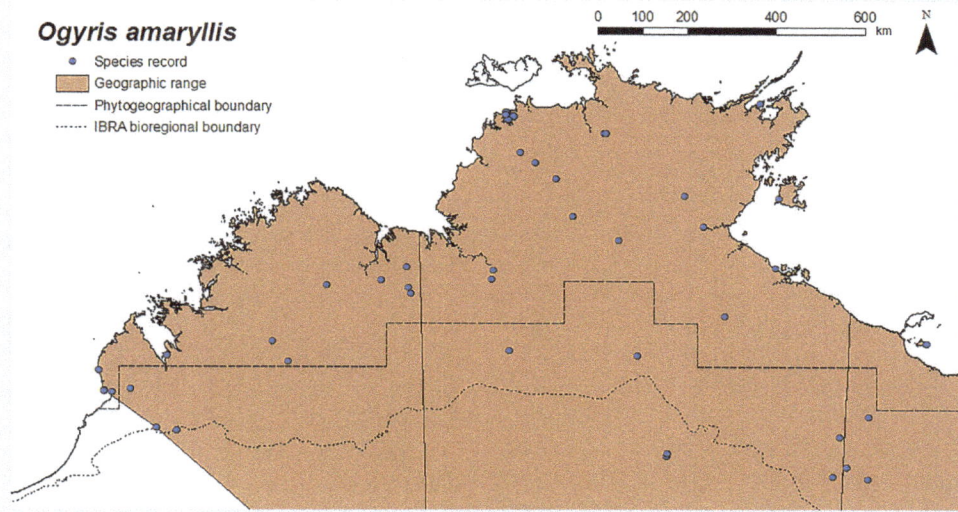

## Ogyris amaryllis (n = 48)

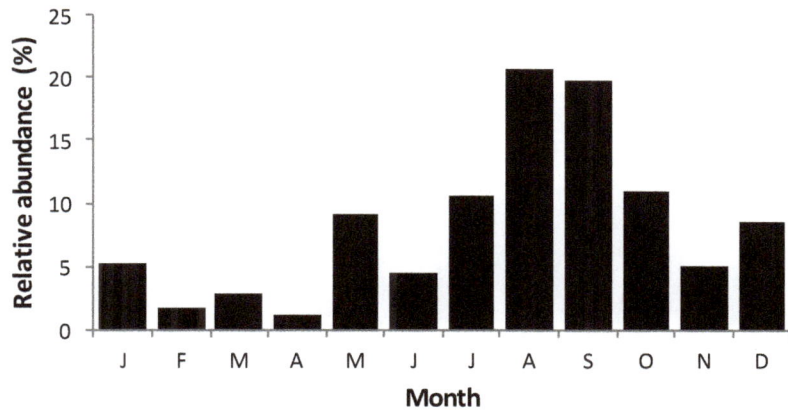

| Month | J | F | M | A | M | J | J | A | S | O | N | D |
|-------|---|---|---|---|---|---|---|---|---|---|---|---|
| Egg   |   |   |   |   |   |   |   |   |   |   |   |   |
| Larva |   |   |   |   |   |   |   |   |   |   |   |   |
| Pupa  |   |   |   |   |   |   |   |   |   |   |   |   |
| Adult |   |   |   |   |   |   |   |   |   |   |   |   |

# Orange-tipped Azure, Dodd's Azure
## *Ogyris iphis* (Waterhouse & Lyell, 1914)

Plate 141 Cobourg Peninsula, NT
Photo: M. F. Braby

Plate 142 Cobourg Peninsula, NT
Photo: M. F. Braby

## Distribution

This species is represented by the subspecies *O. iphis doddi* (Waterhouse & Lyell, 1914), which is endemic to the study region. It occurs in the north of the Northern Territory, where it is restricted to the higher rainfall areas (> 1,400 mm mean annual rainfall) of the north-western corner of the Top End. It has been recorded from only three locations: Darwin (Parap), Melville Island (Pularumpi) and Cobourg Peninsula (near Danger Point), NT (Braby 2000, 2015a; Woinarski et al. 2007b). The species was originally discovered at Darwin (Parap) in 1908–09 by F. P. and W. D. Dodd (Waterhouse and Lyell 1914; Dodd 1935a; Braby 2015a), who collected and reared a series of at least 14 specimens (five males, nine females in AM, ANIC, NMV, SAM, BMNH). Although the known larval food plant is very widely distributed throughout the study region, the extent of the attendant ant is much more restricted (based on records in the TERC and ANIC). However, the known geographic range of *O. iphis* is considerably smaller than the spatial distribution of its associated ant. The ant has also been recorded in the north-western Kimberley (Admiralty Gulf and Kalumburu), WA; Litchfield (Tolmer Falls and Lost City), Kakadu (Jim Jim Falls area) and Limmen (Butterfly Springs) national parks and from several sites on Gove Peninsula, NT. Further field surveys are thus required to determine whether *O. iphis* is present in the near-coastal areas of the north-western Kimberley and northern and north-eastern Arnhem Land (from Murgenella to Nhulunbuy and Birany Birany and Wandawuy homelands). In recent years, targeted searches for *O. iphis* in the Darwin area (Meyer et al. 2006) and in the more inland areas at Robin Falls and Litchfield and Kakadu national parks have not detected the butterfly, which suggests *O. iphis* is absent from these areas. Outside the study region, *O. iphis* occurs in north-eastern Queensland.

## Excluded data

Sands and New (2002) tentatively listed the species from the Mitchell Plateau, WA, and Mt Burrell, NT, based on possible sightings of adults, but given the difficulty of identifying adults in the field (the butterfly flies high and fast in the canopy) and potential confusion with other similar species, especially *Ogyris oroetes*, which regularly hilltops at Mt Burrell, these records need to be confirmed before they can be accepted.

## Habitat

*Ogyris iphis* has been recorded breeding in near-coastal savannah woodland on a gently sloping laterite outcrop/breakaway on the edge of a plateau where a mature clump of the larval food plant grew as a mistletoe (parasitic shrub) on *Eucalyptus tetrodonta* and where extensive colonies of the attendant ant were established (Braby 2015a, 2015e). The co-occurrence of the ant and mistletoe is very local or patchy in distribution. The ant has been recorded more frequently in coastal eucalypt woodland on loamy sand and sand, and the butterfly may well breed in this habitat.

Photo: Tolmer Falls, Litchfield National Park, NT, M.F. Braby

## Larval food plant

*Amyema sanguinea* (Loranthaceae).

## Attendant ant

*Froggattella kirbii* (Formicidae: Dolichoderinae). The larvae and pupae are constantly attended by numerous Froglet Ants (*F. kirbii*) in an obligatory myrmecophilous association (Braby 2015a, 2015e).

## Seasonality

The seasonal abundance and breeding phenology of this species are not well understood. Most adults have been recorded or reared during the second half of the year (July–December), but we have too few records (n = 15) to assess any seasonal changes in abundance. The immature stages have been recorded in the late dry season (August–October) and sporadically during the wet season (November–March). Presumably, the species breeds continuously throughout the year and several generations are completed annually.

## Breeding status

This species is resident in the study region.

## Conservation status

VU, B2ab(ii)(iii). The subspecies *O. iphis doddi* is currently listed as Endangered (EN) under the *TPWCA*. However, available data suggest VU is a more appropriate Red List category. The taxon is a short-range endemic (AOO is likely to be < 2,000 sq km, with spatial buffering of extant records providing a first approximation of 1,400 sq km). Despite targeted searching, it is currently known from only two extant locations, one of which is in a conservation reserve (Garig Gunak Barlu National Park) (Braby 2015a). The population at Darwin (Parap) is no longer extant due to extensive urbanisation. Braby (2000) and Meyer et al. (2006) recorded a possible sighting at Bens Hill near Darwin in 1992, but the habitat surrounding this small hilltop has since been cleared for development. The population on the Tiwi Islands (Melville Island) has no doubt been severely compromised, if not eliminated, by extensive loss of its critical habitat for timber plantation, and there are similar threats facing Bathurst Island to develop potential breeding habitat for agriculture. Inappropriate fire regime is another threat because the mistletoe food plant is sensitive to frequent hot fires; *Amyema sanguinea* is variously killed by fire or it may resprout from haustorial tissue after being burnt (Start 2013). It is likely that *O. iphis* is now restricted to habitats that are infrequently burnt or in which fires are less intense and do not frequently scorch the canopy of the host trees (*Eucalyptus, Corymbia*) supporting mistletoe clumps. It is also suspected that the species does not tolerate habitat fragmentation and it may require relatively large intact landscapes to persist (Braby 2015a). Targeted field surveys to clarify the distribution and ecological requirements should be a high priority for this species.

Photo: Cobourg Peninsula, NT, M.F. Braby

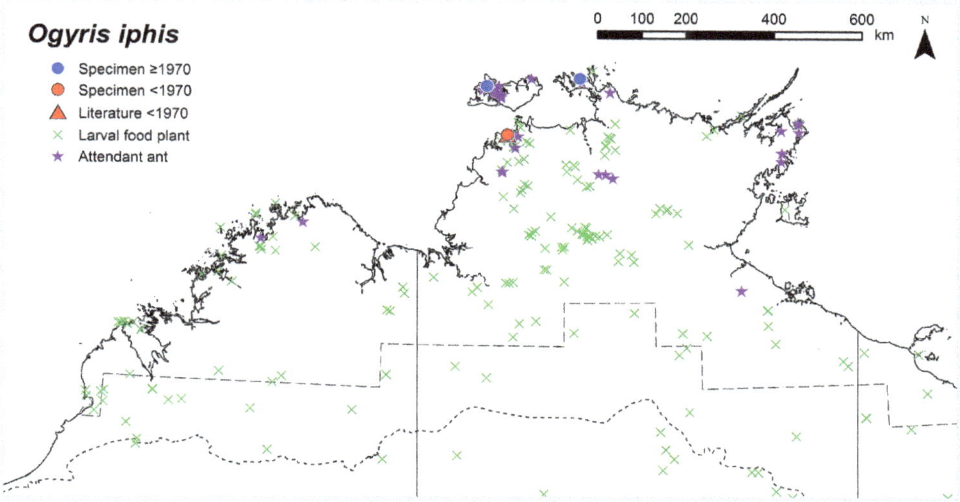

**Ogyris iphis**

- ● Specimen ≥1970
- ● Specimen <1970
- ▲ Literature <1970
- × Larval food plant
- ★ Attendant ant

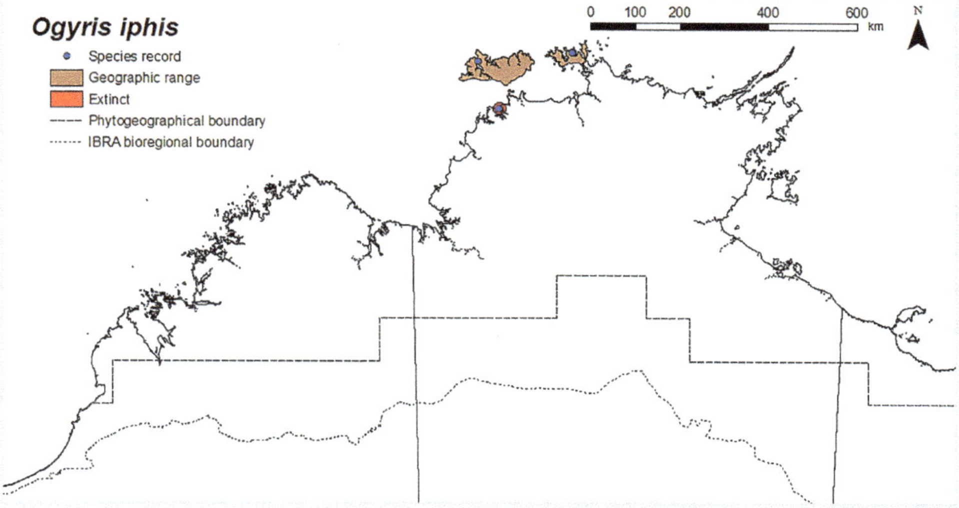

**Ogyris iphis**

- ● Species record
- ▢ Geographic range
- ▢ Extinct
- – – – Phytogeographical boundary
- ······ IBRA bioregional boundary

| Month | J | F | M | A | M | J | J | A | S | O | N | D |
|-------|---|---|---|---|---|---|---|---|---|---|---|---|
| Egg   |   |   |   |   |   |   |   |   |   |   |   |   |
| Larva |   |   |   |   |   |   |   |   |   |   |   |   |
| Pupa  |   |   |   |   |   |   |   |   |   |   |   |   |
| Adult |   |   |   |   |   |   |   |   |   |   |   |   |

# Northern Purple Azure

## *Ogyris zosine* (Hewitson, [1853])

Plate 143 Kakadu National Park, NT
Photo: M. F. Braby

## Distribution

This species is represented in the study region by the subspecies *O. zosine zosine* (Hewitson, [1853]). It occurs very widely in the region, extending from moist coastal areas to drier inland areas of the semi-arid zone (< 500 mm mean annual rainfall), as well as the arid zone of central Australia beyond the southern boundary of the study region. Its southernmost limits include 100 km south of Broome and Fitzroy Crossing, WA (Field 1990b); Elliott, NT (Le Souëf 1971); and 45 km east–south-east of Camooweal, Qld. Its wide geographic range corresponds well with the spatial distribution of its larval food plants. The food plants, however, also occur on the Tiwi Islands; thus, further field surveys are required to determine whether *O. zosine* occurs on Bathurst and Melville islands, NT. Outside the study region, *O. zosine* occurs throughout most of the northern half of Australia, as well as in the semi-arid areas of southern Western Australia.

## Habitat

*Ogyris zosine* breeds locally in savannah woodland, eucalypt heathy woodland, eucalypt open woodland and *Acacia* low open woodland on various substrates where the larval food plants grow as mistletoes (parasitic shrubs) on a range of trees, including *Corymbia*, *Eucalyptus*, *Acacia* and *Alstonia*, and where subterranean nests of the attendant ant are established (Weir et al. 2011; Braby 2011a, 2015e). The co-occurrence of the ant and mistletoe is very local or patchy in distribution.

## Larval food plants

*Amyema benthamii*, *A. bifurcata*, *A. miquelii*, *A. sanguinea*, *A. villiflora*, *Decaisnina signata*, *Diplatia grandibractea* (Loranthaceae).

## Attendant ants

*Camponotus oetkeri*, *Camponotus* sp. (*crozieri* group), *Camponotus* sp. (*humilior* group), *Camponotus* sp. (*novaehollandiae* group) (Formicidae: Formicinae). The larvae and pupae are constantly attended by numerous Sugar Ants (*Camponotus* spp.) in an obligatory myrmecophilous association (Braby 2011a; Weir et al. 2011). Very occasionally the immature stages may be associated with *Iridomyrmex pallidus* ants in disturbed suburban areas (Weir et al. 2011), although the nature of this association has not been established.

## Seasonality

Adults occur throughout the year. They appear to show little seasonal variation in abundance, although they tend to be more numerous during the second half of the year, particularly during the mid to late dry season (July–October). The immature stages have been recorded during most months of the year, especially during the dry season, indicating that the species breeds continuously. Presumably, several generations are completed annually.

## Breeding status

This species is resident in the study region.

## Conservation status

LC. In some areas of the range, inappropriate fire regimes, particularly frequent burning and/or intense landscape fires, are a threat because the mistletoe food plants are highly susceptible to fire.

## Ogyris zosine

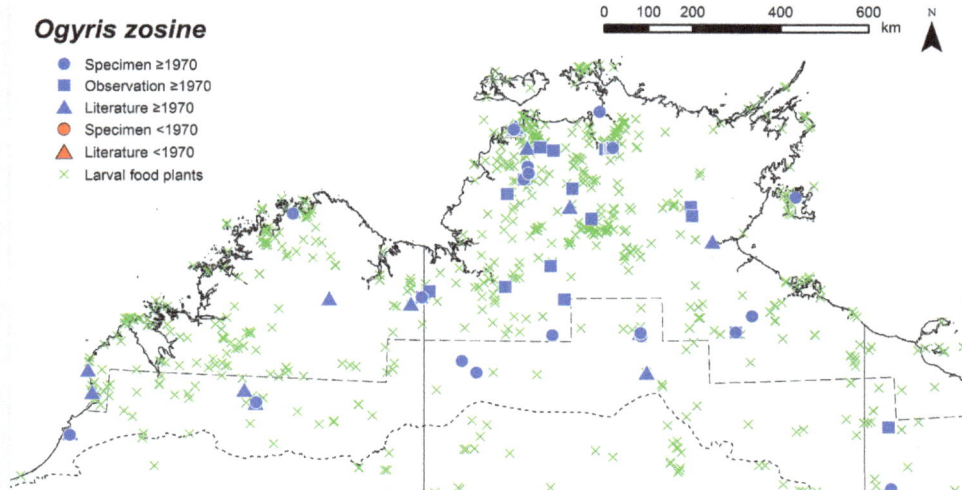

Legend:
- ● Specimen ≥1970
- ■ Observation ≥1970
- ▲ Literature ≥1970
- ● Specimen <1970
- ▲ Literature <1970
- × Larval food plants

## Ogyris zosine

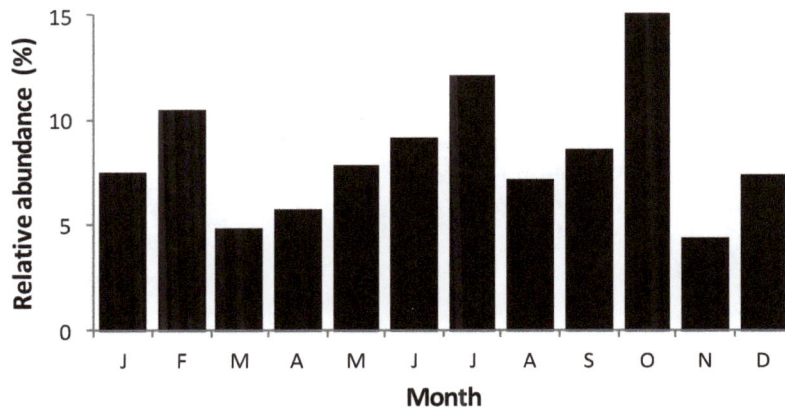

Legend:
- ● Species record
- Geographic range
- --- Phytogeographical boundary
- ···· IBRA bioregional boundary

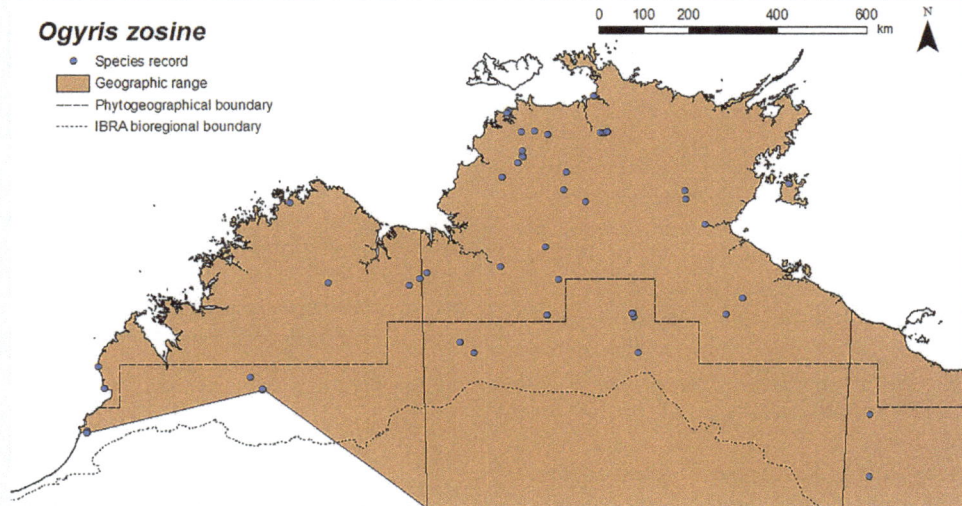

Ogyris zosine (n = 64)

| Month | J | F | M | A | M | J | J | A | S | O | N | D |
|-------|---|---|---|---|---|---|---|---|---|---|---|---|
| Egg | | | | | ■ | | ■ | | | | | |
| Larva | | | | ■ | ■ | ■ | ■ | ■ | | ■ | ■ | |
| Pupa | | ■ | ■ | ■ | ■ | ■ | ■ | ■ | ■ | ■ | ■ | ■ |
| Adult | ■ | ■ | ■ | ■ | ■ | ■ | ■ | ■ | ■ | ■ | ■ | ■ |

# Amethyst Hairstreak

*Jalmenus icilius* Hewitson, [1865]

Plate 144 Alice Springs, NT
Photo: M. F. Braby

## Distribution

This species is restricted to the Northern Deserts and western Gulf Country, where it occurs in the semi-arid and arid zones (< 400 mm mean annual rainfall) in the far south-eastern corner of the study region. It has been recorded 30 km south-east of Barkly Homestead Roadhouse, NT (Braby 2000), and from two locations just outside the southern and eastern boundaries: 120 km east of Argadargada Station, NT (J. Archibald), and 59 km (by road) north–north-east of Burke & Wills Roadhouse, Qld (Dunn 2016). The spatial distribution of the larval food plants is substantially more widespread in the semi-arid zone; thus, further field surveys are required to determine whether *J. icilius* is similarly more widespread than present records indicate. Outside the study region, *J. icilius* occurs widely in the arid and semi-arid areas of central, south-western and south-eastern Australia.

Material from the western Gulf Country (Dunn 2016) is provisionally placed here pending further taxonomic investigation of this population. The specimens are rather distinct and may prove to be *J. daemeli* Semper, 1879 or a species allied to it.

## Habitat

The breeding habitat of *J. icilius* has been recorded only from semi-arid and arid areas just outside the study region, where the species is associated with shrubs of the larval food plants growing in open grassland with scattered shrubs or in mixed open woodland, and where colonies of the attendant ant are established.

## Larval food plants

*Acacia aneura*, *A. tetragonophylla*, *Senna artemisioides* (Fabaceae).

## Attendant ants

*Iridomyrmex chasei*, *Iridomyrmex* sp. (*rufoniger* group) (Formicidae: Dolichoderinae). The larvae and pupae are constantly attended by numerous small black ants in an obligatory myrmecophilous association.

## Seasonality

The seasonal abundance and breeding phenology of this species are not well understood. Adults have been recorded in 'autumn' (March and April) and again in 'spring' (October). In central Australia, they have also been commonly recorded in September (Field 1990a; Grund 2005); however, they have also been collected at other times of the year (for example, in January and February), possibly depending on rainfall. Presumably, the species completes several generations annually and the eggs remain in diapause during the cooler winter months or possibly during prolonged dry periods.

## Breeding status

This species is resident in the study region.

## Conservation status

LC. Although the species *J. icilius* has a narrow range in the study region (EOO = 22,530 sq km), there are no known threats facing the taxon. However, it is currently not known from any conservation reserves.

## Jalmenus icilius

- ● Specimen ≥1970
- ■ Observation ≥1970
- ▲ Literature ≥1970
- × Larval food plants

## Jalmenus icilius

- ● Species record
- ▨ Geographic range
- — — Phytogeographical boundary
- ····· IBRA bioregional boundary

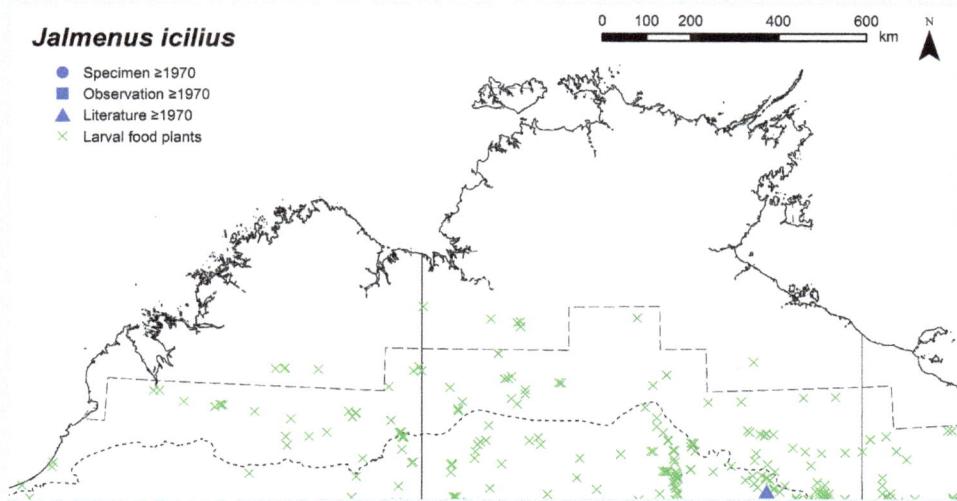

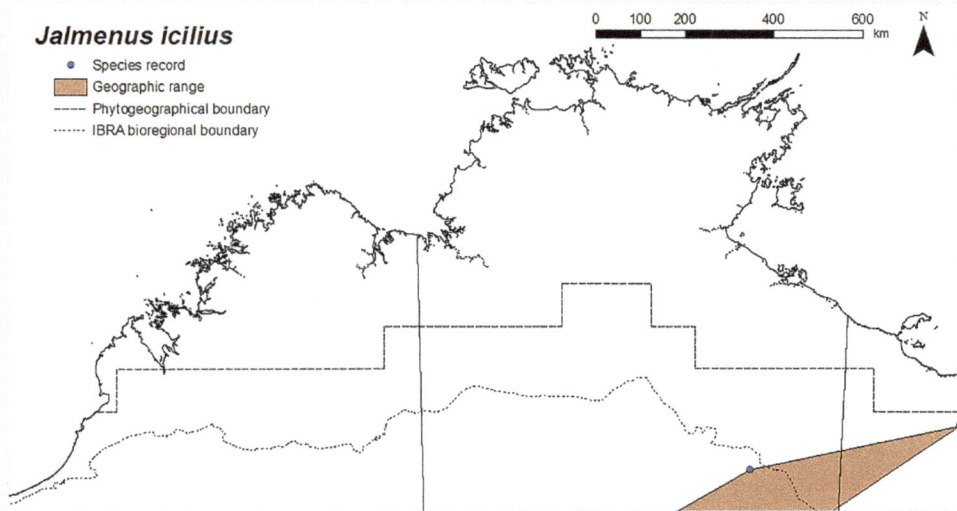

| Month | J | F | M | A | M | J | J | A | S | O | N | D |
|-------|---|---|---|---|---|---|---|---|---|---|---|---|
| Egg   |   |   |   |   |   |   |   |   |   |   |   |   |
| Larva |   |   |   |   |   |   |   |   |   |   |   |   |
| Pupa  |   |   |   |   |   |   |   |   |   |   |   |   |
| Adult |   |   |   |   |   |   |   |   |   |   |   |   |

# Black-spotted Flash
## *Hypolycaena phorbas* (Fabricius, 1793)

Plate 145 Black Point, Cobourg Peninsula, NT
Photo: M. F. Braby

## Distribution

This species is represented in the study region by the subspecies *H. phorbas phorbas* (Fabricius, 1793). It has a disjunct distribution, occurring in the western and northern Kimberley and throughout the Top End. It occurs mainly in the higher rainfall areas (> 1,000 mm mean annual rainfall), but it has been recorded in drier areas receiving an average annual rainfall of 800 mm. Its southernmost limits include Pasco Island, WA (Lambkin 2006); and Manbulloo Station 10 km west of Katherine (A. Carlson), and Ngukurr (S. Normand), NT. The species also occurs in the Gulf of Carpentaria at Walker Creek 36 km east of Karumba, Qld, just outside the eastern boundary of the study region (Braby 2015d). The geographical range of *H. phorbas* broadly corresponds with the spatial distribution of its associated ant, especially within the Top End. Although the larval food plants are widely distributed throughout the region, the attendant ant is absent from most of the Northern Deserts and western Gulf Country and is much more restricted in extent (Lokkers 1986). The species has not been recorded from the Tiwi Islands despite the presence of the attendant ant and larval food plants; thus, further field surveys are required to determine whether *H. phorbas* is present on Bathurst and Melville islands, NT. Outside the study region, *H. phorbas* occurs in mainland New Guinea and adjacent islands, north-eastern Australia and the Bismarck Archipelago.

Previously, the name *H. phorbas ingura* Tindale, 1923 was applied to populations in northern Australia as a subspecies endemic to the region, but Lambkin (2006) concluded that *H. phorbas ingura* did not differ from the nominate subspecies from the northern Torres Strait Islands, Qld.

## Habitat

*Hypolycaena phorbas* breeds in a variety of habitats, including semi-deciduous monsoon vine thicket and mixed eucalypt woodland and riparian woodland with rainforest elements in the understorey, where the larval food plants grow and arboreal nests of the attendant ant are established. The food plants are usually tall shrubs or small trees in the understorey, but are sometimes vines (*Smilax australis*) or mistletoes (*Decaisnina signata*).

## Larval food plants

*Clerodendrum floribundum* (Lamiaceae), *Planchonia careya* (Lecythidaceae), *Decaisnina signata* (Loranthaceae), *Breynia cernua* (Phyllanthaceae), *Smilax australis* (Smilacaceae). The larvae feed on several food plants, but they are frequently found on *C. floribundum*.

## Attendant ant

*Oecophylla smaragdina* (Formicidae: Formicinae). The larvae and pupae are constantly attended by numerous Green Tree Ants (*O. smaragdina*) in an obligatory myrmecophilous association (Eastwood and Fraser 1999; Braby 2000).

## Seasonality

Adults occur during most months of the year, but they are most abundant during the wet season and early dry season (January–May), and are very scarce during the late dry season (August–October). The immature stages have been recorded during the wet season and early dry season (November–May), when population numbers are higher. Presumably, the species breeds continuously throughout the year and several generations are completed annually given there is no evidence of diapause in any of the life history stages.

## Breeding status

This species is resident in the study region.

## Conservation status

LC.

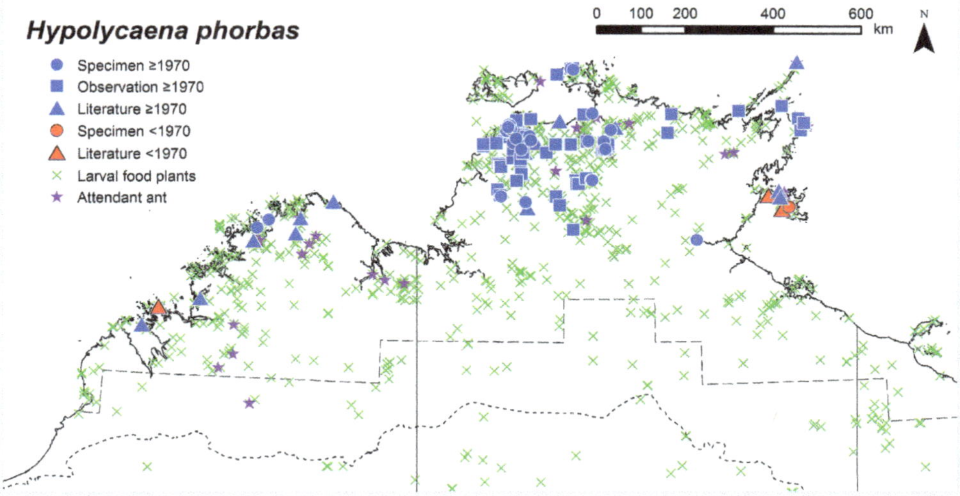

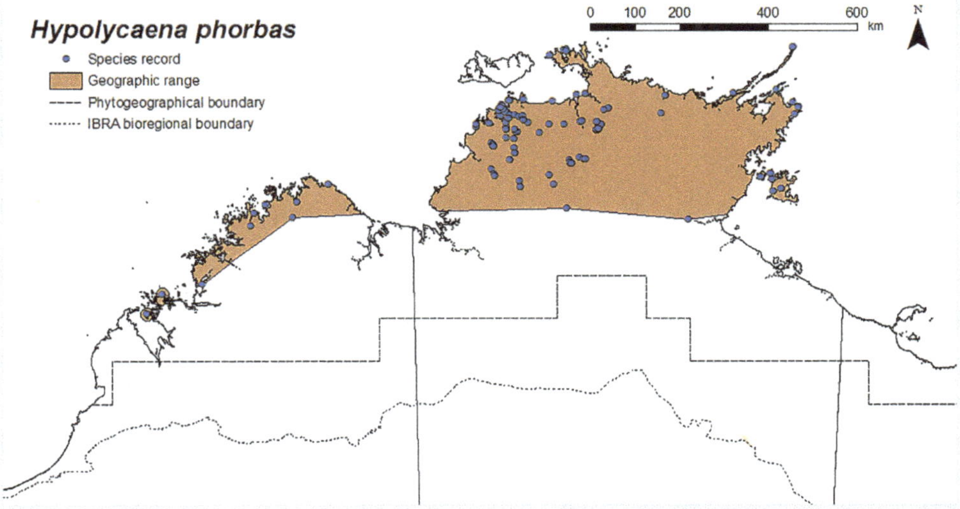

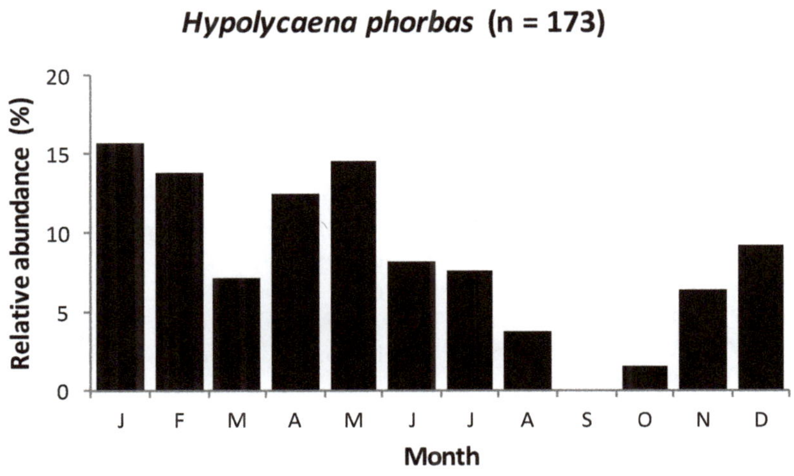

# Hypolycaena phorbas

# Hypolycaena phorbas

## Hypolycaena phorbas (n = 173)

| Month | J | F | M | A | M | J | J | A | S | O | N | D |
|---|---|---|---|---|---|---|---|---|---|---|---|---|
| Egg | | ▓ | | ▓ | | | | | | | ▓ | |
| Larva | ▓ | ▓ | | ▓ | | | | | | | ▓ | ▓ |
| Pupa | | ▓ | ▓ | ▓ | | | | | | | ▓ | |
| Adult | ▓ | ▓ | ▓ | ▓ | ▓ | ▓ | ▓ | ▓ | | ▓ | ▓ | ▓ |

# Princess Flash
## *Deudorix smilis* Hewitson, [1863]

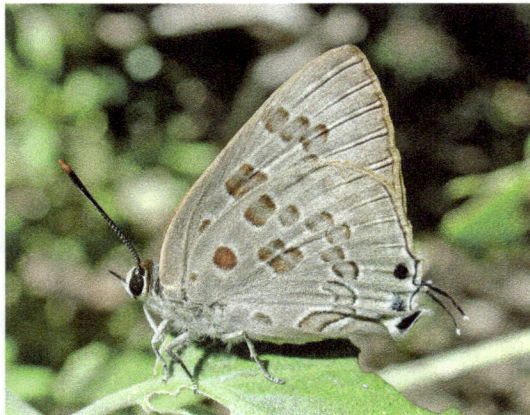

Plate 146 Lee Point, Casuarina Coastal Reserve, NT
Photo: M. F. Braby

## Distribution

This species is represented by the subspecies *D. smilis dalyensis* (Le Souëf & Tindale, 1970), which is endemic to the study region. *Deudorix smilis* does not occur elsewhere in Australia. It occurs mainly in the western half of the Top End, where it is restricted to the higher rainfall areas (> 900 mm mean annual rainfall, but mainly > 1,200 mm). It has been recorded as far south as Daly River (Oolloo Crossing) (Le Souëf and Tindale 1970; Hutchinson 1978) and as far east as Wongalara Wildlife Sanctuary (Echo Gorge) (Braby 2012a), NT. Although the larval food plant is considerably more widespread, extending into the drier areas of the Northern Deserts and the Kimberley, targeted searches in these areas have not detected *D. smilis*. However, the plant does occur in northern and north-eastern Arnhem Land and Groote Eylandt, NT, where the rainfall and habitat are potentially more suitable. Further surveys are therefore required in the eastern half of the Top End and nearby islands to determine whether *D. smilis* is present in these areas. Outside the study region, *D. smilis* occurs from the Andaman Islands, Burma (Myanmar) and the Malay Peninsula to Indonesia, including Maluku.

## Habitat

*Deudorix smilis* breeds in semi-deciduous monsoon vine thicket on coastal sand dunes and lateritic cliffs above the beach, on inland sandstone hills or rocky outcrops, in the edges of evergreen monsoon vine forest associated with permanent freshwater streams or in the ecotone between monsoon forest and eucalypt woodland where the larval food plant grows as a tall deciduous shrub or small tree up to 6 m high (Braby 2016c). Adults also fly in open woodland and males congregate on hilltops to locate females, but they do not breed in these habitats.

## Larval food plant

*Strychnos lucida* (Loganiaceae).

## Attendant ant

*Crematogaster* sp. 3 (Group A) (Formicidae: Myrmicinae). The larvae and pupae are usually not attended by ants, but very occasionally the pupae are attended by a small number of black and tan ants (Eastwood et al. 2008; Braby 2016c).

## Seasonality

Adults occur during most months of the year, but they are most abundant during the mid dry season (May–July), with a peak in abundance in June. The fruits of the larval food plant are generally seasonal, being more numerous during the early to mid dry season, and it is during this period that the immature stages of the butterfly are abundant. However, the immature stages have also been found on young (green) and ripe (orange) fruits sporadically at other times of the year, including during the 'build-up' and wet season (October–April). Presumably, the species breeds continuously throughout the year and several generations are completed annually given there is no evidence of diapause in any of the life history stages.

## Breeding status

This species is resident in the study region.

## Conservation status

LC. The subspecies *D. smilis dalyensis* has a restricted range within which it occurs in several conservation reserves, including Casuarina Coastal Reserve, Black Jungle Conservation Reserve, Mary River National Park, Garig Gunak Barlu National Park, Kakadu National Park and Wongalara Wildlife Sanctuary, NT. It is currently not threatened, but it may be susceptible to habitat loss and fragmentation because it appears to depend on relatively large patches of monsoon forest and/ or a network of smaller interconnected patches of monsoon forest embedded within the savannah matrix (Braby 2016c).

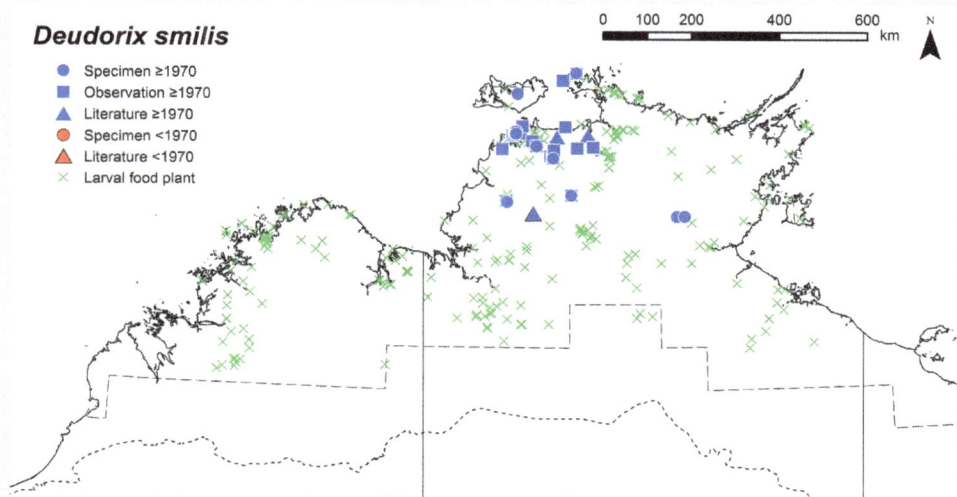

*Deudorix smilis*

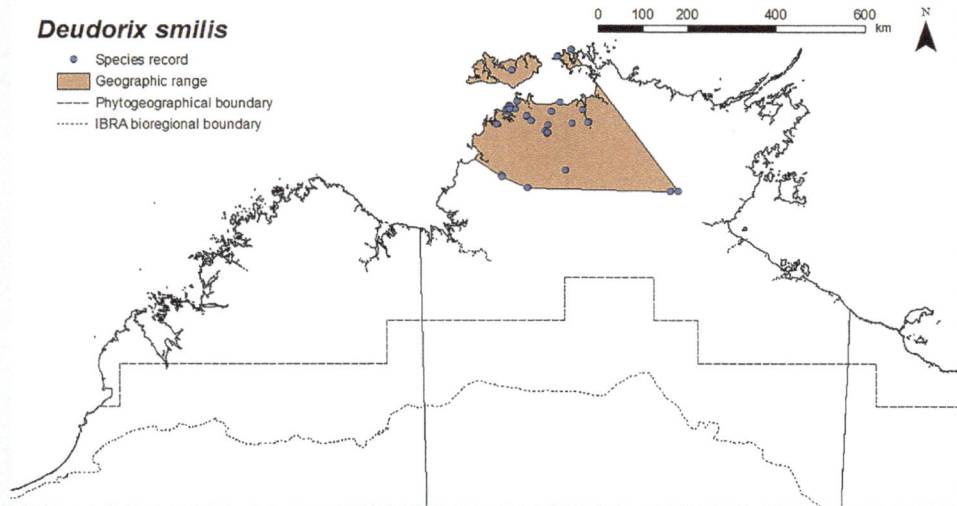

*Deudorix smilis*

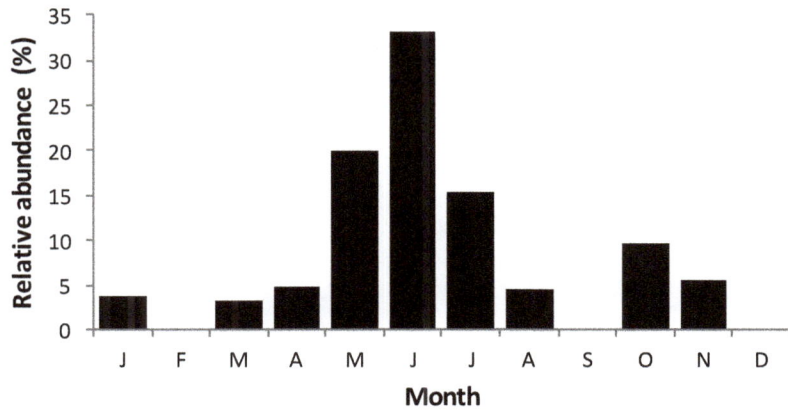

*Deudorix smilis* (n = 29)

| Month | J | F | M | A | M | J | J | A | S | O | N | D |
|-------|---|---|---|---|---|---|---|---|---|---|---|---|
| Egg | | | | | | | | | | | | |
| Larva | | | | | | | | | | | | |
| Pupa | | | | | | | | | | | | |
| Adult | | | | | | | | | | | | |

# Sword-tail Flash
## *Bindahara phocides* (Fabricius, 1793)

Plate 147 Rollingstone, Qld
Photo: M. F. Braby

## Distribution

This species is known only from an observation made in the Gulf of Carpentaria at Limmen National Park (Nathan River Ranger Station), NT, where a specimen was collected and released in January 2009 (M. Kessner). The species has not previously been recorded from the study region, but presumably it is the subspecies *B. phocides yurgama* Couchman, 1965, which occurs in north-eastern Queensland. The putative larval food plant (*Salacia chinensis*) occurs along the northern coastline of the Top End, some 400 km further north of where the specimen was captured. Outside the study region, *B. phocides* occurs widely from India and South-East Asia, through mainland New Guinea and adjacent islands and north-eastern Australia to the Bismarck Archipelago and the Solomon Islands.

## Habitat

The breeding habitat of *B. phocides* has not been recorded in the study region. The specimen was recorded in a residential garden (rangers' quarters) surrounded by natural dry woodland in which the mean annual rainfall is about 700 mm; however, the putative larval food plant (*Salacia chinensis*) grows as a vine in coastal semi-deciduous monsoon vine thicket close to the beach in the higher rainfall areas (> 1,300 mm mean annual rainfall).

## Larval food plants

Not recorded in the study region; probably *Salacia chinensis* (Celastraceae), which is a larval food plant in north-eastern Queensland (Braby 2000).

## Attendant ant

Not recorded in the study region. In Queensland, the larvae and pupae are not attended by ants (Eastwood and Fraser 1999; Braby 2000).

## Seasonality

The seasonal abundance and breeding phenology of this species are not well understood. The adult was recorded during the wet season (January).

## Breeding status

The breeding status of *B. phocides* is uncertain. It may well be resident in the Top End, but the specimen observed at Nathan River was possibly a vagrant from coastal areas further north. The specimen captured was, however, in relatively good condition, with the long hindwing tails intact (M. Kessner, pers. comm.), suggesting it may have bred locally.

## Conservation status

DD. The distribution and critical habitat of the species *B. phocides* are poorly known. Targeted field surveys in coastal areas extending from the Wessell Islands and Gove Peninsula to Limmen Bight in the Gulf of Carpentaria are required to determine whether the species is established with permanent breeding populations.

## *Bindahara phocides*

- ■ Observation ≥1970
- × Putative larval food plant

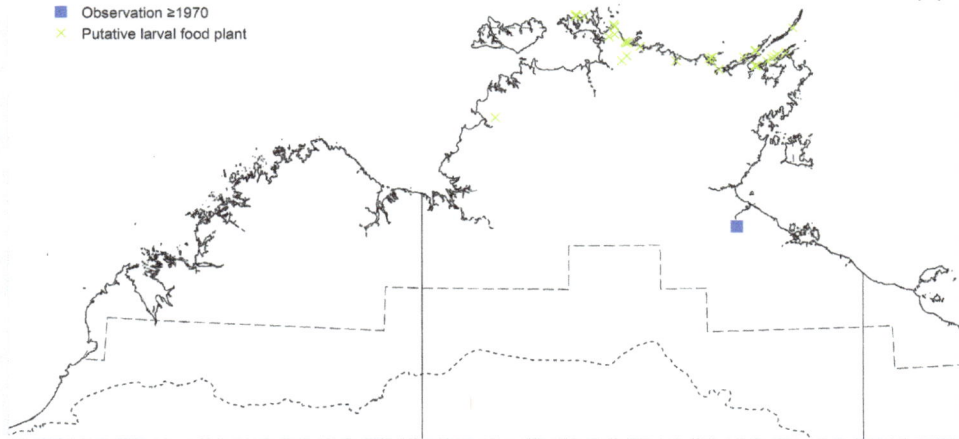

| Month | J | F | M | A | M | J | J | A | S | O | N | D |
|-------|---|---|---|---|---|---|---|---|---|---|---|---|
| Egg   |   |   |   |   |   |   |   |   |   |   |   |   |
| Larva |   |   |   |   |   |   |   |   |   |   |   |   |
| Pupa  |   |   |   |   |   |   |   |   |   |   |   |   |
| Adult |   |   |   |   |   |   |   |   |   |   |   |   |

Photo: Boodjamulla/Lawn Hill National Park, Qld, M.F. Braby

# Trident Pencil-blue, Northern Pencil-blue
## *Candalides margarita* (Semper, 1879)

Plate 148 Wanguri, Darwin, NT
Photo: M. F. Braby

## Distribution

This species is represented by the subspecies *C. margarita gilberti* Waterhouse, 1903, which is endemic to the study region. It occurs widely in the Kimberley and Top End and more sporadically in the western Gulf Country, extending from moist coastal areas of high rainfall (> 1,700 mm mean annual rainfall) to drier inland areas of the semi-arid zone (500–700 mm). Its southernmost limits include Napier Range Pass, WA; Judbarra/Gregory National Park, Savannah Way 75 km west–north-west of the NT/Qld border, and Boodjamulla/Lawn Hill National Park, Qld (Daniels and Edwards 1998; Franklin 2007). The geographical range corresponds well with the spatial distribution of its larval food plants. Although the plants extend further inland, it is considered unlikely that *C. margarita* breeds in these areas. Outside the study region, *C. margarita* occurs in the Aru Islands, mainland New Guinea and north-eastern and eastern Australia.

## Habitat

*Candalides margarita* breeds in savannah woodland and patches of evergreen monsoon vine forest and semi-deciduous monsoon vine thicket where the larval food plants grow as mistletoes (parasitic shrubs) on various trees (Dodd 1935d; Samson and Wilson 1995; Braby 2008b, 2015e). In Darwin, NT, it also occurs in suburban parks and gardens where the food plant (*Decaisnina signata*) frequently parasitises *Alstonia actinophylla* and *Planchonia careya*. Males readily congregate on hilltops to locate females, but they do not breed in these habitats.

## Larval food plants

*Amyema miquelii*, *A. sanguinea*, *A. villiflora*, *Decaisnina signata*, *Decaisnina triflora*, *Dendrophthoe glabrescens*, *Dendrophthoe odontocalyx* (Loranthaceae).

## Attendant ant

*Crematogaster* sp. (Formicidae: Myrmicinae). The larvae and pupae are usually not attended by ants, but very occasionally the larvae are attended by a few ants (Samson and Wilson 1995; Meyer 1996a; Braby 2008b).

## Seasonality

Adults occur throughout the year, but they are most abundant during the mid dry season (June–August), although they may be locally common at other times of the year. The immature stages have been recorded in each month of the year and the life cycle is completed relatively quickly (approximately one month), indicating that breeding occurs continuously and several generations are completed annually.

## Breeding status

This species is resident in the study region.

## Conservation status

LC.

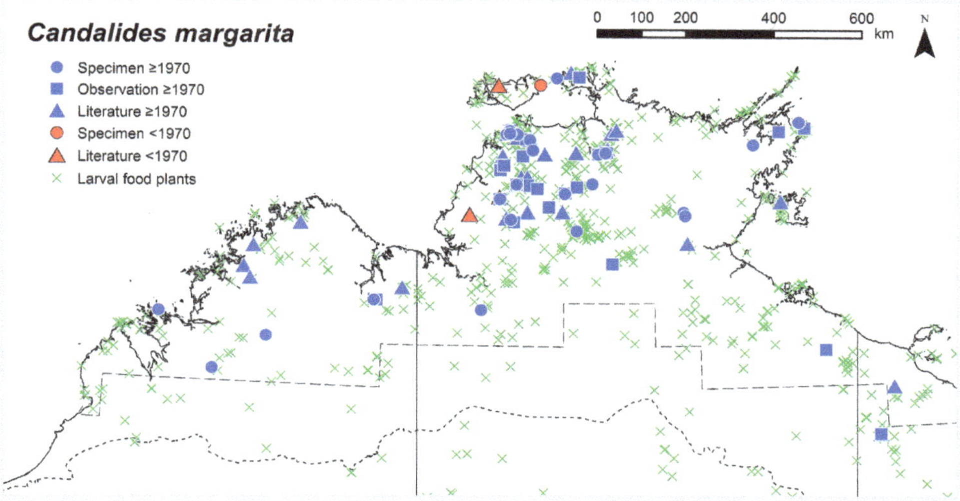

# *Candalides margarita*

- ● Specimen ≥1970
- ■ Observation ≥1970
- ▲ Literature ≥1970
- ● Specimen <1970
- ▲ Literature <1970
- ✕ Larval food plants

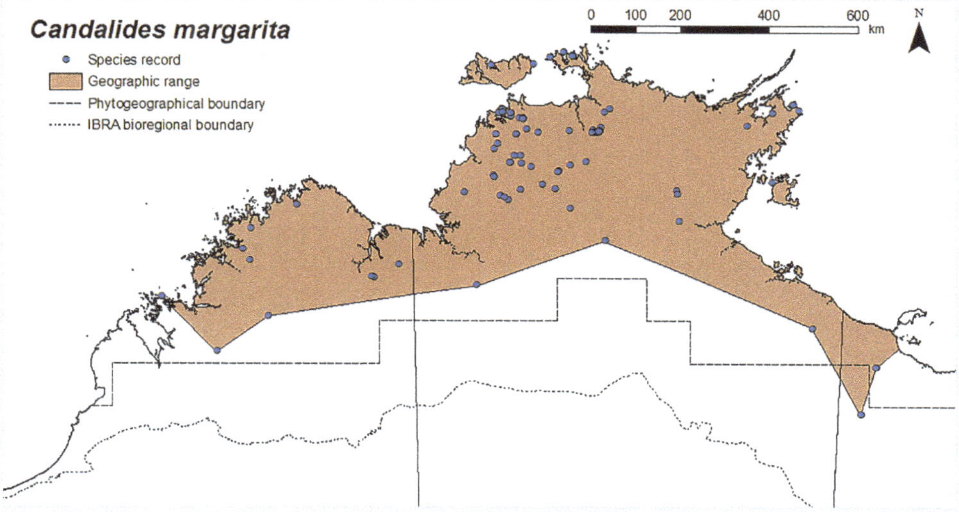

# *Candalides margarita*

- ● Species record
- ▨ Geographic range
- --- Phytogeographical boundary
- ⋯ IBRA bioregional boundary

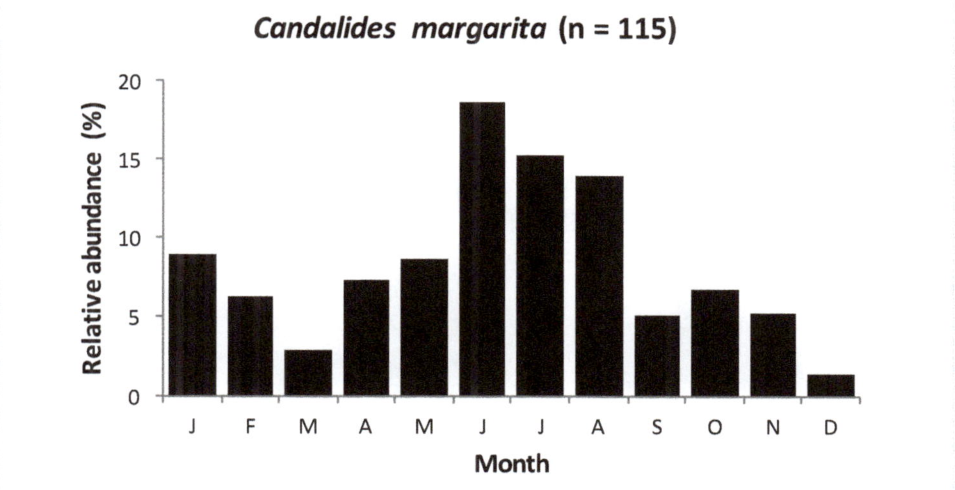

## *Candalides margarita* (n = 115)

| Month | J | F | M | A | M | J | J | A | S | O | N | D |
|-------|---|---|---|---|---|---|---|---|---|---|---|---|
| Egg   |   |   |   |   |   |   |   |   |   |   |   |   |
| Larva |   |   |   |   |   |   |   |   |   |   |   |   |
| Pupa  |   |   |   |   |   |   |   |   |   |   |   |   |
| Adult |   |   |   |   |   |   |   |   |   |   |   |   |

# Twin Dusky-blue
*Candalides geminus* E. D. Edwards & Kerr, 1978

Plate 149 Kakadu National Park, NT
Photo: M. F. Braby

Plate 150 Kakadu National Park, NT
Photo: M. F. Braby

## Distribution

This species is represented by the subspecies *C. geminus gagadju* Braby, 2017, which is endemic to the study region. It occurs in the Top End, its presence in the region first reported by Kikkawa and Monteith (1980) based on a small series of specimens collected in 1979. Examination of material in the ANIC, however, indicated that it was detected seven years earlier, in October 1972, based on a pair collected from western Arnhem Land by I. F. B. Common, E. D. Edwards and M. S. Upton (Braby 2017). It has a patchy and restricted distribution, occurring in Litchfield National Park and on the Arnhem Land Plateau, where it has been recorded at Kakadu and Nitmiluk (Katherine Gorge) national parks and Wongalara Wildlife Sanctuary, NT (Braby 2012a). Although the larval food plants are very widely distributed throughout the study region, *C. geminus* is confined to areas where the plants grow on sandstone in the higher rainfall areas (> 900 mm mean annual rainfall). Outside the study region, *C. geminus* occurs sporadically in north-eastern and eastern Australia.

## Habitat

*Candalides geminus* breeds in eucalypt heathy woodland and open woodland with a spinifex understorey on rocky sandstone escarpments and plateaus where the larval food plants grow as scrambling parasitic vines in the understorey (Braby 2011a, 2017).

## Larval food plants

*Cassytha filiformis*, *C. capillaris* (Lauraceae). The usual food plant is *C. filiformis*, but *C. capillaris* is also used where it is present (Braby 2011a).

## Attendant ant

The larvae and pupae are not attended by ants.

## Seasonality

Adults have been recorded during most months of the year, but they appear to be more abundant during the wet season (November–January) and again in the mid dry season (July). The immature stages have been recorded during most months of the year, except March, and the life cycle is completed relatively quickly (approximately one month), indicating that breeding occurs continuously and several generations are completed annually.

## Breeding status

This species is resident in the study region.

## Conservation status

LC. The subspecies *C. geminus gagadju* is a narrow range endemic (geographic range = 39,940 sq km) and occurs in at least three conservation reserves: Litchfield and Kakadu national parks and Wongalara Wildlife Sanctuary. Despite its restricted occurrence, there are no known threats facing the taxon.

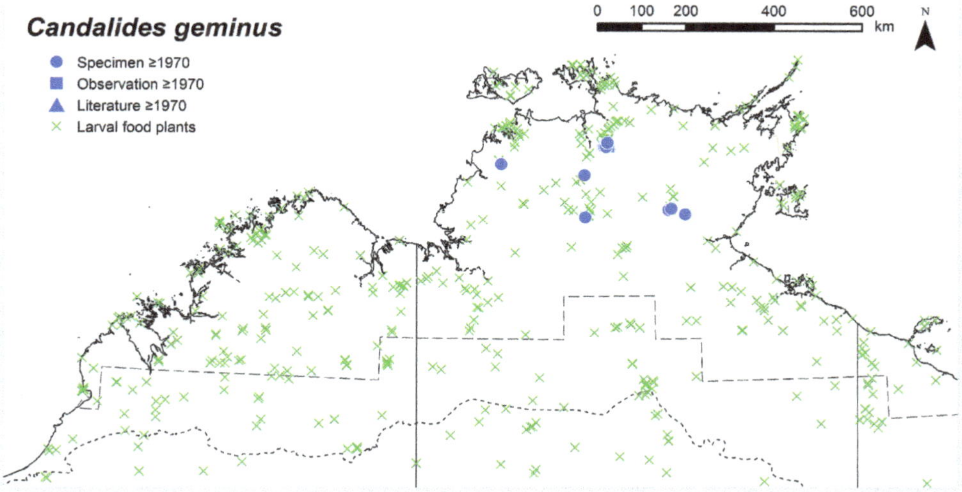

Candalides geminus
0 100 200 400 600 km
N
Specimen ≥1970
Observation ≥1970
Literature ≥1970
Larval food plants

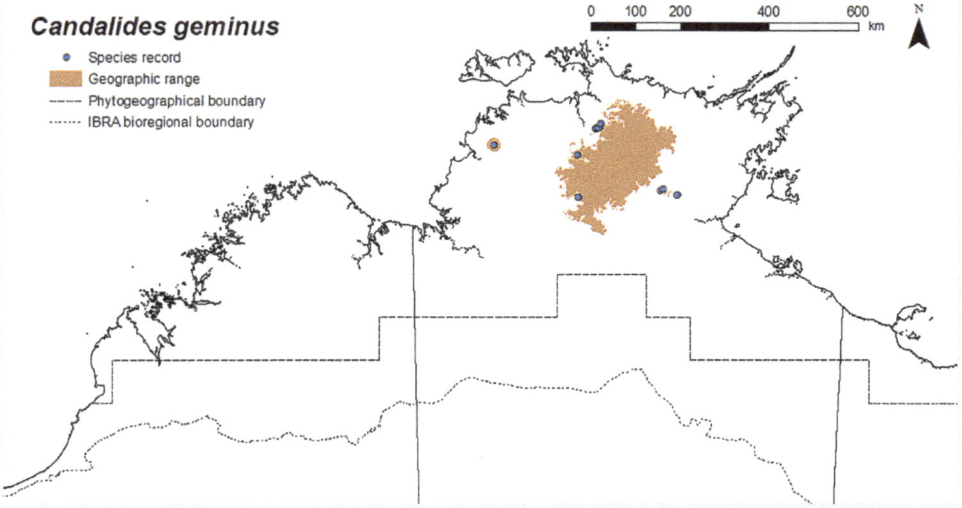

Candalides geminus
0 100 200 400 600 km
N
Species record
Geographic range
Phytogeographical boundary
IBRA bioregional boundary

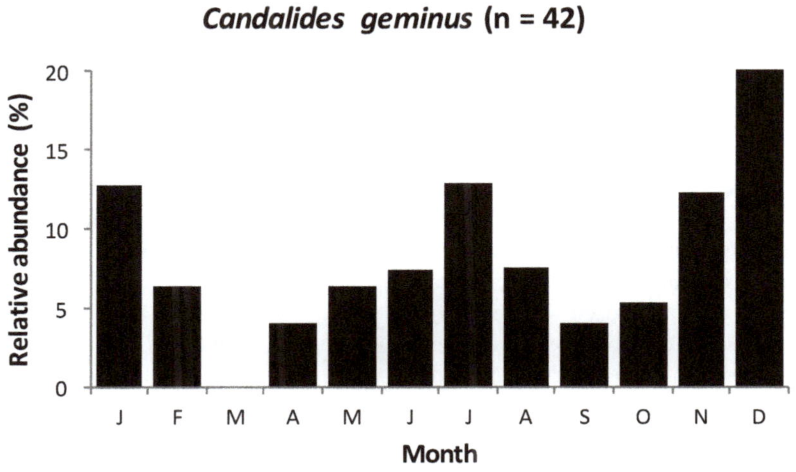

Candalides geminus (n = 42)
Relative abundance (%)
0
5
10
15
20
J F M A M J J A S O N D
Month

| Month | J | F | M | A | M | J | J | A | S | O | N | D |
|---|---|---|---|---|---|---|---|---|---|---|---|---|
| Egg | ■ | | | | ■ | ■ | ■ | ■ | | ■ | ■ | ■ |
| Larva | ■ | ■ | | | ■ | ■ | ■ | ■ | | | ■ | ■ |
| Pupa | | ■ | | ■ | ■ | ■ | ■ | ■ | ■ | | ■ | ■ |
| Adult | ■ | ■ | | ■ | ■ | ■ | ■ | ■ | ■ | ■ | ■ | ■ |

# Small Dusky-blue
*Candalides erinus* (Fabricius, 1775)

Plate 151 Kakadu National Park, NT
Photo: M. F. Braby

Plate 152 Kakadu National Park, NT
Photo: M. F. Braby

## Distribution

This species is represented in the study region by the subspecies *C. erinus erinus* (Fabricius, 1775). It occurs very widely in the Kimberley, Top End, Northern Deserts and western Gulf Country, extending from moist coastal areas to the lower rainfall areas of the semi-arid zone (< 500 mm mean annual rainfall). In the inland it has been recorded as far south as the Edgar Ranges (Common 1981) and Landrigan Creek crossing (Williams et al. 2006), WA; 50 km south–south-east of Elliott, NT; and Boodjamulla/Lawn Hill National Park (Daniels and Edwards 1998) and Rotary Lookout 2 km west of Cloncurry (Dunn 2017a), Qld. The geographical range corresponds well with the spatial distribution of its larval food plants, which are similarly widespread. Outside the study region, *C. erinus* occurs in the Pilbara of Western Australia and in north-eastern and eastern Australia.

## Habitat

*Candalides erinus* breeds in a variety of habitats, including savannah woodland on various substrates, eucalypt woodland with rainforest elements in the understorey, the ecotone between evergreen monsoon vine forest and eucalypt woodland, semi-deciduous monsoon vine thicket, paperbark woodland and paperbark–pandanus swampland, where the larval food plants grow as scrambling parasitic vines in the understorey.

## Larval food plants

*Cassytha filiformis*, *C. capillaris* (Lauraceae).

## Attendant ant

*Iridomyrmex* sp. (*mattiroloi* group) (Formicidae: Dolichoderinae). The larvae and pupae are usually not attended by ants, but very occasionally the larvae are attended by a few small black ants (Braby 2011a).

## Seasonality

Adults occur throughout the year, but they are most abundant during the first half of the year, particularly during the mid dry season (May–July), and are very scarce during the late dry season and early wet season (September–December). The immature stages have been recorded during most months of the year, being particularly prevalent during the wet season and early dry season (February–May), when adults are most numerous. The life cycle is completed relatively quickly (approximately one month), indicating that breeding occurs continuously and several generations are completed annually.

## Breeding status

This species is resident in the study region.

## Conservation status

LC.

## Candalides erinus

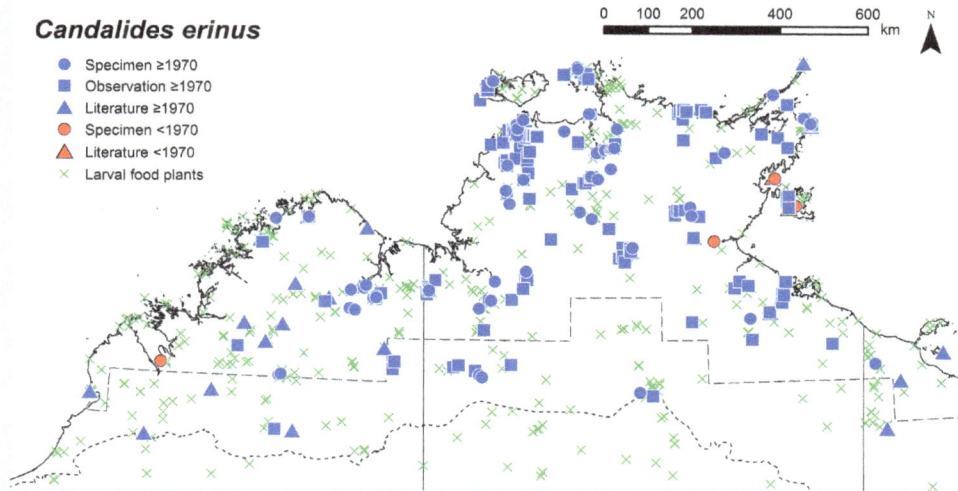

Legend:
- ● Specimen ≥1970
- ■ Observation ≥1970
- ▲ Literature ≥1970
- ● Specimen <1970
- ▲ Literature <1970
- ✕ Larval food plants

## Candalides erinus

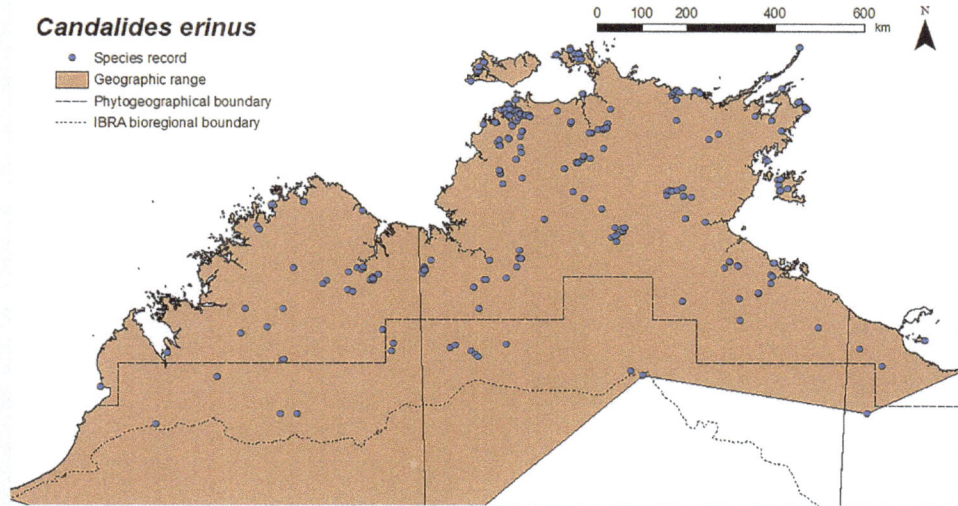

Legend:
- ● Species record
- ▨ Geographic range
- — Phytogeographical boundary
- ··· IBRA bioregional boundary

## Candalides erinus (n =367)

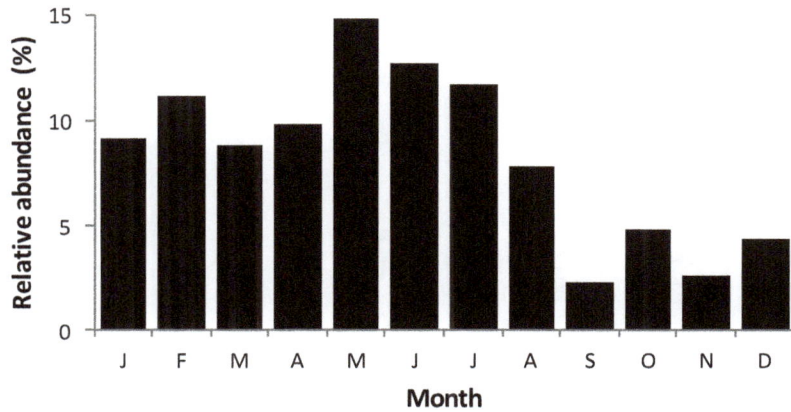

| Month | J | F | M | A | M | J | J | A | S | O | N | D |
|-------|---|---|---|---|---|---|---|---|---|---|---|---|
| Egg   |   | ▨ | ▨ | ▨ | ▨ | ▨ | ▨ |   |   |   | ▨ |   |
| Larva |   | ▨ | ▨ | ▨ | ▨ | ▨ | ▨ | ▨ |   |   | ▨ |   |
| Pupa  |   | ▨ | ▨ |   | ▨ | ▨ | ▨ |   |   |   | ▨ |   |
| Adult | ▨ | ▨ | ▨ | ▨ | ▨ | ▨ | ▨ | ▨ | ▨ | ▨ | ▨ | ▨ |

# Spotted Dusky-blue
*Candalides delospila* (Waterhouse, 1903)

Plate 153 Judbarra/Gregory National Park, NT
Photo: M. F. Braby

## Distribution

This species occurs widely in the lower rainfall areas of the study region, south of latitude 14°S (350–900 mm mean annual rainfall). Its northernmost limits are Queen Island in the western Kimberley, WA (Waterhouse 1938); and Bradshaw Field Training Area (Fitzmaurice River catchment) (Archibald and Braby 2017) and Wongalara Wildlife Sanctuary (12 km south of the homestead) (Braby 2012a) in the Top End, NT. The geographic range broadly corresponds with the spatial distribution of its larval food plant. Although the food plant extends further north into higher rainfall areas of the Top End, targeted searches in this area have not detected *C. delospila*. Outside the study region, *C. delospila* occurs well into the drier inland areas of the arid zone just beyond the southern boundary, where it has been recorded as far south as Pipingarry Station Road 30 km south-east of Port Hedland, Whim Creek (Johnson and Valentine 2004) and Balgo Hills (Braby 2000), WA; Newhaven (Australian Wildlife Conservancy property) 300 km west–north-west of Alice Springs, NT, where males were observed hilltopping on several occasions in July–August 2017 (P. Gilmour); and Johnson Creek 70 km north-west of Mt Isa, Mary Kathleen 55 km east of Mt Isa (Braby 2000), 9 km south east by east of Burke & Wills Roadhouse and Rotary Lookout near Cloncurry (Dunn 2017a), Qld. It also extends further east in the inland areas of northern Queensland.

## Habitat

*Candalides delospila* breeds in savannah woodland and eucalypt low open woodland with a hummock (spinifex) grass understorey and hummock grassland on loamy sand, sandstone escarpments, plateaus and platforms, where the larval food plant grows as a fine scrambling parasitic vine over clumps of spinifex (*Triodia*) and other grasses (Braby 2011a).

## Larval food plant

*Cassytha capillaris* (Lauraceae).

## Attendant ant

*Iridomyrmex* sp. (*gracilis* group) (Formicidae: Dolichoderinae). The larvae and pupae are usually not attended by ants, but very occasionally the larvae are attended by a few small black ants (Braby 2011a).

## Seasonality

The seasonal occurrence of adults, breeding phenology and seasonal history of the immature stages are not well understood. Adults have been recorded sporadically from January to October, with apparent peaks in abundance in March, July and August, but none has been recorded during the late dry season (November and December), when conditions are typically very hot. Similarly, the immature stages have been recorded sporadically from February to September. The life cycle is completed relatively quickly (approximately seven weeks), but it is not known whether the species undergoes pupal diapause. Presumably, the seasonal pattern is similar to *Candalides erinus*, in which adults are most abundant during the first half of the year, with a possible peak during the mid dry season (July), during which several generations are completed, but it remains to be established whether the species breeds during the late dry season.

## Breeding status

This species is resident in the study region.

## Conservation status

LC.

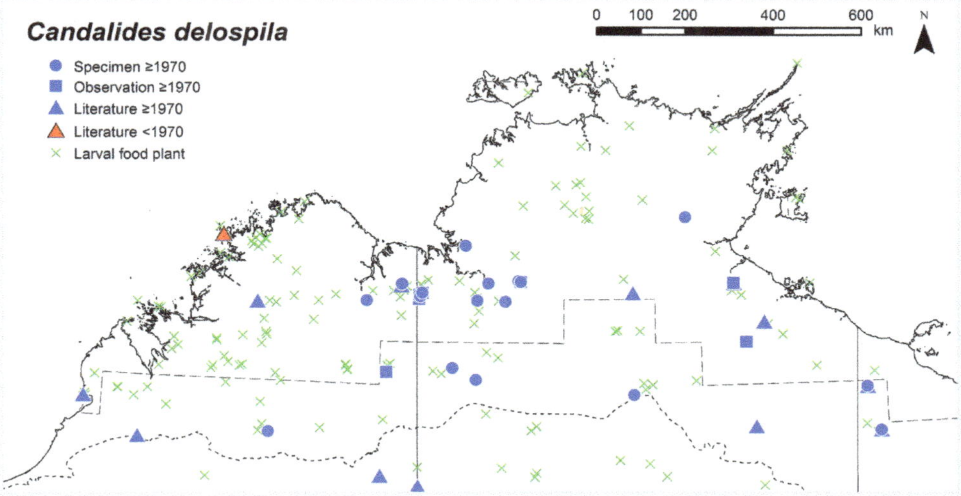

Candalides delospila
0 100 200 400 600
km
N
Specimen ≥1970
Observation ≥1970
Literature ≥1970
Literature <1970
Larval food plant


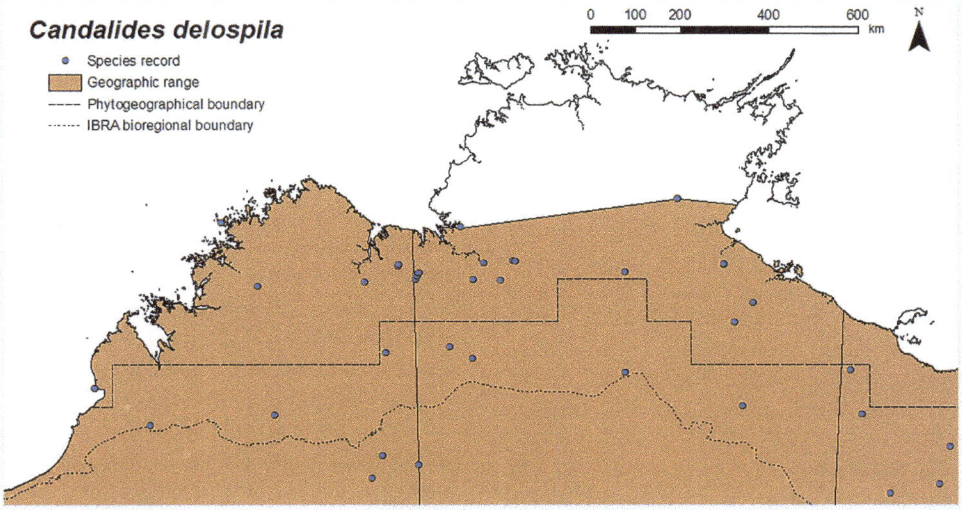

Candalides delospila
0 100 200 400 600
km
N
Species record
Geographic range
Phytogeographical boundary
IBRA bioregional boundary


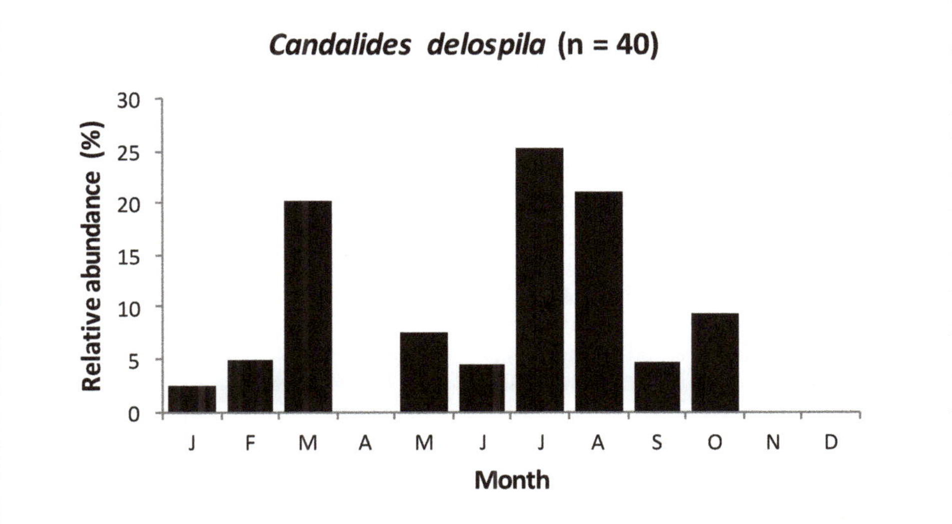

Candalides delospila (n = 40)
Relative abundance (%)
30
25
20
15
10
5
0
J F M A M J J A S O N D
Month


| Month | J | F | M | A | M | J | J | A | S | O | N | D |
|---|---|---|---|---|---|---|---|---|---|---|---|---|
| Egg | | | ● | | | | ● | | | | | |
| Larva | | ● | ● | | | | ● | ● | | | | |
| Pupa | | ● | ● | | | | | | ● | | | |
| Adult | ● | ● | ● | | ● | ● | ● | ● | ● | ● | | |

# Spotted Opal

## *Nesolycaena urumelia* (Tindale, 1922)

Plate 154 Kakadu National Park, NT
Photo: M. F. Braby

## Distribution

This species is endemic to the study region. It occurs widely in the Top End and western Gulf Country, generally in the higher rainfall areas (> 800 mm mean annual rainfall), although it has been recorded in the semi-arid zone in western Queensland (c. 500 mm). Its southernmost limits include Spirit Hills (42 km north-east of Keep River National Park ranger station) and McArthur River Homestead (Edwards 1980), NT; and Boodjamulla/Lawn Hill National Park, Qld (Braby 1996; Daniels and Edwards 1998). The geographic range corresponds well with the spatial distribution of its larval food plants in the Northern Territory. Although the food plants extend to the Kimberley, the species is replaced with *Nesolycaena caesia* in northern Western Australia. However, *N. urumelia* does not appear to have been recorded from the Tiwi Islands despite the presence of the larval food plants; thus, further field surveys are required to determine whether *N. urumelia* is established on Bathurst and Melville islands, NT.

## Habitat

*Nesolycaena urumelia* breeds in eucalypt heathy woodland and open woodland on sandstone outcrops, escarpments, plateaus and breakaways or scree slopes below cliffs and in mixed riparian woodland along sandstone gorges where the larval food plants grow as shrubs on sand or loamy sand (Edwards 1980; Braby 2011a, 2015e). It also breeds in coastal eucalypt heathy woodland on white sand and lowland sandsheets.

## Larval food plants

*Boronia lanceolata, B. lanuginosa, B. laxa, B. wilsonii* (Rutaceae). The main food plants are *B. lanceolata* and *B. lanuginosa* (Edwards 1980; Meyer 1996a; Braby 2000, 2011a); however, larvae also feed on *B. laxa* in western Arnhem Land (Braby 2011a) and *B. wilsonii* at Spirit Hills, NT (Braby 2015e).

## Attendant ant

*Polyrhachis gab* (Formicidae: Formicinae), *Monomorium* sp. 8 (*carinatum* group) (Formicidae: Myrmicinae). The larvae and pupae are usually not attended by ants, but very occasionally the larvae are attended by a few ants (Edwards 1980; Meyer 1996a; Eastwood et al. 2008).

## Seasonality

Adults occur during most months of the year, but they are generally most abundant during the wet season (January–April) after average or above average rainfall, with a peak in abundance in April. Their numbers diminish as the dry season progresses, and they are scarce towards the end of the dry season (August–October). The immature stages (eggs or larvae) have been recorded during most months of the year except during the late dry season (August–October). Pupae from larvae reared in April or May remain in diapause for up to six months or more during the dry season (Edwards 1980; M. F. Braby, unpublished data). These observations suggest breeding is seasonal and restricted largely to the wet season and early dry season, during which several generations are completed.

## Breeding status

This species is resident in the study region.

## Conservation status

LC. Although the species *N. urumelia* has a wide geographical range, some areas are potentially threatened by inappropriate fire regimes. Sands and New (2002) suggested the species and/or its larval food plants are sensitive to too frequent burning of sandstone plant communities, and this may be a threatening process in some areas of the range such as the Arnhem Land Plateau (see also Russell-Smith et al. 1998, 2002). The food plants are obligate seeders and, if the fire regime is too frequent, with short interfire intervals (every one to two years), there may be little or no recruitment. That is, in some areas of the range, the population

of *N. urumelia* may be reduced in future based on a projected decline in the AOO and/or quality of its habitat. Although two of the larval food plants (*B. lanceolata*, *B. lanuginosa*) are currently listed as Least Concern (LC), the two other species (*B. laxa*, *B. wilsonii*) are listed as Near Threatened (NT) under the *TPWCA*. Thus, *N. urumelia* may qualify as Near Threatened (NT) in future. Monitoring of the abundance or occupancy of the butterfly and its food plants is required to clarify the effect of fire as a key threatening process. The fire regime on the sandstone plateau–breakaway country needs to be carefully managed (e.g. at least more than five years) to ensure the food plants have sufficient time to mature, flower and set seed.

Photo: Nourlangie Rock, Kakadu National Park, NT, M.F. Braby

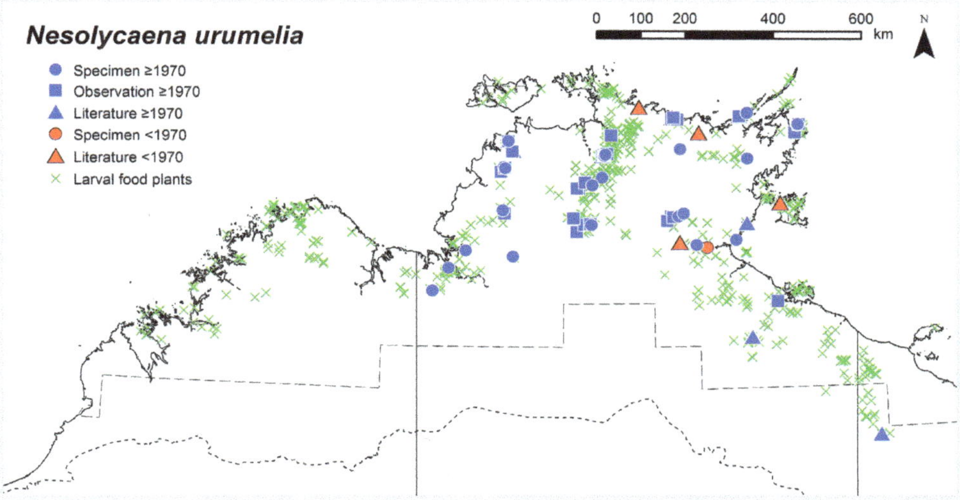

Nesolycaena urumelia
0 100 200 400 600
km
N
Specimen ≥1970
Observation ≥1970
Literature ≥1970
Specimen <1970
Literature <1970
Larval food plants

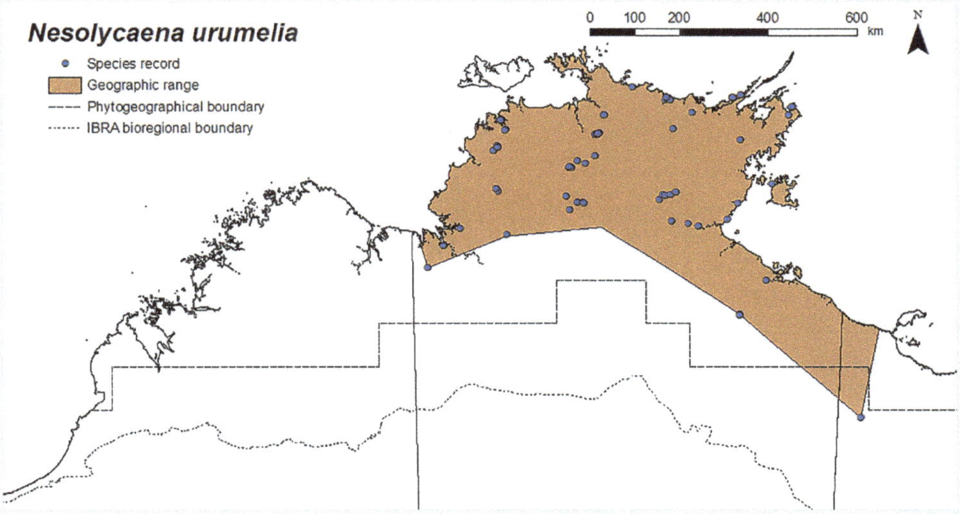

Nesolycaena urumelia
0 100 200 400 600
km
N
Species record
Geographic range
Phytogeographical boundary
IBRA bioregional boundary

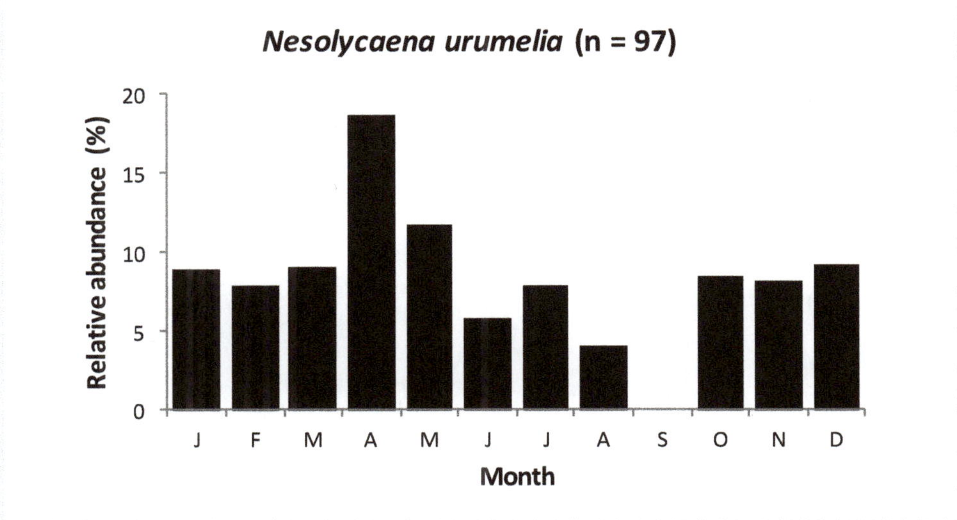

Nesolycaena urumelia (n = 97)
Relative abundance (%)
0
5
10
15
20
J F M A M J J A S O N D
Month

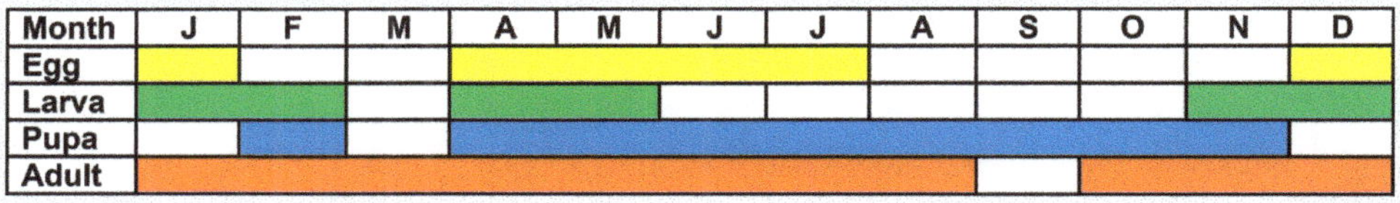

Month J F M A M J J A S O N D
Egg
Larva
Pupa
Adult

# Kimberley Opal

## *Nesolycaena caesia* d'Apice & Miller, 1992

Plate 155 Kalumburu, WA
Photo: M. F. Braby

## Distribution

This species is endemic to the study region. It was described only as recently as 1992, two years after it was first discovered, in June 1990 (d'Apice and Miller 1992). It is restricted to northern Western Australia, where it has been recorded in the higher rainfall areas of the northern Kimberley at and near Kalumburu (d'Apice and Miller 1992) and on Jar and Steep Head islands (J. E. and A. Koeyers), and in the lower rainfall areas of the eastern Kimberley (El Questro Wilderness Park, WA) (Meyer 1996c). Its main larval food plant (*Boronia wilsonii*) is substantially more widespread, particularly in the western Kimberley south-west to King Sound; thus, further field surveys are required to determine whether *N. caesia* is established in this area.

## Habitat

*Nesolycaena caesia* breeds in eucalypt heathy low woodland on sandstone outcrops, ridges and breakaways where the larval food plants grow as small shrubs on sandy soil (d'Apice and Miller 1992; Meyer et al. 2013).

## Larval food plants

*Boronia kalumburuensis*, *B. wilsonii* (Rutaceae). The main food plant is *B. wilsonii* (d'Apice and Miller 1992; Grund and Hunt 2001; Meyer et al. 2013).

## Attendant ant

*Iridomyrmex* sp. (Formicidae: Dolichoderinae). The larvae and pupae are usually not attended by ants, but very occasionally the larvae are attended by a few ants (Eastwood and Fraser 1999; Meyer et al. 2013).

## Seasonality

Adults have been recorded during most months of the year (November–July), but there are too few records (n = 15) to assess any seasonal changes in abundance. In general, they have been noted to be more numerous during the wet season (January–April) (Meyer et al. 2013; J. E. and A. Koeyers, pers. comm.). The breeding phenology and seasonal history of the immature stages are not well understood. The immature stages (eggs or larvae) have been recorded in November and from March to July. The pupal stage is known to remain in diapause for up to eight months during the dry season (Braby 2000). Presumably, the life cycle strategy is similar to *Nesolycaena urumelia* in which breeding is seasonal and restricted to the wet season and early dry season, during which several generations are completed, and adults are scarce or absent during the late dry season (August–October).

## Breeding status

This species is resident in the study region.

## Conservation status

DD. The species *N. caesia* is a short-range endemic (EOO = 9,170 sq km). It currently lacks adequate protection because most sites occur outside conservation reserves, although it does occur in the Uunguu IPA. However, it may be impacted by inappropriate fire regimes. The larval food plants are obligate seeders and if the fire regime on sandstone plant communities is too frequent, with short interfire intervals (every one–two years), there may be little or no recruitment. That is, the population of *N. caesia* may be reduced in future based on a projected decline in the AOO and/or quality of its habitat. Thus, the taxon may qualify as Near Threatened (NT) once adequate data are available. Monitoring of the abundance and occupancy of the butterfly and its food plants is required to clarify the effect of fire frequency as a key threatening process.

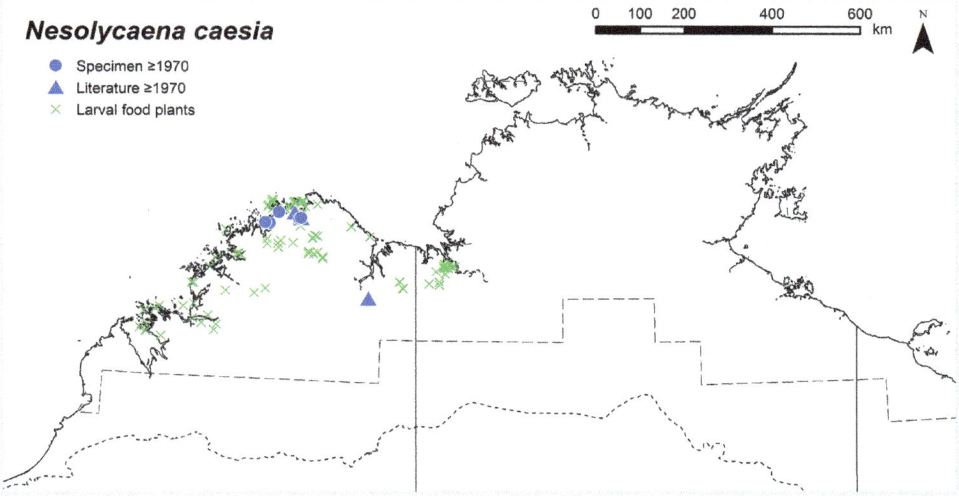

**Nesolycaena caesia**

- Specimen ≥1970
- Literature ≥1970
- Larval food plants

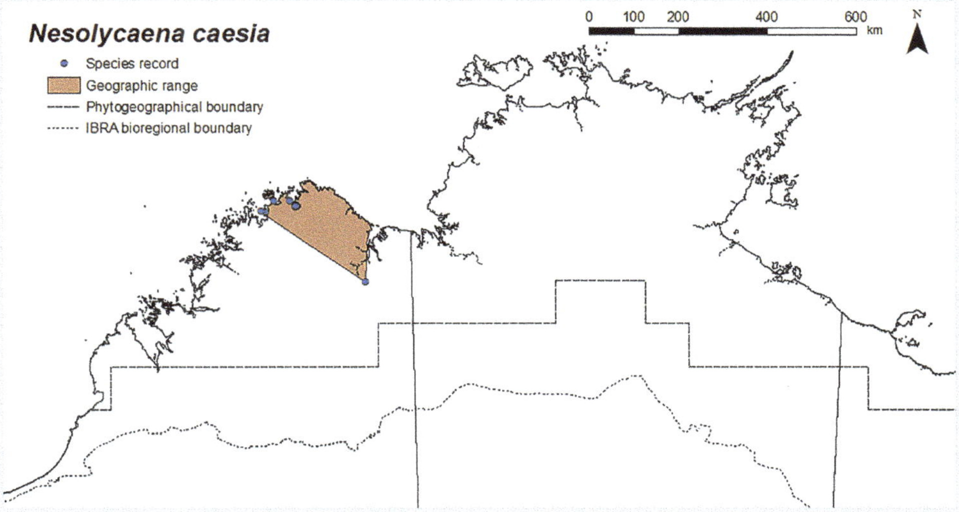

**Nesolycaena caesia**

- Species record
- Geographic range
- — — Phytogeographical boundary
- ········ IBRA bioregional boundary

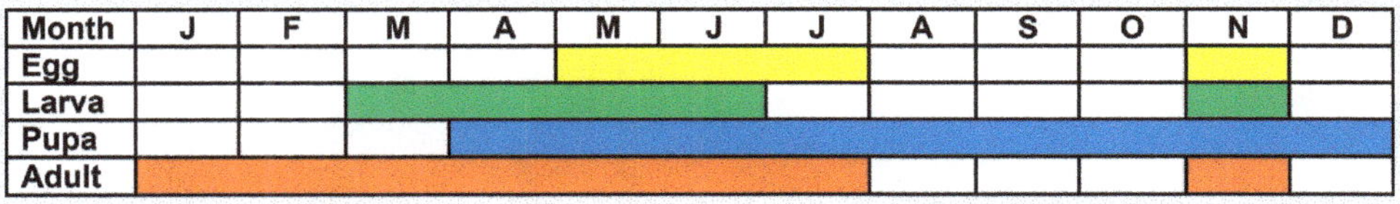

| Month | J | F | M | A | M | J | J | A | S | O | N | D |
|-------|---|---|---|---|---|---|---|---|---|---|---|---|
| Egg   |   |   |   |   |   |   |   |   |   |   |   |   |
| Larva |   |   |   |   |   |   |   |   |   |   |   |   |
| Pupa  |   |   |   |   |   |   |   |   |   |   |   |   |
| Adult |   |   |   |   |   |   |   |   |   |   |   |   |

# Pale Ciliate-blue
## *Anthene lycaenoides* (C. Felder, 1860)

Plate 156 Herberton, Qld
Photo: Don Franklin

## Distribution

This species is represented in the study region by the subspecies *A. lycaenoides godeffroyi* (Semper, [1879]). It occurs in the northern and eastern Kimberley and more widely in the Top End. It extends from moist coastal areas to drier inland areas of the semi-arid zone (c. 700 mm mean annual rainfall), where it has been recorded as far south as El Questro Wilderness Park (El Questro Gorge and Pentecost River crossing), WA (Braby 2012b); and Timber Creek, NT. The larval food plants are considerably more widespread in the Kimberley, the eastern half of the Top End and western Gulf Country, as well as on the Tiwi Islands and Groote Eylandt, NT. The spatial distribution of the larval food plants, together with the occurrence of the butterfly in the lower rainfall areas of the study region, suggests the geographic range of *A. lycaenoides* is incomplete. Further field surveys are thus required to determine whether *A. lycaenoides* occurs in the western Kimberley, Bathurst and Melville islands, NT, the Limmen Bight area in the Gulf of Carpentaria and the western Gulf Country. Outside the study region, *A. lycaenoides* occurs in Maluku, including the Kai Islands, mainland New Guinea and adjacent islands and north-eastern Australia.

## Habitat

*Anthene lycaenoides* breeds mainly in monsoon forest, including riparian evergreen monsoon vine forest along permanent freshwater streams, semi-deciduous monsoon vine thicket and riparian paperbark open forest with rainforest elements in the understorey, where the larval food plants usually grow as small trees. It also occurs in suburban parks and gardens where the food plants are propagated as ornamental trees.

## Larval food plants

*Millettia pinnata*, *Senna surattensis* (Fabaceae), *Flagelleria indica* (Flagellariaceae), *Barringtonia acutangula* (Lecythidaceae), *Bridelia tomentosa* (Phyllanthaceae), *Cupaniopsis anacardioides* (Sapindaceae).

## Attendant ants

*Crematogaster* spp. (Formicidae: Myrmicinae). The larvae and pupae are usually not attended by ants, but very occasionally the larvae are attended by a few ants (Meyer 1996a).

## Seasonality

Adults occur during most months of the year, but the patterns of their seasonal changes in abundance are not entirely clear. Adults are rarely observed in large numbers. Our limited data suggest they are more numerous during the late dry season, wet season and early dry season, with a possible peak in abundance in April and May, but are scarce during the mid dry season (June–August). The breeding phenology is not well understood. The larvae feed on flower buds and flowers, and the immature stages have been recorded sporadically during the wet season (December–April) and also in July. Presumably, the species breeds continuously throughout the year and several generations are completed annually by switching its food plants on a seasonal basis.

## Breeding status

This species is resident in the study region.

## Conservation status

LC.

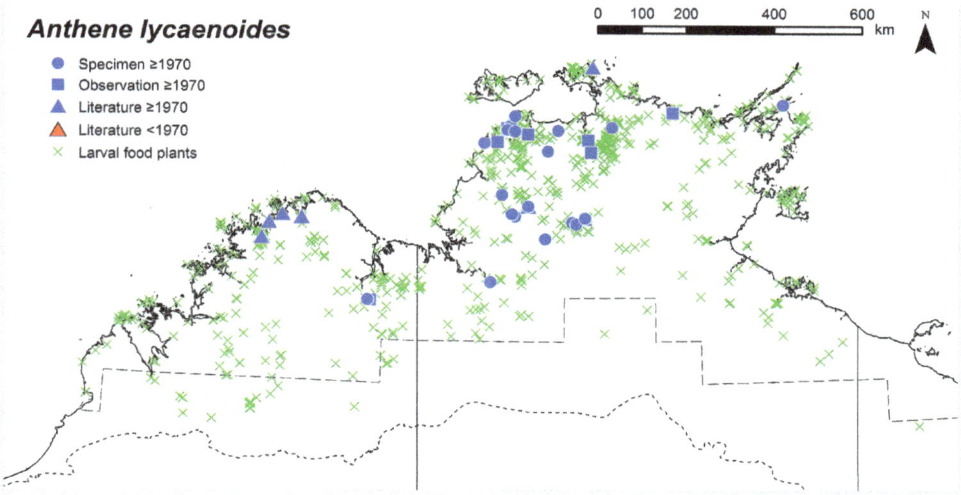

Anthene lycaenoides
0 100 200 400 600 km
N
Specimen ≥1970
Observation ≥1970
Literature ≥1970
Literature <1970
Larval food plants

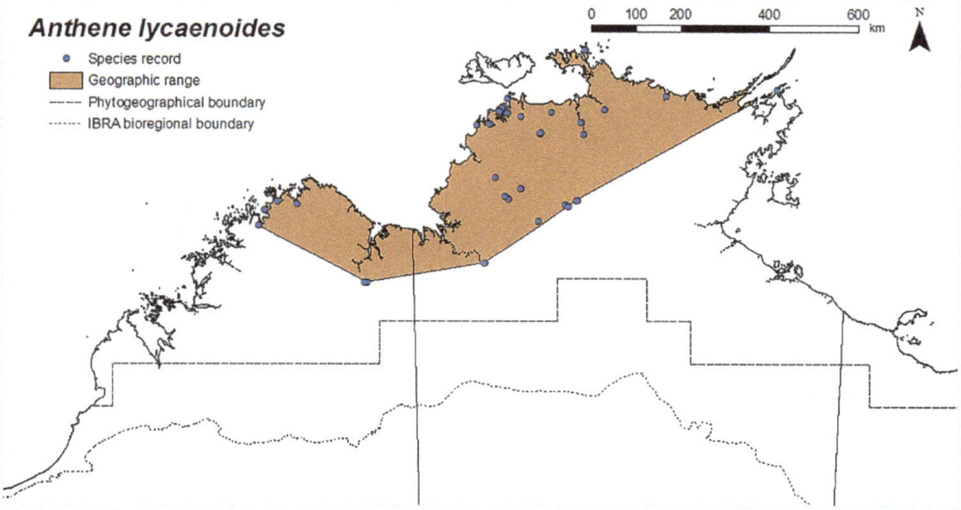

Anthene lycaenoides
0 100 200 400 600 km
N
Species record
Geographic range
Phytogeographical boundary
IBRA bioregional boundary

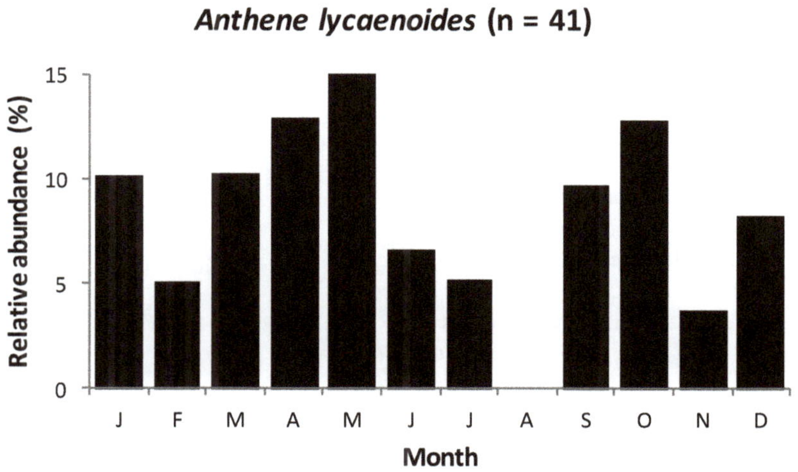

Anthene lycaenoides (n = 41)
Relative abundance (%)
0
5
10
15
J F M A M J J A S O N D
Month

| Month | J | F | M | A | M | J | J | A | S | O | N | D |
|---|---|---|---|---|---|---|---|---|---|---|---|---|
| Egg | | | | ■ | | | | | | | | |
| Larva | ■ | | ■ | ■ | | | ■ | | | | | ■ |
| Pupa | ■ | | ■ | ■ | | | | | | | | |
| Adult | ■ | ■ | ■ | ■ | ■ | ■ | ■ | | ■ | ■ | ■ | ■ |

# Dark Ciliate-blue
## *Anthene seltuttus* (Röber, 1886)

Plate 157 Beatrice Hill, NT
Photo: M. F. Braby

## Distribution

This species is represented in the study region by the subspecies *A. seltuttus affinis* (Waterhouse & R. E. Turner, 1905). It is restricted to the Top End, where it occurs mainly in the higher rainfall areas (> 800 mm mean annual rainfall, but generally > 900 mm). Its southernmost extent includes Katherine (Angel 1951), Roper River and Groote Eylandt (Tindale 1923), NT. The species also occurs in the Gulf of Carpentaria at Walker Creek 36 km east of Karumba, Qld, just outside the eastern boundary of the study region (Braby 2015d). The geographical range of *A. seltuttus* closely corresponds with the spatial distribution of its associated ant within the Top End. Although the larval food plants and attendant ant also occur widely in the Kimberley, searches have not detected *A. seltuttus* in this area. Similarly, although the larval food plants are more widely distributed in the Northern Deserts and the western Gulf Country, the ant is absent from these areas and is much more restricted in extent (Lokkers 1986). Outside the study region, *A. seltuttus* occurs from Maluku, including the Kai and Aru islands, through mainland New Guinea and adjacent islands and north-eastern Australia to the Louisiade Archipelago.

## Habitat

*Anthene seltuttus* breeds mainly in the edges of semi-deciduous monsoon vine thicket, in both coastal and inland areas, and in mixed riparian open forest with rainforest elements in the understorey where the larval food plants (*Millettia pinnata* and *Cupaniopsis anacardioides*) grow as understorey trees and where arboreal nests of the attendant ant are established. It also breeds in the ecotone between savannah woodland and monsoon forest where saplings of an alternative food plant (*Corymbia disjuncta*) grow (Braby 2015e).

## Larval food plants

*Millettia pinnata* (Fabaceae), *Corymbia disjuncta* (Myrtaceae), *Cupaniopsis anacardioides* (Sapindaceae); also *\*Delonix regia* (Fabaceae), *\*Litchi chinensis* (Sapindaceae).

## Attendant ant

*Oecophylla smaragdina* (Formicidae: Formicinae). The larvae and pupae are constantly attended by numerous Green Tree Ants (*O. smaragdina*) in an obligatory myrmecophilous association (Eastwood and Fraser 1999; Braby 2000).

## Seasonality

Adults occur throughout the year, but they are generally more abundant during the wet season (November–April). The breeding phenology is not well understood. The larvae feed on new soft leaves, and the immature stages have been recorded during the second half of the year (August–December) and also in April. Presumably, the species breeds continuously throughout the year and several generations are completed annually by switching its larval food plants on a seasonal basis.

## Breeding status

This species is resident in the study region.

## Conservation status

LC.

# *Anthene seltuttus*

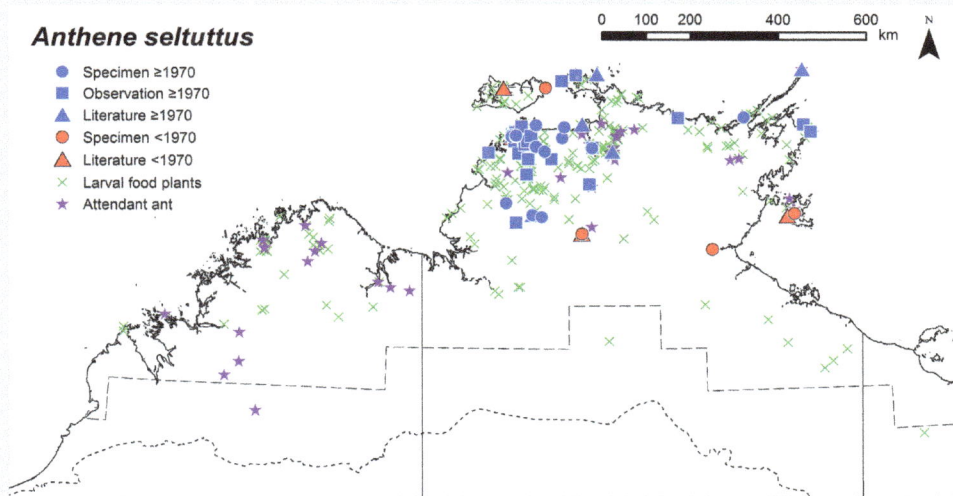

# *Anthene seltuttus*

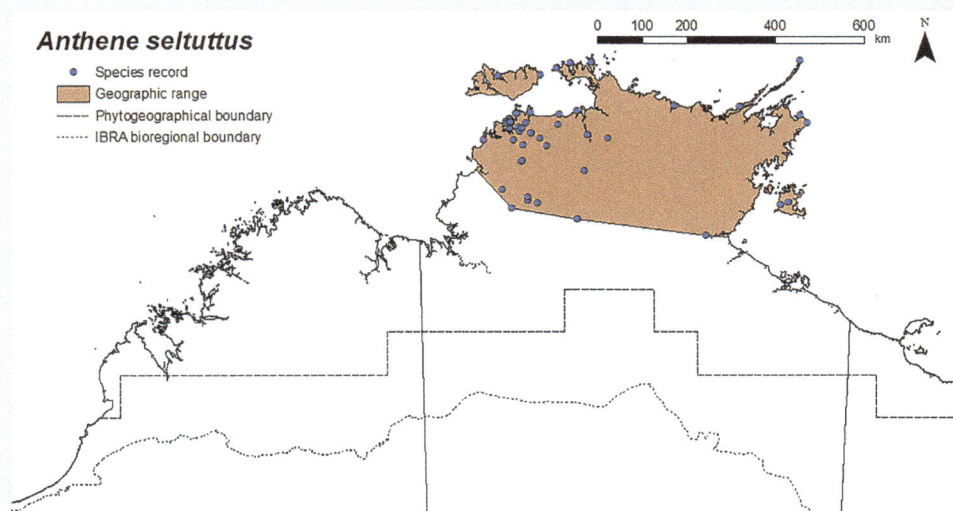

## *Anthene seltuttus* (n = 68)

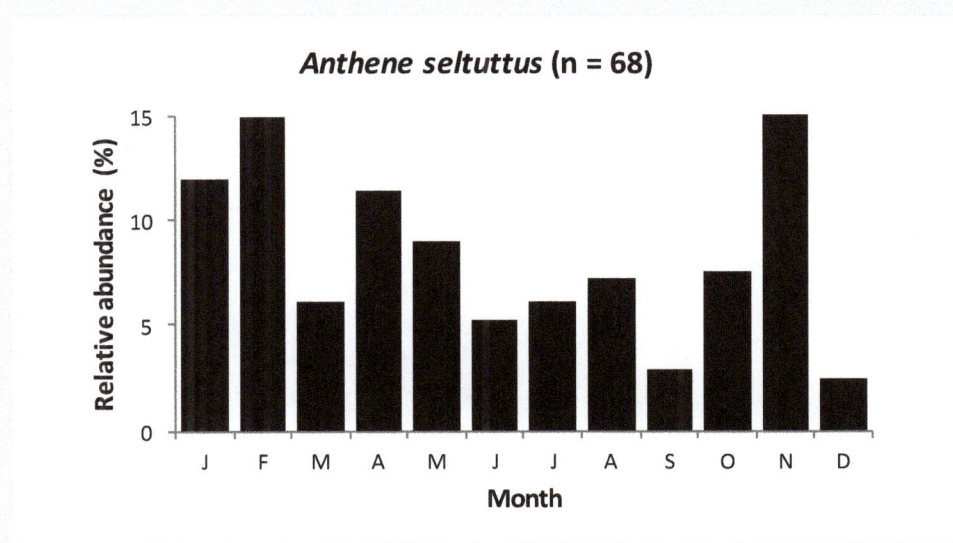

| Month | J | F | M | A | M | J | J | A | S | O | N | D |
|---|---|---|---|---|---|---|---|---|---|---|---|---|
| Egg | | | | | | | | | | | | |
| Larva | | | | | | | | | | | | |
| Pupa | | | | | | | | | | | | |
| Adult | | | | | | | | | | | | |

# Mauve Line-blue
## *Petrelaea tombugensis* (Röber, [1886])

Plate 158 Carson River crossing, WA
Photo: M. F. Braby

Plate 159 Carson River crossing, WA
Photo: M. F. Braby

## Distribution

This species is known from only two locations within the study region: in the northern Kimberley on the Carson River crossing 18 km south–south-east of Kalumburu, WA, based on six specimens collected in May 2015 (Braby 2015b); and in the Top End at Black Point on Cobourg Peninsula, NT, based on two males collected in January 1977 (Common and Waterhouse 1981; Braby 2015b). The putative larval food plants (*Terminalia* spp.) have a disjunct distribution that largely corresponds with the known geographic range of *P. tombugensis*; *Terminalia muelleri* is known only from the northern Kimberley (Vansittart Bay, WA) and Cobourg Peninsula (Coral Bay, NT), whereas *T. catappa* is known further east on the Wessell Islands and Gove Peninsula, NT. Further field surveys are thus required to determine whether *P. tombugensis* is present in north-eastern Arnhem Land. Outside the study region, *P. tombugensis* occurs from the Andaman Islands and Maluku, through mainland New Guinea and north-eastern Australia to New Britain and the Solomon Islands.

## Habitat

The breeding habitat of *P. tombugensis* has not been recorded in the study region. Adults have been collected in coastal semi-deciduous monsoon vine thicket and riparian evergreen monsoon vine forest (Braby 2015b)—habitats in which the species no doubt breeds.

## Larval food plants

Not recorded in the study region; probably *Terminalia catappa* and *T. muelleri* (Combretaceae), which are larval food plants in northern Queensland (Samson and Lambkin 2003; Hopkinson and Bourne 2018).

## Attendant ant

Not recorded in the study region. In north-eastern Queensland, the larvae and pupae are not attended by ants (Samson and Lambkin 2003; Hopkinson and Bourne 2018).

## Seasonality

The seasonal abundance and breeding phenology of this species are not well understood. Adults have been recorded only during the wet season (January) and early dry season (May), but there are too few records (n = 3) to assess any seasonal changes in abundance. In northern Queensland, the larvae feed on flowers of the food plant; adults are seasonal and present only from September to May, which broadly coincides with the flowering period of *Terminalia catappa* and *T. muelleri*, which extends from the 'build-up' through the wet season to the start of the dry season (October–May) (Müller et al. 1998; Samson and Lambkin 2003; Hopkinson and Bourne 2018). Presumably, the breeding phenology of *P. tombugensis* in northern Australia is similar to that in northern Queensland. However, it is not known how the species survives the dry season (June–September).

## Breeding status

This species is resident in the study region.

## Conservation status

DD. Despite the wide geographical range of *P. tombugensis* in the study region, available data suggest it has a disjunct distribution with a limited AOO (is likely to be < 2,000 sq km, with spatial buffering of records providing a first approximation of 1,400 sq km). One of the two known locations occurs in a conservation reserve (Garig Gunak Barlu National Park). Targeted field surveys to clarify the extent of its distribution and ecological requirements should be a priority for this species.

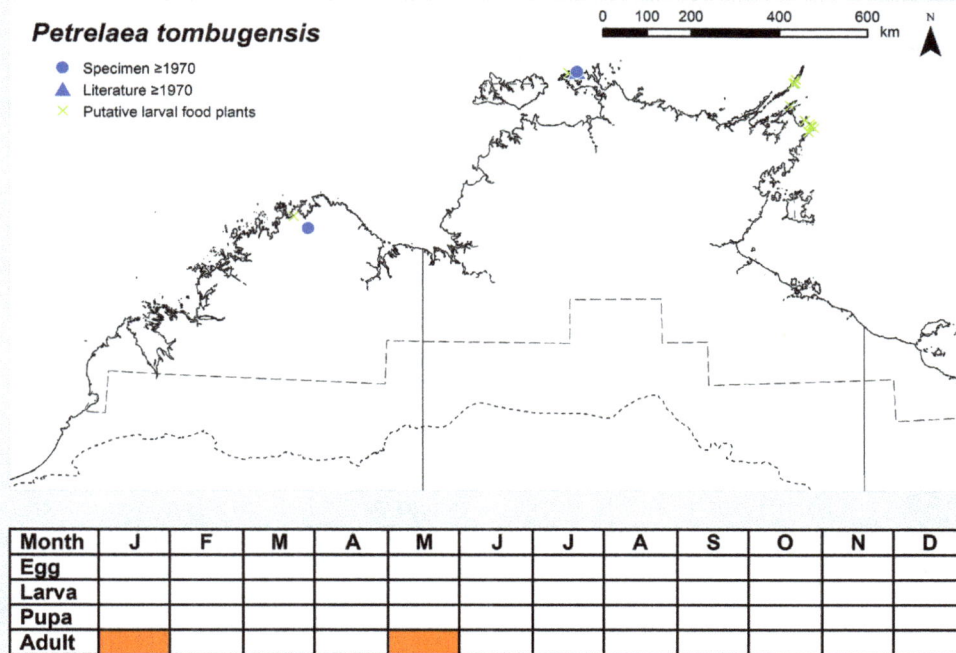

*Petrelaea tombugensis*

- ● Specimen ≥1970
- ▲ Literature ≥1970
- ✕ Putative larval food plants

| Month | J | F | M | A | M | J | J | A | S | O | N | D |
|---|---|---|---|---|---|---|---|---|---|---|---|---|
| Egg | | | | | | | | | | | | |
| Larva | | | | | | | | | | | | |
| Pupa | | | | | | | | | | | | |
| Adult | ▨ | | | | ▨ | | | | | | | |

# White-banded Line-blue
## *Nacaduba kurava* (Moore, [1858])

Plate 160 Marrakai Road, NT
Photo: M. F. Braby

## Distribution

This species is represented by the subspecies *N. kurava felsina* Waterhouse & Lyell, 1914, which is endemic to the study region. It occurs in the north of the Northern Territory, where it is restricted to the higher rainfall areas of the north-western corner of the Top End (> 800 mm mean annual rainfall, but generally > 900 mm). The geographic range broadly corresponds with the spatial distribution of its known larval food plant. The food plant, however, extends further east to Kakadu National Park (East Alligator River), NT; thus, further field surveys are required to determine whether *N. kurava* occurs in the vicinity of western Arnhem Land. Specimens of *N. kurava* have been collected from several locations where the food plant is absent, including Bullocky Point in Darwin (Braby 2014a) and further inland at Katherine (Common and Waterhouse 1981) and Mataranka (Dunn and Dunn 1991), NT, indicating that alternative (as yet unreported) food plants are used at these locations. Outside the study region, *N. kurava* occurs widely from India, Japan and South-East Asia, through mainland New Guinea and north-eastern Australia to the Bismarck Archipelago.

## Habitat

*Nacaduba kurava* breeds mainly in riparian evergreen monsoon vine forest associated with permanent freshwater streams where the known larval food plant grows as a rambling woody vine (Meyer 1996b). Adults have also been recorded in semi-deciduous monsoon vine thicket on laterite in coastal areas and on limestone outcrops in inland areas, where they no doubt breed on alternative food plants.

## Larval food plants

*Embelia curvinervia* (Primulaceae). The recorded food plant is the vine *E. curvinervia* (Meyer 1996b); however, Common and Waterhouse (1981: 550) reported an additional species, noting at Katherine that a female 'was reared from a larva feeding on the young foliage of a rain-forest tree growing on the bank of the Katherine River'. The identity of this second food plant has not been determined.

## Attendant ant

The larvae and pupae are not attended by ants.

## Seasonality

Adults occur throughout the year, and they appear to be more numerous during the second half (July–December, and January), but there are too few records (n = 27) to assess any seasonal changes in abundance. The larvae feed on new leaf growth and flowers. We have recorded the immature stages during the early dry season (April–July), although Meyer (1996b: 74) noted that the 'larvae occur throughout the year but are numerous from May to July'. Presumably, the species breeds throughout the year and several generations are completed annually.

## Breeding status

This species is resident in the study region.

## Conservation status

LC. The subspecies *N. kurava felsina* is a narrow-range endemic (EOO = 25,250 sq km) and occurs in at least one conservation reserve (Fish River Station). Despite its restricted occurrence and poor level of protection, there are no known threats facing the taxon. The larval food plant is currently listed as Least Concern (LC) under the *TPWCA*.

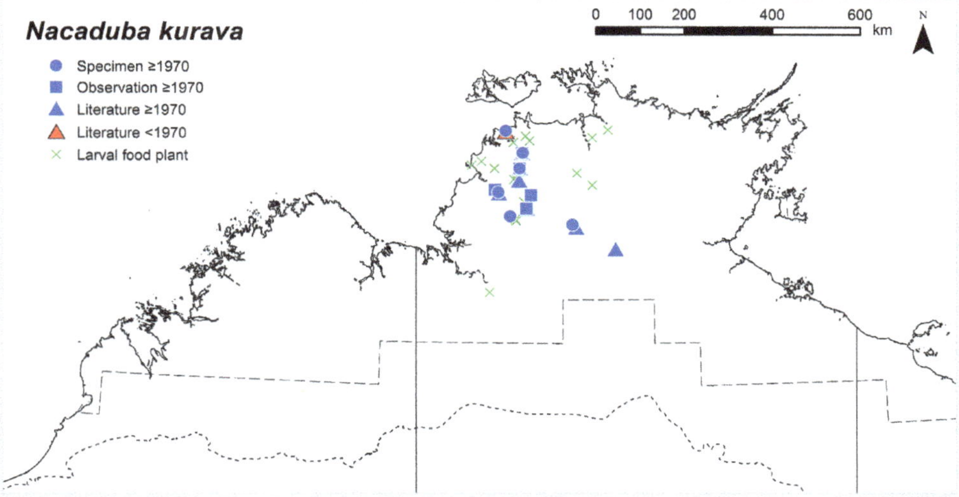

Nacaduba kurava
0 100 200 400 600 km
N
Specimen ≥1970
Observation ≥1970
Literature ≥1970
Literature <1970
Larval food plant

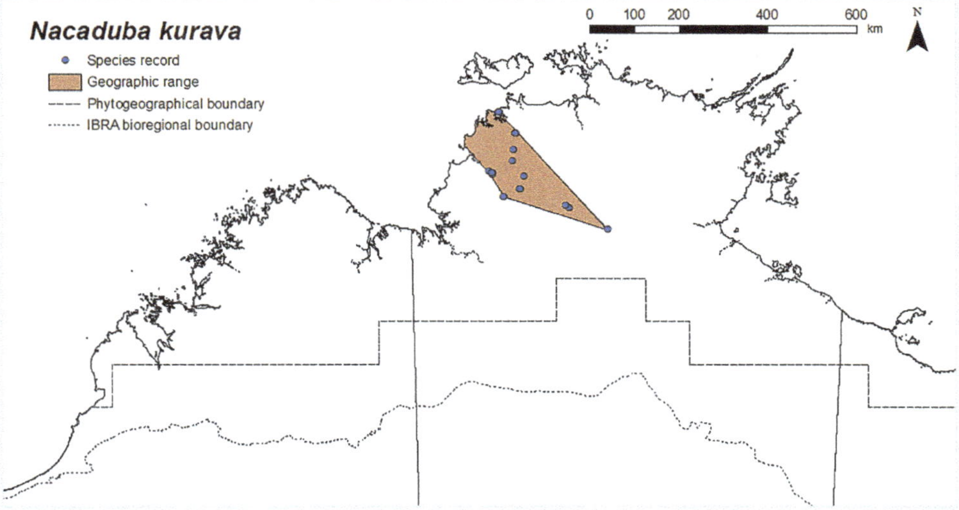

Nacaduba kurava
0 100 200 400 600 km
N
Species record
Geographic range
Phytogeographical boundary
IBRA bioregional boundary

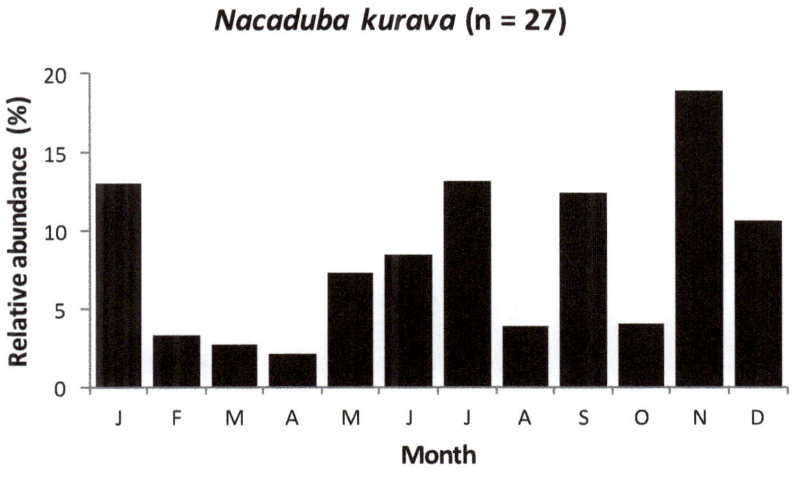

Nacaduba kurava (n = 27)
Relative abundance (%)
20
15
10
5
0
J F M A M J J A S O N D
Month

| Month | J | F | M | A | M | J | J | A | S | O | N | D |
|---|---|---|---|---|---|---|---|---|---|---|---|---|
| Egg | | | | | | | | | | | | |
| Larva | | | | | | | | | | | | |
| Pupa | | | | | | | | | | | | |
| Adult | | | | | | | | | | | | |

# Two-spotted Line-blue
## *Nacaduba biocellata* (C. & R. Felder, 1865)

Plate 161 Nhulunbuy, NT
Photo: M. F. Braby

Plate 162 Wongalara Wildlife Sanctuary, NT
Photo: M. F. Braby

## Distribution

This species is represented in the study region by the subspecies *N. biocellata biocellata* (C. & R. Felder, 1865). It occurs very widely in the region and also extends into the arid zone of central Australia beyond the southern boundary of the study region. It is generally more prevalent in the low rainfall areas of the semi-arid zone. The geographic range corresponds well with the spatial distribution of its larval food plants, which are similarly widespread. Interestingly, both the butterfly and its known food plants are absent from the Tiwi Islands, NT. Outside the study region, *N. biocellata* occurs in the Lesser Sunda Islands and throughout Australia, Vanuatu and New Caledonia.

## Habitat

*Nacaduba biocellata* breeds in savannah woodland and open woodland where the larval food plants grow as small trees (Braby 2015e).

## Larval food plants

*Acacia plectocarpa*, *A. torulosa*, *A. tumida* (Fabaceae). The immature stages and male mate-location behaviour have been recorded associated with at least three species of *Acacia* (Braby 2015e), but presumably *N. biocellata* utilises a much wider range of *Acacia* spp. than currently recorded (Dunn 2017c).

## Attendant ant

Not recorded in the study region. Elsewhere, the larvae and pupae are usually not attended by ants, but very occasionally they are attended by a few ants (Eastwood and Fraser 1999; Braby 2000).

## Seasonality

Adults have been recorded during most months of the year, but they are most abundant during the mid and early dry seasons (May–August), with a pronounced peak in abundance in July, and are generally rare or absent in most years during the 'build-up' and wet season (October–April). The breeding phenology and seasonal history of the immature stages are not well understood. Larvae have been collected and reared on flowers of the larval food plant in May. There are no published records of migration of the species, but it is strongly suspected that populations are highly mobile and can disperse considerable distances (F. Douglas, pers. comm.). Presumably, *N. biocellata* colonises much of the study region—or at least the higher rainfall areas of the region—on a seasonal basis; breeding is limited to the dry season, when their larval food plants are in flower, and it then vacates much of the region or contracts to the arid zone during the hot humid months. The few records during the wet season possibly represent vagrants or temporary local breeding populations in some years.

## Breeding status

*Nacaduba biocellata* appears to be a regular immigrant in the study region, breeding temporarily during the dry season and then vacating the region or contracting to the drier inland areas before the onset of the wet.

## Conservation status

LC.

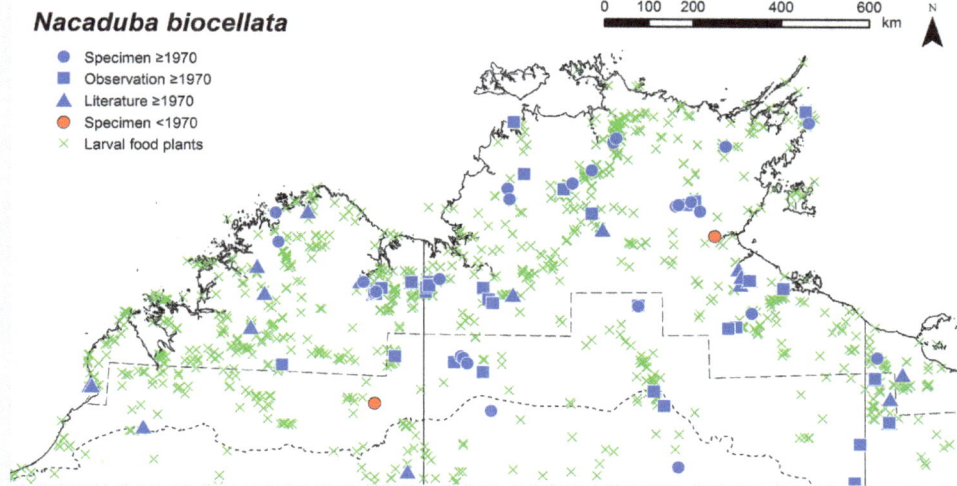

*Nacaduba biocellata*

- ● Specimen ≥1970
- ■ Observation ≥1970
- ▲ Literature ≥1970
- ● Specimen <1970
- × Larval food plants

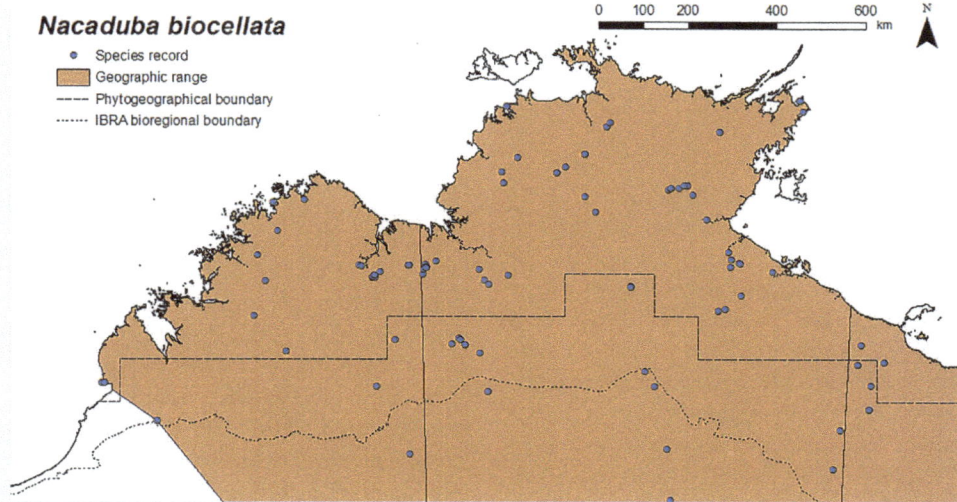

*Nacaduba biocellata*

- ● Species record
- ▬ Geographic range
- – – – Phytogeographical boundary
- ········ IBRA bioregional boundary

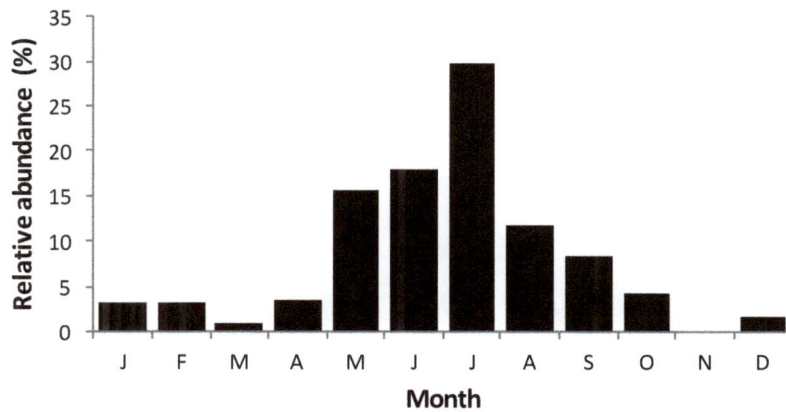

*Nacaduba biocellata* (n = 94)

| Month | J | F | M | A | M | J | J | A | S | O | N | D |
|-------|---|---|---|---|---|---|---|---|---|---|---|---|
| Egg   |   |   |   |   |   |   |   |   |   |   |   |   |
| Larva |   |   |   |   |   |   |   |   |   |   |   |   |
| Pupa  |   |   |   |   |   |   |   |   |   |   |   |   |
| Adult |   |   |   |   |   |   |   |   |   |   |   |   |

# Purple Line-blue

*Prosotas dubiosa* (Semper, [1879])

Plate 163 Gunn Point, NT
Photo: M. F. Braby

## Distribution

This species is represented in the study region by the subspecies *P. dubiosa dubiosa* (Semper, [1879]). It occurs in the Kimberley, Top End and western Gulf Country, extending from moist coastal areas to drier inland areas of the semi-arid zone (c. 700 mm mean annual rainfall). Its southern limits include Broome (Common and Waterhouse 1981) and Lake Argyle (Dunn 1980), WA; and Jasper Gorge about 58 km south-west of Victoria River Roadhouse (Franklin et al. 2005) and Bessie Spring (Dunn and Dunn 1991), NT. It has also been recorded further east in the Gulf of Carpentaria at Walker Creek 36 km east of Karumba, Qld, just outside the eastern boundary of the study region (Dunn 2017c). Comparison of the geographic range with the spatial distribution of its larval food plants indicates that the known food plants are absent from much of the range of *P. dubiosa* in the Kimberley, indicating that other (as yet unreported) food plants are used in this area. Outside the study region, *P. dubiosa* occurs widely from India and South-East Asia, through mainland New Guinea and adjacent islands and north-eastern and eastern Australia to the Solomon Islands.

## Habitat

*Prosotas dubiosa* breeds in a wide variety of habitats, including semi-deciduous monsoon vine thicket, the edges of riparian evergreen monsoon vine forest, savannah woodland and eucalypt heathy woodland, where the larval food plants usually grow as tall shrubs or small trees (Braby 2011a, 2015e).

## Larval food plants

*Semecarpus australiensis* (Anacardiaceae), *Acacia auriculiformis*, *A. scopulorum*, *Millettia pinnata* (Fabaceae), *Cupaniopsis anacardioides* (Sapindaceae); also *Dalbergia sissoo* (Fabaceae).

## Attendant ant

The larvae and pupae are not attended by ants.

## Seasonality

Adults occur throughout the year, but they are generally more abundant during the mid to late dry season (June–October), with peaks in abundance in July and October, but they may also be locally abundant at other times of the year (e.g. April), depending on the seasonal availability of flower buds. The immature stages have been recorded sporadically from April to December, but undoubtedly occur at other times of the year. Presumably, *P. dubiosa* breeds continuously throughout the year and several generations are completed annually by switching its food plants on a seasonal basis according to the flowering period of each species.

## Breeding status

This species is resident in the study region.

## Conservation status

LC.

## *Prosotas dubiosa*

- ● Specimen ≥1970
- ■ Observation ≥1970
- ▲ Literature ≥1970
- ● Specimen <1970
- ▲ Literature <1970
- × Larval food plants

## *Prosotas dubiosa*

- ● Species record
- ▬ Geographic range
- — — Phytogeographical boundary
- ······ IBRA bioregional boundary

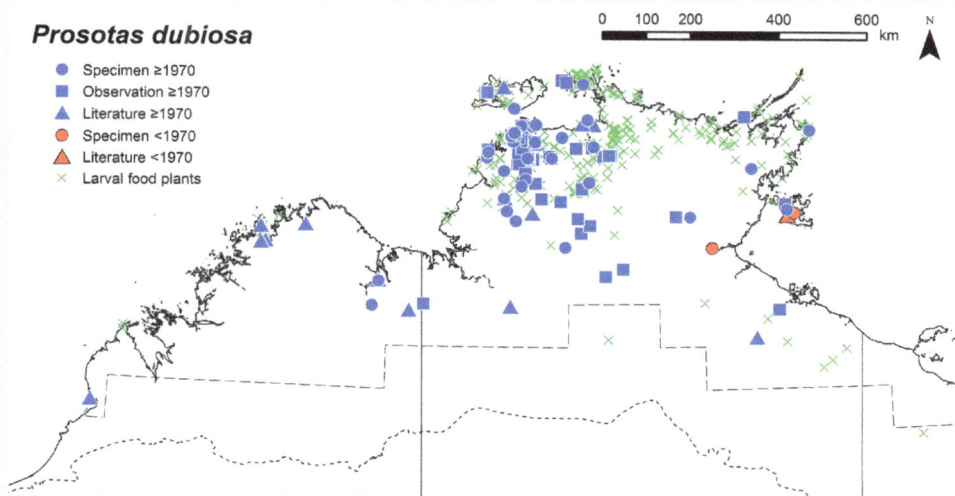

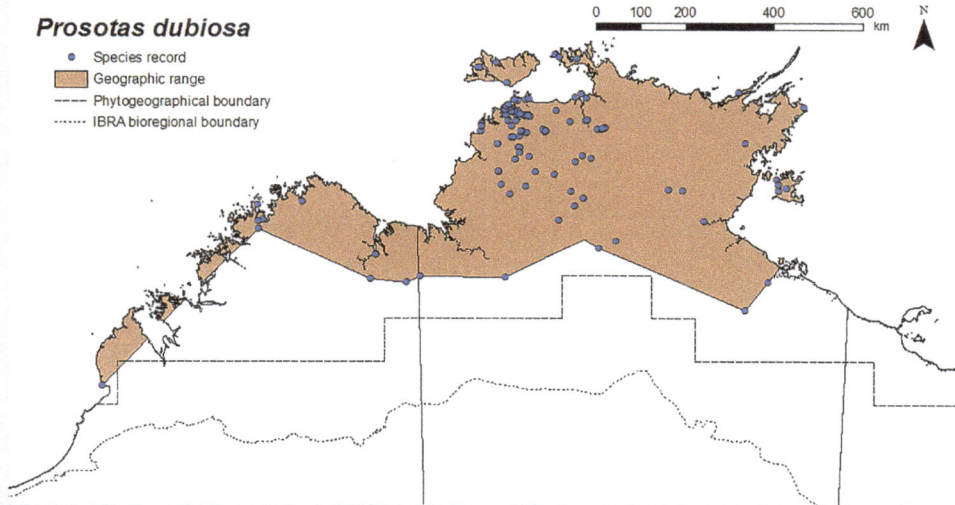

## *Prosotas dubiosa* (n = 185)

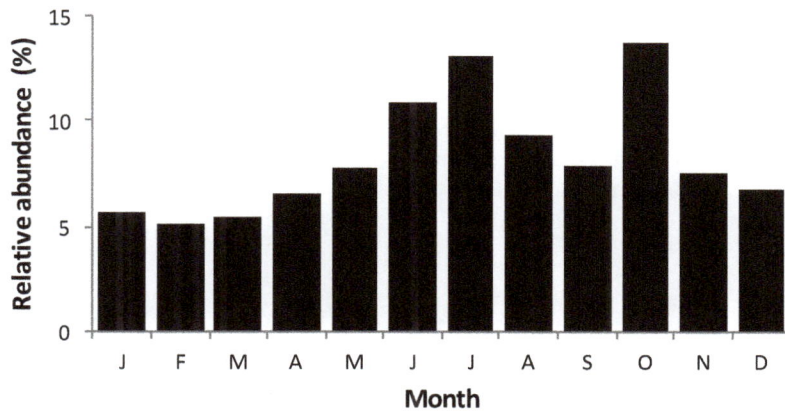

| Month | J | F | M | A | M | J | J | A | S | O | N | D |
|-------|---|---|---|---|---|---|---|---|---|---|---|---|
| Egg | | | | ▨ | | | ▨ | | ▨ | | | ▨ |
| Larva | | | | | | | ▨ | | | | | |
| Pupa | | | | | | | | ▨ | | | | |
| Adult | ▬ | | | | | | | | | | | ▬ |

# Speckled Line-blue
## *Catopyrops florinda* (Butler, 1877)

Plate 164 Kununurra, WA
Photo: Mark Golding

Plate 165 Wanguri, Darwin, NT
Photo: M. F. Braby

## Distribution

This species is represented in the study region by the subspecies *C. florinda estrella* (Waterhouse & Lyell, 1914). It occurs very widely in the region, from the Kimberley, through the Top End and Northern Deserts to the western Gulf Country, extending from moist coastal areas to drier inland areas of the semi-arid zone (c. 500 mm mean annual rainfall). It has been recorded as far south as the Edgar Ranges (Common 1981) and Halls Creek (Grund and Hunt 2001), WA; Newcastle Waters, NT (Braby 2000); and Boodjamulla/Lawn Hill National Park, Qld (Daniels and Edwards 1998). It has also been recorded at Walker Creek 36 km east of Karumba, Qld, just outside the eastern boundary of the study region (Dunn 2017d). The geographic range corresponds well with the spatial distribution of its larval food plants, indicating that *C. florinda* has been well sampled in the region. Outside the study region, *C. florinda* occurs from Timor, through mainland New Guinea and north-eastern and eastern Australia to New Britain and New Caledonia.

## Habitat

*Catopyrops florinda* breeds in a variety of habitats, including semi-deciduous monsoon vine thicket, savannah woodland, eucalypt open woodland on rocky outcrops and along rocky seasonal gullies, and mixed riparian woodland with rainforest elements along riverbanks or gorges in the drier inland areas, where the larval food plants grow as shrubs (Braby 2011a).

## Larval food plants

*Mallotus nesophilus* (Euphorbiaceae), *Dodonaea hispidula* (Sapindaceae). The immature stages are found frequently on *D. hispidula* (Braby 2011a).

## Attendant ant

The larvae and pupae are not attended by ants.

## Seasonality

Adults occur throughout the year, but they are most abundant during the dry season (April–September). The immature stages have been recorded sporadically from February to October, but no doubt also occur at other times. Presumably, the species breeds continuously throughout the year and several generations are completed annually.

## Breeding status

This species is resident in the study region.

## Conservation status

LC.

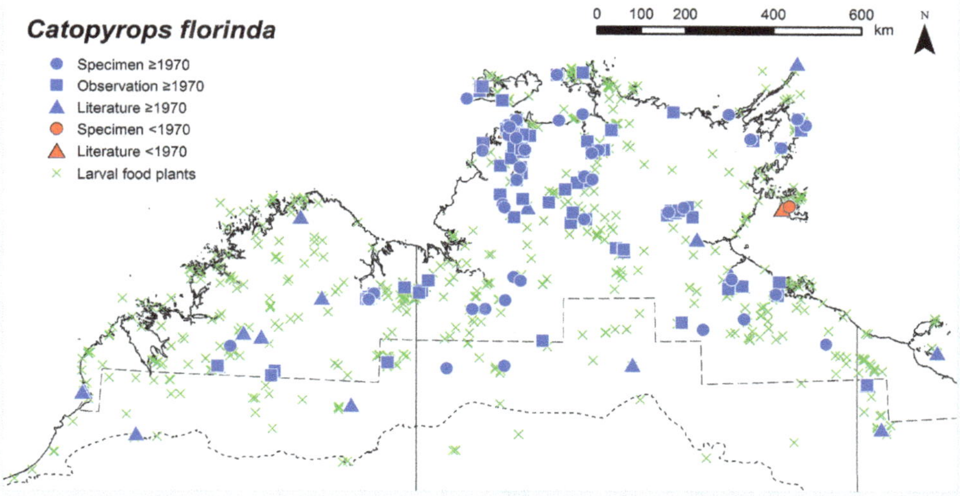

Catopyrops florinda
0 100 200 400 600 km
N
Specimen ≥1970
Observation ≥1970
Literature ≥1970
Specimen <1970
Literature <1970
Larval food plants

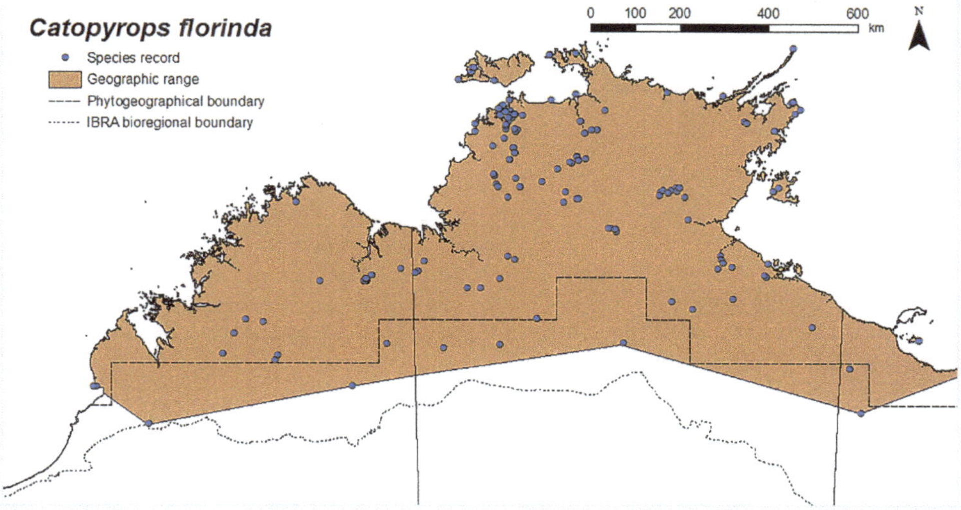

Catopyrops florinda
0 100 200 400 600 km
N
Species record
Geographic range
Phytogeographical boundary
IBRA bioregional boundary

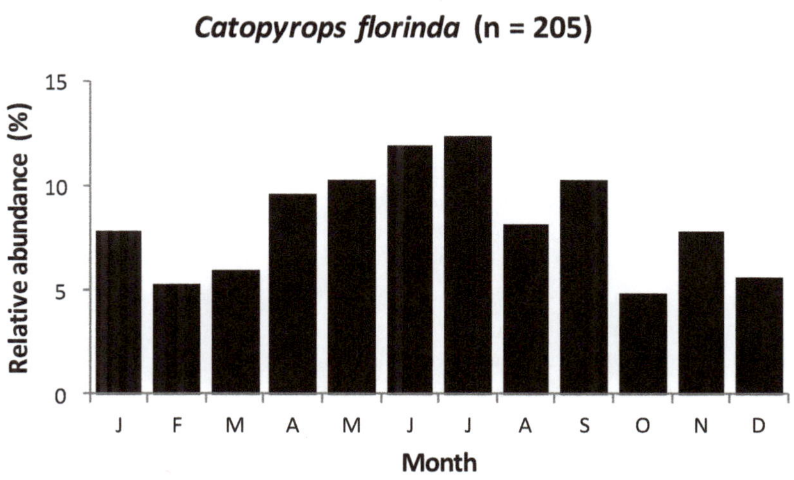

Catopyrops florinda (n = 205)
Relative abundance (%)
15
10
5
0
J F M A M J J A S O N D
Month

| Month | J | F | M | A | M | J | J | A | S | O | N | D |
|---|---|---|---|---|---|---|---|---|---|---|---|---|
| Egg | | | | ■ | ■ | ■ | | | ■ | ■ | | |
| Larva | | ■ | ■ | ■ | ■ | ■ | | | ■ | ■ | | |
| Pupa | | ■ | ■ | | | | | | | ■ | | |
| Adult | ■ | ■ | ■ | ■ | ■ | ■ | ■ | ■ | ■ | ■ | ■ | ■ |

# Glistening Line-blue
## *Sahulana scintillata* (T. P. Lucas, 1889)

Plate 166 Mary River Reserve, NT
Photo: M. F. Braby

## Distribution

This species occurs in the north of the Northern Territory, where it is restricted to the higher rainfall areas of the Top End (> 1,200 mm mean annual rainfall, but mostly > 1,400 mm). It appears to be rare, known only from seven locations (Meyer et al. 2006; Franklin et al. 2007a; Braby 2014a). It has been recorded from Darwin to the Mary River crossing (Arnhem Highway) and further east, on the Koolatong River crossing (Gapuwiyak–Balma Track) (Braby 2014a), NT. The putative larval food plants (*Acacia aulacocarpa* and *Cupaniopsis anacardioides*) occur more widely in the Top End; thus, further field surveys are required to determine more precisely the southern extent of *S. scintillata*. Outside the study region, *S. scintillata* occurs in mainland New Guinea and north-eastern and eastern Australia.

## Habitat

The breeding habitat of *S. scintillata* has not been recorded in the study region. Adults have been collected mostly in riparian forest, riparian woodland or riparian paperbark open woodland (Braby 2014a), but they have also been recorded from coastal areas adjacent to mangroves or semi-deciduous monsoon vine thicket (Meyer et al. 2006; Franklin et al. 2007a). Adults have been observed feeding at flowers of *Melaleuca*, *Lophostemon* and *Calytrix*.

## Larval food plants

Not recorded in the study region; possibly *Acacia aulacocarpa* (Fabaceae) and *Cupaniopsis anacardioides* (Sapindaceae), which are two larval food plants in eastern Australia (Braby 2000, 2016a).

## Attendant ant

Not recorded in the study region. In Queensland, the larvae and pupae are not attended by ants (Eastwood and Fraser 1999; Braby 2000).

## Seasonality

The seasonal abundance and breeding phenology of this species are not well understood. Adults are seasonal, occurring only in the dry season (June–September), but there are too few records (n = 12) to assess temporal changes in abundance. Most records are from June and July. In Queensland, the larvae feed on the flower buds and flowers of their food plants. Presumably, the seasonal appearance of *S. scintillata* in northern Australia is tightly synchronised with the flowering period of its larval food plant(s). However, it is not clear how the species survives the wet season.

## Breeding status

This species is resident in the study region.

## Conservation status

LC. The species *S. scintillata* has a restricted range in the study region within which it occurs in several conservation reserves, including Casuarina Coastal Reserve, Djukbinj National Park and Mary River National Park. Despite its restricted occurrence, there are no known threats facing the taxon. Further investigations are needed to determine the ecological requirements of this species.

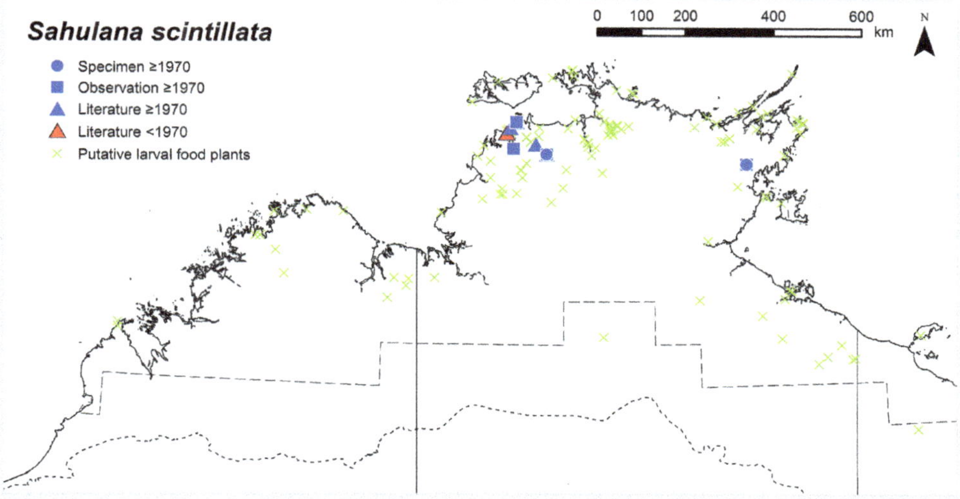

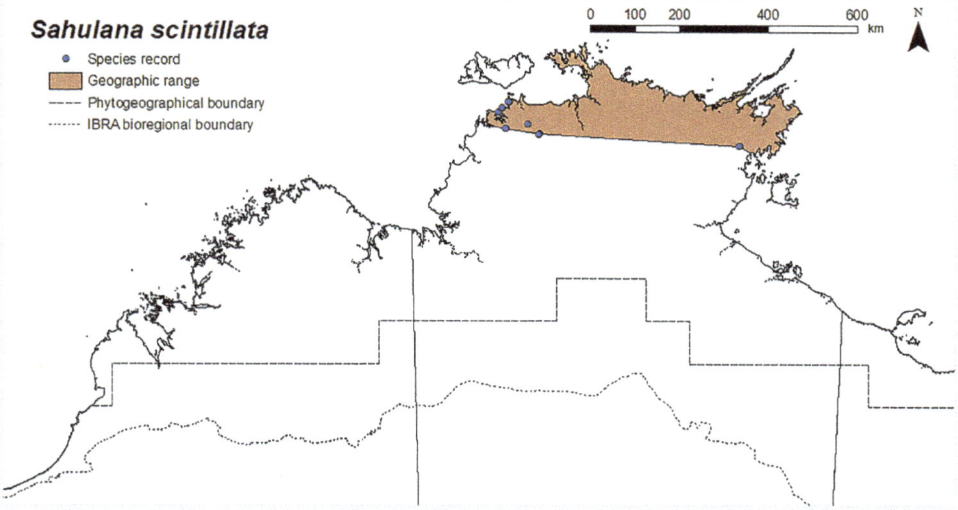

## Sahulana scintillata

- ● Specimen ≥1970
- ■ Observation ≥1970
- ▲ Literature ≥1970
- ▲ Literature <1970
- × Putative larval food plants

## Sahulana scintillata

- ● Species record
- ▨ Geographic range
- --- Phytogeographical boundary
- ···· IBRA bioregional boundary

| Month | J | F | M | A | M | J | J | A | S | O | N | D |
|-------|---|---|---|---|---|---|---|---|---|---|---|---|
| Egg   |   |   |   |   |   |   |   |   |   |   |   |   |
| Larva |   |   |   |   |   |   |   |   |   |   |   |   |
| Pupa  |   |   |   |   |   |   |   |   |   |   |   |   |
| Adult |   |   |   |   |   |   |   |   |   |   |   |   |

# Samphire Blue
## *Theclinesthes sulpitius* (Miskin, 1890)

Plate 167 Near Palmerston, NT
Photo: M. F. Braby

## Distribution

This species occurs sporadically in coastal areas of the Kimberley, Top End and western Gulf Country of the study region, its presence detected only as recently as 1991 (Meyer and Wilson 1995). It has been recorded from six widely dispersed locations: Broome (Peters 2006; Williams et al. 2006; Braby 2012b) and Wyndham (Braby 2000; Williams et al. 2006), WA; the Darwin area (Shoal Bay, Elizabeth River, Cox Peninsula) (Meyer and Wilson 1995), from Maningrida to Milingimbi (Bisa 2013), Groote Eylandt (D. Webb) and Bing Bong 48 km north of Borroloola (Braby 2012b), NT; as well as Karumba–Normanton, Qld (Pierce 2008, 2010, 2011; Braby 2015d), just outside the eastern boundary of the region. The larval food plants are distributed more or less continuously between the known locations; thus, further field studies are required in the intervening areas to determine whether *T. sulpitius* occurs throughout the coastal areas of the study region, particularly in the Kimberley. Outside the study region, *T. sulpitius* occurs in the Torres Strait Islands, Qld, and widely along the eastern Australian coast.

The published records of *Theclinesthes serpentatus* (Herrich-Schäffer, 1869) for Broome (Peters 2006) and Karumba (Pierce 2008, 2010) are erroneous and refer to *T. sulpitius* (see Braby 2012b, 2015d).

## Habitat

*Theclinesthes sulpitius* breeds in estuarine saltmarsh where the larval food plants grow as samphires in the intertidal zone of rivers and creeks (Meyer and Wilson 1995; Braby 2011a, 2014a, 2015d).

## Larval food plants

*Tecticornia australasica, T. halocnemoides, T. indica* (Amaranthaceae).

## Attendant ant

The larvae and pupae are usually not attended by ants, but very occasionally the larvae are attended by a few small black ants of a species yet to be determined (Meyer and Wilson 1995).

## Seasonality

Adults have been recorded mainly in the dry season (April–August), as well as in the mid wet season (December and January), but there are too few records to assess any seasonal changes in abundance. We have observed adults to be locally abundant during the mid dry season (June–August). The immature stages (eggs or larvae) have been recorded sporadically from December to June. Presumably, the species breeds continuously throughout the year and several generations are completed annually.

## Breeding status

This species is resident in the study region.

## Conservation status

LC. Although the species *T. sulpitius* has a wide geographical range in the study region, it is known to occur in only one conservation reserve (Djelk IPA).

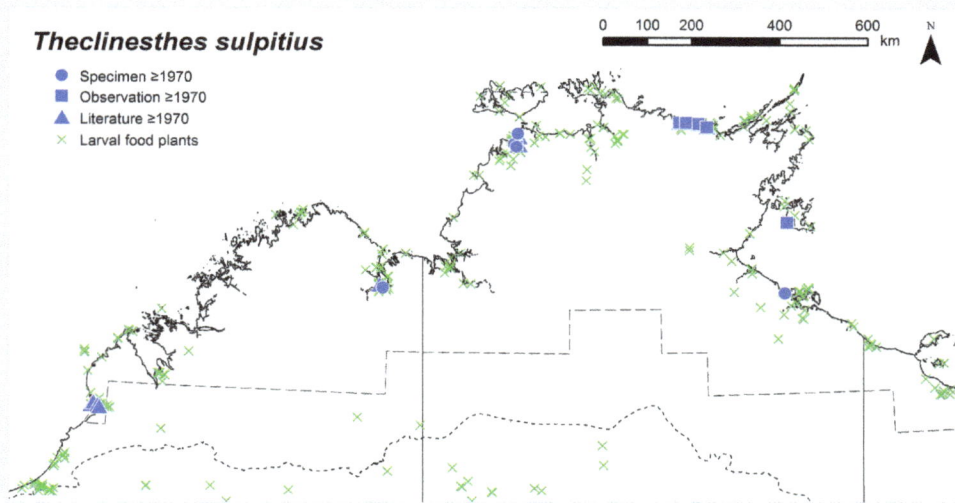

*Theclinesthes sulpitius*

- ● Specimen ≥1970
- ■ Observation ≥1970
- ▲ Literature ≥1970
- ✕ Larval food plants

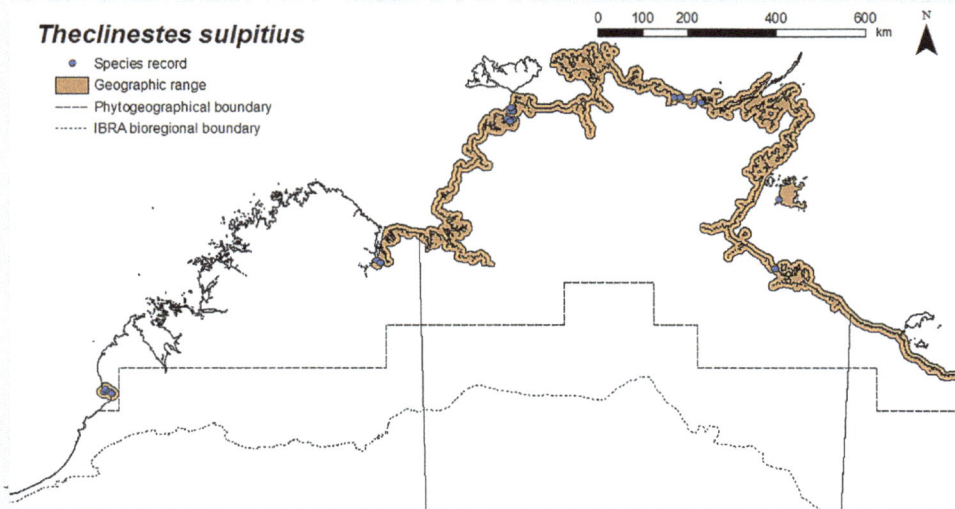

*Theclinestes sulpitius*

- ● Species record
- ▨ Geographic range
- — Phytogeographical boundary
- ⋯ IBRA bioregional boundary

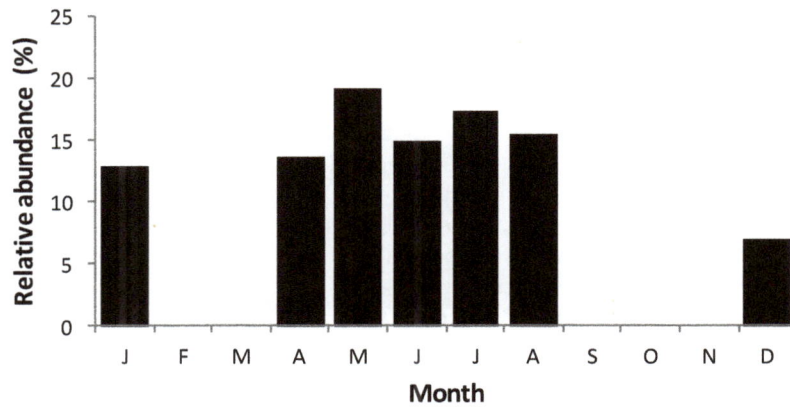

*Theclinesthes sulpitius* (n = 26)

| Month | J | F | M | A | M | J | J | A | S | O | N | D |
|-------|---|---|---|---|---|---|---|---|---|---|---|---|
| Egg | | | | ▥ | ▥ | ▥ | | | | | | |
| Larva | | ▥ | | | | | | | | | | ▥ |
| Pupa | | | | | | | | | | | | |
| Adult | ▥ | | | ▥ | ▥ | ▥ | ▥ | ▥ | | | | ▥ |

# Wattle Blue

## *Theclinesthes miskini* (T. P. Lucas, 1889)

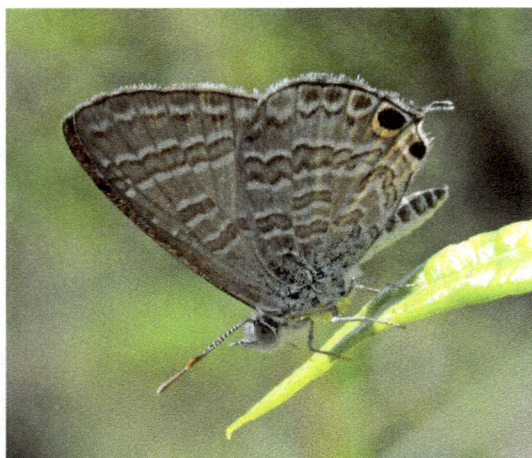

Plate 168 Mt Burrell, NT
Photo: M. F. Braby

## Distribution

This species is represented in the study region by the subspecies *T. miskini miskini* (T. P. Lucas, 1889). It occurs almost throughout the entire region, from moist coastal areas to drier inland areas of the semi-arid zone; it also extends into the arid zone of central Australia beyond the southern boundary of the study region. The geographic range corresponds with the spatial distribution of its larval food plants, which are similarly very widespread. Outside the study region, *T. miskini* occurs from the Lesser Sunda Islands, Tanimbar and the Kai and Aru islands, through mainland New Guinea and adjacent islands to the Admiralty Islands, the Bismarck Archipelago and the D'Entrecasteaux and Goodenough islands. It also occurs throughout the Australian continent.

## Habitat

*Theclinesthes miskini* breeds mainly in savannah woodland, open woodland and riparian woodland, favouring open disturbed areas where the larval food plants grow as pioneer shrubs, seedlings or saplings regenerating after fire, flood or unnatural events such as roadside earthworks (Braby 2011a, 2015e). It also breeds in the edges of monsoon forest.

## Larval food plants

*Acacia auriculiformis, A. difficilis, A. holosericea, A. platycarpa, Cathormion umbellatum, Sesbania cannabina* (Fabaceae), *Corymbia bella, Corymbia disjuncta, Corymbia ferruginea* (Myrtaceae), *Santalum lanceolatum* (Santalaceae), *Atalaya hemiglauca, Atalaya variifolia* (Sapindaceae).

## Attendant ants

*Iridomyrmex reburrus, I. sanguineus, Iridomyrmex* sp. (*anceps* group), *Iridomyrmex* sp. (*gracilis* group), *Iridomyrmex* sp. (*mattiroloi* group), *Iridomyrmex* sp. near *minor, I. pallidus* (Formicidae: Dolichoderinae), *Camponotus* sp. (*crozieri* group), *Camponotus* sp. (*denticulatus* group), *Camponotus* sp. (*novaehollandiae* group) (Formicidae: Formicinae). The larvae and pupae are usually attended by varying numbers of ants representing several genera, including Meat Ants (*Iridomyrmex* spp.), small black ants (*Iridomyrmex* spp.) and Sugar Ants (*Camponotus* spp.), in a facultative myrmecophilous association (Braby 2011a, 2015e).

## Seasonality

Adults occur throughout the year, but they are most abundant during the first half (January–June), which coincides with the wet season and early dry season, although they may be locally abundant at other times. The immature stages have been recorded during most months of the year, but are observed more frequently during the wet season (December–April), when adults are abundant. The young larvae feed on the new soft leaf growth of the food plants. Presumably, the species breeds continuously throughout the year and several generations are completed annually.

## Breeding status

This species is resident in the study region.

## Conservation status

LC.

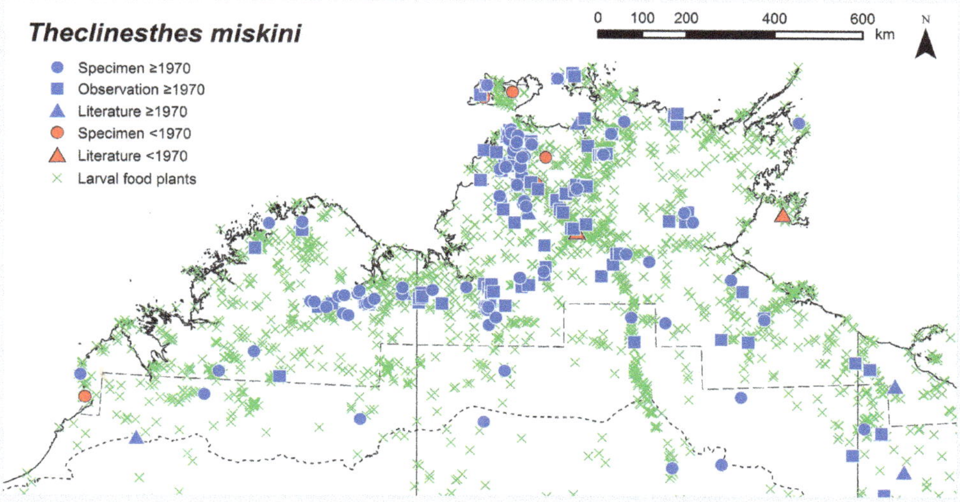

Theclinesthes miskini
0 100 200 400 600 km
N
Specimen ≥1970
Observation ≥1970
Literature ≥1970
Specimen <1970
Literature <1970
Larval food plants

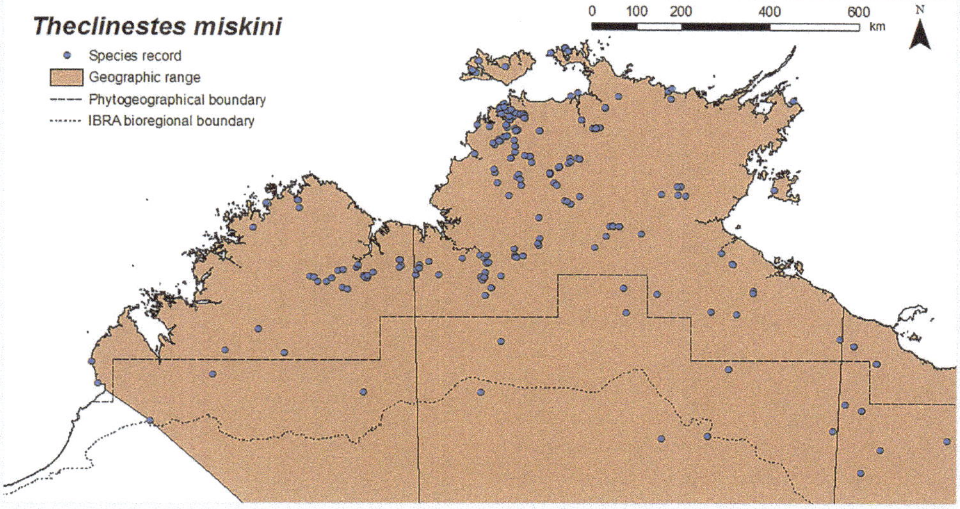

Theclinestes miskini
0 100 200 400 600 km
N
Species record
Geographic range
Phytogeographical boundary
IBRA bioregional boundary

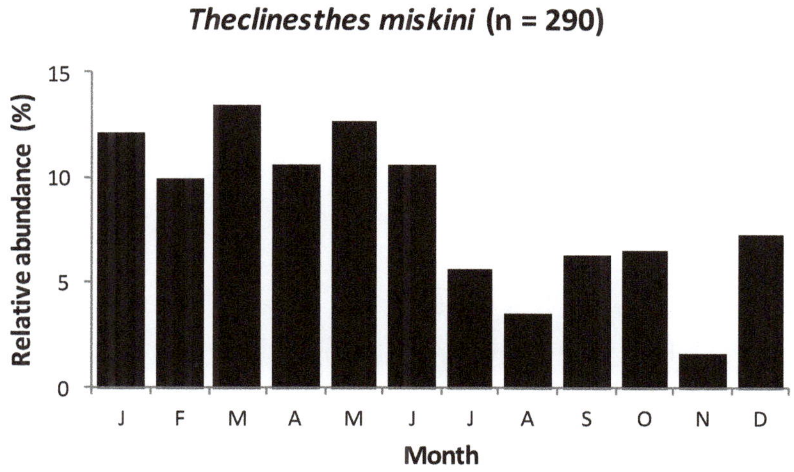

Theclinesthes miskini (n = 290)
Relative abundance (%)
15
10
5
0
J F M A M J J A S O N D
Month

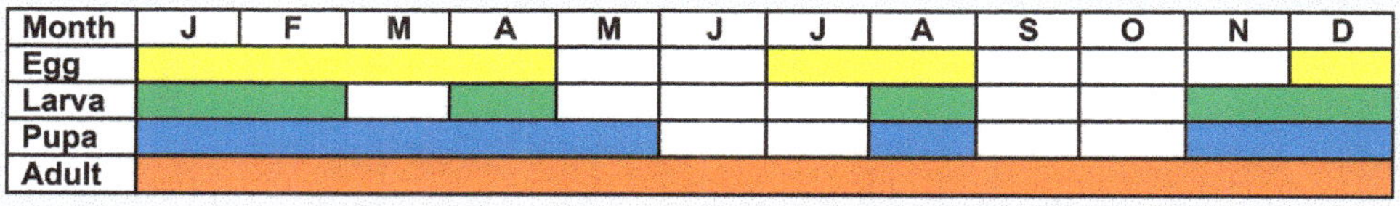

Month
J F M A M J J A S O N D
Egg
Larva
Pupa
Adult

# Bitter-bush Blue
## *Theclinesthes albocinctus* (Waterhouse, 1903)

Plate 169 Wyperfeld National Park, Vic
Photo: M. F. Braby

## Distribution

This species is known only from the low rainfall areas of the south-western Kimberley of the study region. It has been recorded at Cable Beach, Broome, WA (Grund 1998; Johnson and Valentine 2004), its presence detected only as recently as 1997. The larval food plant occurs further east in drier inland areas of the semi-arid zone; thus, further field surveys are required to determine whether *T. albocinctus* occurs in these areas, particularly in the southern and eastern Kimberley. Grund (1996) reported the possible presence of the species at Geikie Gorge National Park and Ord River Gorge, WA, based on putative eggs on vouchered herbarium specimens; however, subsequent searches along the Fitzroy River at Geikie Gorge did not detect the butterfly (Grund 1998). Outside the study region, *T. albocinctus* occurs mainly in arid and semi-arid areas of western, central and southern Australia.

## Habitat

*Theclinesthes albocinctus* breeds in low open shrubland on the landward side of coastal sand dunes where the larval food plant grows as a shrub up to 2 m high.

## Larval food plant

*Adriana tomentosa* (Euphorbiaceae).

## Attendant ant

Not recorded in the study region. Elsewhere, the larvae and pupae are usually attended by a few ants representing several genera, in a facultative myrmecophilous association (Eastwood and Fraser 1999; Braby 2000).

## Seasonality

The seasonal abundance and breeding phenology of this species are not well understood. Adults have been recorded in the late dry season (September and October). Grund (1998: 67) noted that the larval food plants 'were generally infested with the early stages of *T. albocincta*', but provided few details.

## Breeding status

This species is resident in the study region.

## Conservation status

DD. The species *T. albocinctus* appears to have a limited range in the study region (AOO is likely to be < 2,000 sq km, with spatial buffering of records providing a first approximation of 700 sq km), and is currently known from only one location, which lacks adequate conservation protection and management. Thus, it may qualify as Near Threatened (NT) once adequate data are available because of its restricted AOO and limited number of locations. Targeted field surveys to clarify the extent of its distribution and identify key threatening processes should be a high priority for this species.

## Theclinesthes albocinctus

▲ Literature ≥1970
✕ Larval food plant

0  100  200     400     600
km

N

| Month | J | F | M | A | M | J | J | A | S | O | N | D |
|-------|---|---|---|---|---|---|---|---|---|---|---|---|
| Egg   |   |   |   |   |   |   |   |   |   |   |   |   |
| Larva |   |   |   |   |   |   |   |   |   |   |   |   |
| Pupa  |   |   |   |   |   |   |   |   |   |   |   |   |
| Adult |   |   |   |   |   |   |   |   |   |   |   |   |

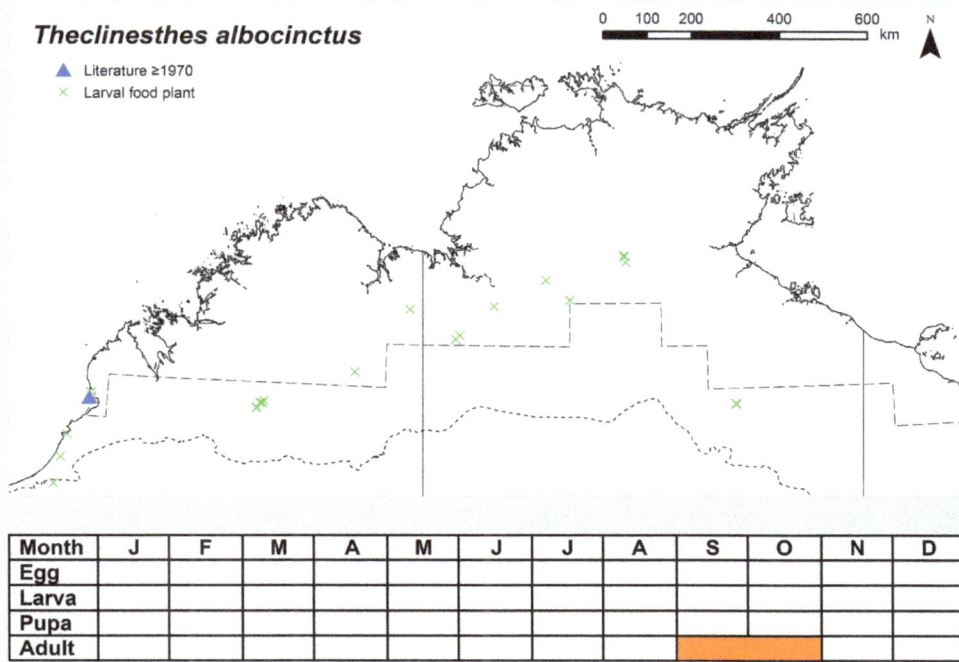

Photo: Mornington Wildlife Sanctuary, WA, M.F. Braby

# Purple Cerulean

*Jamides phaseli* (Mathew, 1889)

**Plate 170 Cairns, Qld**
Photo: M. F. Braby

## Distribution

This species occurs widely in the Kimberley, Top End and western Gulf Country of the study region. It extends from moist coastal areas to drier inland areas of the semi-arid zone (500 mm mean annual rainfall), reaching its southernmost limits at Broome (Grund 1998; Peters 2006, 2008) and Halls Creek (Grund and Hunt 2001), WA; and Boodjamulla/Lawn Hill National Park, Qld (Daniels and Edwards 1998). The geographic range corresponds well with the spatial distribution of its larval food plants. The food plants, however, also occur on the Tiwi Islands and in the drier areas of the Northern Deserts; thus, further surveys are required to determine whether *J. phaseli* occurs in these areas. Outside the study region, *J. phaseli* occurs in the Torres Strait Islands, Qld, and north-eastern and eastern Australia.

## Habitat

*Jamides phaseli* breeds in a variety of habitats, including savannah woodland and open woodland on rocky sandstone country, where several of the larval food plants (*Bossiaea, Cajanus, Tephrosia*) grow as understorey shrubs, and coastal sand dunes where an alternative food plant (*Canavalia rosea*) grows as a trailing prostrate creeper (Braby 2011a, 2015e). It also occurs in the edges of semi-deciduous monsoon vine thicket and suburban parks and nature strips, where it commonly breeds on trees of *Millettia pinnata*.

## Larval food plants

*Bossiaea bossiaeoides, Canavalia rosea, Cajanus aromaticus, Millettia pinnata, Sesbania simpliciuscula, Tephrosia spechtii* (Fabaceae).

## Attendant ant

The larvae and pupae are not attended by ants.

## Seasonality

Adults occur throughout the year, but they are most abundant during the wet season and early dry season (February–June), depending on the local and seasonal availability of their larval food plants. The immature stages have been recorded mostly from February to May, when adults are abundant, but also in the late dry season (September and October). The larvae feed on the flower buds and new leaf growth of their food plants. Presumably, *J. phaseli* breeds continuously throughout the year and several generations are completed annually by switching its food plants on a seasonal basis according to the flowering period of each species.

## Breeding status

This species is resident in the study region.

## Conservation status

LC.

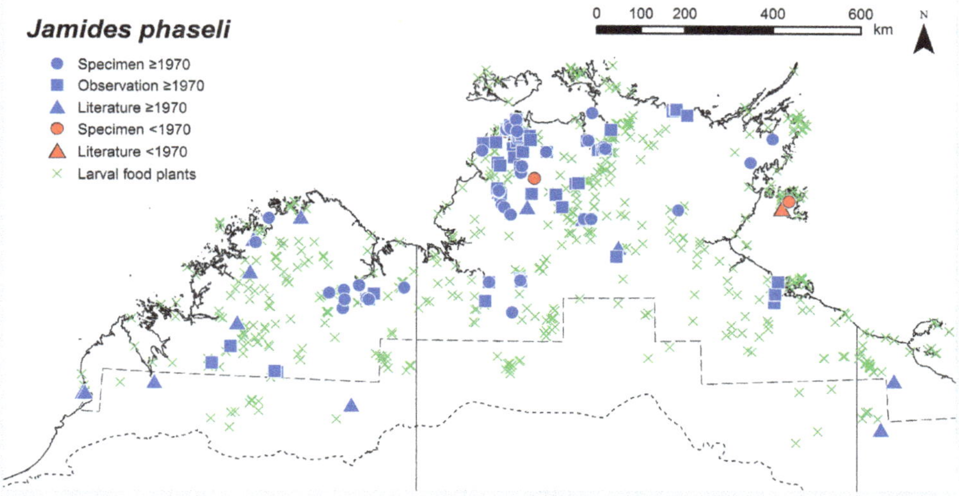

Jamides phaseli
0 100 200 400 600 km
N
Specimen ≥1970
Observation ≥1970
Literature ≥1970
Specimen <1970
Literature <1970
Larval food plants

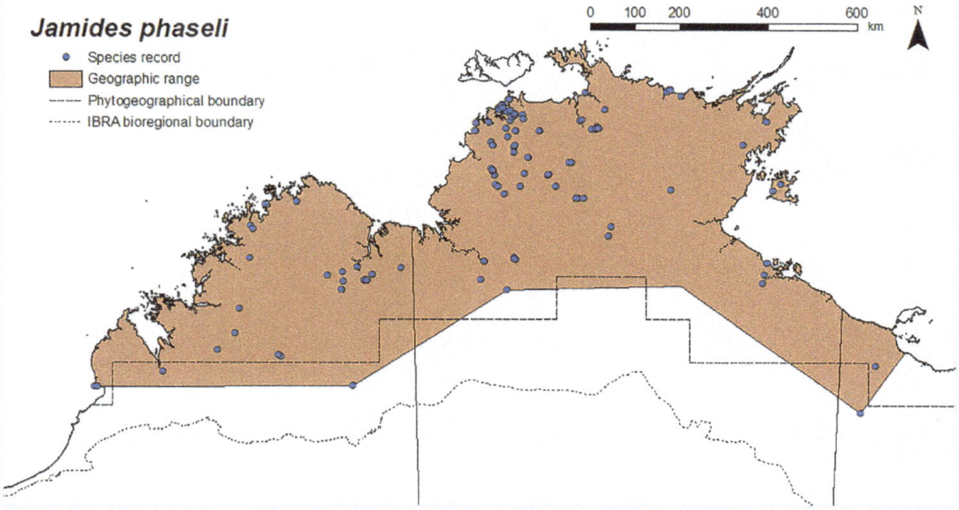

Jamides phaseli
0 100 200 400 600 km
N
Species record
Geographic range
Phytogeographical boundary
IBRA bioregional boundary

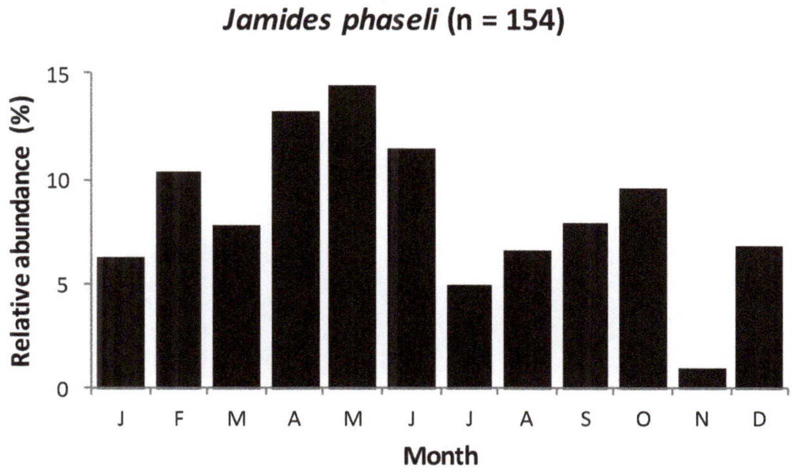

Jamides phaseli (n = 154)
Relative abundance (%)
15
10
5
0
J F M A M J J A S O N D
Month

| Month | J | F | M | A | M | J | J | A | S | O | N | D |
|---|---|---|---|---|---|---|---|---|---|---|---|---|
| Egg | | ■ | ■ | | ■ | | | | ■ | ■ | | |
| Larva | | ■ | ■ | | ■ | | | | | | | |
| Pupa | | ■ | ■ | ■ | | | | | | | | |
| Adult | ■ | ■ | ■ | ■ | ■ | ■ | ■ | ■ | ■ | ■ | ■ | ■ |

# Pale Pea-blue
## *Catochrysops panormus* (C. Felder, 1860)

Plate 171 Herberton, Qld
Photo: Don Franklin

## Distribution

This species is represented in the study region by the subspecies *C. panormus platissa* (Herrich-Schäffer, 1869). It occurs widely in the Kimberley, Top End, Northern Deserts and western Gulf Country, extending from moist coastal areas to drier inland areas of the semi-arid zone (500 mm mean annual rainfall). It reaches its southernmost limits at Broome (Peters 2006) and Halls Creek (Grund and Hunt 2001), WA; Newcastle Creek near Newcastle Waters, NT (Dunn and Dunn 1991); and Boodjamulla/Lawn Hill National Park, Qld (Daniels and Edwards 1998). The geographic range corresponds well with the spatial distribution of its larval food plants. The food plants, however, also occur on the Tiwi Islands; thus, further surveys are required to determine whether *C. panormus* is present on Bathurst and Melville islands, NT. Outside the study region, *C. panormus* occurs widely from India and South-East Asia, through mainland New Guinea and adjacent islands and north-eastern and eastern Australia to the Solomon Islands, the Loyalty Islands and Vanuatu.

## Habitat

*Catochrysops panormus* breeds mainly in savannah woodland and open woodland where the larval food plants grow as shrubs, often in open rocky areas (Braby 2011a, 2015e).

## Larval food plants

*Cajanus aromaticus, C. pubescens, Flemingia lineata, Sesbania simpliciuscula* (Fabaceae).

## Attendant ant

The larvae and pupae are not attended by ants.

## Seasonality

Adults occur throughout the year, but they are generally more abundant during the late wet season (February–April), with a peak in abundance in March. Their numbers appear to diminish as the dry season progresses, and they are very scarce at the start of the wet season. We have limited data on the phenology of the immature stages, which have been recorded in the mid wet season and mid dry season. The larvae feed on the flower buds and flowers of their larval food plants. Presumably, *C. panormus* breeds continuously throughout the year and several generations are completed annually by switching its food plants on a seasonal basis according to the flowering period of each species.

## Breeding status

This species is resident in the study region.

## Conservation status

LC.

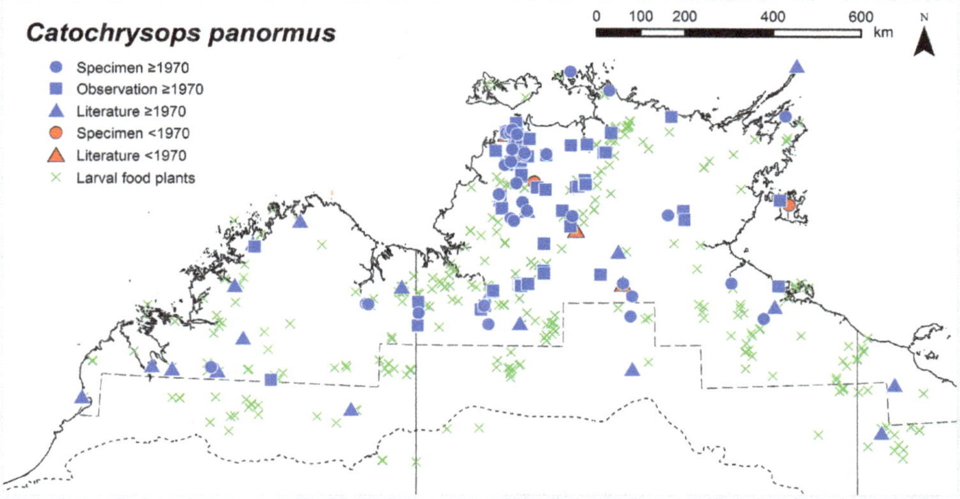

Catochrysops panormus
0 100 200 400 600
km
N
Specimen ≥1970
Observation ≥1970
Literature ≥1970
Specimen <1970
Literature <1970
Larval food plants

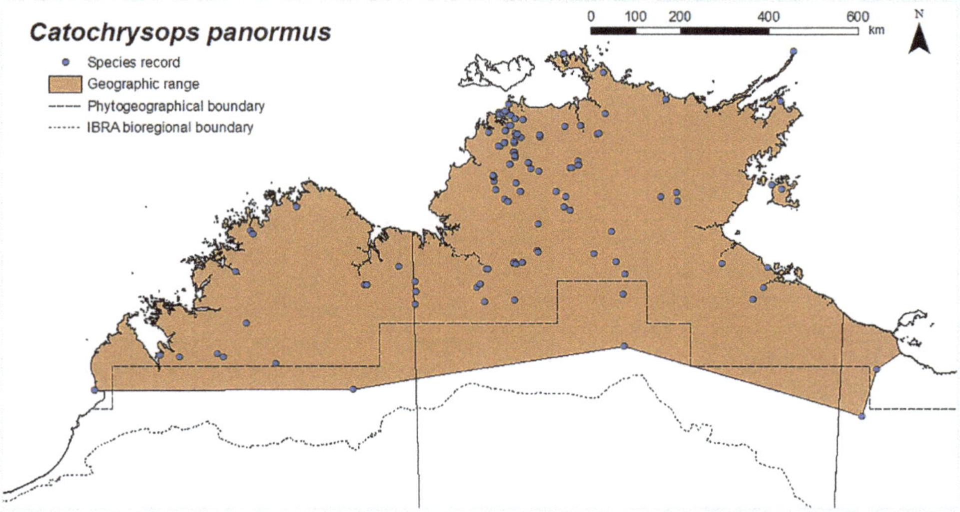

Catochrysops panormus
0 100 200 400 600
km
N
Species record
Geographic range
Phytogeographical boundary
IBRA bioregional boundary

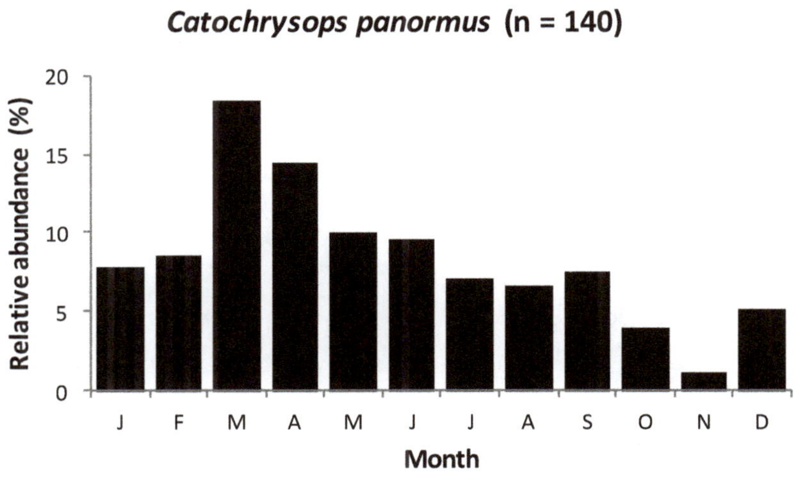

Catochrysops panormus (n = 140)
Relative abundance (%)
20
15
10
5
0
J F M A M J J A S O N D
Month

| Month | J | F | M | A | M | J | J | A | S | O | N | D |
|---|---|---|---|---|---|---|---|---|---|---|---|---|
| Egg | | | | | | | | | | | | |
| Larva | | | | | | | | | | | | |
| Pupa | | | | | | | | | | | | |
| Adult | | | | | | | | | | | | |

# Long-tailed Pea-blue
## *Lampides boeticus* (Linnaeus, 1767)

Plate 172 Hervey Range, Qld
Photo: M. F. Braby

## Distribution

This species occurs very widely in the study region. It also extends into the arid zone of central Australia beyond the southern boundary of the study region. The species is more prevalent in the low rainfall areas of the semi-arid zone. There are very few records from the Top End north of latitude 13°S (Waterhouse and Lyell 1914; Le Souëf 1971; Franklin et al. 2005; Meyer et al. 2006). The geographic range broadly corresponds with the spatial distribution of the known larval food plants. Outside the study region, *L. boeticus* occurs very widely from Western Europe, Africa and South-East Asia, through mainland New Guinea to the islands of the South Pacific. It also occurs throughout the Australian continent.

## Habitat

The breeding habitat and larval food plants of *L. boeticus* are not well documented from the study region. Grund (1998) recorded the species breeding in the Kimberley (on *Crotalaria* spp.) at Broome and El Questro Wilderness Park, WA. At Mornington Wildlife Sanctuary, WA, it was recorded breeding in open grassland (E. P. Williams, pers. comm.). Adults are usually encountered in savannah woodland and riparian woodland, often in open disturbed areas, and presumably they breed in these habitats.

## Larval food plants

*Crotalaria cunninghamii*, *C. novae-hollandiae* (Fabaceae). In Queensland, the larvae feed on a wide range of legumes (Braby 2000). Presumably, the number of species of food plants utilised is much larger than present records indicate.

## Attendant ant

*Froggattella* sp. (Formicidae: Dolichoderinae). The larvae are usually attended by a few small black ants in a facultative myrmecophilous association (Grund 1998).

## Seasonality

Adults have been recorded during most months of the year, but they are most abundant during the dry season (July–October), with a pronounced peak in abundance in July. They are rare or absent during the wet season (November–April) in most years. The breeding phenology and seasonal history of the immature stages are not well understood. In the Kimberley, Grund (1998) noted that the immature stages were recorded in October, but provided few details. The species is migratory (Smithers 1985; Braby 2000) and adults are capable of dispersing considerable distances. Presumably, *L. boeticus* colonises much of the study region—or at least the higher rainfall areas of the study region—on a seasonal basis; breeding is limited to the mid to late dry season, and it then vacates much of the region or contracts to the arid zone during the hot humid months. The few records during the wet season possibly represent vagrants or temporary local breeding populations in some years.

## Breeding status

*Lampides boeticus* appears to be a regular immigrant in the study region, breeding temporarily during the dry season and then vacating the region or contracting to the drier inland areas before the onset of the wet.

## Conservation status

LC.

## Lampides boeticus

- ● Specimen ≥1970
- ■ Observation ≥1970
- ▲ Literature ≥1970
- ▲ Literature <1970
- ✕ Larval food plants

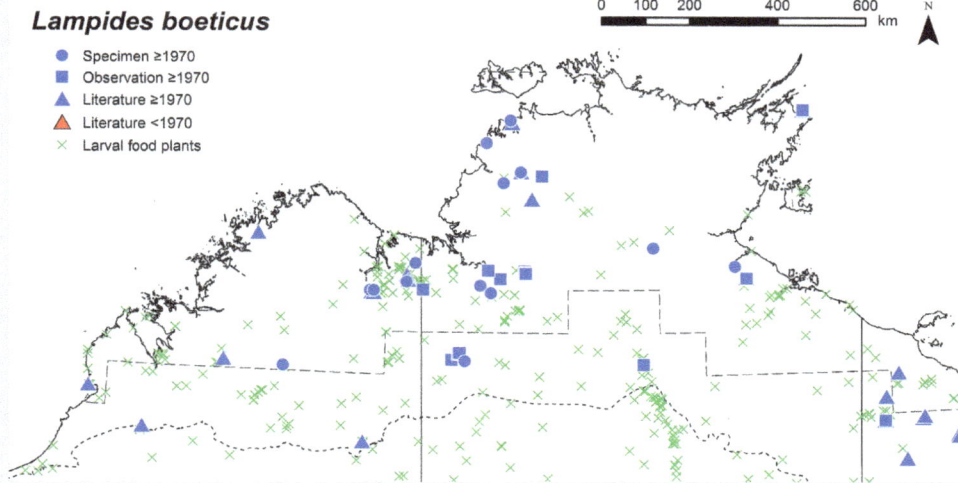

## Lampides boeticus

- ● Species record
- ▨ Geographic range
- --- Phytogeographical boundary
- ···· IBRA bioregional boundary

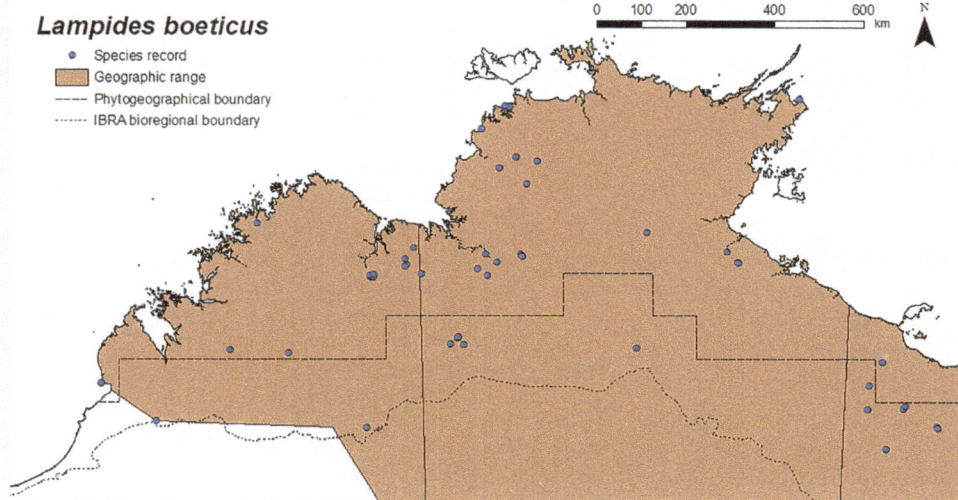

### Lampides boeticus (n = 47)

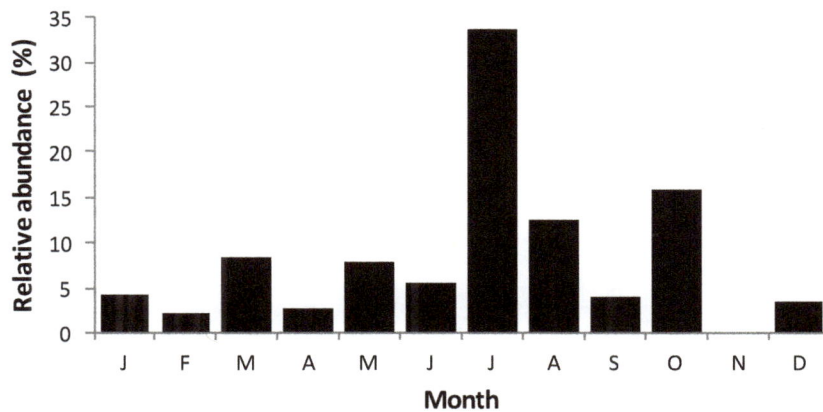

| Month | J | F | M | A | M | J | J | A | S | O | N | D |
|-------|---|---|---|---|---|---|---|---|---|---|---|---|
| Egg   |   |   |   |   |   |   |   |   |   |   |   |   |
| Larva |   |   |   |   |   |   |   |   |   |   |   |   |
| Pupa  |   |   |   |   |   |   |   |   |   |   |   |   |
| Adult |   |   |   |   |   |   |   |   |   |   |   |   |

# Spotted Grass-blue
## *Zizeeria karsandra* (Moore, 1865)

Plate 173 Cobourg Peninsula, NT
Photo: M. F. Braby

Plate 174 Cobourg Peninsula, NT
Photo: M. F. Braby

## Distribution

This species occurs very widely in the study region, extending from moist coastal areas to drier inland areas of the semi-arid zone (< 500 mm mean annual rainfall). It has been recorded as far south as Broome and Mary River Pool, approximately 100 km south-west of Halls Creek, WA (Williams et al. 2006); Elliott, NT, and Boodjamulla/Lawn Hill National Park, Qld (Daniels and Edwards 1998). It also occurs in the Davenport Ranges, NT (Pierce 2008), just outside the southern boundary of the study region. The geographic range corresponds well with the spatial distribution of its known and putative (*Glinus* spp.) larval food plants. The food plants also occur on the Tiwi Islands; thus, further surveys are required to determine whether *Z. karsandra* occurs on Bathurst and Melville islands, NT. Outside the study region, *Z. karsandra* occurs widely from northern Africa, northern Arabia and South-East Asia to the Torres Strait Islands, Qld, and central, north-eastern and eastern Australia.

## Habitat

*Zizeeria karsandra* breeds in beach scrubland on sand dunes and sand above the high-tide mark in coastal areas, and in open woodland in the more inland areas, favouring open disturbed areas where the larval food plants grow as prostrate herbs (Braby 2011a). It also occurs in open sandy areas along riverbanks and dry river beds in the more inland areas, where the species undoubtedly breeds on alternative food plants (*Glinus* spp.).

## Larval food plants

*Tribulopis bicolor*, *Tribulus cistoides* (Zygophyllaceae); probably *Glinus lotoides* and *G. oppositifolius* (Molluginaceae), which are two food plants in eastern Australia (Braby 2000). *Tribulus cistoides* is the main food plant in coastal areas (Braby 2011a).

## Attendant ant

The larvae and pupae are not attended by ants.

## Seasonality

Adults occur throughout the year, but they are generally more abundant during the wet season (December–April), although they may be locally abundant at other times such as the mid dry season. The immature stages have been recorded sporadically during both the wet and the dry seasons. The larval food plants have varying life history strategies: *Tribulus cistoides* is a short-lived perennial that is seasonally dependent on rainfall, whereas *Tribulopis bicolor* and *Glinus* spp. are annuals, although *G. lotoides* may also be a short-lived perennial. It is not clear how *Z. karsandra* survives the late dry season, when their food plants are frequently not available. Presumably, the species breeds continuously throughout the year by dispersing and tracking the local availability of their food plants over a very wide area.

## Breeding status

This species is resident in the study region, but populations appear to be nomadic and are possibly temporary in many areas.

## Conservation status

LC.

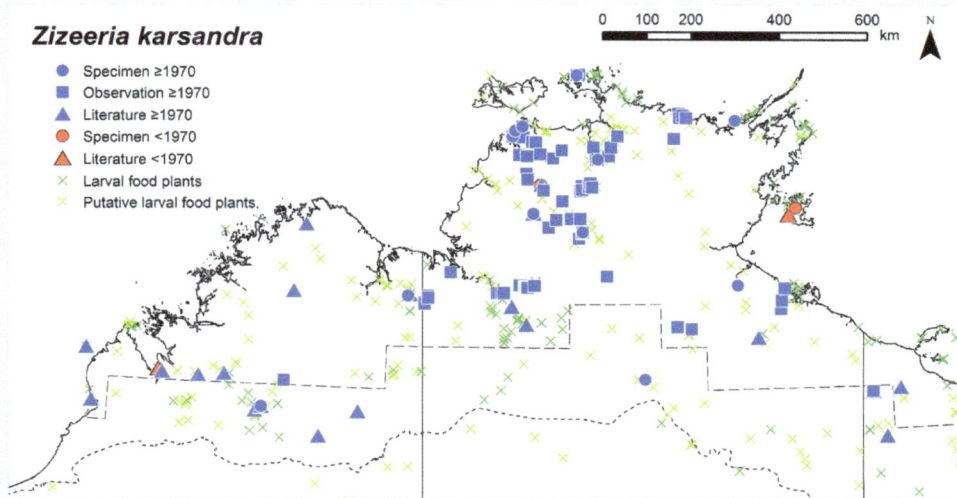

*Zizeeria karsandra*

- ● Specimen ≥1970
- ■ Observation ≥1970
- ▲ Literature ≥1970
- ● Specimen <1970
- ▲ Literature <1970
- ✕ Larval food plants
- ✕ Putative larval food plants.

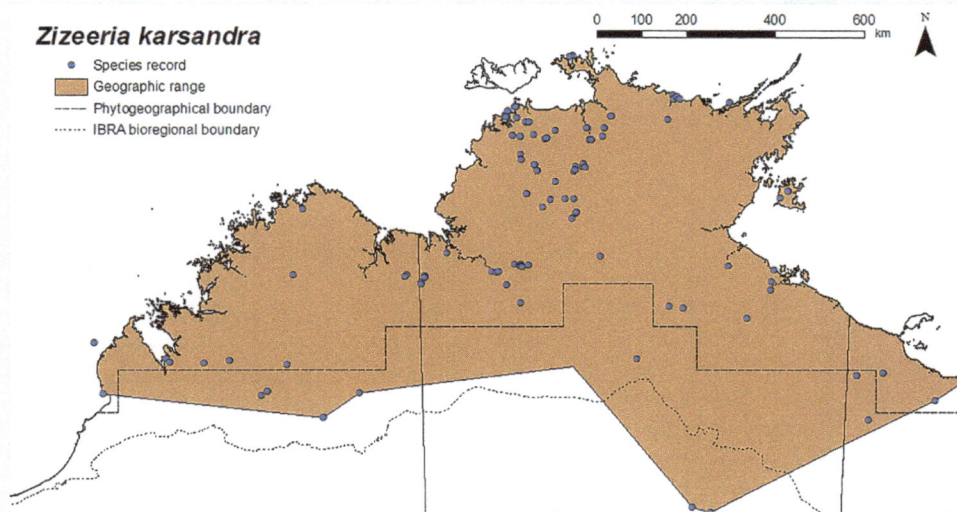

*Zizeeria karsandra*

- ● Species record
- ▨ Geographic range
- – – – Phytogeographical boundary
- ······ IBRA bioregional boundary

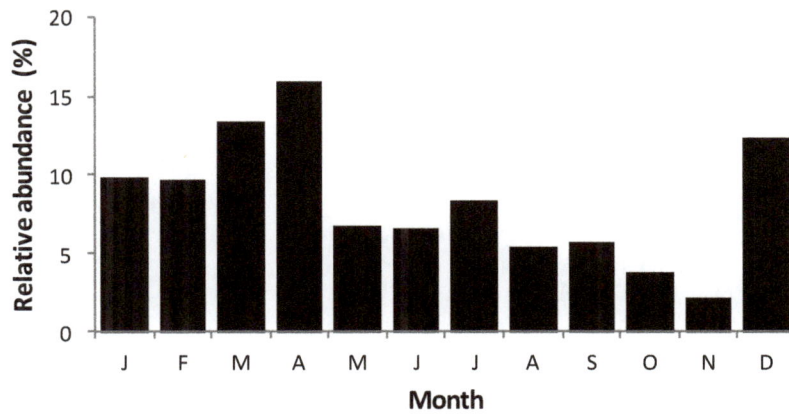

*Zizeeria karsandra* (n = 140)

| Month | J | F | M | A | M | J | J | A | S | O | N | D |
|-------|---|---|---|---|---|---|---|---|---|---|---|---|
| Egg | | | | | | | | | | | | |
| Larva | | | | | | | | | | | | |
| Pupa | | | | | | | | | | | | |
| Adult | | | | | | | | | | | | |

# Common Grass-blue
## *Zizina otis* (Fabricius, 1787)

Plate 175 Paluma, Qld
Photo: M. F. Braby

## Distribution

This species is represented in the study region by the subspecies *Z. otis labradus* (Godart, [1824]). It occurs very widely in the region, extending from moist coastal areas to drier inland areas of the semi-arid zone (< 400 mm mean annual rainfall), as well as the arid zone of central Australia beyond the southern boundary of the study region. It has been recorded as far south as 160 km north-west of Rabbit Flat, WA (Pierce 2008); 7 km north–north-east of Helen Springs Homestead, NT; and Boodjamulla/ Lawn Hill National Park, Qld (Daniels and Edwards 1998). Outside the study region, *Z. otis* occurs widely from India, Japan and South-East Asia, through mainland New Guinea and adjacent islands to New Caledonia, Vanuatu, Fiji, Samoa and New Zealand. It also occurs throughout the Australian continent.

## Habitat

The natural breeding habitat of *Z. otis* has not been recorded in the study region, but adults occur in a very wide range of habitats in which they no doubt breed. They also occur in suburban parks and gardens, where they breed on an introduced larval food plant.

## Larval food plants

*Desmodium triflorum* (Fabaceae). The native food plants have not been recorded in the study region, but elsewhere the food plants include a wide range of native and introduced low-spreading legumes (Braby 2000).

## Attendant ant

Not recorded in the study region. Elsewhere, the larvae are usually attended by a few ants representing several genera in a facultative myrmecophilous association (Eastwood and Fraser 1999; Braby 2000).

## Seasonality

Adults occur throughout the year, but they are most abundant during the wet season and early dry season (January–June), with a peak in abundance in May, and scarce during the remainder of the year, particularly during the late dry season and 'build-up' (August–November). We have few data on the phenology of the immature stages. Presumably, *Z. otis* breeds continuously from at least the early wet season to the mid dry season and several generations are completed annually. However, it is not clear how the species survives the late dry season.

## Breeding status

This species is resident in the study region.

## Conservation status

LC.

## Zizina otis

- ● Specimen ≥1970
- ■ Observation ≥1970
- ▲ Literature ≥1970
- ▲ Literature <1970

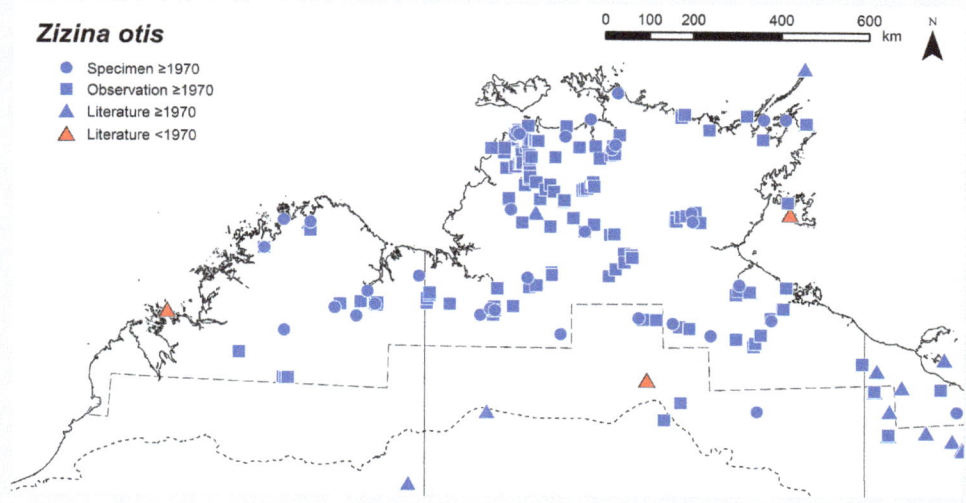

## Zizina otis

- ● Species record
- ▨ Geographic range
- — — Phytogeographical boundary
- ······ IBRA bioregional boundary

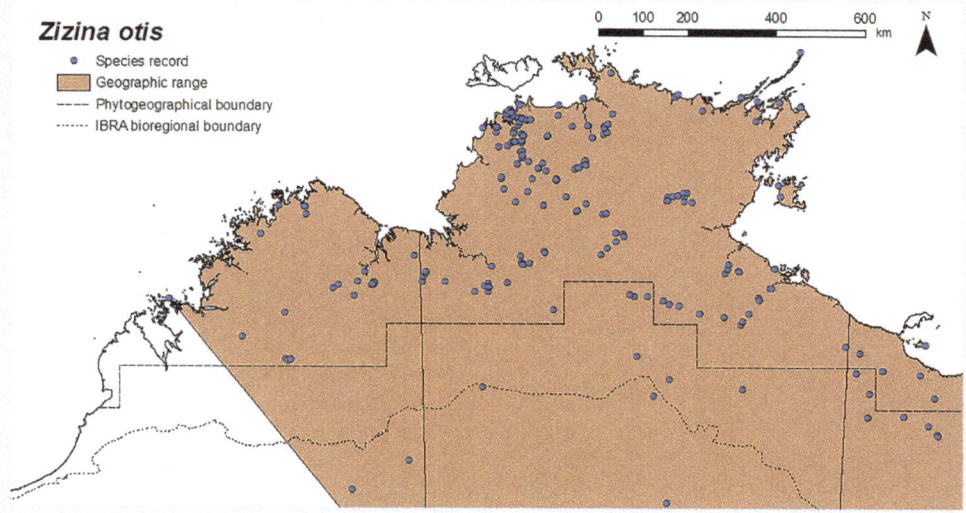

### Zizina otis (n = 344)

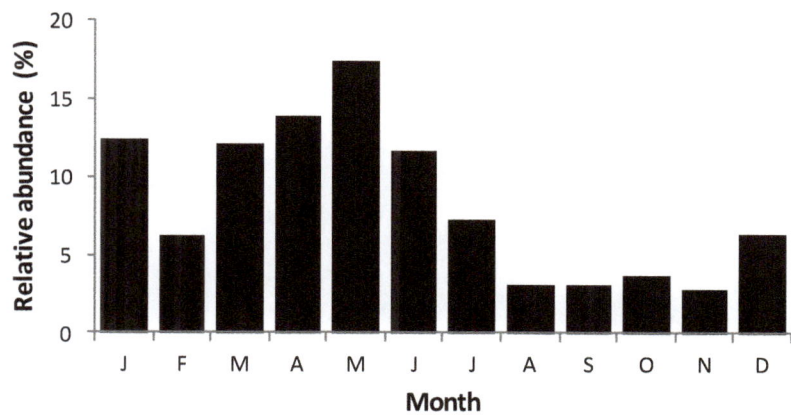

| Month | J | F | M | A | M | J | J | A | S | O | N | D |
|-------|---|---|---|---|---|---|---|---|---|---|---|---|
| Egg | | | | | | | | | | | | |
| Larva | | | | | | | | | | | | |
| Pupa | | | | | | | | | | | | |
| Adult | | | | | | | | | | | | |

# Black-spotted Grass-blue
*Famegana alsulus* (Herrich-Schäffer, 1869)

Plate 176 Buntine Highway, NT
Photo: Deb Bisa

## Distribution

This species is represented in the study region by the subspecies *F. alsulus alsulus* (Herrich-Schäffer, 1869). It occurs very widely in the study region, extending from moist coastal areas to drier inland areas of the semi-arid zone (500 mm mean annual rainfall), as well as the arid zone of central Australia beyond the southern boundary of the study region. It has been recorded as far south as the Edgar Ranges (Common 1981) and Old Halls Creek (Williams et al. 2006), WA; Devils Marbles, NT (Dunn 1980); and Boodjamulla/Lawn Hill National Park, Qld (Daniels and Edwards 1998). It has also been recorded at Terry Smith Lookout 80 km north–north-west of Cloncurry, Qld, just outside the south-eastern corner of the study region (Dunn 2017b). The geographic range corresponds well with the spatial distribution of its larval food plants, indicating that *F. alsulus* has been well sampled in the region. Outside the study region, *F. alsulus* occurs in southern China, Taiwan and the Philippines, and from central and eastern Australia to Vanuatu, Fiji, Samoa and Tonga. It also occurs in the Pilbara of Western Australia.

## Habitat

*Famegana alsulus* breeds in savannah woodland where the main larval food plants (*Vigna* spp.) typically grow as seasonal perennial herbaceous twiners (vines) in the grassy understorey (Braby 2011a, 2015e). It also breeds in riparian open woodland and open sandy areas along riverbanks in the more inland areas.

## Larval food plants

*Tephrosia* sp., *Vigna lanceolata*, *V. radiata*, *V. vexillata* (Fabaceae); also *Macroptilium atropurpureum* (Fabaceae). The usual larval food plants are *Vigna* spp., but at Leichhardt Falls in the Gulf of Carpentaria, Qld, a female was observed ovipositing several eggs on the flower buds of a *Tephrosia* sp. growing as a small shrub in dry river sand. This species may prove to be *T. remotiflora*; adults have been observed attracted to the flowers of this species at nearby Boodjamulla/Lawn Hill National Park, Qld (Dunn 2017b).

## Attendant ant

Not recorded in the study region. Elsewhere, the larvae are usually attended by a few ants representing several genera, in a facultative myrmecophilous association (Eastwood and Fraser 1999; Braby 2000).

## Seasonality

Adults occur throughout the year, but they are most abundant during the late wet season (March and April), when immense numbers of males may be observed puddling at damp sand near creeks. Their numbers diminish as the dry season progresses, and they are very scarce during the 'build-up' at the start of the wet season. However, in some areas, they may be locally abundant during the dry season depending on the availability of flower buds and flowers of their larval food plants. The immature stages (eggs or larvae) have been recorded frequently from March to May, when adults are most abundant, but also in September. The main larval food plants (*Vigna* spp.) are seasonal perennials that lose their aerial stems and leaves during the dry season; they regenerate from tuberous roots (i.e. facultative basal resprouters) depending on rainfall/moisture and post fire. Presumably, *F. alsulus* breeds continuously throughout much of the year and several generations are completed annually, but it is not clear how the species survives the late dry season, when the food plants are frequently not available.

## Breeding status

This species is resident in the study region.

## Conservation status

LC.

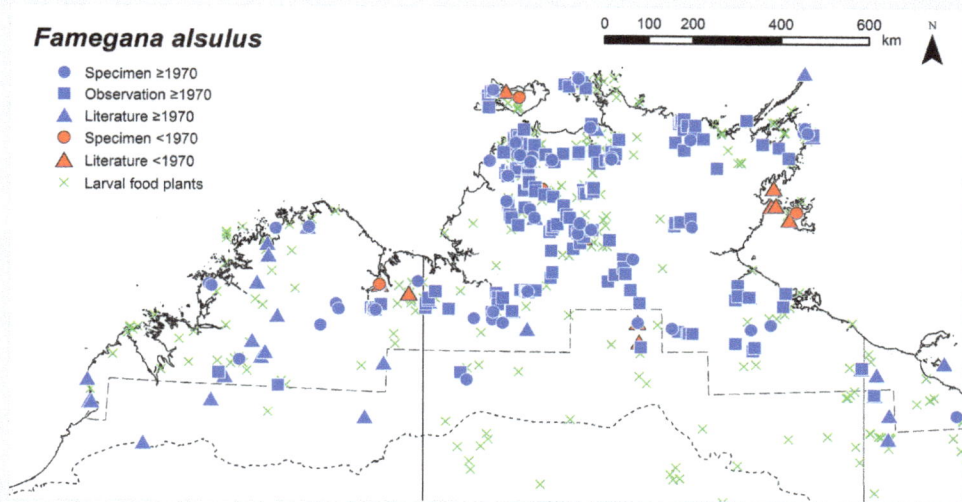

*Famegana alsulus*

- ● Specimen ≥1970
- ■ Observation ≥1970
- ▲ Literature ≥1970
- ● Specimen <1970
- ▲ Literature <1970
- ✕ Larval food plants

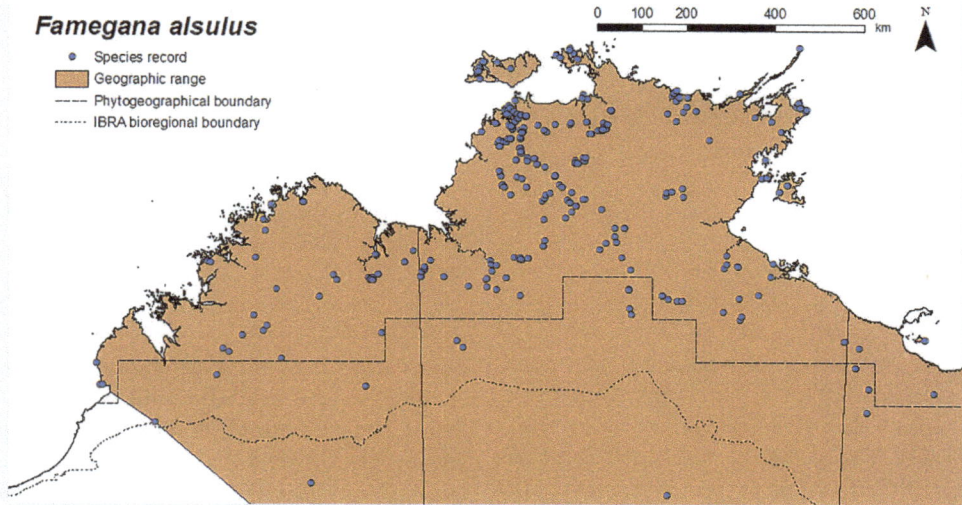

*Famegana alsulus*

- ● Species record
- ▬ Geographic range
- --- Phytogeographical boundary
- ···· IBRA bioregional boundary

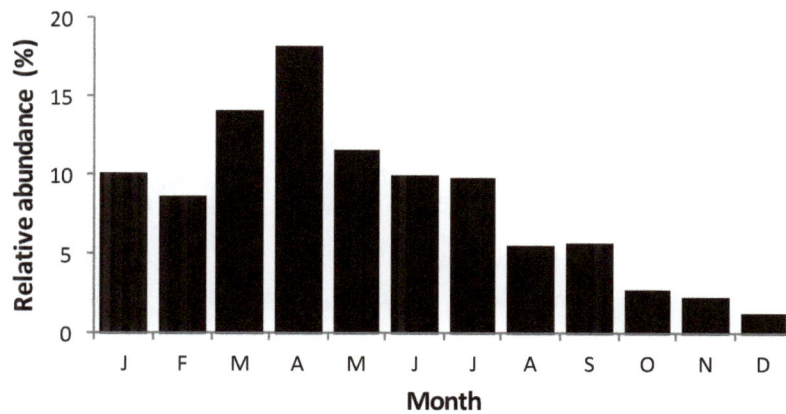

*Famegana alsulus* (n = 418)

| Month | J | F | M | A | M | J | J | A | S | O | N | D |
|-------|---|---|---|---|---|---|---|---|---|---|---|---|
| Egg   |   |   |   |   |   |   |   |   |   |   |   |   |
| Larva |   |   |   |   |   |   |   |   |   |   |   |   |
| Pupa  |   |   |   |   |   |   |   |   |   |   |   |   |
| Adult |   |   |   |   |   |   |   |   |   |   |   |   |

# Dainty Grass-blue
## *Zizula hylax* (Fabricius, 1775)

Plate 177 Kakadu National Park, NT
Photo: Ian Morris

## Distribution

This species is represented in the study region by the subspecies *Z. hylax attenuata* (T. P. Lucas, 1890). Le Souëf (1971) first reported its presence in the region, detecting the species at Darwin, NT, in 1971. It occurs in the Kimberley, Top End and western Gulf Country, extending from moist coastal areas to drier inland areas of the semi-arid zone (< 700 mm mean annual rainfall). It has been recorded as far south as Windjana Gorge National Park, WA (Pierce 2010); Judbarra/Gregory National Park (Humbert River crossing) (J. Archibald) and Limmen National Park (Pandanus Creek crossing), NT; and Boodjamulla/Lawn Hill National Park, Qld (Daniels and Edwards 1998). The geographic range corresponds well with the spatial distribution of its putative larval food plant (*Hygrophila angustifolia*), although the food plant extends slightly further inland. Outside the study region, *Z. hylax* occurs widely from Africa, India and South-East Asia, through mainland New Guinea and north-eastern and eastern Australia to the Solomon Islands and Vanuatu. It also occurs in the Pilbara of Western Australia and the arid zone of central Australia.

## Habitat

The breeding habitat of *Z. hylax* has not been recorded in the study region. Adults usually occur in pandanus swamps and swampy areas adjacent to spring-fed monsoon forest or within riparian woodland/forest where the putative larval food plant grows as an annual herb, but they have also been recorded in grassy areas and savannah woodland well away from water, where they probably breed on alternative food plants.

## Larval food plants

Not recorded in the study region; probably *Hygrophila angustifolia* (Acanthaceae), which is a food plant in Queensland (Braby 2000).

## Attendant ant

Not recorded in the study region. Elsewhere, the larvae and pupae are usually not attended by ants, but very occasionally the larvae are attended by a few small black ants (Eastwood and Fraser 1999; Braby 2000).

## Seasonality

Adults have been recorded during most months of the year. Our field observations, together with the limited number of monthly records available, suggest they are more abundant during the mid dry season (May–August), with a peak in abundance in July. The breeding phenology and seasonal history of the immature stages have not been recorded. Larvae are known to feed on the young fruits of the putative larval food plant, which germinates during the wet season and is available only during the late wet season and early dry season; it then withers and dies off in the mid to late dry season. Presumably, *Z. hylax* breeds from at least the mid wet season to the early dry season and one or more generations are completed annually. However, it is not clear how the species survives the late dry season and early wet season.

## Breeding status

This species is resident in the study region.

## Conservation status

LC.

## Zizula hylax

- ● Specimen ≥1970
- ■ Observation ≥1970
- ▲ Literature ≥1970
- ✕ Putative larval food plant

## Zizula hylax

- ● Species record
- ▨ Geographic range
- – – – Phytogeographical boundary
- ········· IBRA bioregional boundary

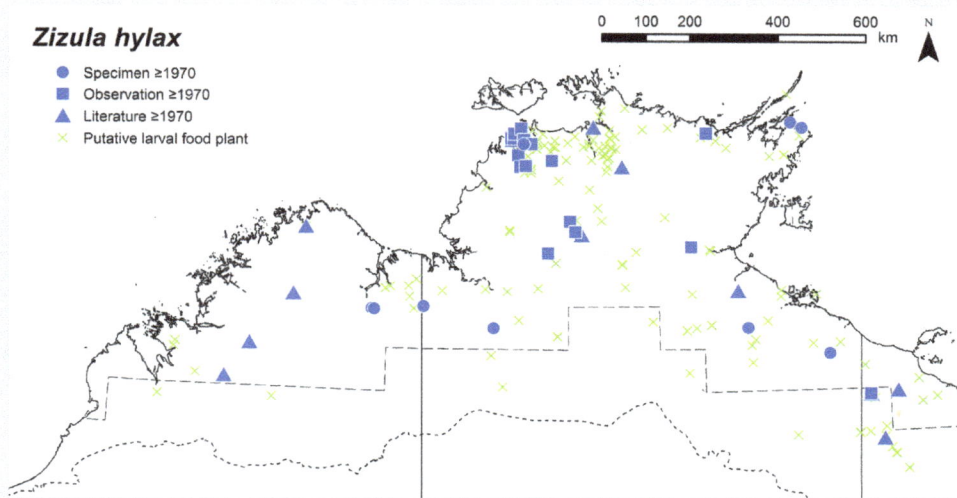

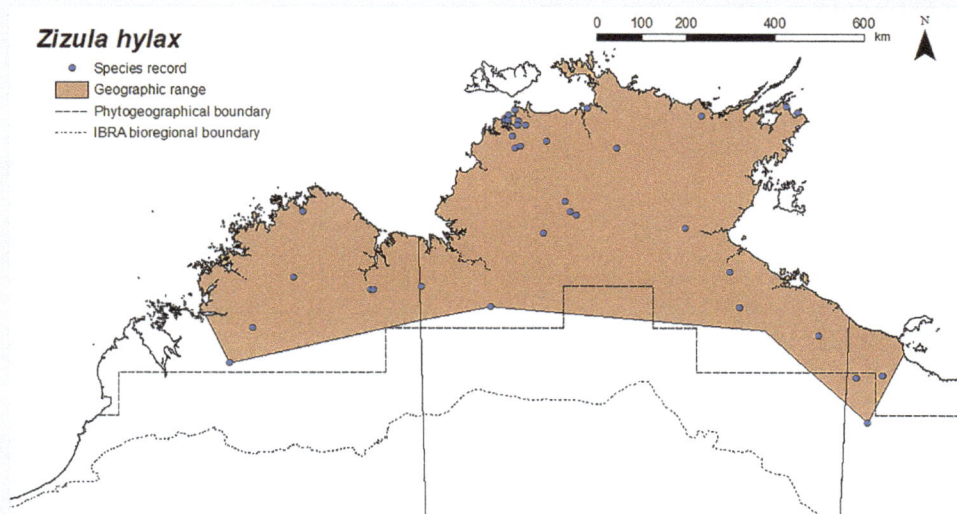

## *Zizula hylax* (n = 41)

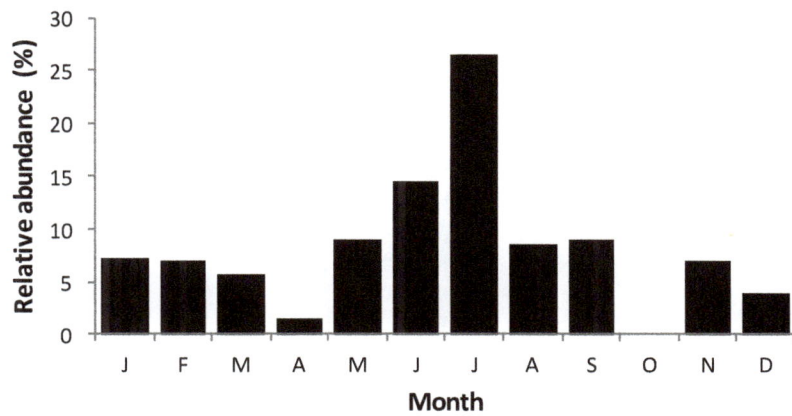

| Month | J | F | M | A | M | J | J | A | S | O | N | D |
|-------|---|---|---|---|---|---|---|---|---|---|---|---|
| Egg   |   |   |   |   |   |   |   |   |   |   |   |   |
| Larva |   |   |   |   |   |   |   |   |   |   |   |   |
| Pupa  |   |   |   |   |   |   |   |   |   |   |   |   |
| Adult |   |   |   |   |   |   |   |   |   |   |   |   |

# Orange-tipped Pea-blue
## *Everes lacturnus* (Godart, [1824])

Plate 178 Fish River Station, NT
Photo: M. F. Braby

## Distribution

This species is represented in the study region by the subspecies *E. lacturnus australis* Couchman, 1962. It has a disjunct distribution, occurring in the northern Kimberley (Johnson 1993) and more widely in the Top End. It is restricted to the higher rainfall areas (mostly > 1,000 mm mean annual rainfall), reaching its southernmost limit at Katherine (Angel 1951) and easternmost limits on Gove Peninsula and Groote Eylandt (Tindale 1923), NT. The geographic range closely corresponds with the spatial distribution of the putative larval food plant (*Desmodium heterocarpon*). The food plant, however, also occurs further south-west in the western Kimberley and on the Tiwi Islands and Sir Edward Pellew Group, NT; thus, further field surveys are required to determine whether *E. lacturnus* occurs in these areas. Outside the study region, *E. lacturnus* occurs widely from India, southern China and South-East Asia, through mainland New Guinea and adjacent islands and north-eastern and eastern Australia to New Britain and the Solomon Islands.

## Habitat

The breeding habitat of *E. lacturnus* has not been recorded in the study region. Adults are usually found in riparian areas along permanent freshwater creeks or streams, where they typically occur in damp open areas or grassy swampy areas along the edge of evergreen monsoon vine forest or within mixed *Lophostemon–Melaleuca* woodland or in open forest adjacent to monsoon forest.

## Larval food plants

Not recorded in the study region; probably *Desmodium heterocarpon* (Fabaceae), which is the food plant in eastern Queensland (Braby 2000).

## Attendant ant

Not recorded in the study region. Elsewhere, the larvae and pupae are usually not attended by ants, but very occasionally the larvae are attended by a few small ants (Eastwood and Fraser 1999; Braby 2000).

## Seasonality

Adults have been recorded during most months of the year, but they are most abundant during the late wet season (March and April) and scarce during the dry season. The breeding phenology and seasonal history of the immature stages have not been recorded. The putative larval food plant is a seasonal perennial herb that is facultatively deciduous, losing its leaves during the dry season when water-stressed and resprouting from rootstock during the wet season or post fire. It is not clear how *E. lacturnus* survives the dry season and early wet season, when the plant is frequently not available. Presumably, the species remains in larval diapause, similar to populations in eastern Queensland (Samson 1991).

## Breeding status

This species is resident in the study region.

## Conservation status

LC.

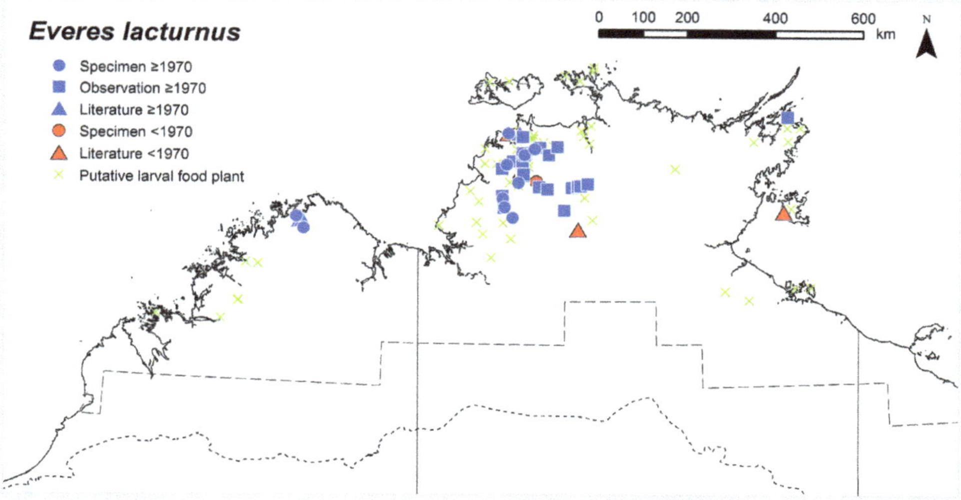

## Everes lacturnus

Legend:
- Specimen ≥1970
- Observation ≥1970
- Literature ≥1970
- Specimen <1970
- Literature <1970
- Putative larval food plant

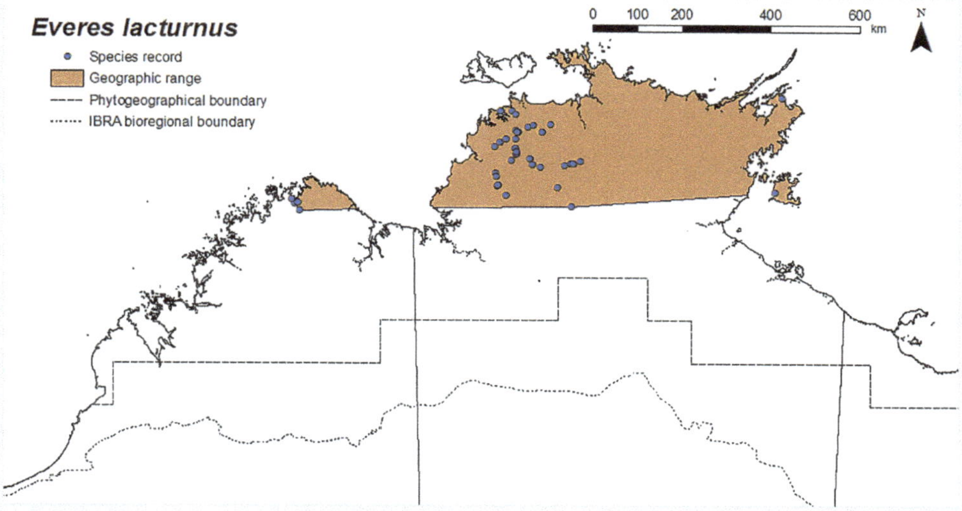

0 100 200 400 600 km

## Everes lacturnus

Legend:
- Species record
- Geographic range
- Phytogeographical boundary
- IBRA bioregional boundary

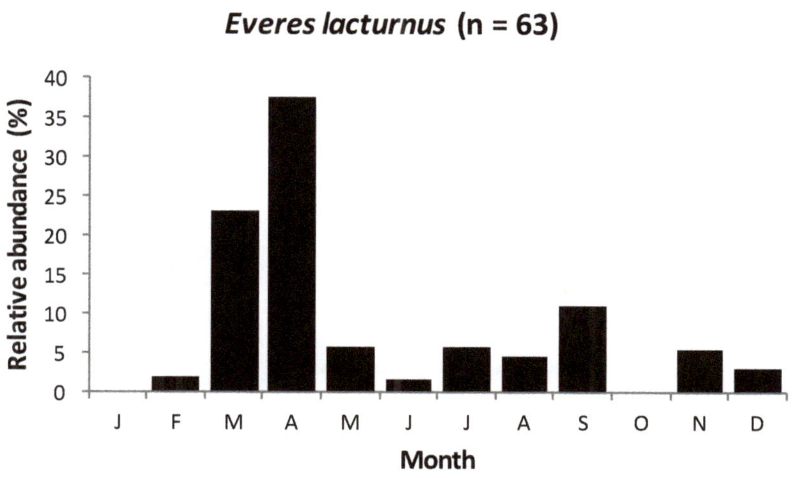

0 100 200 400 600 km

Everes lacturnus (n = 63)

| Month | J | F | M | A | M | J | J | A | S | O | N | D |
|-------|---|---|---|---|---|---|---|---|---|---|---|---|
| Egg | | | | | | | | | | | | |
| Larva | | | | | | | | | | | | |
| Pupa | | | | | | | | | | | | |
| Adult | | ■ | ■ | ■ | ■ | ■ | ■ | ■ | ■ | | ■ | ■ |

# Spotted Pea-blue
## *Euchrysops cnejus* (Fabricius, 1798)

Plate 179 Berry Springs, NT
Photo: M. F. Braby

## Distribution

This species is represented in the study region by the subspecies *E. cnejus cnidus* Waterhouse & Lyell, 1914. It occurs very widely in the Kimberley, Top End and western Gulf Country, extending from moist coastal areas to drier inland areas of the semi-arid zone (500 mm mean annual rainfall). It has been recorded as far inland as Halls Creek, WA (Grund and Hunt 2001); Daly Waters, NT (A. Allwood and T. Weir); and Boodjamulla/Lawn Hill National Park (Daniels and Edwards 1998) and Lake Moondarra (Dunn 2017d), Qld, as well as Corella River crossing 45 km west by south of Cloncurry, Qld, just outside the south-eastern corner of the study region. The geographic range is broadly similar to the spatial distribution of its larval food plants, although the food plants extend further inland to the arid zone. Outside the study region, *E. cnejus* occurs widely from India and South-East Asia, through mainland New Guinea and adjacent islands and north-eastern and eastern Australia to the Bismarck Archipelago and Samoa.

## Habitat

*Euchrysops cnejus* breeds mainly in savannah woodland where the native larval food plants (*Vigna* spp.) typically grow as seasonal perennial herbaceous twiners (vines) in the grassy understorey (Braby 2011a, 2015e). It also breeds in paperbark swamps and in open disturbed areas where the introduced food plants (*Macroptilium* spp.) grow in abundance in the ground layer.

## Larval food plants

*Vigna lanceolata*, *V. marina*, *V. radiata*, *V. vexillata* (Fabaceae); also *\*Macroptilium atropurpureum*, *\*M. lathyroides*, *\*Vigna unguiculata* (Fabaceae).

## Attendant ant

Not recorded in the study region. Elsewhere, the larvae are usually attended by a few ants representing several genera in a facultative myrmecophilous association (Eastwood and Fraser 1999; Braby 2000).

## Seasonality

Adults occur throughout the year, but they are most abundant during the late wet season and early dry season (March–June), with a peak in abundance in April and May, and least abundant during the late dry season (September–November). The immature stages have been recorded sporadically from January to August. The larvae feed on the flower buds and young pods of the larval food plants. The native food plants (*Vigna* spp.) are seasonal perennials that lose their aerial stems and leaves during the dry season; they regenerate from tuberous roots (i.e. facultative basal resprouters) depending on rainfall/moisture and post fire. Presumably, *E. cnejus* breeds continuously from at least the mid wet season to the mid dry season and several generations are completed annually. However, is not clear how the species survives the dry season, when its larval food plants are frequently not available.

## Breeding status

This species is resident in the study region.

## Conservation status

LC.

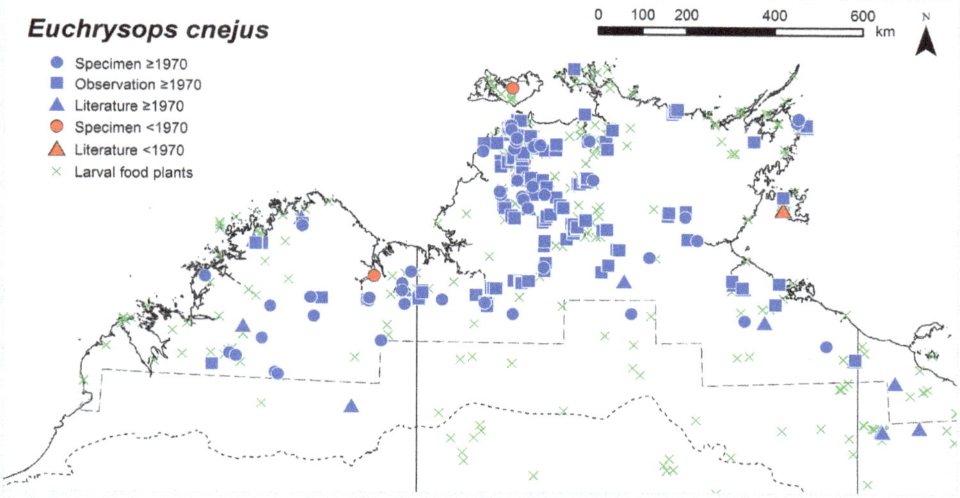

## Euchrysops cnejus

Legend:
- ● Specimen ≥1970
- ■ Observation ≥1970
- ▲ Literature ≥1970
- ● Specimen <1970
- ▲ Literature <1970
- ✕ Larval food plants

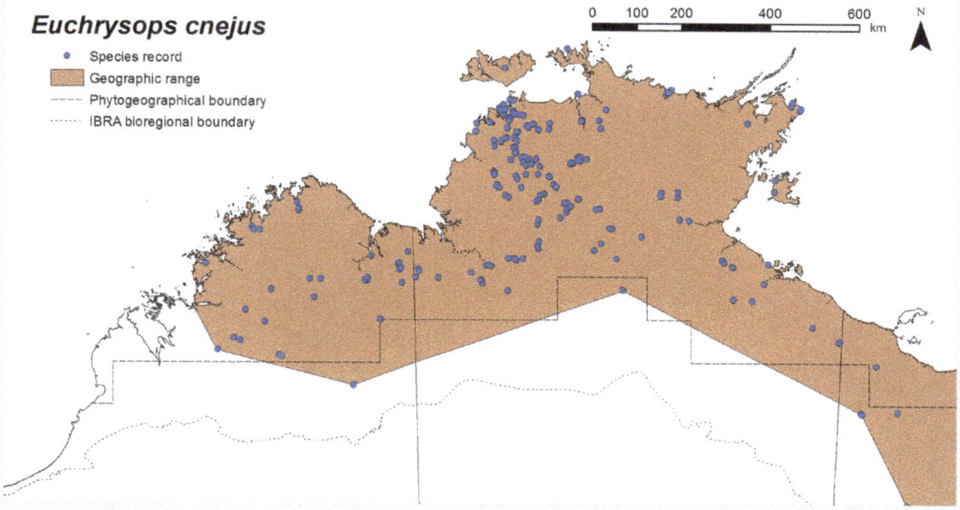

## Euchrysops cnejus

Legend:
- • Species record
- ▨ Geographic range
- — · — Phytogeographical boundary
- ⋯⋯ IBRA bioregional boundary

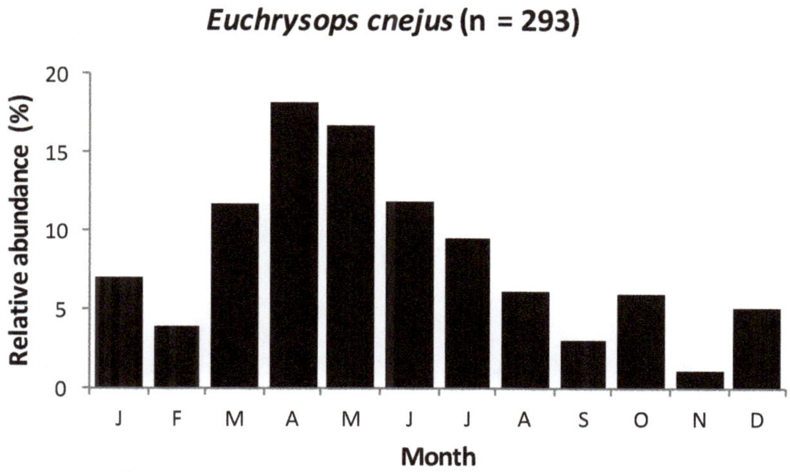

### Euchrysops cnejus (n = 293)

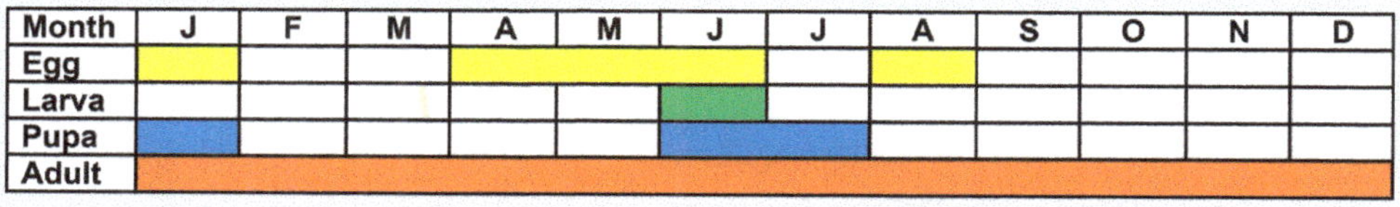

| Month | J | F | M | A | M | J | J | A | S | O | N | D |
|-------|---|---|---|---|---|---|---|---|---|---|---|---|
| Egg   | ▓ |   |   | ▓ | ▓ | ▓ |   | ▓ |   |   |   |   |
| Larva |   |   |   |   |   | ▓ |   |   |   |   |   |   |
| Pupa  | ▓ |   |   |   |   | ▓ | ▓ |   |   |   |   |   |
| Adult | ▓ | ▓ | ▓ | ▓ | ▓ | ▓ | ▓ | ▓ | ▓ | ▓ | ▓ | ▓ |

# Jewelled Grass-blue
## *Freyeria putli* (Kollar, [1844])

Plate 180 Kununurra, WA
Photo: Mark Golding

## Distribution

This species is represented in the study region by the subspecies *F. putli putli* (Kollar, [1844]). It occurs very widely in the Kimberley, Top End, Northern Deserts and western Gulf Country. It extends from moist coastal areas to the semi-arid zone (< 500 mm mean annual rainfall), where it has been recorded as far south as Derby (Warham 1957) and Halls Creek (Grund and Hunt 2001), WA; 7 km north–north-east of Helen Springs Homestead near Renner Springs, NT; and Boodjamulla/Lawn Hill National Park, Qld (Daniels and Edwards 1998). The geographic range broadly corresponds with the spatial distribution of its larval food plants, although the food plants extend further inland to the arid zone. The food plants also occur on the Tiwi Islands; thus, further surveys are required to determine whether *F. putli* is present on Bathurst and Melville islands, NT. Outside the study region, *F. putli* occurs widely from Nepal, India and South-East Asia, through mainland New Guinea to north-eastern and eastern Australia.

## Habitat

*Freyeria putli* breeds in savannah woodland, favouring open disturbed areas where the main larval food plant grows as an annual herb (Braby 2015e).

## Larval food plants

*Flemingia lineata*, *Indigofera linifolia* (Fabaceae). The main food plant appears to be *I. linifolia* (Braby 2015e).

## Attendant ant

*Iridomyrmex* sp. (1 *anceps* group), *Ochetellus* sp. near *glaber* (Formicidae: Dolichoderinae), *Polyrhachis schenkii* (Formicidae: Formicinae). The larvae are usually attended by a few ants representing several genera in a facultative myrmecophilous association (Meyer 1996a; Braby 2015e).

## Seasonality

Adults have been recorded during most months of the year, but they are most abundant during the late wet season and early dry season (March–May). They are generally absent or very scarce during the late dry season (August–November). We have recorded the immature stages in March and April, when adults are abundant, but undoubtedly they occur throughout the wet season and early dry season. The main larval food plant is a seasonal herb with an annual lifespan; the plants die off during the dry season and then reproduce from seed after rainfall. Presumably, *F. putli* breeds continuously from at least the mid wet season to the early dry season and several generations are completed annually. However, it is not clear how the species survives the late dry season, when the main food plant is not available; it is possible the immature stages enter diapause or it switches to an alternative food plant, such as *Flemingia lineata*, which is a small perennial shrub that maintains its foliage throughout the year.

## Breeding status

This species is resident in the study region.

## Conservation status

LC.

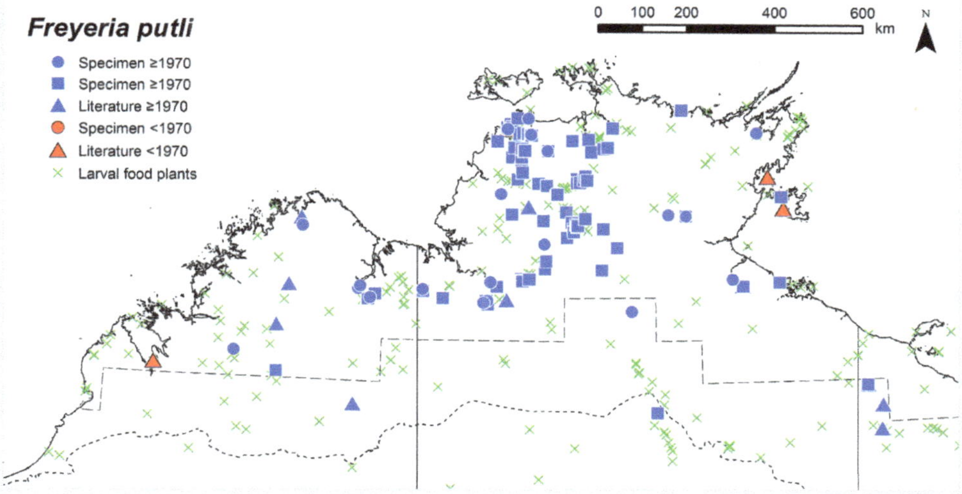

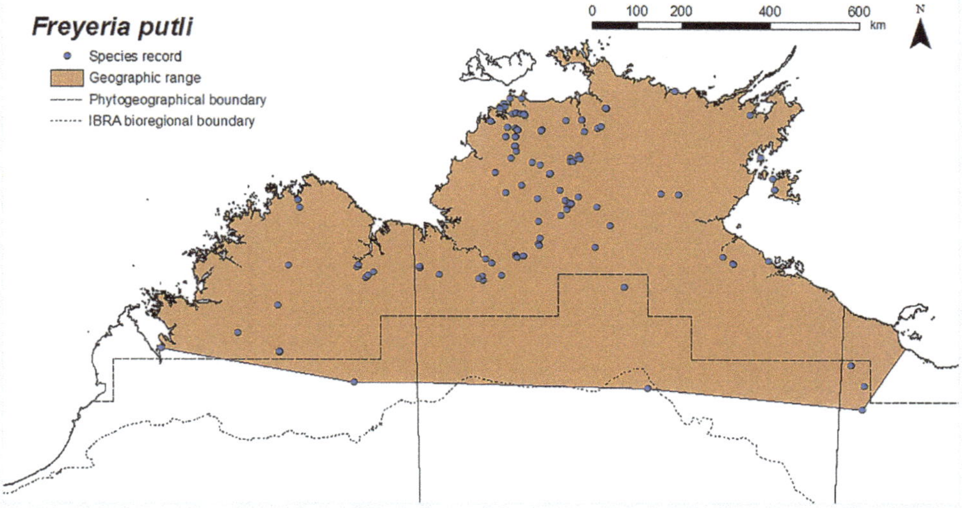

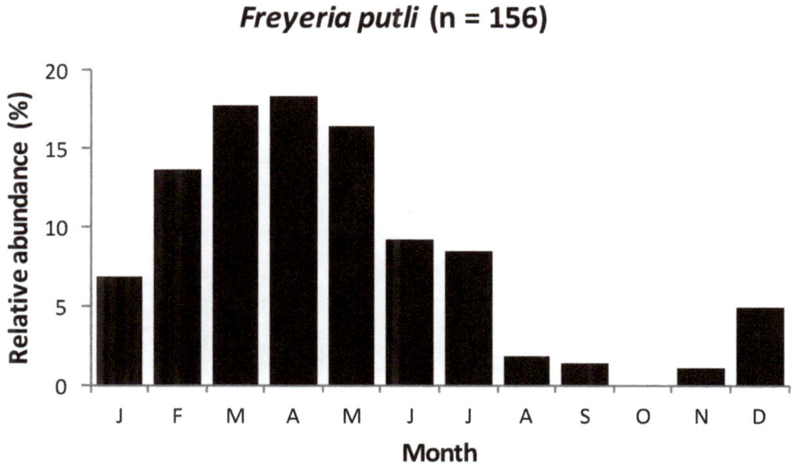

Freyeria putli
Specimen ≥1970
Specimen ≥1970
Literature ≥1970
Specimen <1970
Literature <1970
Larval food plants
0 100 200 400 600 km
N

Freyeria putli
Species record
Geographic range
Phytogeographical boundary
IBRA bioregional boundary
0 100 200 400 600 km
N

Freyeria putli (n = 156)
Relative abundance (%)
20
15
10
5
0
J F M A M J J A S O N D
Month

| Month | J | F | M | A | M | J | J | A | S | O | N | D |
|---|---|---|---|---|---|---|---|---|---|---|---|---|
| Egg | | | | | | | | | | | | |
| Larva | | | | | | | | | | | | |
| Pupa | | | | | | | | | | | | |
| Adult | | | | | | | | | | | | |

# Day-flying moths

(Sesiidae, Castniidae, Zygaenidae, Immidae, Geometridae, Uraniidae, Erebidae and Noctuidae)

10

# Clearwing Moth
## *Pseudosesia oberthuri* (Le Cerf, 1916)

Plate 181 Lee Point, Casuarina Coastal Reserve, NT
Photo: Axel Kallies

Plate 182 Lee Point, Casuarina Coastal Reserve, NT
Photo: Axel Kallies

## Distribution

This species is endemic to the study region. It occurs in the north of the Northern Territory, where it is restricted to the higher rainfall areas of the north-western corner of the Top End (> 1,300 mm mean annual rainfall). It extends from the Tiwi Islands (Melville Island) (G. F. Hill) south to Mt Burrell on Tipperary Station (Braby 2011a) and east to Kakadu National Park (Nourlangie Rock) (A. Kallies). The larval food plants are considerably more widespread than the known geographic range, occurring in the Kimberley and throughout the Top End and coastal islands of the western Gulf Country. Further surveys are thus required to determine whether *P. oberthuri* occurs in these areas and is more widespread than present records indicate. Males of the species are usually detected by attracting them to artificial pheromone lures (Kallies 2001).

## Habitat

*Pseudosesia oberthuri* breeds along the edges of semi-deciduous monsoon vine thicket and in savannah woodland where the larval food plants grow either as vines (*Ampelocissus acetosa*) or as shrubs (*A. frutescens*) (Braby 2011a).

## Larval food plants

*Ampelocissus acetosa*, *A. frutescens* (Vitaceae). The main food plant appears to be *A. acetosa* (Kallies 2001; Braby 2011a).

## Seasonality

Adults are seasonal, occurring only during the wet season and early dry season (January–May), but there are too few records (n = 10) to assess temporal changes in abundance. The life cycle is incompletely known. The immature stages (larvae or pupae) have been recorded sporadically from January to May. Near Darwin, NT, large numbers of fully grown larvae and pupae have been found in March (A. Kallies, pers. comm.). The larvae feed in the woody stems of their food plants (*Ampelocissus* spp.), which are short-lived seasonal perennial vines or shrubs. The larvae cause significant swelling of the stems and sometimes multiple larvae (> 10) can be found in a single large gall. During the dry season, the food plants senesce and remain dormant as underground tubers; they then resprout and regenerate during the 'build-up' following the pre-monsoon storms. Presumably, only a single generation is completed annually, but it is not clear whether the species survives the dry season as eggs or early instar larvae.

## Breeding status

This species is resident in the study region.

## Conservation status

LC. The species *P. oberthuri* is a narrow-range endemic (geographic range = 29,990 sq km) and occurs in several conservation reserves, including East Point Reserve, Casuarina Coastal Reserve and Kakadu National Park. Despite its restricted occurrence, there are no known threats facing the taxon.

*Pseudosesia oberthuri*

- ● Specimen ≥1970
- ■ Observation ≥1970
- ▲ Literature ≥1970
- ● Specimen <1970
- ✕ Larval food plants

*Pseudosesia oberthuri*

- ● Species record
- ▨ Geographic range
- -- Phytogeographical boundary
- ⋯ IBRA bioregional boundary

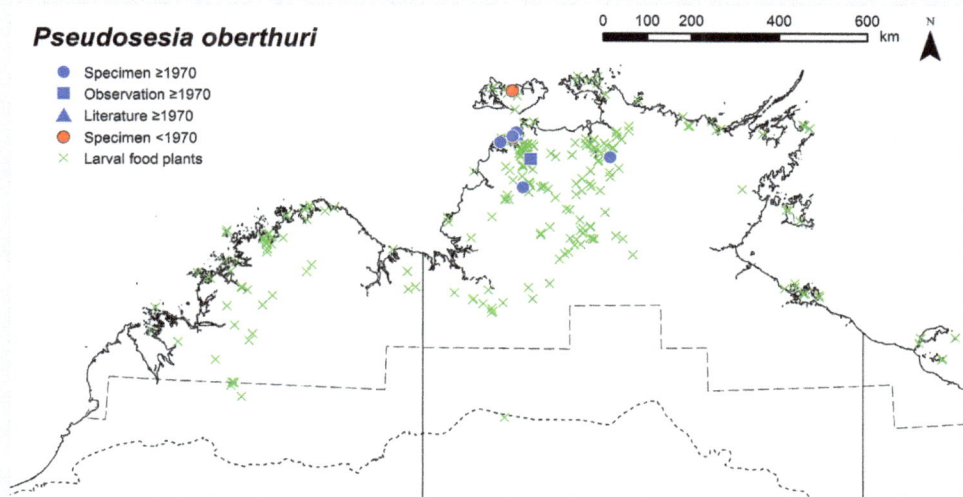

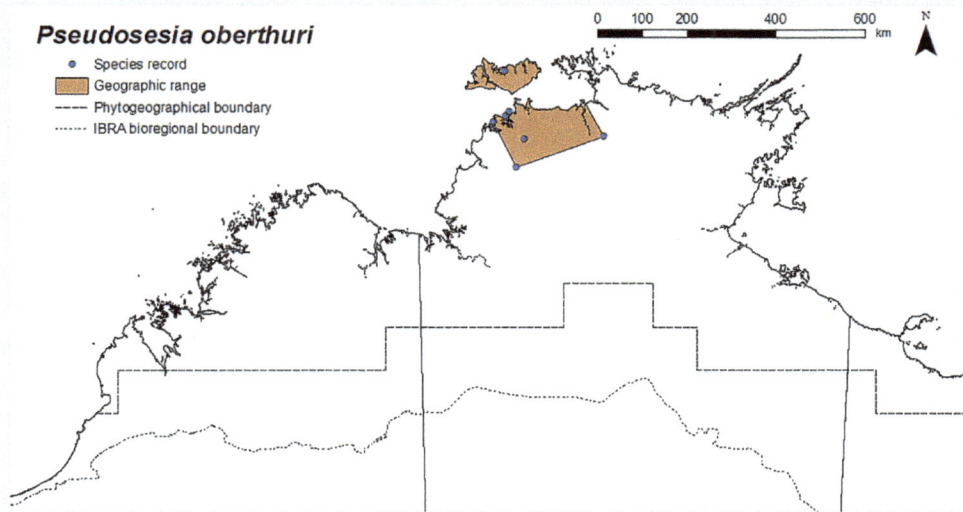

| Month | J | F | M | A | M | J | J | A | S | O | N | D |
|-------|---|---|---|---|---|---|---|---|---|---|---|---|
| Egg   |   |   |   |   |   |   |   |   |   |   |   |   |
| Larva |   |   |   |   |   |   |   |   |   |   |   |   |
| Pupa  |   |   |   |   |   |   |   |   |   |   |   |   |
| Adult |   |   |   |   |   |   |   |   |   |   |   |   |

# Northern White-spotted Sun-moth
## *Synemon phaeoptila* Turner, 1906

Plate 183 Cobourg Peninsula, NT
Photo: M. F. Braby

Plate 184 Pinkerton Range, NT
Photo: M. F. Braby

## Distribution

This species occurs in the western half of the Top End of the study region. It extends from moist coastal areas of high rainfall (> 1,600 mm mean annual rainfall) to drier inland areas, reaching its southernmost limit on the Pinkerton Range, NT (Braby 2011a). It generally occurs in higher rainfall areas (> 800 mm) than *Synemon wulwulam*, although the ranges of the two species overlap in the area between Adelaide River, Katherine and the Victoria River District, NT. The geographic range corresponds well with the spatial distribution of its larval food plant, although the food plant also occurs in the Kimberley and on the Tiwi Islands, NT. Further field surveys are thus required to determine whether *S. phaeoptila* occurs in these areas. Outside the study region, *S. phaeoptila* occurs in northern Queensland.

## Habitat

*Synemon phaeoptila* breeds in savannah woodland, favouring open grassy areas where the larval food plant grows as a perennial grass (Braby 2011a).

## Larval food plant

*Chrysopogon latifolius* (Poaceae).

## Seasonality

Adults are seasonal, occurring only during the wet season (November–February), before the flight season of *Synemon wulwulam*. They are most abundant during December and January following the first substantial rainfall events, when the larval food plants resprout or regenerate after fire. In some locations or seasons they may also be common in February, depending on the timing of the monsoon rains. The immature stages (eggs and fresh pupal shells) have been recorded in February. The larvae are believed to feed underground on the thick rhizome of the food plant. Presumably, one generation is completed annually and the larvae develop during the late wet season and dry season.

## Breeding status

This species is resident in the study region.

## Conservation status

LC.

## Synemon phaeoptila

- ● Specimen ≥1970
- ● Specimen <1970
- ✕ Larval food plant

## Synemon phaeoptila

- ● Species record
- ▨ Geographic range
- —— Phytogeographical boundary
- ······ IBRA bioregional boundary

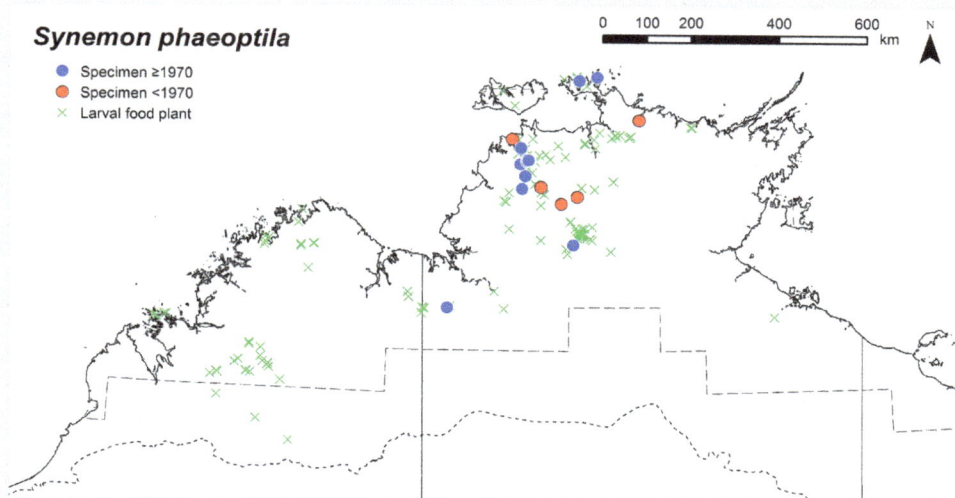

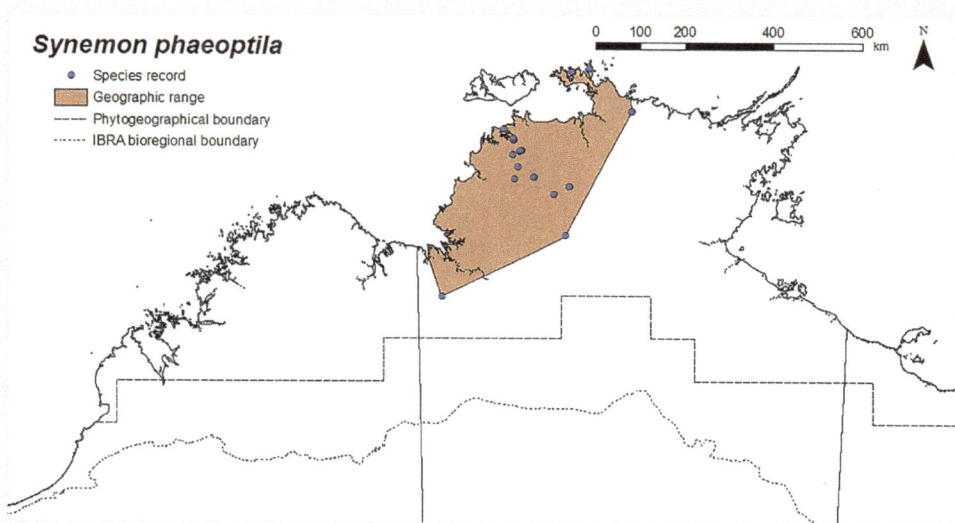

## Synemon phaeoptila (n = 29)

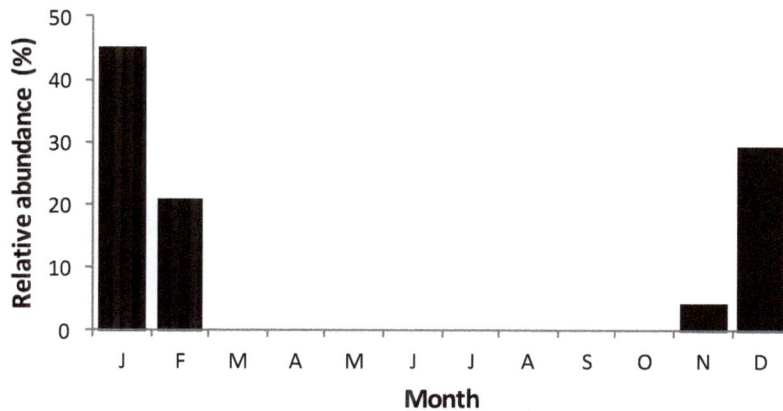

| Month | J | F | M | A | M | J | J | A | S | O | N | D |
|-------|---|---|---|---|---|---|---|---|---|---|---|---|
| Egg | | | | | | | | | | | | |
| Larva | | | | | | | | | | | | |
| Pupa | | | | | | | | | | | | |
| Adult | | | | | | | | | | | | |

# Kimberley Sun-moth
## *Synemon* sp. 'Kimberley'

Plate 185 East of Borda Island, WA
Photo: You Ning Su

Plate 186 East of Borda Island, WA
Photo: You Ning Su

## Distribution

This undescribed species is endemic to the northern Kimberley, where it occurs in the higher rainfall areas (> 1,000 mm mean annual rainfall), its presence in the region first detected in 2000 by G. Swann. It is currently known from only two coastal locations 150 km apart: on the mainland east of Borda Island (Williams et al. 2016) and Bertram Cove, south of Cape Bernier (G. Swann), WA. Further field surveys are required to determine whether *Synemon* sp. 'Kimberley' occurs elsewhere in the Kimberley, particularly in the more inland areas.

This species was provisionally placed under *Synemon phaeoptila* by Williams et al. (2016: 146) based on limited material (one female and one male), but with the caveat that '[i]t is possible that this is an undescribed species … More specimens, and better quality material from the Kimberley is required before a thorough taxonomic assessment can be made'. Examination and comparison of additional material (14 males, two females) recently collected by J. E. and A. Koeyers indicate that it is indeed a distinct species. *Synemon* sp. 'Kimberley' differs from *S. phaeoptila* in several characters—namely, the adults are substantially smaller in size; the wings are narrower and a different shape, with the forewing obovate and apex more rounded, and the hindwing termen straighter and less rounded; the pattern on the upperside of the forewing is less complex, being uniformly grey with black veins, the white cell spot consists of a small delicate dull white rounded spot (rather than a larger, brighter white oblique bar enclosed by a variable black patch in *S. phaeoptila*), and the broad iridescent silver curved streak along

the cubital vein is usually absent or poorly developed; and the prominent orange-brown terminal band on the upperside of the hindwing in males is absent or, when present, reduced to a series of indistinct spots separated by dark veins. In addition, on the underside of the forewing, females have four distinct black elongate subapical spots, which are absent in *S. phaeoptila*, and males have the ground colour predominantly dark brown, with the orange-brown area substantially reduced to a narrow margin near the termen, whereas in *S. phaeoptila* only the inner half is dark brown, which contrasts against the outer portion, which is bright orange-brown.

## Habitat

The breeding habitat of *Synemon* sp. 'Kimberley' has not been recorded. Adults have been collected in coastal *Acacia–Grevillea* low open woodland, where the putative larval food plant grows as a dominant perennial grass on sandstone (Williams et al. 2016).

## Larval food plant

Not recorded in the study region; possibly *Chrysopogon fallax* (Poaceae) (Williams et al. 2016).

## Seasonality

The seasonal abundance and breeding phenology of this species are not well understood. Adults have been recorded only in February, but we have too few records (n = 3) to assess any seasonal changes in abundance. It is likely the flight period is limited to the wet season, similar to that of *S. phaeoptila*. Presumably, one generation is completed annually and the larvae develop during the late wet season and dry season.

## Breeding status

This species is resident in the study region.

## Conservation status

DD. The species *Synemon* sp. 'Kimberley' is currently known from only two locations in the study region (AOO is likely to be < 2,000 sq km, with spatial buffering of records providing a first approximation of 1,400 sq km). Although both sites are located within conservation reserves (Balanggarra and Uunguu IPAs), further investigations are required to clarify the ecological requirements, critical habitat, distribution and conservation status of this species.

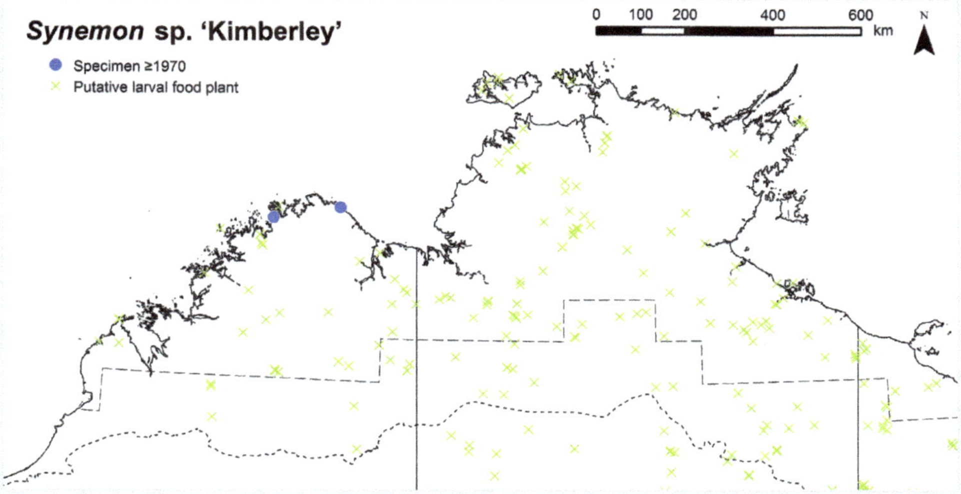

*Synemon* sp. 'Kimberley'

- Specimen ≥1970
- Putative larval food plant

| Month | J | F | M | A | M | J | J | A | S | O | N | D |
|-------|---|---|---|---|---|---|---|---|---|---|---|---|
| Egg   |   |   |   |   |   |   |   |   |   |   |   |   |
| Larva |   |   |   |   |   |   |   |   |   |   |   |   |
| Pupa  |   |   |   |   |   |   |   |   |   |   |   |   |
| Adult |   | ■ |   |   |   |   |   |   |   |   |   |   |

# Golden-flash Sun-moth

*Synemon wulwulam* Angel, 1951

Plate 187 Pine Creek, NT
Photo: M. F. Braby

Plate 188 Mt Burrell, NT
Photo: M. F. Braby

## Distribution

This species occurs widely in the study region, extending from the Kimberley through the Top End and Northern Deserts to the western Gulf Country. In the Kimberley, it has been recorded from Derby, the Mitchell Plateau, Kununurra and the Lake Argyle area (Williams et al. 2016), WA. In the Northern Territory, it extends from the higher rainfall areas of the Top End, reaching its northernmost limit at Adelaide River (Angel 1951), to drier inland areas of the semi-arid zone (< 400 mm mean annual rainfall), where it occurs as far south as the Barkly Tableland (Barkly Homestead Roadhouse and 22 km east–north-east of Soudan Homestead) (E. D. Edwards and M. Mathews). It generally occurs in lower rainfall areas (< 1,400 mm) than *Synemon phaeoptila*, although the ranges of the two species overlap in the area between Adelaide River, Katherine and the Victoria River District, NT. The larval food plants are more widespread than the known geographic range; thus, further field surveys are required to determine whether *S. wulwulam* is present in the southern Kimberley and Northern Deserts (Tanami Desert). Outside the study region, *S. wulwulam* occurs in western Queensland, reaching its easternmost limit at Cloncurry just outside the boundary of the study region.

## Habitat

*Synemon wulwulam* breeds in savannah woodland, favouring open grassy areas, where the larval food plants grow as perennial grasses (Braby 2015e; Williams et al. 2016).

## Larval food plants

*Chrysopogon setifolius*, *Sorghum plumosum* (Poaceae).

## Seasonality

Adults are seasonal, occurring only during the early dry season (March–June), after the flight season of *Synemon phaeoptila*. They are most abundant in May. The immature stages (eggs and a fresh pupal shell) have also been recorded in May. The larvae are believed to feed underground on the thick rhizome of the food plants. Presumably, one generation is completed annually and the larvae develop during the late dry season and wet season.

## Breeding status

This species is resident in the study region.

## Conservation status

LC.

## Synemon wulwulam

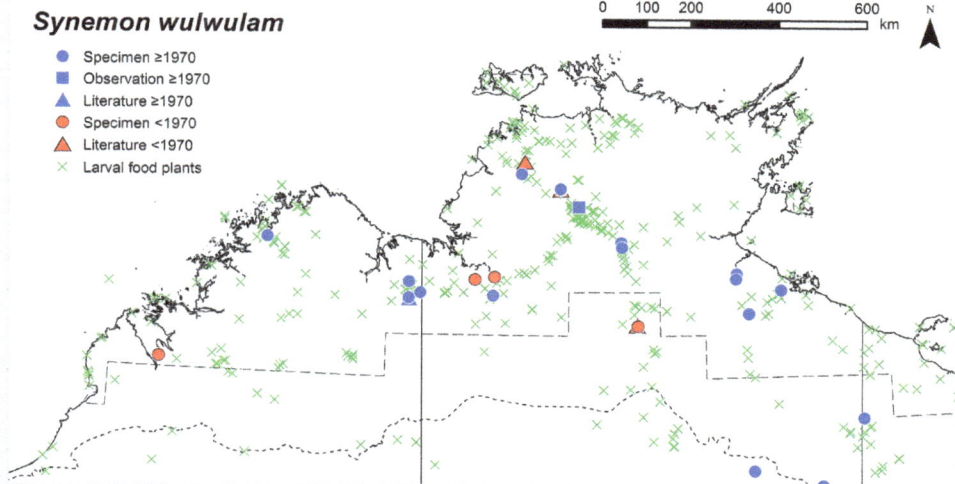

Legend:
- ● Specimen ≥1970
- ■ Observation ≥1970
- ▲ Literature ≥1970
- ● Specimen <1970
- ▲ Literature <1970
- × Larval food plants

## Synemon wulwulam

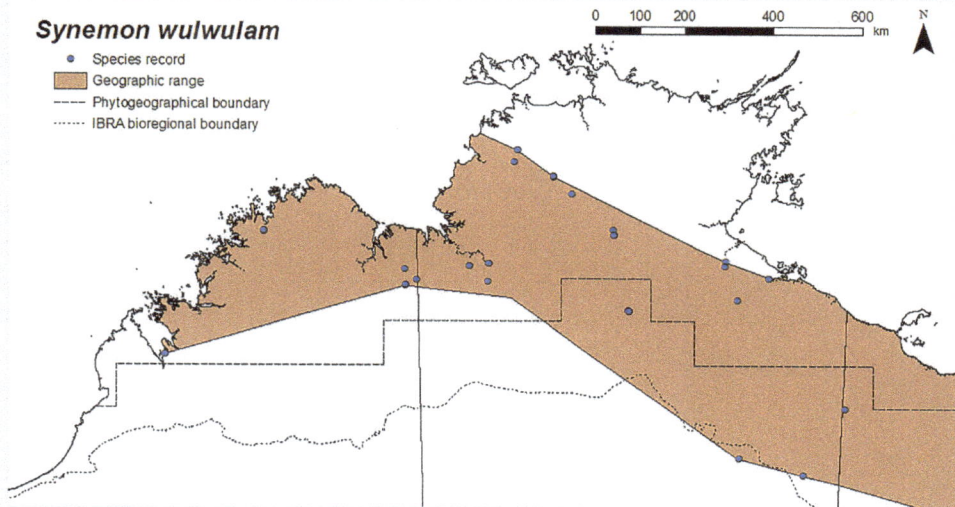

Legend:
- • Species record
- Geographic range
- – – Phytogeographical boundary
- ⋯⋯ IBRA bioregional boundary

### Synemon wulwulam (n = 30)

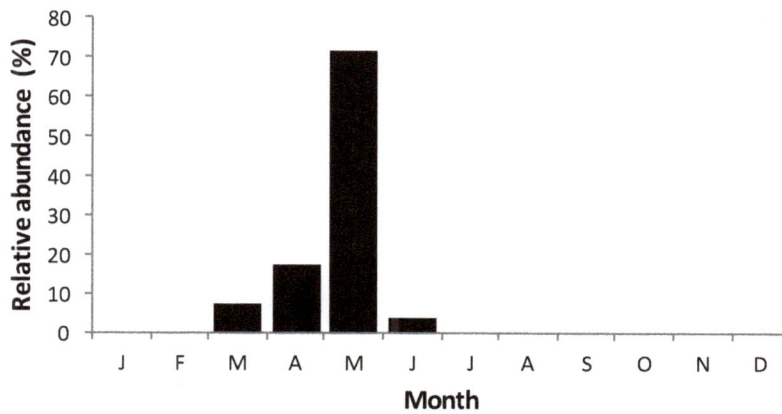

| Month | J | F | M | A | M | J | J | A | S | O | N | D |
|-------|---|---|---|---|---|---|---|---|---|---|---|---|
| Egg | | | | | ▮ | | | | | | | |
| Larva | | | | | | | | | | | | |
| Pupa | | | | | ▮ | | | | | | | |
| Adult | | | ▮ | ▮ | ▮ | ▮ | | | | | | |

# White-veined Sun-moth

## *Synemon* sp. 'Roper River'

Plate 189 Nathan River, Limmen National Park, NT
Photo: M. F. Braby

Plate 190 Nathan River, Limmen National Park, NT
Photo: M. F. Braby

## Distribution

This undescribed species is endemic to the study region. It is restricted mainly to the western Gulf Country, where it occurs in the lower rainfall areas (< 800 mm mean annual rainfall). There is also an isolated record further west in the southern Kimberley (16 km south-east of Halls Creek, WA) based on a historical specimen collected in 1944 (in WADA) (Williams et al. 2016). It extends as far south as the Barkly Tableland (Ranken Road 1 km north of the Barkly Highway, NT) in the semi-arid zone (< 400 mm). Its northernmost limit is Roper River (N. B. Tindale) and the easternmost limit is 15 km west of Musselbrook Resource Centre (G. Daniels and M. A. Schneider), NT. The larval food plants are considerably more widespread than the known geographic range; thus, further field surveys are required to determine whether *Synemon* sp. 'Roper River' is more widespread in the southern Kimberley and Northern Deserts, particularly in the intervening area between Halls Creek and the NT–Qld border.

## Habitat

*Synemon* sp. 'Roper River' breeds in savannah woodland, favouring flat open grassy areas where the larval food plants grow as perennial grasses (Braby 2011a; Williams et al. 2016). Although the geographic range falls within that of *Synemon wulwulam* and although it has a similar flight period to that species, the two species are rarely found together.

## Larval food plants

*Chrysopogon fallax*, *C. pallidus* (Poaceae).

## Seasonality

Adults are seasonal, occurring only in the early dry season (May). The immature stages (fresh pupal shells) have also been recorded in May. The larvae are believed to feed underground on the thick rhizome of the food plants. Presumably, one generation is completed annually and the larvae develop during the late dry season and wet season.

## Breeding status

This species is resident in the study region.

## Conservation status

LC. *Synemon* sp. 'Roper River' appears to be poorly represented in conservation reserves, and is currently known only from Limmen National Park.

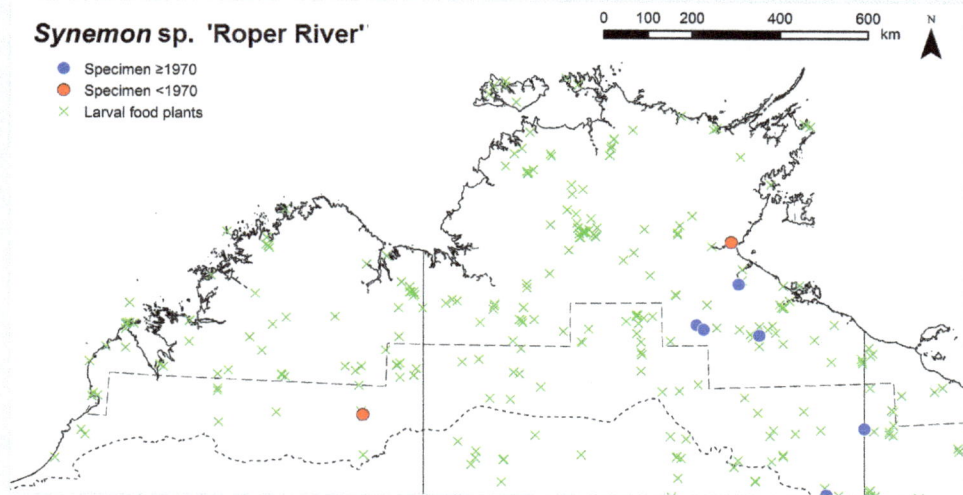

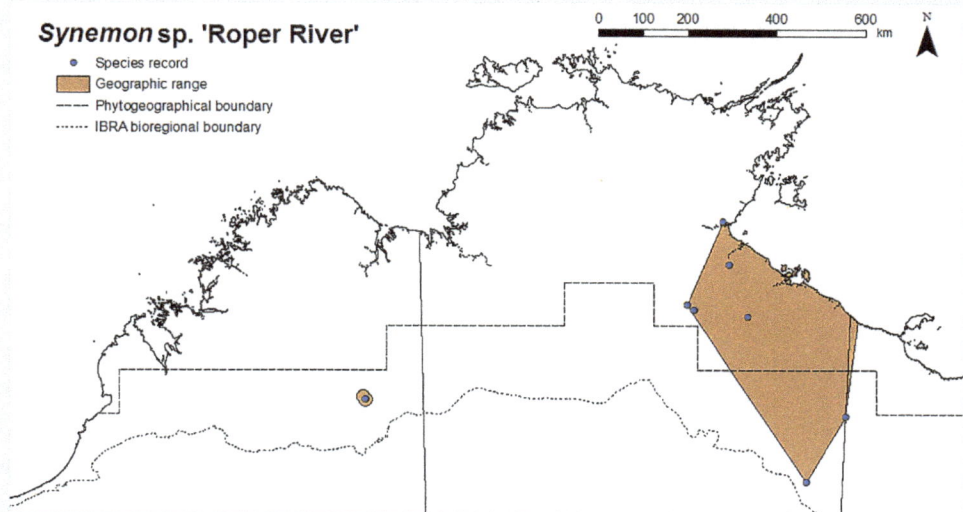

| Month | J | F | M | A | M | J | J | A | S | O | N | D |
|--------|---|---|---|---|---|---|---|---|---|---|---|---|
| Egg | | | | | | | | | | | | |
| Larva | | | | | | | | | | | | |
| Pupa | | | | | | | | | | | | |
| Adult | | | | | | | | | | | | |

# Northern Forester Moth

## *Pollanisus* sp. 7

Plate 191 Darwin, NT
Photo: M. F. Braby

## Distribution

This undescribed species is endemic to the study region, where it is known from only a single specimen collected from Darwin, NT, in 1908 by F. P. Dodd (Tarmann 2004). The putative larval food plant (*Pipturus argenteus*) occurs mainly in the higher rainfall areas (> 1,000 mm mean annual rainfall) in the eastern half of the Top End. Further field surveys are required to determine whether *Pollanisus* sp. 7 is still extant in the Darwin region (e.g. Black Jungle) or occurs elsewhere in the Top End.

## Habitat

The breeding habitat of *Pollanisus* sp. 7 has not been recorded. The putative larval food plant grows as a pioneer shrub or small tree up to 6 m high along the edges of wet and dry monsoon forests.

## Larval food plants

Not recorded in the study region; possibly *Pipturus argenteus* (Urticaceae), which is a food plant for the closely related *Pollanisus eumetopus* Turner, 1926 in north-eastern Queensland (Tarmann 2004; Mollet and Tarmann 2010).

## Seasonality

The seasonal abundance and breeding phenology of this species are not known.

## Breeding status

This species is assumed to be resident in the study region.

## Conservation status

DD. *Pollanisus* sp. 7 has not been recorded for more than 110 years since it was first discovered in 1908. Targeted field surveys to clarify the distribution and status of this species should be a high priority.

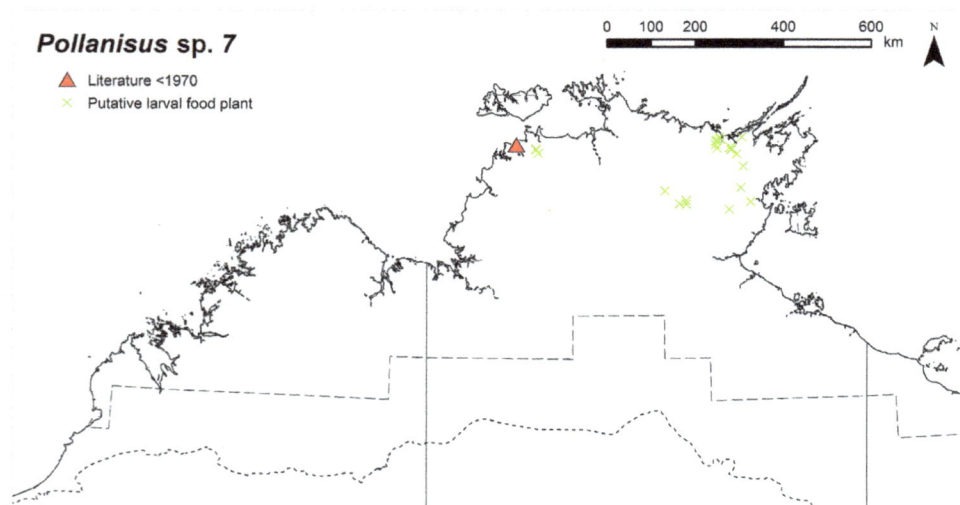

**Pollanisus sp. 7**

▲ Literature <1970
× Putative larval food plant

0   100   200     400     600   km   N

Photo: Western Lost City, Limmen National Park, NT, M.F. Braby

# Yellow Forester Moth

## *Hestiochora xanthocoma* Meyrick, 1886

Plate 192 Wongalara Wildlife Sanctuary, NT
Photo: M. F. Braby

## Distribution

This species occurs in the northern and eastern Kimberley and Top End of the study region. It occurs mainly in higher rainfall areas, although it has been recorded in drier areas (c. 800 mm mean annual rainfall). It has been recorded from several widely dispersed localities, including Jack Creek Bay and Palm Island, WA, in the northern Kimberley (J. E. & A. Koeyers); and Keep River National Park (Jarrnarm), Darwin, Howard Springs, Katherine (Tarmann 2004), East Alligator River in Kakadu National Park (A. Kallies), Wongalara Wildlife Sanctuary (Braby 2012a) and further east on Groote Eylandt (Tarmann 2004), NT. Further field surveys are required to determine whether *H. xanthocoma* occurs on the Tiwi Islands and in northern and north-eastern Arnhem Land of the Top End. Outside the study region, *H. xanthocoma* occurs in northern and south-eastern Queensland.

## Habitat

The breeding habitat of *H. xanthocoma* has not been recorded in the study region. Most locations at which adults have been collected comprise woodland on sandstone or sandy soil. At Wongalara Wildlife Sanctuary, NT, several adults were observed on a hill supporting low open woodland with a spinifex understorey on sandstone; they were feeding locally on flowers growing in the ground layer (Braby 2012a).

## Larval food plants

Not recorded in the study region.

## Seasonality

The seasonal abundance and breeding phenology of this species are not well understood. Adults have been recorded mainly in the late wet season and early dry season (March–June), as well as in the late dry season (October), but there are too few records (n = 10) to assess any seasonal changes in abundance. The seasonal history of the immature stages is not known.

## Breeding status

This species is resident in the study region.

## Conservation status

LC.

## Hestiochora xanthocoma

- ● Specimen ≥1970
- ▲ Literature ≥1970
- ▲ Literature <1970

0  100  200    400    600 km

N

## Hestiochora xanthocoma

- ● Species record
- ▓ Geographic range
- ‐‐‐ Phytogeographical boundary
- ⋯⋯ IBRA bioregional boundary

0  100  200    400    600 km

N

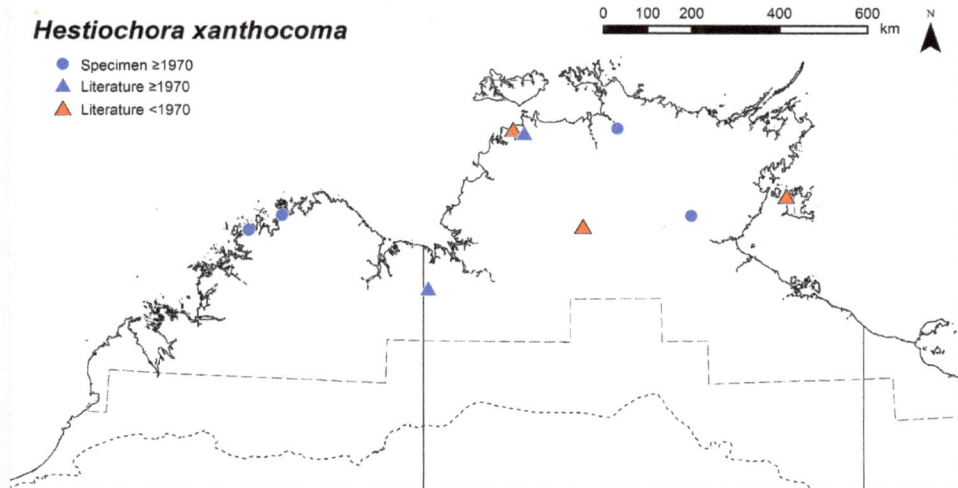

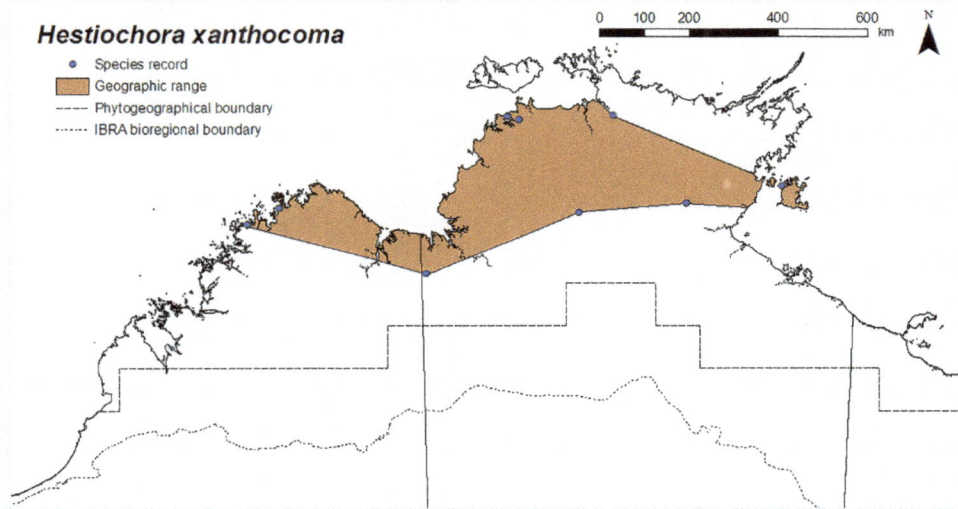

| Month | J | F | M | A | M | J | J | A | S | O | N | D |
|-------|---|---|---|---|---|---|---|---|---|---|---|---|
| Egg   |   |   |   |   |   |   |   |   |   |   |   |   |
| Larva |   |   |   |   |   |   |   |   |   |   |   |   |
| Pupa  |   |   |   |   |   |   |   |   |   |   |   |   |
| Adult |   |   |   |   |   |   |   |   |   |   |   |   |

# Orange-banded Velvet Moth

*Birthana cleis* (R. Felder & Rogenhofer, 1875)

Plate 193 Leanyer, Darwin, NT
Photo: M. F. Braby

Plate 194 Leanyer, Darwin, NT
Photo: M. F. Braby

## Distribution

This species has a disjunct distribution, occurring in the northern Kimberley and more widely in the Top End, where it occurs in the higher rainfall areas (> 900 mm mean annual rainfall). The southern limits of its range include Mitchell Falls National Park (Mertens Creek), WA (E. P. Williams); and Fish River Station (6 km south-east of Fish River homestead) (Braby and Kessner 2012) and Wongalara Wildlife Sanctuary (Wongalara homestead) (Braby 2012a), NT. The geographic range falls within the spatial distribution of its larval food plants, which occur very widely throughout the study region. Further surveys are thus required to determine more precisely the southern limits of *B. cleis* and whether it is present on the Tiwi Islands and Groote Eylandt, NT. Outside the study region, *B. cleis* occurs in Ambon and north-eastern Australia.

## Habitat

*Birthana cleis* breeds in savannah woodland and riparian open forest where the larval food plants grow as mistletoes (parasitic shrubs) on various trees, including *Eucalyptus*, *Planchonia*, *Alstonia*, *Erythrophleum* and *Grevillea* (Anderson and Braby 2009; Braby 2011a, 2015e). It also breeds in suburban parks and gardens.

## Larval food plants

*Amyema sanguinea*, *Decaisnina signata*, *Dendrophthoe glabrescens*, *Dendrophthoe odontocalyx* (Loranthaceae).

## Seasonality

Adults occur during most months of the year. In general, adults tend to be more numerous during the wet season (January and March–April), but they may be locally common at other times of the year depending on the synchronous development of the immature stages, of which the larvae feed gregariously on the foliage of their mistletoe food plants. The immature stages have been recorded during all months of the year, indicating that breeding occurs continuously and several generations are completed annually.

## Breeding status

This species is resident in the study region.

## Conservation status

LC.

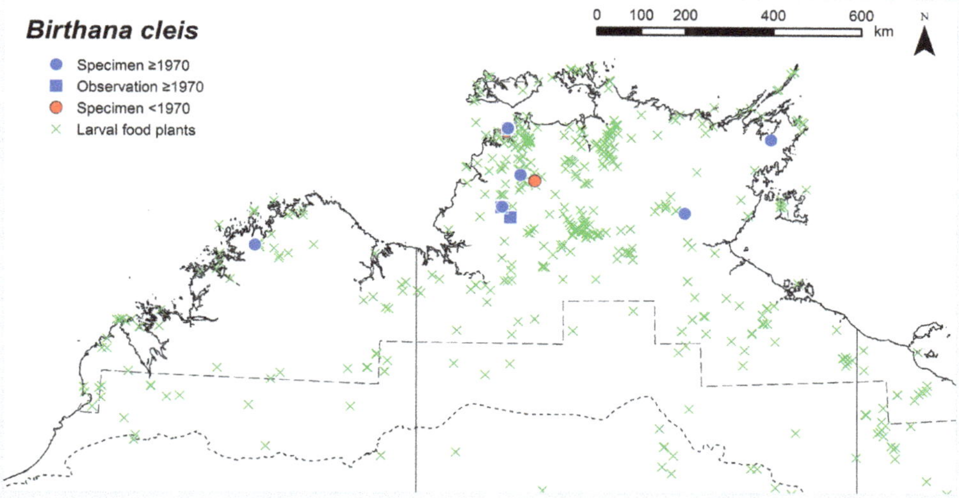

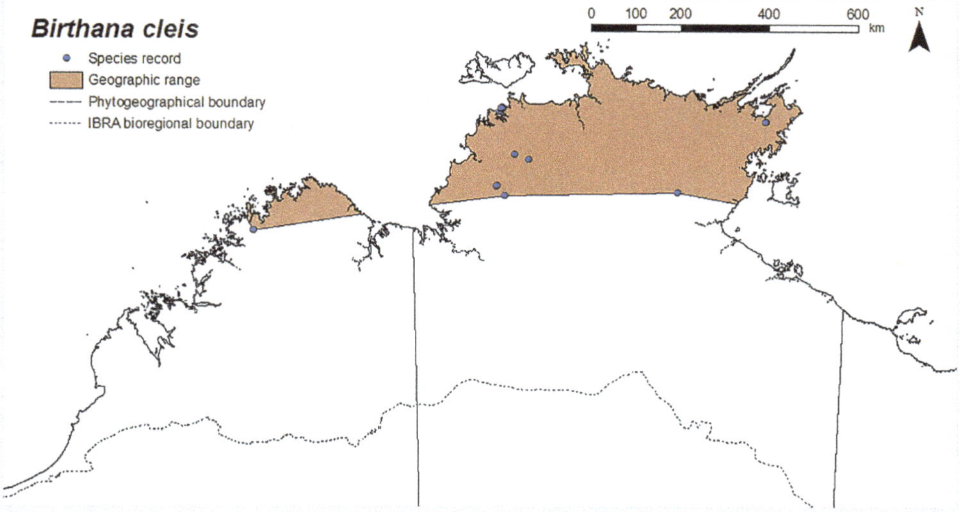

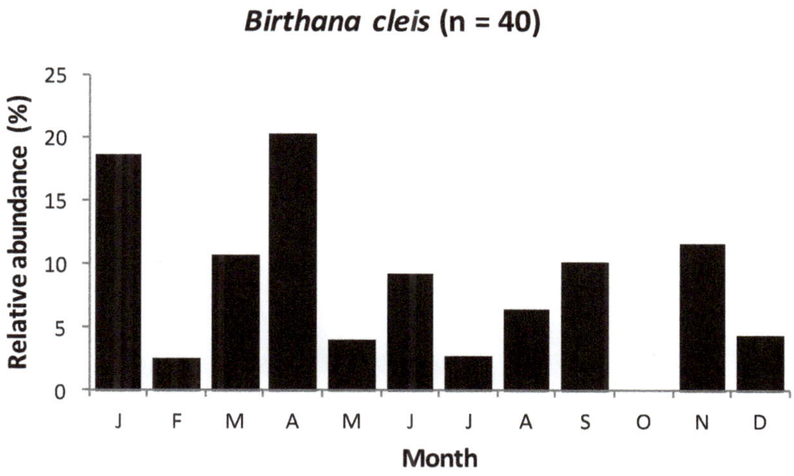

Birthana cleis
Specimen ≥1970
Observation ≥1970
Specimen <1970
Larval food plants
0 100 200 400 600 km
N

Birthana cleis
Species record
Geographic range
Phytogeographical boundary
IBRA bioregional boundary
0 100 200 400 600 km
N

Birthana cleis (n = 40)
Relative abundance (%)
25
20
15
10
5
0
J F M A M J J A S O N D
Month

| Month | J | F | M | A | M | J | J | A | S | O | N | D |
|---|---|---|---|---|---|---|---|---|---|---|---|---|
| Egg | | | | | | | | | | | | |
| Larva | | | | | | | | | | | | |
| Pupa | | | | | | | | | | | | |
| Adult | | | | | | | | | | | | |

# Flame Geometer Moth

## *Ctimene* sp. 'Top End'

Plate 195 Mosquito Creek, Gove Peninsula, NT
Photo: M. F. Braby

Plate 196 Riyala, Elizabeth Valley, NT
Photo: Ian Morris

## Distribution

This undescribed species is endemic to the study region. It is restricted to the north of the Northern Territory, where it occurs in the higher rainfall areas (> 1,200 mm mean annual rainfall) of the Top End.

## Habitat

The breeding habitat of *Ctimene* sp. 'Top End' has not been recorded in the study region. Adults have been recorded only in semi-deciduous monsoon vine thicket and riparian evergreen monsoon vine forest in which they no doubt breed. The species also occurs in the suburbs and rural areas of Darwin, NT, where suitable patches of monsoon forest habitat persists.

## Larval food plants

Not recorded in the study region.

## Seasonality

The seasonal abundance and breeding phenology of this species are not well understood. Adults have been recorded during most months of the year, but there are too few records (n = 14) to assess any seasonal changes in abundance. In general, adults tend to be more numerous during the mid dry season (July–September). The seasonal history of the immature stages is not known.

## Breeding status

This species is resident in the study region.

## Conservation status

LC. Although the species *Ctimene* sp. 'Top End' has a restricted range, there are no known threats facing the taxon.

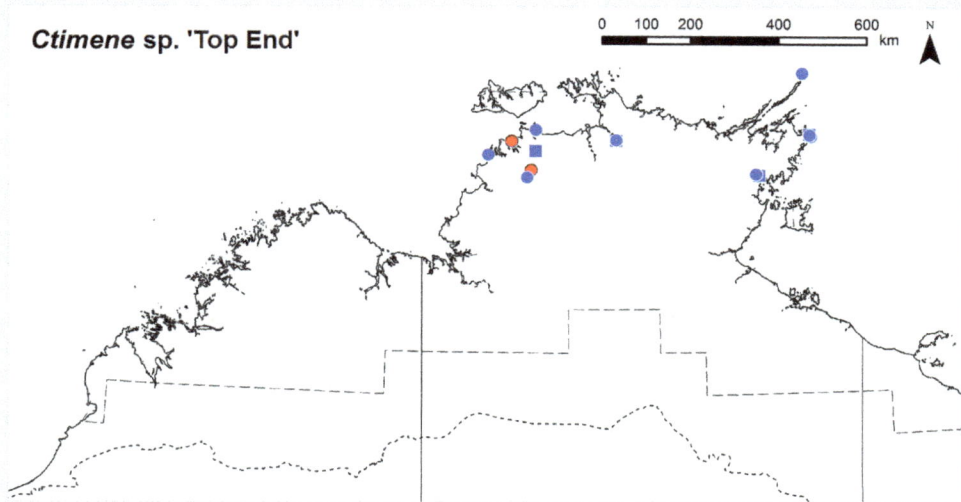

***Ctimene*** sp. 'Top End'

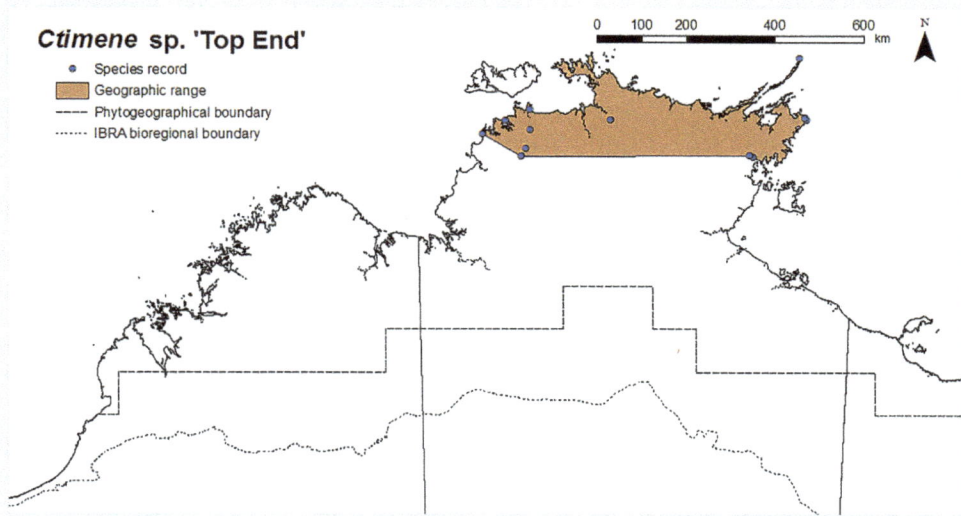

***Ctimene*** sp. 'Top End'

- Species record
- Geographic range
- Phytogeographical boundary
- IBRA bioregional boundary

| Month | J | F | M | A | M | J | J | A | S | O | N | D |
|---|---|---|---|---|---|---|---|---|---|---|---|---|
| Egg | | | | | | | | | | | | |
| Larva | | | | | | | | | | | | |
| Pupa | | | | | | | | | | | | |
| Adult | | | | | | | | | | | | |

# Six O'clock Moth

*Dysphania numana* (Cramer, 1779)

Plate 197 Marrara, Darwin, NT
Photo: M. F. Braby

## Distribution

This species occurs in the northern Kimberley and in the Top End of the study region. It generally occurs in the higher rainfall areas (> 900 mm mean annual rainfall). It has been recorded as far south as the Prince Regent Nature Reserve (Youwanjela Creek and Roe River), WA (L. Scott-Virtue and S. Black); and the Bradshaw Field Training Area (Fitzmaurice River catchment) (Archibald and Braby 2017) and Ngukurr (S. Normand), NT. Its geographical range corresponds well with the spatial distribution of its larval food plant, although the food plant extends slightly further inland into lower rainfall areas. Outside the study region, *D. numana* occurs from Maluku, Tanimbar and the Kai and Aru islands, through mainland New Guinea and adjacent islands and north-eastern Australia to the Bismarck Archipelago, Bougainville and the Solomon Islands.

The taxonomic status of *Dysphania numana* in northern Australia requires closer scrutiny. The adults and immature stages (larvae) are phenotypically distinct from populations elsewhere in the range, and it is likely more than one species is involved in the complex, with the population in the Kimberley and Top End probably comprising a separate species (Prout 1921–34).

## Habitat

*Dysphania numana* breeds in a variety of wet and dry monsoon forests where the larval food plant grows as a small to medium-sized evergreen tree. Typical habitats include semi-deciduous monsoon vine thicket on coastal lateritic cliffs and in drier inland rocky sandstone escarpments and gullies, evergreen monsoon vine forest associated with permanent freshwater streams and perennial springs, mixed riparian monsoon forest–woodland along creeks and mixed tall paperbark swampland with rainforest elements in the understorey. Adults regularly disperse into savannah woodland, but they rarely breed in this habitat.

## Larval food plant

*Carallia brachiata* (Rhizophoraceae).

## Seasonality

Adults occur throughout the year. They appear to show little seasonal variation in abundance, although they tend to be more numerous during the early to mid dry season (May–August), but they may also be common during the late wet season. The immature stages have been recorded during most months of the year except during the 'build-up' (October and November); they may be numerous during the mid to late wet season and early dry season (January–June), when the food plant produces new leaf growth. The species breeds continuously throughout the year and several generations are completed annually.

## Breeding status

This species is resident in the study region.

## Conservation status

LC.

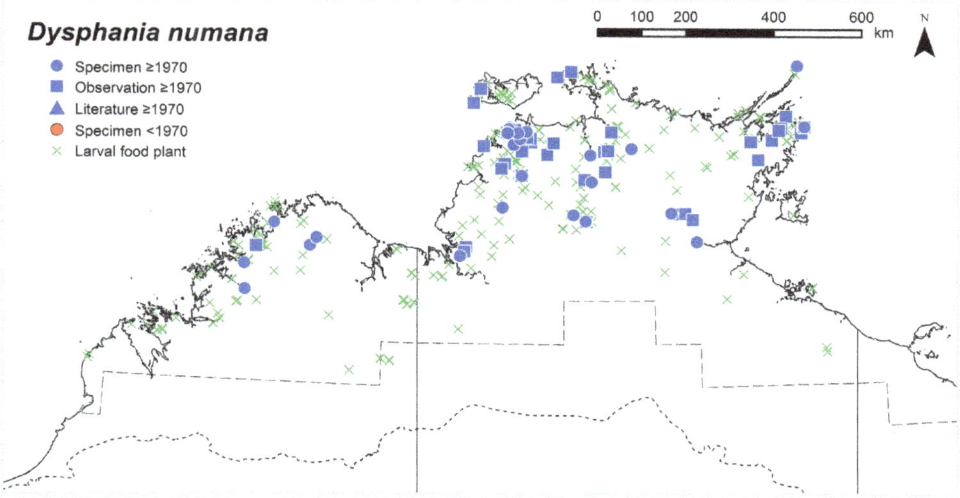

*Dysphania numana*
0 100 200 400 600 km
N
Specimen ≥1970
Observation ≥1970
Literature ≥1970
Specimen <1970
Larval food plant

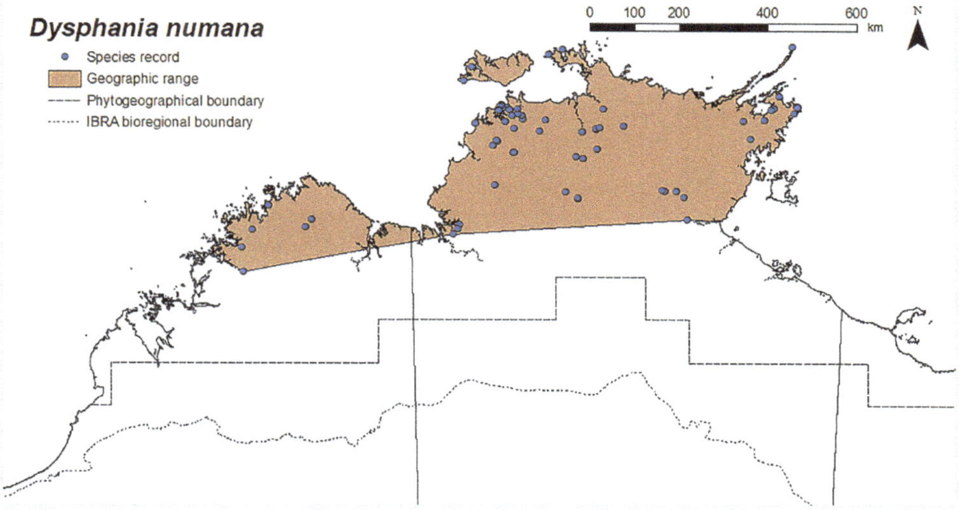

*Dysphania numana*
0 100 200 400 600 km
N
Species record
Geographic range
Phytogeographical boundary
IBRA bioregional boundary

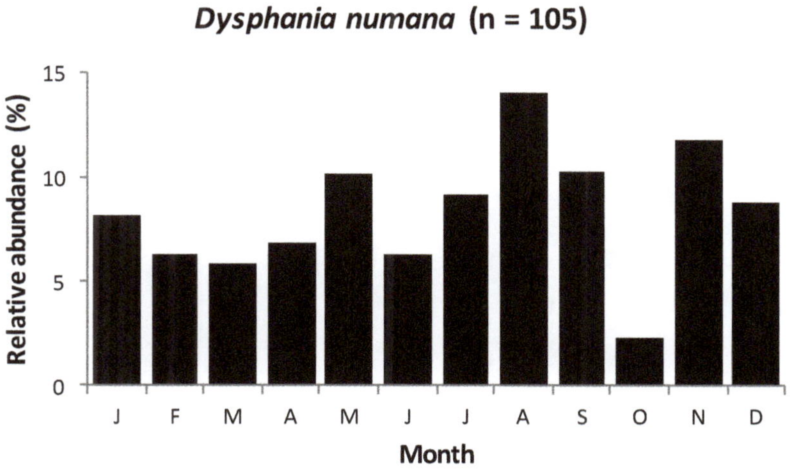

*Dysphania numana* (n = 105)
Relative abundance (%)
15
10
5
0
J F M A M J J A S O N D
Month

| Month | J | F | M | A | M | J | J | A | S | O | N | D |
|---|---|---|---|---|---|---|---|---|---|---|---|---|
| Egg | | | | | | | | | | | | |
| Larva | | | | | | | | | | | | |
| Pupa | | | | | | | | | | | | |
| Adult | | | | | | | | | | | | |

# Zodiac Moth
## *Alcides metaurus* (Hopffer, 1856)

Plate 198 Mackay, Qld
Photo: M. F. Braby

## Distribution

This species is known only from the Top End of the study region. It has been recorded at two locations in eastern Arnhem Land, NT: Elcho Island (Galiwinku) in the Wessel Islands, based on several observations and photographs during the 'build-up' between 1971 and 1976 by I. Morris; and at Nhulunbuy (Drimmie Head) on Gove Peninsula, based on a specimen collected in October 2006 by L. Wilson (Braby 2014a). Outside the study region, *A. metaurus* occurs in northern and north-eastern Queensland.

## Habitat

The breeding habitat of *A. metaurus* has not been recorded in the study region.

## Larval food plants

Not recorded in the study region; possibly *Endospermum myrmecophilum* (Euphorbiaceae). In north-eastern Queensland, the food plants include *Omphalea* spp. and *Endospermum* spp., all of which are tropical rainforest vines or tall trees (Coleman and Monteith 1981; Monteith and Wood 1987; Harrison 2010; Moss 2010). None of these species occurs on Gove Peninsula or the Wessel Islands, NT, and only one (*E. myrmecophilum*) occurs in the Northern Territory, where it is restricted to the Tiwi Islands and north-western corner of the Top End.

## Seasonality

The seasonal abundance and breeding phenology of this species are not well understood. Adults have been recorded only during the late dry season (October and November). In Queensland, the species is well known for its overwintering aggregations and migratory flights (Smithers and Peters 1977; Coleman and Monteith 1981).

## Breeding status

*Alcides metaurus* does not appear to have become permanently established in the study region. It is likely the records from north-eastern Arnhem Land represent vagrants that dispersed from northern Queensland across the Gulf of Carpentaria to the Northern Territory associated with large-scale migration of adults following the breakup of their winter dry season aggregations.

## Conservation status

NA.

## Alcides metaurus

● Specimen ≥1970
■ Observation ≥1970

| Month | J | F | M | A | M | J | J | A | S | O | N | D |
|-------|---|---|---|---|---|---|---|---|---|---|---|---|
| Egg   |   |   |   |   |   |   |   |   |   |   |   |   |
| Larva |   |   |   |   |   |   |   |   |   |   |   |   |
| Pupa  |   |   |   |   |   |   |   |   |   |   |   |   |
| Adult |   |   |   |   |   |   |   |   |   | ■ | ■ |   |

Photo: NE Arnhem Land, NT, M.F. Braby

# Scarlet Tiger Moth

## *Euchromia creusa* (Linnaeus, 1758)

Plate 199 Oenpelli, NT
Photo: Ian Morris

## Distribution

This species is known from only a single record in the Top End of the study region, where I. Morris photographed a specimen near Oenpelli, NT, in December 1993 (Braby 2014a). Outside the study region, *E. creusa* occurs from Ambon and the Kai Islands, through mainland New Guinea and north-eastern Australia to New Caledonia, Vanuatu and Fiji.

## Habitat

The breeding habitat of *E. creusa* has not been recorded in the study region. The single individual encountered was perched on a rock in sandstone country within the Arnhem Land Plateau.

## Larval food plants

Not recorded in the study region.

## Seasonality

The seasonal abundance and breeding phenology of this species are not well understood. The single individual was recorded during the early wet season (December). Presumably, adults are highly seasonal and local in occurrence.

## Breeding status

*Euchromia creusa* is assumed to be resident in the study region.

## Conservation status

DD. The species *E. creusa* is currently known from only one location in the study region (AOO is likely to be < 2,000 sq km, with spatial buffering of records providing a first approximation of 700 sq km). Although it has been recorded from a conservation reserve (Warddeken IPA), further investigations are required to clarify the ecological requirements, critical habitat and distribution of this species.

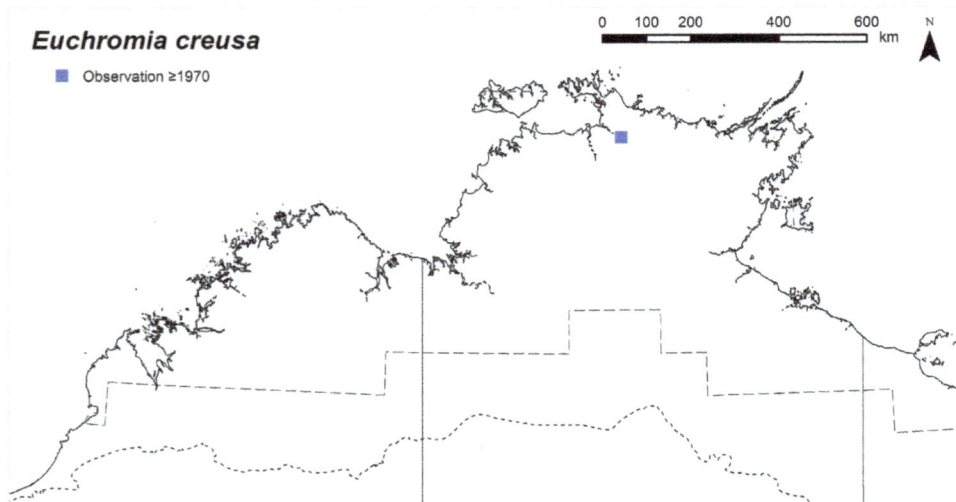

*Euchromia creusa*

■ Observation ≥1970

| Month | J | F | M | A | M | J | J | A | S | O | N | D |
|-------|---|---|---|---|---|---|---|---|---|---|---|---|
| Egg   |   |   |   |   |   |   |   |   |   |   |   |   |
| Larva |   |   |   |   |   |   |   |   |   |   |   |   |
| Pupa  |   |   |   |   |   |   |   |   |   |   |   |   |
| Adult |   |   |   |   |   |   |   |   |   |   |   | ■ |

# Galaxy Day-moth
## Genus 1 sp. 'Sandstone'

Plate 200 Kakadu National Park, NT
Photo: M. F. Braby

## Distribution

This undescribed species, which has not yet been assigned to a genus according to material curated in the ANIC, has a very wide but sporadic distribution within the study region. It occurs in both high and low rainfall areas (600–1,500 mm mean annual rainfall), extending from moist coastal areas to the semi-arid zone. It has been recorded as far south-west as Broome, WA (M. S. Moulds); as far north as the Arnhem Land Plateau, NT (15 km south-west by south of Nimbuwah Rock) (E. D. Edwards and M. S. Upton); and as far south-east as Boodjamulla/Lawn Hill National Park, Qld (Amphitheatre Springs 27 km north of Musselbrook Camp) (E. D. Edwards). Further field surveys are required to determine whether Genus 1 sp. 'Sandstone' is more widely distributed than available records indicate, particularly in the Kimberley and the Limmen Bight area in the Gulf of Carpentaria. Outside the study region, Genus 1 sp. 'Sandstone' occurs in northern Queensland.

## Habitat

The breeding habitat of Genus 1 sp. 'Sandstone' has not been recorded. All specimens have been collected in eucalypt heathy woodland associated with sandstone breakaways, escarpments and rocky outcrops, in which they no doubt breed.

## Larval food plants

Not recorded in the study region.

## Seasonality

Adults are seasonal, occurring mainly during the warmer months of high humidity and high temperatures (October–May), with a peak in abundance during the early wet season (November–January), depending on the timing of the pre-monsoon storms. The breeding phenology and seasonal history of the immature stages have not been recorded, but it is possible there is only a single generation annually, followed by a partial second generation in the late wet season. Presumably, the species survives the dry season in pupal diapause.

## Breeding status

This species is resident in the study region.

## Conservation status

LC.

## Genus 1 sp. 'Sandstone'

● Specimen ≥1970
■ Observation ≥1970

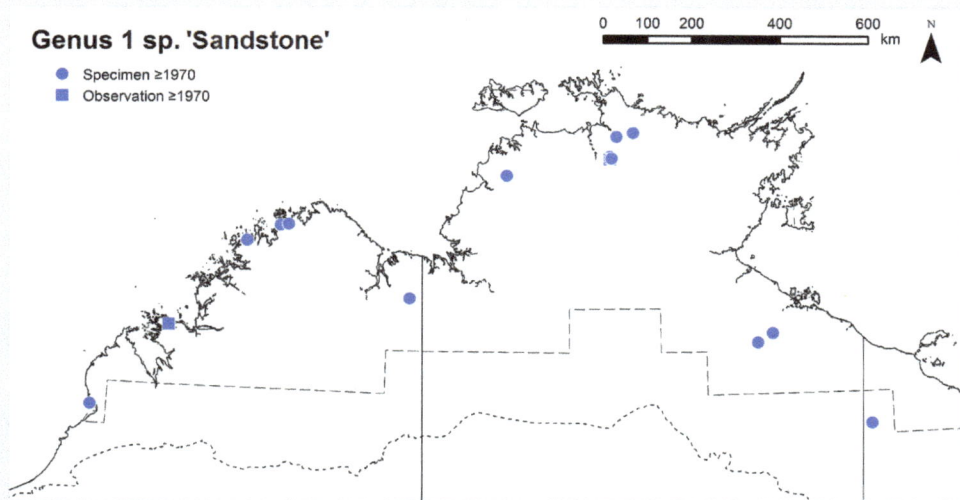

## Genus 1 sp. 'Sandstone'

● Species record
▨ Geographic range
--- Phytogeographical boundary
······ IBRA bioregional boundary

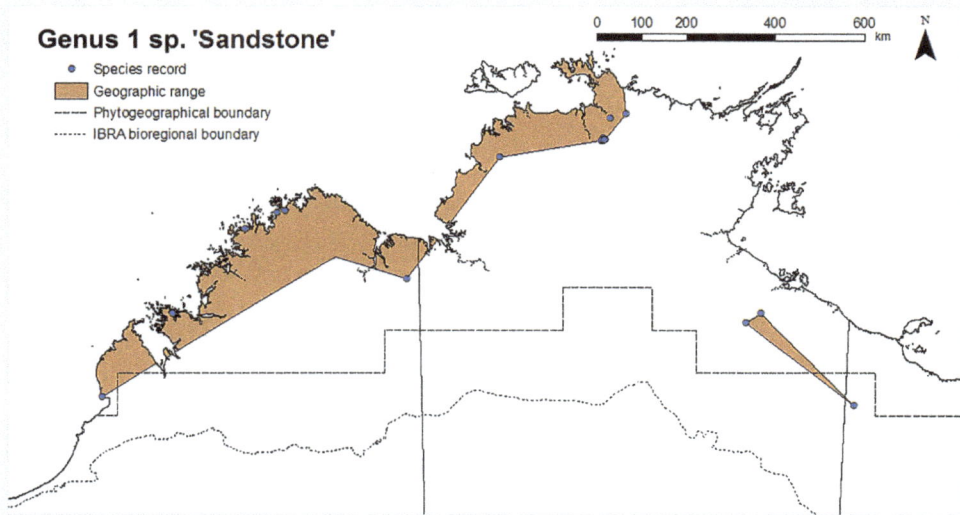

## Genus 1 sp. 'Sandstone' (n = 21)

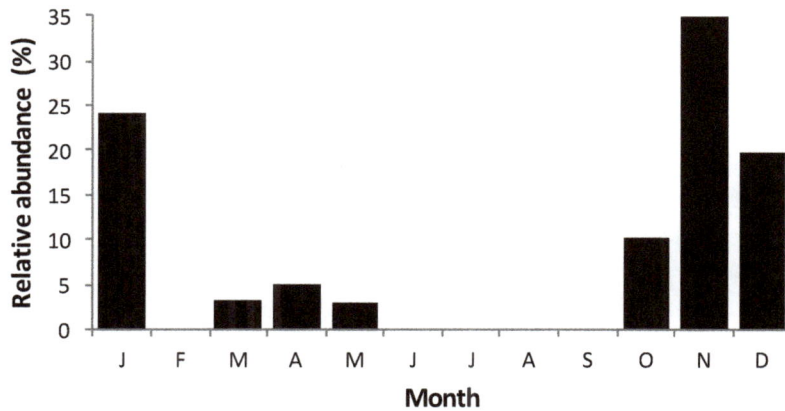

| Month | J | F | M | A | M | J | J | A | S | O | N | D |
|-------|---|---|---|---|---|---|---|---|---|---|---|---|
| Egg | | | | | | | | | | | | |
| Larva | | | | | | | | | | | | |
| Pupa | | | | | | | | | | | | |
| Adult | | | | | | | | | | | | |

# Pale Banded Day-moth
## *Leucogonia cosmopis* (Lower, 1897)

Plate 201 Dundee Beach, NT
Photo: M. F. Braby

Plate 202 Black Point, Cobourg Peninsula, NT
Photo: M. F. Braby

## Distribution

This species occurs in the northern and eastern Kimberley and western half of the Top End of the study region. It occurs mainly in higher rainfall areas, although it has been recorded in drier areas (c. 800 mm mean annual rainfall) at Keep River National Park, NT (Cockatoo Creek) (G. Cocking, Y. N. Su and A. Zwick). Further field surveys are required to determine whether *L. cosmopis* occurs in the eastern half of the Top End. Outside the study region, *L. cosmopis* occurs in northern Queensland.

## Habitat

The breeding habitat of *L. cosmopis* has not been recorded in the study region. Adults have been collected mostly at night by light traps set in savannah woodland and semi-deciduous monsoon vine thicket.

## Larval food plants

Not recorded in the study region.

## Seasonality

Adults are seasonal, occurring mainly during the wet season (November–March), but there are too few records (n = 19) to assess temporal changes in abundance. In general, adults tend to be more numerous during the early wet season (December and January). The breeding phenology and seasonal history of the immature stages have not been recorded, but it is possible there is only a single generation annually, followed by a partial second generation. Presumably, the species survives the dry season in pupal diapause.

## Breeding status

This species is resident in the study region.

## Conservation status

LC.

## Leucogonia cosmopis

- Specimen ≥1970
- Specimen <1970

## Leucogonia cosmopis

- Species record
- Geographic range
- — — Phytogeographical boundary
- ······ IBRA bioregional boundary

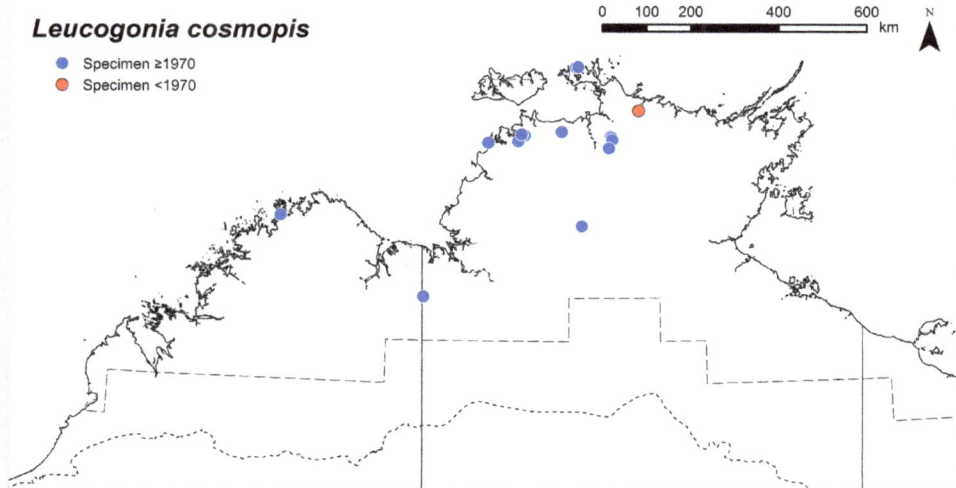

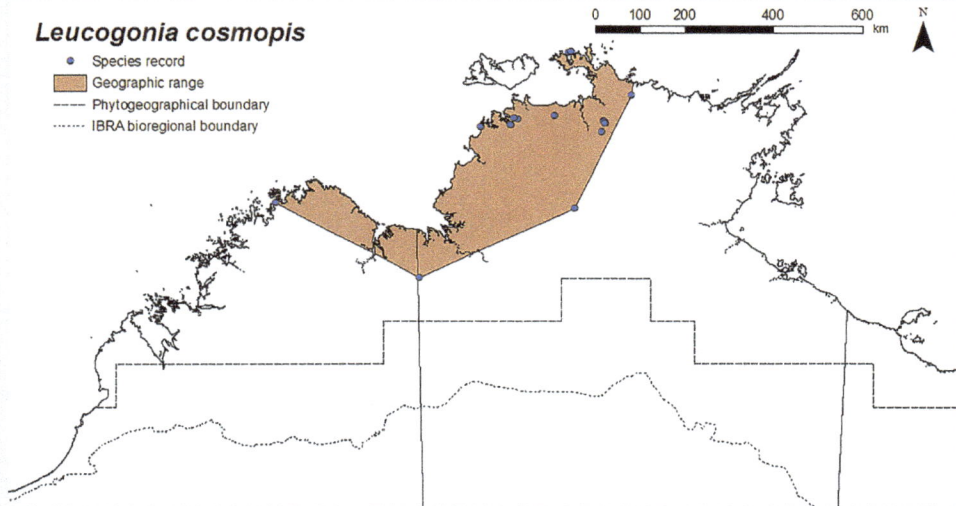

| Month | J | F | M | A | M | J | J | A | S | O | N | D |
|-------|---|---|---|---|---|---|---|---|---|---|---|---|
| Egg   |   |   |   |   |   |   |   |   |   |   |   |   |
| Larva |   |   |   |   |   |   |   |   |   |   |   |   |
| Pupa  |   |   |   |   |   |   |   |   |   |   |   |   |
| Adult |   |   |   |   |   |   |   |   |   |   |   |   |

# Laced Day-moth
*Ipanica cornigera* (Butler, 1886)

Plate 203 Black Point, Cobourg Peninsula, NT
Photo: M. F. Braby

Plate 204 Black Point, Cobourg Peninsula, NT
Photo: M. F. Braby

## Distribution

This species occurs in the eastern Kimberley, the Top End, Northern Deserts and western Gulf Country of the study region. It extends very widely from moist coastal areas to drier inland areas of the semi-arid zone (< 400 mm mean annual rainfall), where it has been recorded as far south as Three Ways Roadhouse 25 km north of Tennant Creek, NT (M. S. Upton). It also occurs in the arid zone of central Australia beyond the southern boundary of the study region. Further field surveys are required to determine whether *I. cornigera* occurs more widely in the Kimberley, eastern Arnhem Land and western Queensland in the Gulf Country. Outside the study region, *I. cornigera* occurs from Indonesia, through mainland New Guinea to central, north-eastern and eastern Australia.

## Habitat

The breeding habitat of *I. cornigera* has not been recorded in the study region. Adults have been collected mostly at night by light traps set in savannah woodland and semi-deciduous monsoon vine thicket, often in disturbed areas.

## Larval food plants

Not recorded in the study region.

## Seasonality

Adults are seasonal, occurring during the late dry season and wet season (August–April), but they are most abundant during the mid wet season (January and February) following the first monsoon rains. The breeding phenology and seasonal history of the immature stages have not been recorded, but it is possible there are only one or two generations annually. Presumably, the species survives the dry season in pupal diapause.

## Breeding status

This species is resident in the study region.

## Conservation status

LC.

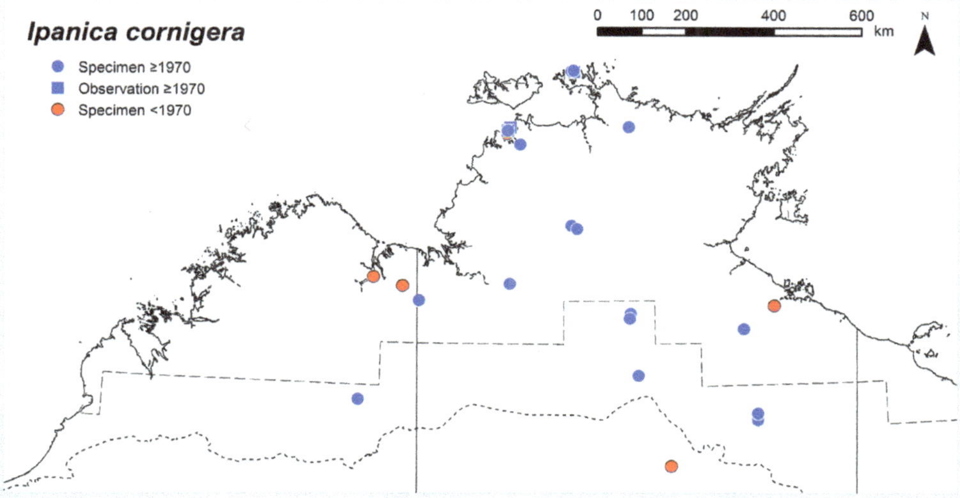

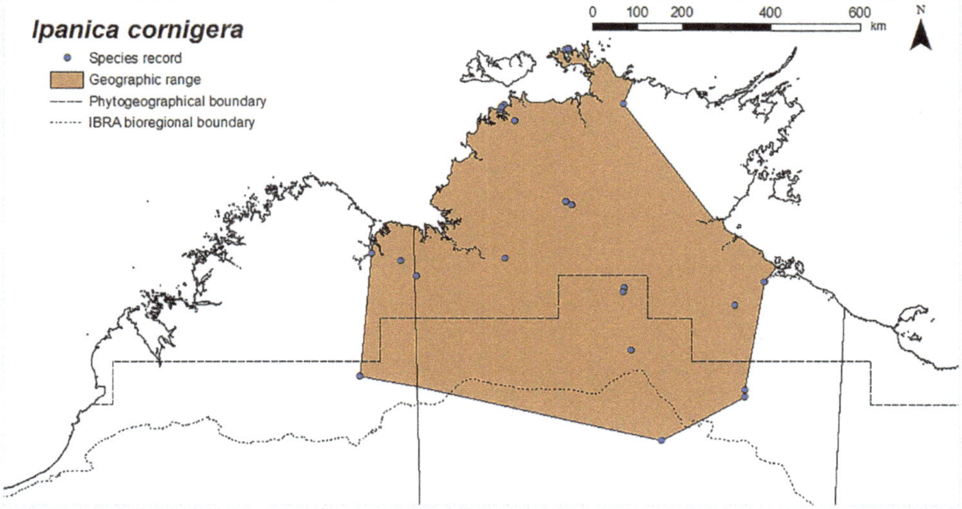

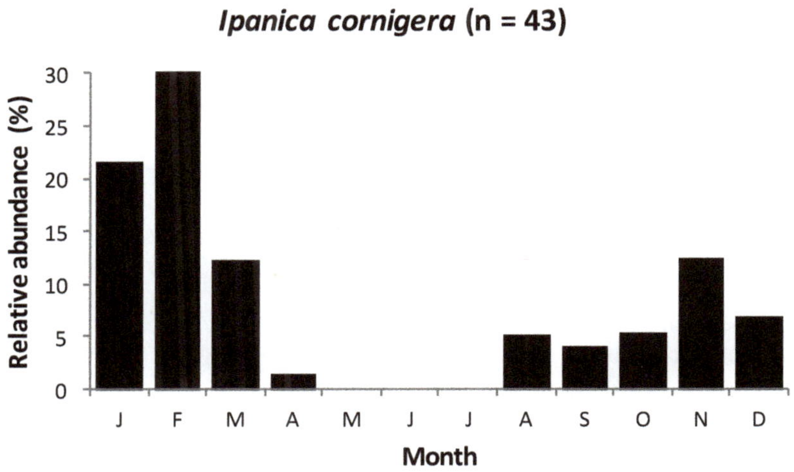

Ipanica cornigera
Specimen ≥1970
Observation ≥1970
Specimen <1970
0 100 200 400 600
km
N

Ipanica cornigera
Species record
Geographic range
Phytogeographical boundary
IBRA bioregional boundary
0 100 200 400 600
km
N

*Ipanica cornigera* (n = 43)
Relative abundance (%)
30
25
20
15
10
5
0
J F M A M J J A S O N D
Month

| Month | J | F | M | A | M | J | J | A | S | O | N | D |
|---|---|---|---|---|---|---|---|---|---|---|---|---|
| Egg | | | | | | | | | | | | |
| Larva | | | | | | | | | | | | |
| Pupa | | | | | | | | | | | | |
| Adult | | | | | | | | | | | | |

# Orange-banded Day-moth

## *Periopta diversa* (Walker, [1865])

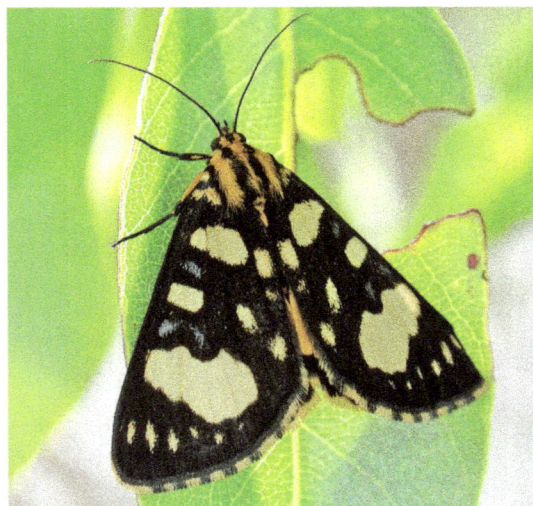

Plate 205 Leanyer, Darwin, NT
Photo: M. F. Braby

Plate 206 Leanyer, Darwin, NT
Photo: M. F. Braby

## Distribution

This species occurs in the northern and eastern Kimberley, the Top End, Northern Deserts and western Gulf Country of the study region. It extends widely from moist coastal areas to drier inland areas of the semi-arid zone (< 500 mm mean annual rainfall), where it has been recorded as far south as the Barkly Tableland, NT (c. 140 km north of Barkly Homestead Roadhouse) (M. S. Moulds). The geographic range falls within the spatial distribution of the known and putative larval food plants (*Spermacoce* spp.), which occur very widely throughout the study region. Further field surveys are thus required to determine whether *P. diversa* occurs more widely in the southern Kimberley, the eastern half of the Top End and western Queensland in the Gulf Country. Outside the study region, *P. diversa* occurs in north-eastern Queensland.

## Habitat

*Periopta diversa* breeds in grassland and open woodland, favouring open disturbed areas of bare ground where the larval food plants grow as pioneer annual herbs (Braby 2011a, 2015e).

## Larval food plants

*Spermacoce phalloides* (Rubiaceae); also *Spermacoce articularis*, *Oldenlandia corymbosa* (Rubiaceae); probably other *Spermacoce* spp.

## Seasonality

Adults are seasonal, occurring only during the wet season (November–April), with a peak in abundance in December and January. The start of the flight season is triggered by the first substantial pre-monsoon storms, but the timing of adult emergence varies from year to year, depending on when the wet season commences. The immature stages (eggs or larvae) have been recorded only from December to February. The larvae feed on the foliage of their food plants, which are short-lived annuals, and it is likely only one or two generations are completed during the flight season, depending on the duration of the wet season. Presumably, the species, like *Periopta ardescens* and other agaristines, survives the dry season in pupal diapause.

## Breeding status

This species is resident in the study region.

## Conservation status

LC.

## Periopta diversa

- ● Specimen ≥1970
- ■ Observation ≥1970
- ● Specimen <1970
- × Larval food plants

## Periopta diversa

- ● Species record
- ▧ Geographic range
- – – Phytogeographical boundary
- ······ IBRA bioregional boundary

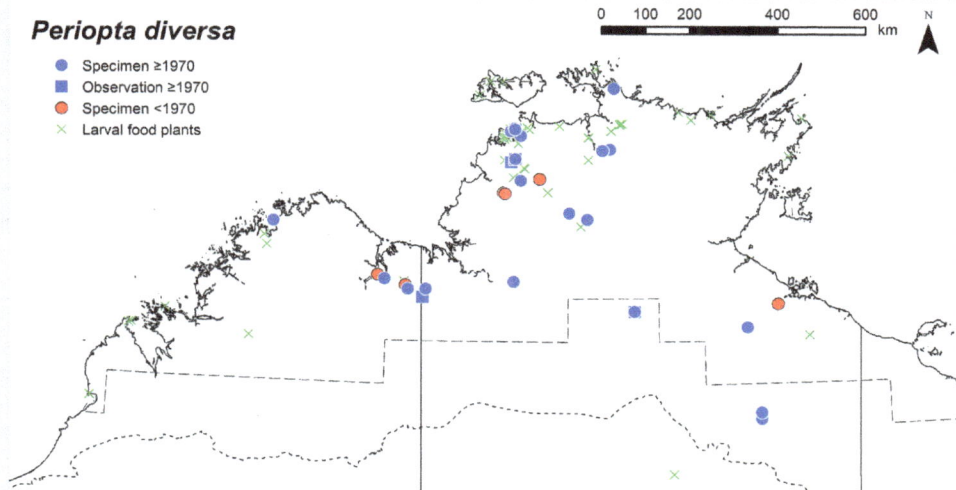

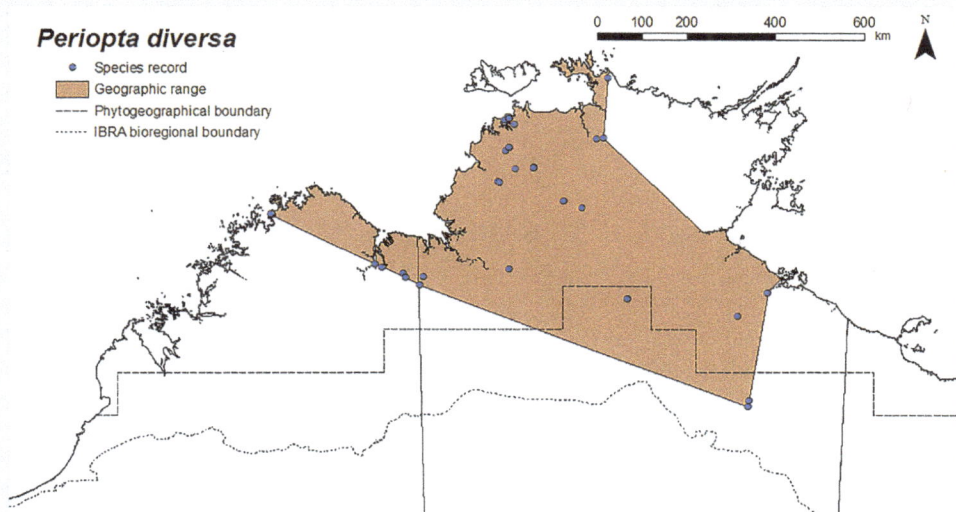

### Periopta diversa (n = 42)

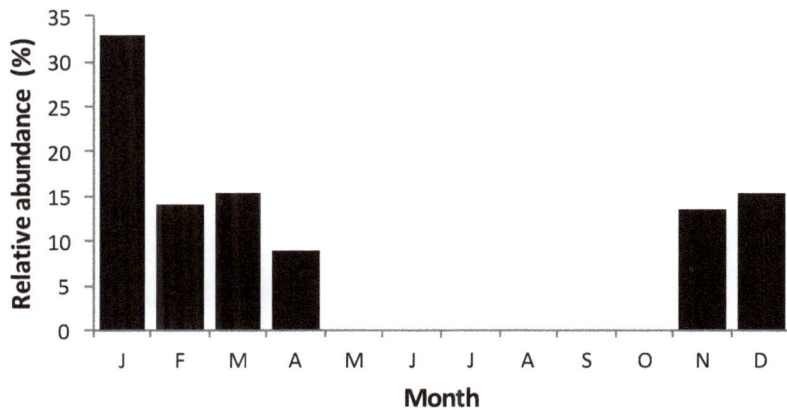

| Month | J | F | M | A | M | J | J | A | S | O | N | D |
|-------|---|---|---|---|---|---|---|---|---|---|---|---|
| Egg | | | | | | | | | | | | |
| Larva | | | | | | | | | | | | |
| Pupa | | | | | | | | | | | | |
| Adult | | | | | | | | | | | | |

# Chestnut Day-moth
## *Periopta ardescens* (Butler, 1884)

Plate 207 Timber Creek, NT
Photo: M. F. Braby

Plate 208 Dundee Beach, NT
Photo: M. F. Braby

## Distribution

This species occurs in the northern and eastern Kimberley and the western half of the Top End of the study region. It occurs mainly in the higher rainfall areas, although it has been recorded in drier inland areas (c. 800 mm mean annual rainfall), where it has been collected as far south as Kununurra, WA (M. S. Moulds and B. J. Moulds); and 11 km south–south-east of Timber Creek, NT. The geographic range falls within the spatial distribution of its larval food plants, which occur more widely in the southern Kimberley, Tiwi Islands, eastern half of the Top End, Groote Eylandt and coastal islands of the western Gulf Country. Further field surveys are thus required to determine whether *P. ardescens* occurs in these areas. Outside the study region, *P. ardescens* occurs in north-eastern Queensland.

## Habitat

*Periopta ardescens* breeds along the edges of semi-deciduous monsoon vine thicket and in savannah woodland where the larval food plants grow either as vines (*Ampelocissus acetosa*) or as shrubs (*A. frutescens*) (Braby 2011a).

## Larval food plants

*Ampelocissus acetosa*, *A. frutescens* (Vitaceae).

## Seasonality

Adults are seasonal, occurring only during the early to mid wet season (October–February), with a peak in abundance in December. The immature stages (eggs or larvae) occur mainly in November and December. The larvae feed on the new soft leaf growth of the food plants, which regenerate from underground tubers during the 'build-up' following the pre-monsoon storms and/or an increase in humidity. The food plants are short-lived seasonal perennial vines or shrubs and they senesce towards the end of the wet season and early dry season. The duration of the pupal stage is variable and there is probably only a single generation annually. The species survives the dry season in the pupal stage, which may remain in diapause for one to two years (Braby 2011a).

## Breeding status

This species is resident in the study region.

## Conservation status

LC.

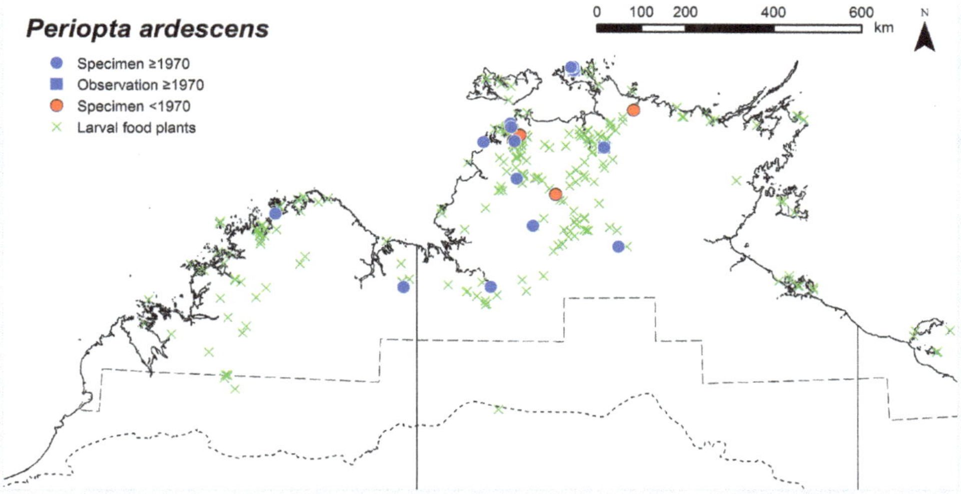

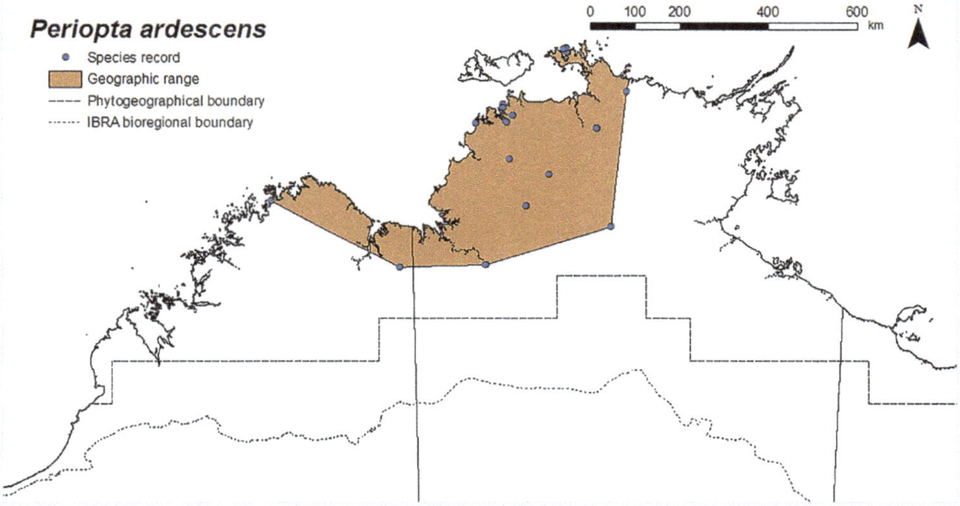

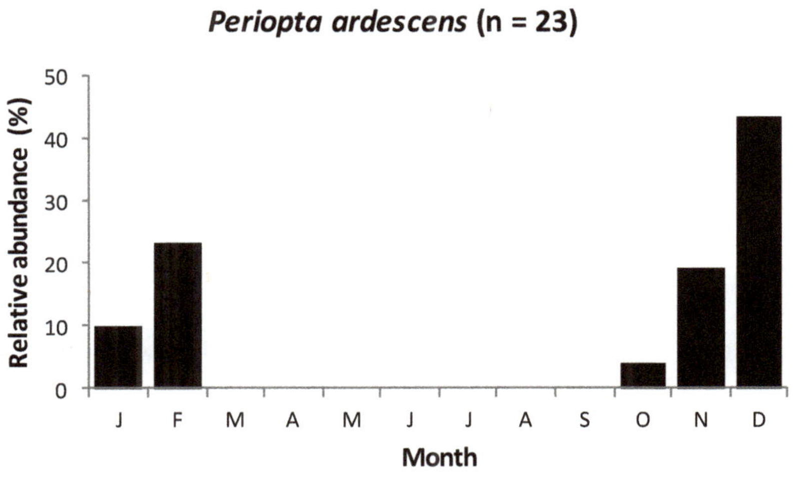

## Periopta ardescens

## Periopta ardescens

## Periopta ardescens (n = 23)

| Month | J | F | M | A | M | J | J | A | S | O | N | D |
|-------|---|---|---|---|---|---|---|---|---|---|---|---|
| Egg | | | | | | | | | | | | |
| Larva | | | | | | | | | | | | |
| Pupa | | | | | | | | | | | | |
| Adult | | | | | | | | | | | | |

# Pearl Day-moth

## *Radinocera vagata* (Walker, 1865)

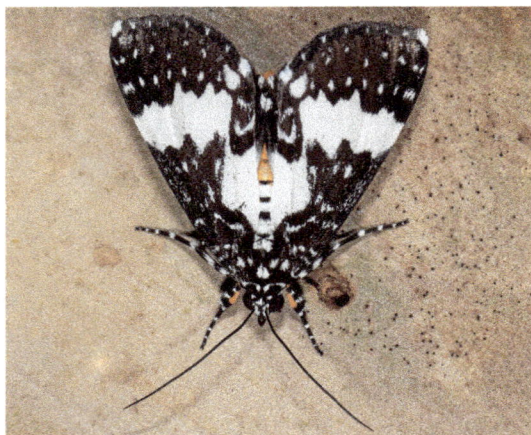

Plate 209 Dundee Beach, NT
Photo: M. F. Braby

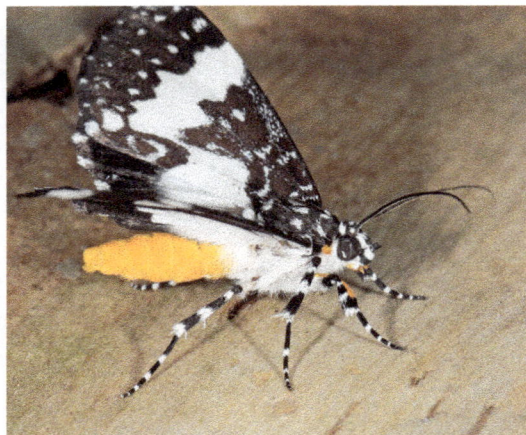

Plate 210 Dundee Beach, NT
Photo: M. F. Braby

## Distribution

This species occurs in the northern and eastern Kimberley, the Top End, Northern Deserts and western Gulf Country of the study region. It extends from moist coastal areas to drier inland areas of the semi-arid zone (c. 500 mm mean annual rainfall), where it has been recorded as far south as Elliott, NT (I. Archibald). The geographic range falls within the spatial distribution of its larval food plants. The food plants, however, occur more widely in the southern Kimberley, Tiwi Islands, eastern half of the Top End, Groote Eylandt and western Queensland in the Gulf Country; thus, further field surveys are required to determine whether *R. vagata* occurs in these areas. Outside the study region, *R. vagata* occurs from the Lesser Sunda Islands, through mainland New Guinea to north-eastern Queensland.

## Habitat

*Radinocera vagata* breeds mainly in savannah woodland and coastal paperbark woodland or swampland where the larval food plants grow either as vines (*Ampelocissus acetosa*, *Cayratia* spp.) or as shrubs (*A. frutescens*) (Braby 2011a, 2015e). It also breeds along the edges of coastal semi-deciduous monsoon vine thicket.

## Larval food plants

*Ampelocissus acetosa*, *A. frutescens*, *Cayratia trifolia*, *C. maritima* (Vitaceae).

## Seasonality

Adults are seasonal, occurring only during the 'build-up' and wet season (October–April), but they are most abundant in November and December following the pre-monsoon storms. The immature stages (larvae) have been recorded from October to March, but they are more prevalent earlier in the wet season (November–January), depending on the timing of rainfall. The larvae feed on the new soft leaf growth of the food plants, which regenerate from underground tubers during the 'build-up' following the pre-monsoon storms and/or an increase in humidity. The food plants are short-lived seasonal perennial vines or shrubs and they senesce towards the end of the wet season and early dry season. The duration of the pupal stage is variable and there is probably only a single generation annually, followed by a partial second generation in the late wet season. The species survives the dry season in the pupal stage, which may remain in diapause for up to 10 months.

## Breeding status

This species is resident in the study region.

## Conservation status

LC.

## Radinocera vagata

- ● Specimen ≥1970
- ■ Observation ≥1970
- ● Specimen <1970
- × Larval food plants

## Radinocera vagata

- ● Species record
- Geographic range
- – – – Phytogeographical boundary
- ········ IBRA bioregional boundary

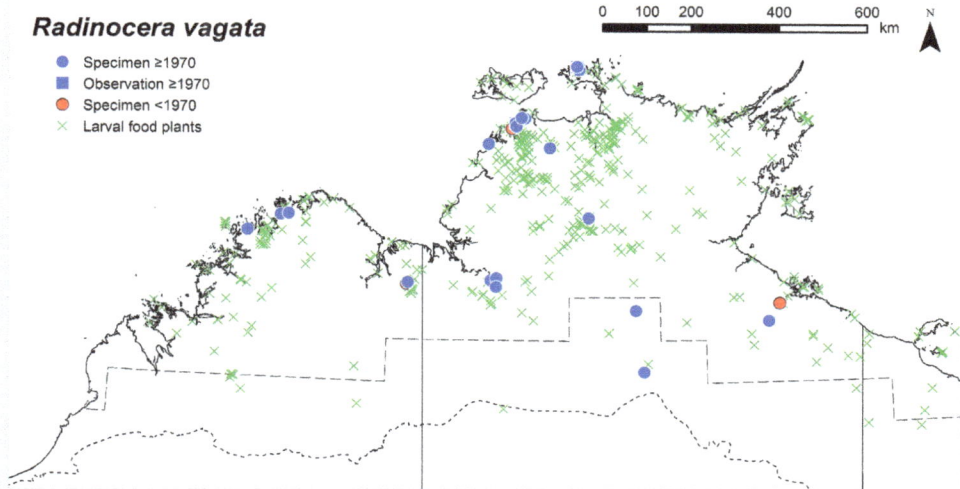

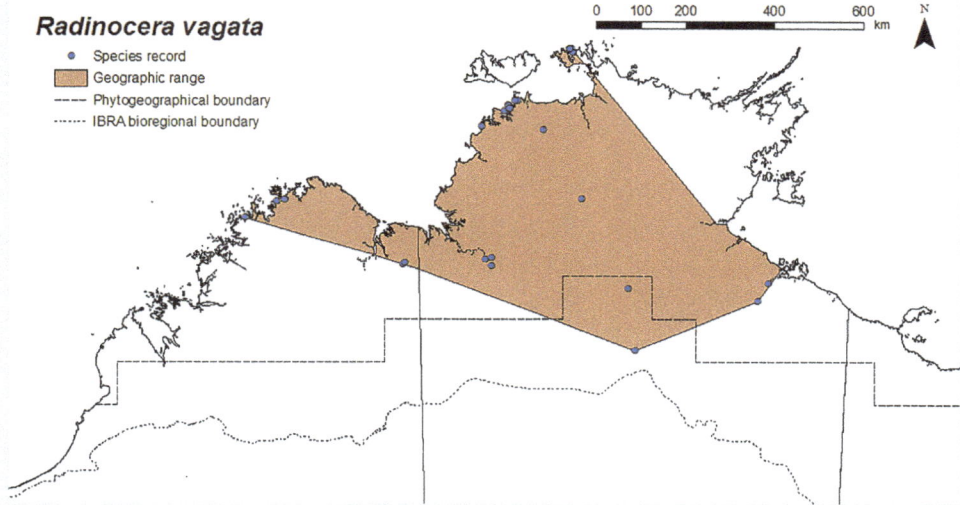

## Radinocera vagata (n = 34)

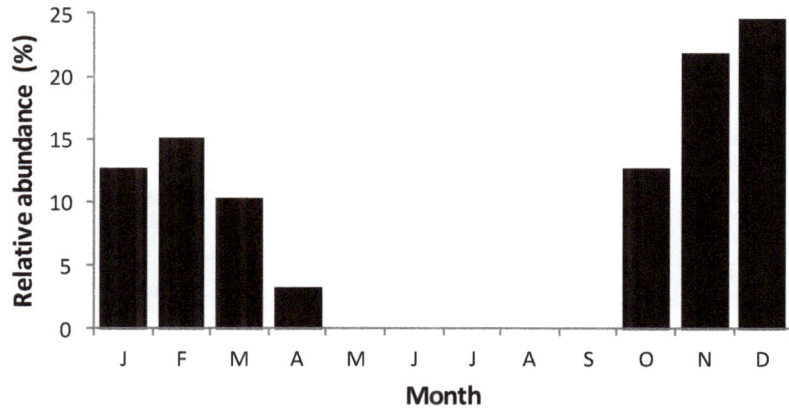

| Month | J | F | M | A | M | J | J | A | S | O | N | D |
|-------|---|---|---|---|---|---|---|---|---|---|---|---|
| Egg | | | | | | | | | | | | |
| Larva | | | | | | | | | | | | |
| Pupa | | | | | | | | | | | | |
| Adult | | | | | | | | | | | | |

# Boulder Day-moth
## *Radinocera* sp. 'Sandstone'

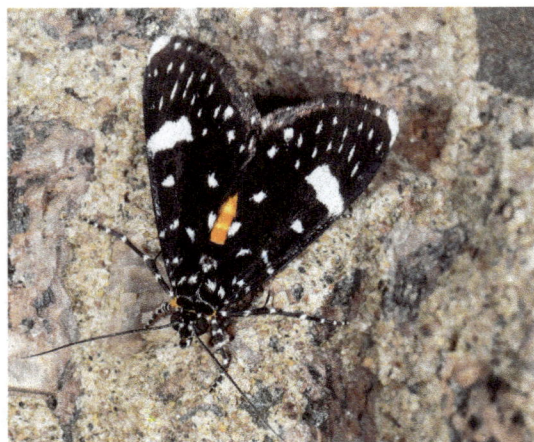

Plate 211 Kakadu National Park, NT
Photo: M. F. Braby

Plate 212 Kakadu National Park, NT
Photo: Justin Armstrong

## Distribution

This undescribed species is endemic to the study region, its presence detected only as recently as 2009 (Braby 2015e). It has a disjunct distribution and is known from only two locations: Jack Creek Bay, WA, in the northern Kimberley, where a single female was collected in 2016 (J. E. and A. Koeyers); and Kakadu National Park, NT (Nourlangie Rock) (Braby 2015e). Although the larval food plant is considerably more widespread in the study region, *Radinocera* sp. 'Sandstone' is confined to areas where the plant grows on sandstone in the higher rainfall areas (> 900 mm mean annual rainfall). Further field surveys are thus required to determine whether it occurs elsewhere in the Kimberley.

## Habitat

*Radinocera* sp. 'Sandstone' breeds in sandstone foot-slope boulders at the base of escarpments where the larval food plant grows as a vine (Braby 2015e).

## Larval food plant

*Ampelocissus acetosa* (Vitaceae).

## Seasonality

Adults are seasonal, occurring only during the 'build-up' and early wet season (October–January), but there are too few records (n = 9) to assess temporal changes in abundance. In general, adults tend to be more prevalent in November and December, depending on the timing of the pre-monsoon storms. The immature stages (eggs or larvae) have been recorded only in November and December. The larvae feed on the new soft leaf growth of the food plant, which regenerates from an underground tuber during the 'build-up' following the pre-monsoon storms and/or an increase in humidity. The food plant is a short-lived seasonal perennial vine and it senesces towards the end of the wet season and early dry season. The duration of the pupal stage is variable and there is probably only a single generation annually, followed by a partial second generation in the mid wet season. The species survives the dry season in the pupal stage, which may remain in diapause for up to 12 months.

## Breeding status

This species is resident in the study region.

## Conservation status

LC. Despite the wide geographical range of the species, available data suggest *Radinocera* sp. 'Sandstone' has a disjunct distribution with a limited AOO. The two known locations both occur in conservation reserves: Uunguu IPA and Kakadu National Park. Despite its restricted occurrence, there are no known threats facing the taxon.

## Radinocera sp. 'Sandstone'

● Specimen ≥1970
■ Observation ≥1970
× Larval food plant

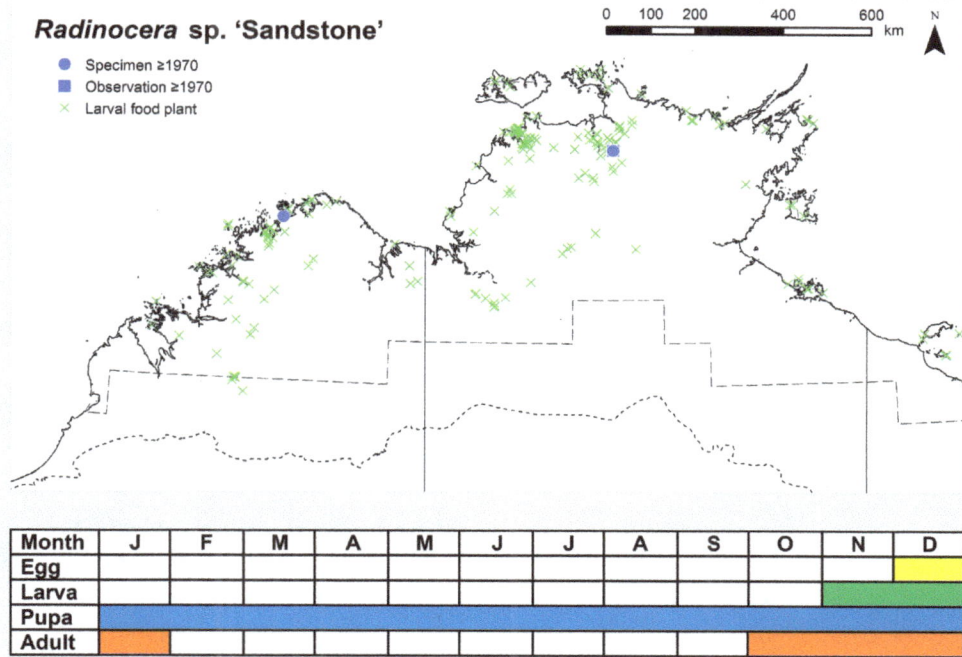

| Month | J | F | M | A | M | J | J | A | S | O | N | D |
|-------|---|---|---|---|---|---|---|---|---|---|---|---|
| Egg   |   |   |   |   |   |   |   |   |   |   |   |   |
| Larva |   |   |   |   |   |   |   |   |   |   |   |   |
| Pupa  |   |   |   |   |   |   |   |   |   |   |   |   |
| Adult |   |   |   |   |   |   |   |   |   |   |   |   |

Photo: Nourlangie Rock, Kakadu National Park, NT, M.F. Braby

# Speckled Day-moth
## *Idalima metasticta* Hampson, 1910

Plate 213 Litchfield National Park, NT
Photo: M. F. Braby

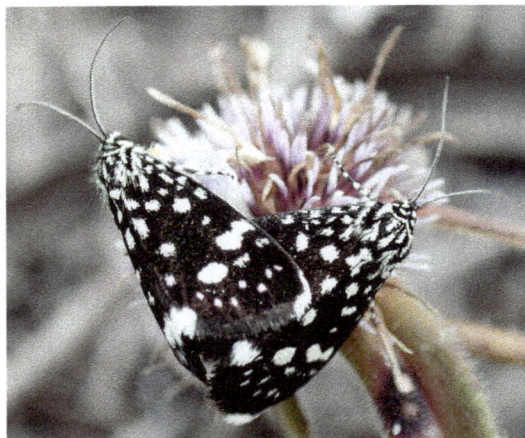

Plate 214 Nhulunbuy, NT
Photo: M. F. Braby

## Distribution

This species is endemic to the study region. It is restricted to the Top End, where it occurs mainly in areas that receive more than 800 mm mean annual rainfall. Its southernmost limits include Mataranka Homestead and Birdum near Larrimah, NT. The geographic range corresponds well with the spatial distribution of its larval food plants, indicating that *I. metasticta* has been well sampled in the region.

## Habitat

*Idalima metasticta* breeds in savannah woodland where the larval food plant (*Hibbertia dilatata*) grows as a low perennial shrub on lateritic outcrops and rocky hill slopes. It also breeds in eucalypt heathy woodland where an alternative food plant (*H. juncea*) grows as a low perennial shrub on white coastal sand and relatively flat sandy terrain between sandstone outcrops (Braby 2015e).

## Larval food plants

*Hibbertia dilatata*, *H. juncea* (Dilleniaceae).

## Seasonality

Adults occur during most months of the year, but they are most abundant during the early wet season (November and December) following the pre-monsoon storms. However, they may also be locally abundant at other times of the year (e.g. October), depending on the timing of the first significant rainfall events. They are generally very scarce or absent during the dry season (May–September), except on Gove Peninsula, where the dry season is less pronounced and adults may be locally abundant in July. The immature stages (eggs or larvae) have been recorded sporadically from November to June. Freshly emerged females have been observed mating in July and October, indicating that breeding also occurs at these times of the year. The larvae feed on the soft new cladodes and flowers of the food plants. Presumably, several generations are completed annually and the species survives the dry periods in pupal diapause.

## Breeding status

This species is resident in the study region.

## Conservation status

LC.

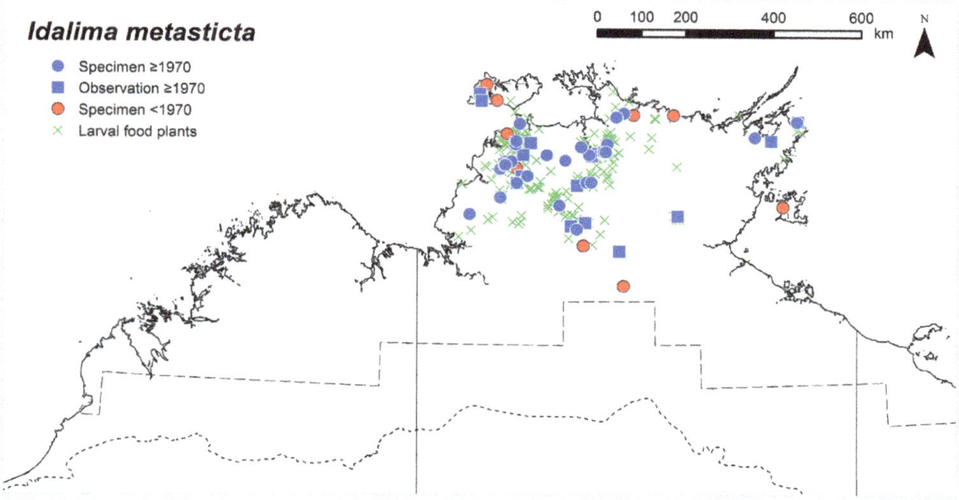
Idalima metasticta
Specimen ≥1970
Observation ≥1970
Specimen <1970
Larval food plants
0 100 200 400 600
km
N

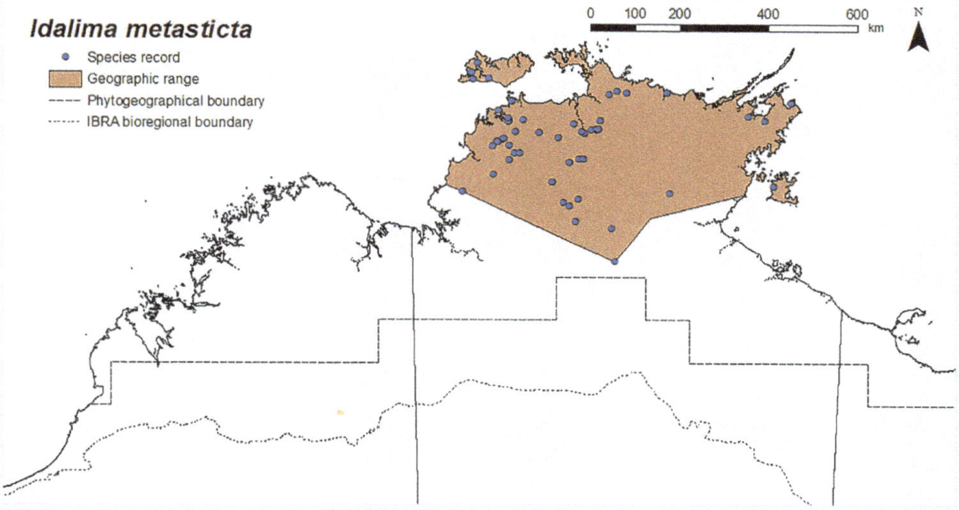
Idalima metasticta
Species record
Geographic range
Phytogeographical boundary
IBRA bioregional boundary
0 100 200 400 600
km
N

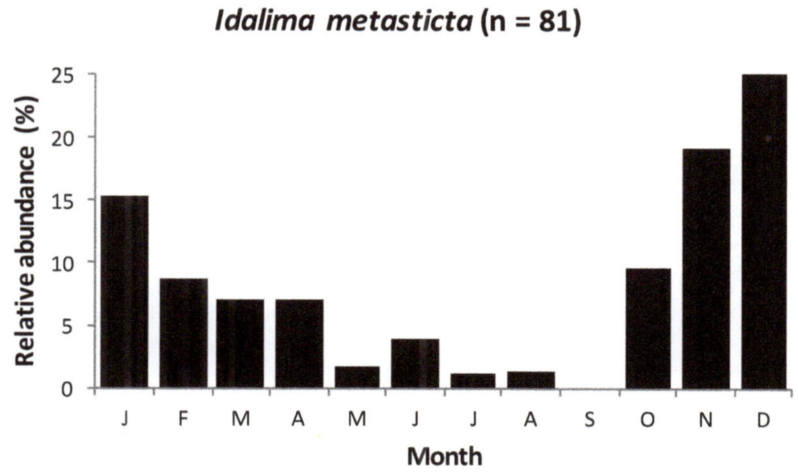
Idalima metasticta (n = 81)
Relative abundance (%)
25
20
15
10
5
0
J F M A M J J A S O N D
Month

| Month | J | F | M | A | M | J | J | A | S | O | N | D |
|---|---|---|---|---|---|---|---|---|---|---|---|---|
| Egg | ■ | | | | | | | | | | | |
| Larva | ■ | ■ | ■ | ■ | | ■ | | | | | ■ | ■ |
| Pupa | | ■ | | | | | | | | | ■ | ■ |
| Adult | ■ | ■ | ■ | ■ | ■ | ■ | ■ | ■ | | ■ | ■ | ■ |

# Indigo Day-moth
## *Idalima leonora* (Doubleday, 1846)

Plate 215 Cobourg Peninsula, NT
Photo: M. F. Braby

Plate 216 Cobourg Peninsula, NT
Photo: M. F. Braby

## Distribution

This species is endemic to the study region. It has a disjunct distribution, occurring in the northern Kimberley (Mitchell Plateau, WA) and in the Top End. It generally occurs in the higher rainfall areas (> 1,300 mm mean annual rainfall), although it has been recorded as far south as Manbullo, NT (c. 900 mm), based on a historical record in the AM in which a specimen was collected in 1916. The geographic range in the Top End corresponds reasonably well with the spatial distribution of its two known larval food plants (*Hibbertia* spp.); however, these plants are absent from the Kimberley, the Darwin area and Litchfield National Park, NT, indicating that several other (as yet unreported) food plants are used in these areas.

The taxonomic status and relationships of *Idalima leonora* require further investigation. Kiriakoff (1977) treated the taxon as a polytypic species, with two allopatric subspecies: *I. leonora leonora* from the Northern Territory and *I. leonora tasso* (Jordan, 1912) from northern Queensland. However, Edwards (1996) treated *I. leonora tasso* as a distinct species, *I. tasso*. The classification of Edwards (1996) is tentatively followed in this work.

## Habitat

*Idalima leonora* breeds in savannah woodland and eucalypt heathy open woodland where the larval food plants grow as perennial shrubs on sandy loam or rocky sandstone outcrops (Braby 2011a). Adults have also been recorded in eucalypt heathy woodland on white coastal sand, where they no doubt breed.

## Larval food plants

*Hibbertia brownii*, *H. candicans* (Dilleniaceae).

## Seasonality

Adults occur during most months of the year, but they are most abundant during the early wet season (November–February), depending on the timing of the pre-monsoon storms or monsoon rains. They are generally scarce or absent during the late wet season and dry season. The immature stages (eggs or larvae) have been recorded mainly during the wet season (November–March); however, on Cobourg Peninsula, NT, adults were noted to be locally common in August and females were observed ovipositing on the larval food plant (*Hibbertia brownii*), which was resprouting from rootstock following a dry season burn two months earlier. Presumably, several generations are completed annually and the species normally survives the dry periods in pupal diapause.

## Breeding status

This species is resident in the study region.

## Conservation status

LC. Although the species *I. leonora* has a wide geographical range, the impact of inappropriate fire regimes—particularly an increase in the frequency of dry season burns in some areas of the range, such as sandstone communities—requires further investigation. Although the larval food plants resprout from rootstock, if the fire regime is too frequent, with short interfire intervals, there may be little or no recruitment. Thus, *I. leonora* may qualify as Near Threatened (NT) in future. Monitoring of the abundance or occupancy of the moth and its food plants is required to clarify the effect of fire frequency as a key threatening process. The fire regime needs to be carefully managed (e.g. an interfire interval of more than five years) to ensure the food plants have sufficient time to mature, flower and set seed.

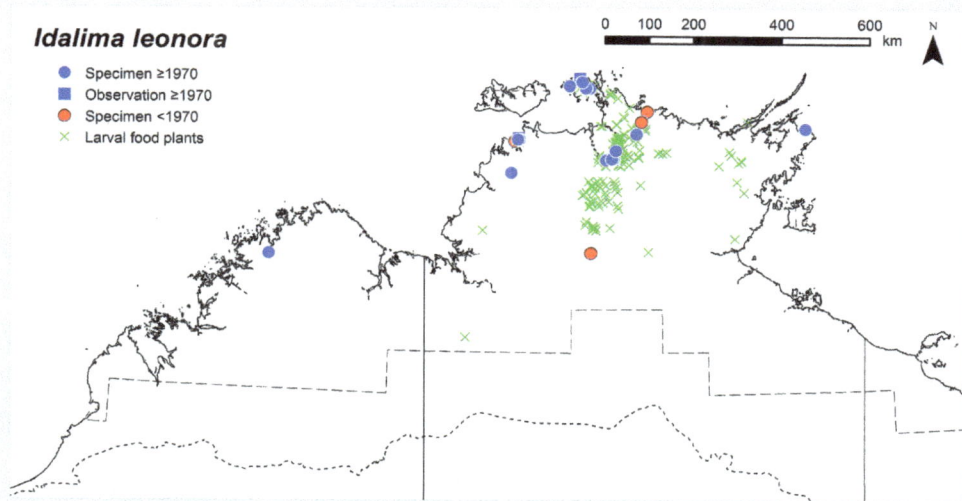

*Idalima leonora*

- ● Specimen ≥1970
- ■ Observation ≥1970
- ● Specimen <1970
- × Larval food plants

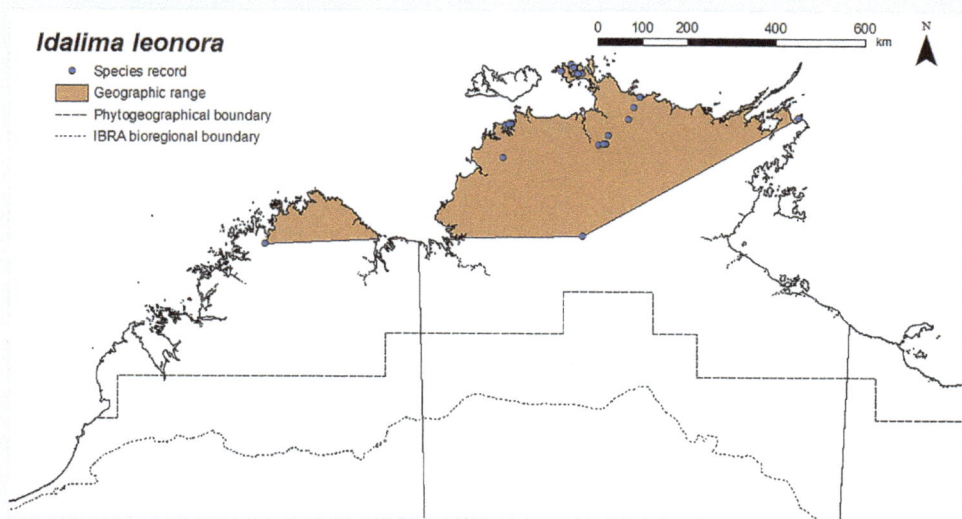

*Idalima leonora*

- ● Species record
- ▬ Geographic range
- --- Phytogeographical boundary
- ···· IBRA bioregional boundary

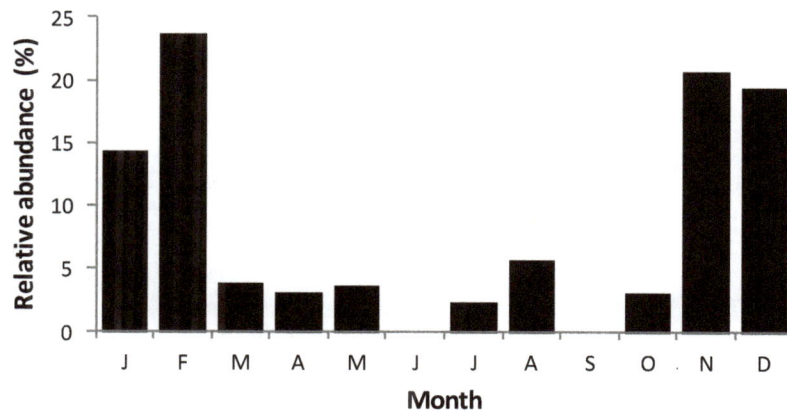

*Idalima leonora* (n = 37)

| Month | J | F | M | A | M | J | J | A | S | O | N | D |
|-------|---|---|---|---|---|---|---|---|---|---|---|---|
| Egg | | | | | | | | | | | | |
| Larva | | | | | | | | | | | | |
| Pupa | | | | | | | | | | | | |
| Adult | | | | | | | | | | | | |

# Splendid Day-moth
## *Idalima aethrias* (Turner, 1908)

Plate 217 Berry Springs, NT
Photo: M. F. Braby

Plate 218 Mt Burrell, NT
Photo: M. F. Braby

## Distribution

This species occurs in the north of the Northern Territory of the study region. It is restricted to the higher rainfall areas (> 1,300 mm mean annual rainfall) of the north-western corner of the Top End. It has been recorded as far south as Daly River Mission (J. F. Hutchinson) and Kakadu National Park (near Twin Falls) and as far east as the King River, NT. The geographic range closely corresponds with the spatial distribution of its larval food plants. The food plants, however, also occur slightly further south, at Fish River Station and Katherine Gorge, and extend further east to Gove Peninsula and Groote Eylandt; thus, further field surveys are required to determine whether *I. aethrias* occurs in these areas, particularly in the eastern half of the Top End. Outside the study region, *I. aethrias* occurs in north-eastern Queensland.

## Habitat

*Idalima aethrias* breeds in savannah woodland where the larval food plants grow as small perennial shrubs (Braby 2011a).

## Larval food plants

*Hibbertia brevipedunculata*, *H. cistifolia* (Dilleniaceae).

## Seasonality

Adults are seasonal, occurring during the 'build-up', wet season and early dry season (October–May), but they are most abundant during the early wet season (November and December) following the first substantial pre-monsoon storms. The immature stages (eggs or larvae) have been recorded during the wet season (November–February). The duration of the pupal stage is variable and there is probably only a single generation annually, followed by a partial second generation in the late wet season. The species survives the dry season in the pupal stage, which may remain in diapause for up to 11 months.

## Breeding status

This species is resident in the study region.

## Conservation status

LC. The species *I. aethrias* has a restricted range in the study region within which it occurs in several conservation reserves, including Litchfield National Park, Garig Gunak Barlu National Park and Kakadu National Park. However, the impact of inappropriate fire regimes in some areas of the range requires further investigation. Although the larval food plants are currently listed as Least Concern (LC) under the *TPWCA* and resprout from rootstock, if the fire regime is too frequent, with short interfire intervals, there may be little or no recruitment. Thus, *I. aethrias* may qualify as Near Threatened (NT) in future. Monitoring of the abundance or occupancy of the moth and its food plants is required to clarify the effect of fire frequency as a key threatening process. The fire regime needs to be carefully managed (e.g. an interfire interval of more than five years) to ensure the food plants have sufficient time to mature, flower and set seed.

## Idalima aethrias

- ● Specimen ≥1970
- ■ Observation ≥1970
- ● Specimen <1970
- × Larval food plants

## Idalima aethrias

- ● Species record
- ▨ Geographic range
- – – Phytogeographical boundary
- ···· IBRA bioregional boundary

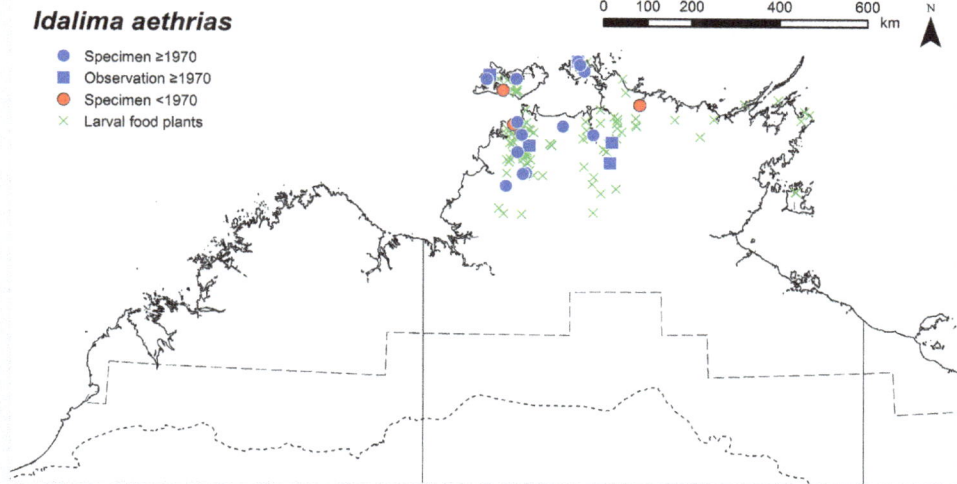

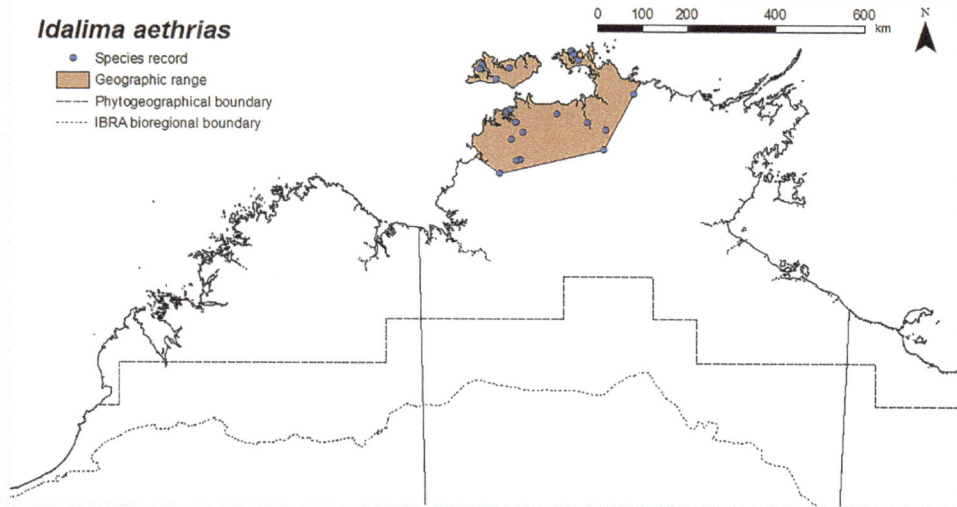

### Idalima aethrias (n = 34)

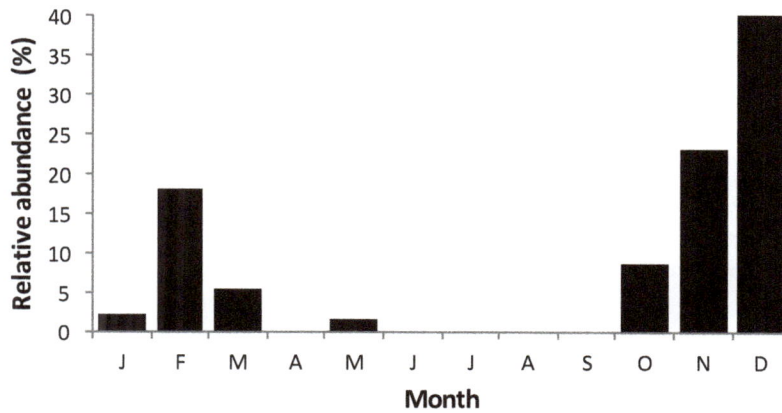

| Month | J | F | M | A | M | J | J | A | S | O | N | D |
|-------|---|---|---|---|---|---|---|---|---|---|---|---|
| Egg   |   |   |   |   |   |   |   |   |   |   |   |   |
| Larva |   |   |   |   |   |   |   |   |   |   |   |   |
| Pupa  |   |   |   |   |   |   |   |   |   |   |   |   |
| Adult |   |   |   |   |   |   |   |   |   |   |   |   |

# Rock-art Day-moth
## *Idalima* sp. 'Arnhem Land'

Plate 219 Kakadu National Park, NT
Photo: M. F. Braby

## Distribution

This undescribed species is endemic to the study region. It occurs in the Top End, where it is restricted to the Arnhem Land Plateau, its presence in the region first detected in 1972. It has been recorded from two locations: 15 km south-west by south of Nimbuwah Rock (E. D. Edwards and M. S. Upton) and Kakadu National Park (Nourlangie Rock), NT (Braby 2015e). The larval food plant occurs throughout the Arnhem Land Plateau, as well as in eastern Arnhem Land; thus, further field surveys are required to determine whether *Idalima* sp. 'Arnhem Land' is more widely distributed in these areas than present records indicate.

## Habitat

*Idalima* sp. 'Arnhem Land' breeds in eucalypt heathy woodland on broken sandstone hill slopes where the larval food plant grows as a perennial shrub in the understorey (Braby 2015e).

## Larval food plant

*Hibbertia candicans* (Dilleniaceae).

## Seasonality

Adults are seasonal, occurring only during the wet season (November–April), but they are most abundant during the early wet season (November–January) following the pre-monsoon storms, although they are never observed in large numbers or high densities. The immature stages (eggs or larvae) have been recorded during the wet season (December–April). There is probably only a single generation annually, followed by a partial second generation in the late wet season. Presumably, the species survives the dry season in pupal diapause.

## Breeding status

This species is resident in the study region.

## Conservation status

Near Threatened (NT). The species *Idalima* sp. 'Arnhem Land' is a short-range endemic (AOO is likely to be < 2,000 sq km, with spatial buffering of records providing a first approximation of 1,400 sq km), with the entire range occurring within a single conservation reserve (Kakadu National Park). However, it may qualify for a threatened category in the near future because the taxon is threatened by inappropriate fire regimes, particularly too frequent fires (every one–two years) on the Arnhem Land Plateau (see also Russell-Smith et al. 1998, 2002). Although the larval food plant is currently listed as Least Concern (LC) under the *TPWCA* and it resprouts from rootstock, if the fire regime on sandstone plant communities is too frequent, with short interfire intervals, there may be little or no recruitment. That is, the population of *Idalima* sp. 'Arnhem Land' is likely to be reduced in future based on a projected decline in the AOO, quality of its habitat and/or number of locations (criteria A3c, D2). Monitoring of the abundance and occupancy of the moth and its food plant is required for this species. The fire regime on the sandstone plateau–breakaway country needs to be carefully managed (e.g. an interfire interval of more than five years) to ensure the food plant has sufficient time to mature, flower and set seed.

## Idalima sp. 'Arnhem Land'

- ● Specimen ≥1970
- ■ Observation ≥1970
- ✕ Larval food plant

## Idalima sp. 'Arnhem Land'

- ● Species record
- ▨ Geographic range
- – – Phytogeographical boundary
- ···· IBRA bioregional boundary

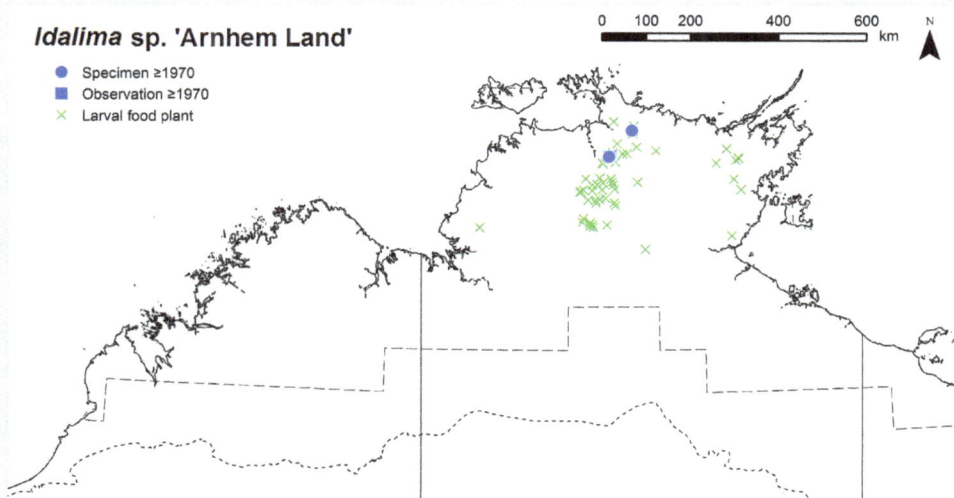

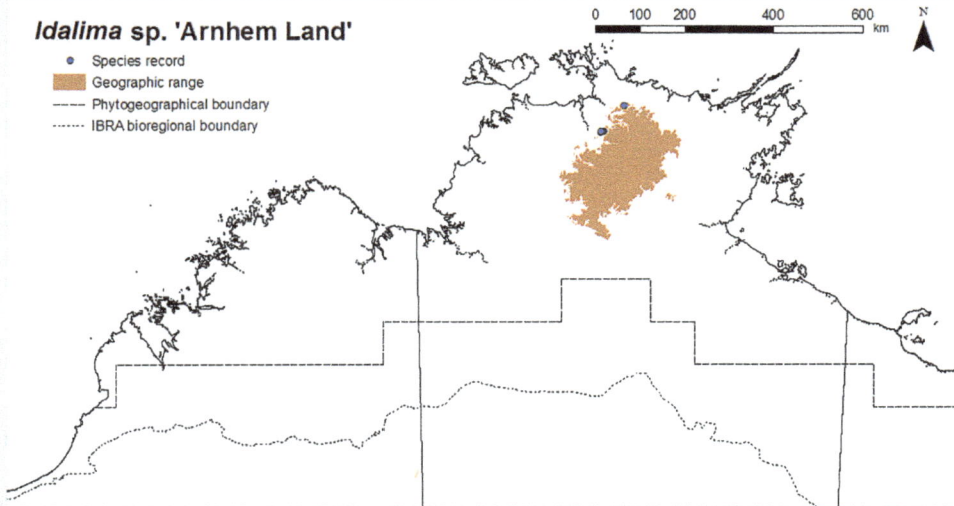

## Idalima sp. 'Arnhem Land' (n = 27)

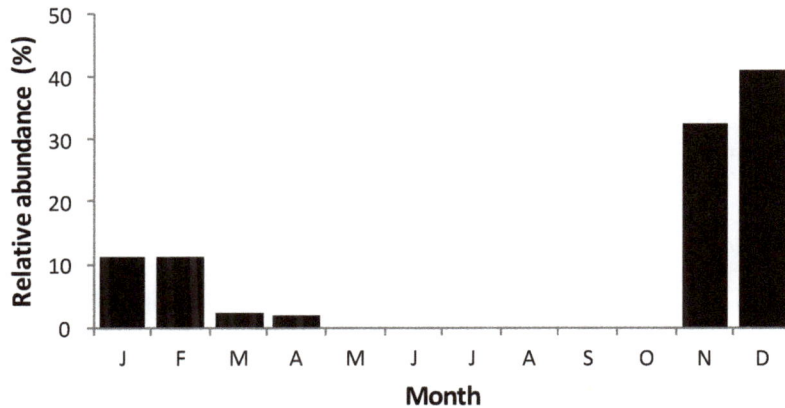

| Month | J | F | M | A | M | J | J | A | S | O | N | D |
|-------|---|---|---|---|---|---|---|---|---|---|---|---|
| Egg   |   | ▨ |   |   |   |   |   |   |   |   |   | ▨ |
| Larva | ▨ | ▨ | ▨ | ▨ |   |   |   |   |   |   |   |   |
| Pupa  | ▨ | ▨ | ▨ | ▨ |   |   |   |   |   |   |   | ▨ |
| Adult | ▨ | ▨ | ▨ | ▨ |   |   |   |   |   |   | ▨ | ▨ |

# Kakadu Whistling Moth

## *Hecatesia* sp. 'Arnhem Land'

Plate 220 Kakadu National Park, NT
Photo: M. F. Braby

Plate 221 Kakadu National Park, NT
Photo: M. F. Braby

## Distribution

This undescribed species is endemic to the study region. It occurs in the Top End, where it is restricted to the Arnhem Land Plateau, its presence in the region first detected in 1973. It is currently known from only two locations: 6 km south–south-west of Oenpelli (J. C. Cardale and M. S. Upton) and Kakadu National Park (Nourlangie Rock) (Braby 2015e), NT. Although the larval food plant is very widely distributed throughout the study region, *Hecatesia* sp. 'Arnhem Land' is confined to areas where the plant grows on sandstone in the higher rainfall areas (> 900 mm mean annual rainfall).

## Habitat

*Hecatesia* sp. 'Arnhem Land' breeds in long-unburnt eucalypt heathy woodland on steep broken sandstone hill slopes where patches of the larval food plant grow as a scrambling parasitic vine in superabundance over understorey shrubs (Braby 2015e).

## Larval food plant

*Cassytha filiformis* (Lauraceae).

## Seasonality

Adults appear to be rare and seasonal, having been recorded only during the wet season (November–April) and early dry season (June), but there are too few records (n = 7) to assess temporal changes in abundance. The immature stages (eggs or larvae) have been recorded from December to March. All pupae reared during the wet season developed directly with no evidence of diapause, suggesting that several generations are completed at this time. However, it is likely the species survives the late dry season in pupal diapause and that adult emergence is triggered by the pre-monsoon storms during the 'build-up'.

## Breeding status

This species is resident in the study region.

## Conservation status

Near Threatened (NT). The species *Hecatesia* sp. 'Arnhem Land' is a narrow-range endemic (AOO is likely to be < 2,000 sq km, with spatial buffering of records providing a first approximation of 1,400 sq km), with the entire range occurring within a single conservation reserve (Kakadu National Park). However, it may qualify for a threatened category in the near future because the taxon is threatened by inappropriate fire regimes, particularly too frequent fires, with short interfire intervals (every one–two years), on the Arnhem Land Plateau (see also Russell-Smith et al. 1998, 2002). That is, the population of *Hecatesia* sp. 'Arnhem Land' is likely to be reduced in future based on a projected decline in the AOO, quality of its habitat and/or number of locations (criteria A3c, D2). Monitoring of the abundance and occupancy of the moth and its food plant is required for this species. The fire regime on the sandstone plateau–breakaway country needs to be carefully managed to ensure that long-unburnt patches of habitat (with an interfire interval of more than five years) persist, which will encourage high local densities of the food plant.

## *Hecatesia* sp. 'Arnhem Land'

- ● Specimen ≥1970
- ■ Observation ≥1970
- ✕ Larval food plant

## *Hecatesia* sp. 'Arnhem Land'

- • Species record
- ▨ Geographic range
- — — Phytogeographical boundary
- ···· IBRA bioregional boundary

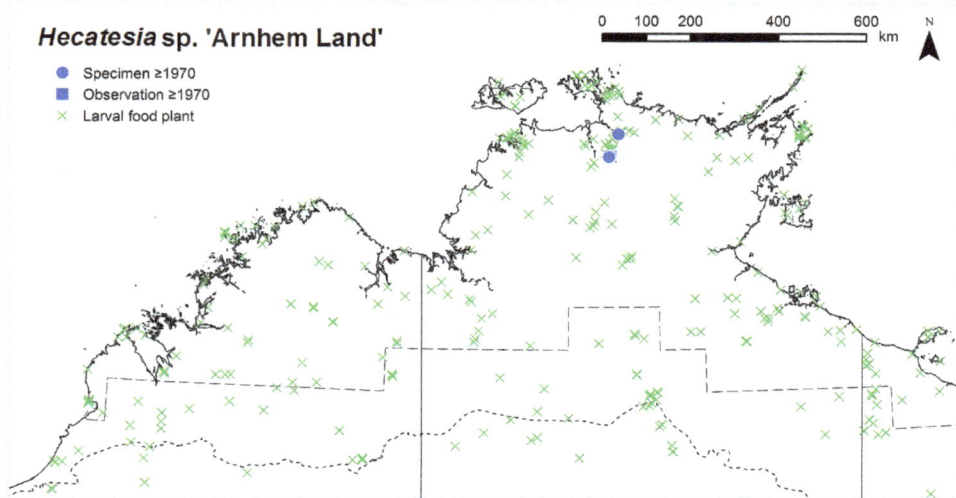

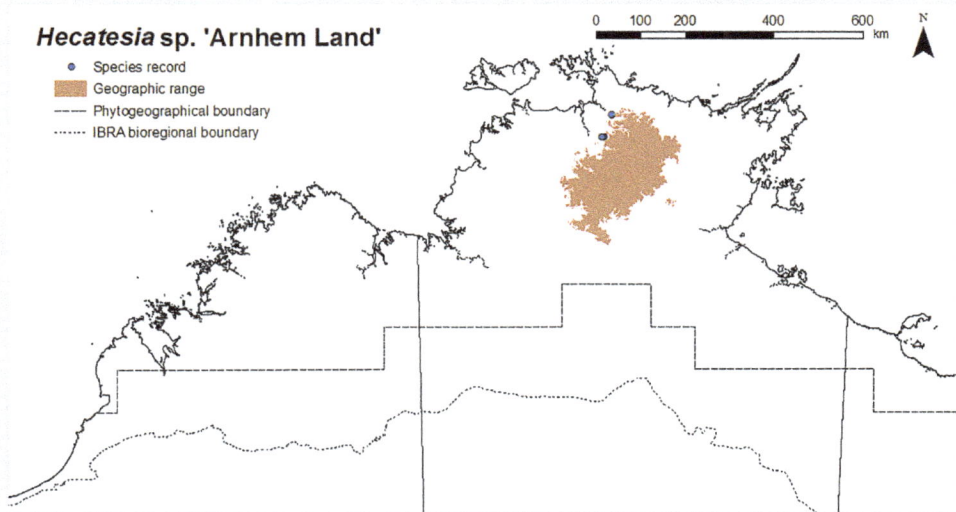

| Month | J | F | M | A | M | J | J | A | S | O | N | D |
|-------|---|---|---|---|---|---|---|---|---|---|---|---|
| Egg   |   |   |   |   |   |   |   |   |   |   |   |   |
| Larva |   |   |   |   |   |   |   |   |   |   |   |   |
| Pupa  |   |   |   |   |   |   |   |   |   |   |   |   |
| Adult |   |   |   |   |   |   |   |   |   |   |   |   |

# Mimetic Whistling Moth
## *Hecatesia* sp. 'Amata'

Plate 222 Keep River National Park, NT
Photo: M. F. Braby

Plate 223 Keep River National Park, NT
Photo: M. F. Braby

## Distribution

This undescribed species is restricted to the lower rainfall areas (600–800 mm mean annual rainfall) of the Kimberley and Northern Deserts, where it has been recorded from three locations: 'Barrier Range' near Napier Range, WA, based on a male specimen in the ANIC collected some time in 1887 (see Froggatt 1934); and Keep River National Park (Braby 2015e) and Daly Waters, NT. The larval food plant is considerably more widespread than the known geographic range; thus, further field surveys are required to determine whether *Hecatesia* sp. 'Amata' is more widespread in the semi-arid zone than present records indicate. Outside the study region, *Hecatesia* sp. 'Amata' occurs in north-eastern Queensland (M. F. Braby, unpublished data).

## Habitat

*Hecatesia* sp. 'Amata' breeds in long-unburnt eucalypt open woodland on flat sandy loam where the larval food plant grows as a fine scrambling parasitic vine in superabundance on various grasses, particularly *Triodia* (Braby 2015e).

## Larval food plant

*Cassytha capillaris* (Lauraceae).

## Seasonality

Adults appear to be seasonal, having been recorded only during the wet season (January–April), but there are too few records (n = 4) to assess temporal changes in abundance. The immature stages (larvae) have been recorded in February and March. All pupae reared during the wet season developed directly with no evidence of diapause, suggesting that at least two generations are completed at this time. However, it is likely the species survives the late dry season in pupal diapause and the first monsoon rains trigger adult emergence.

## Breeding status

This species is resident in the study region.

## Conservation status

LC. The species *Hecatesia* sp. 'Amata' has a restricted range within which it occurs in at least one conservation reserve (Keep River National Park). However, the effect of inappropriate fire regimes in some areas of the range requires further investigation because breeding colonies are likely to be impacted by too frequent fires, with short interfire intervals (every one–two years), such as prescribed control burns. Thus, *Hecatesia* sp. 'Amata' may qualify as Near Threatened (NT) in future. The fire regime needs to be carefully managed to ensure that long-unburnt patches of habitat (with an interfire interval of more than five years) persist, which will encourage high local densities of the food plant.

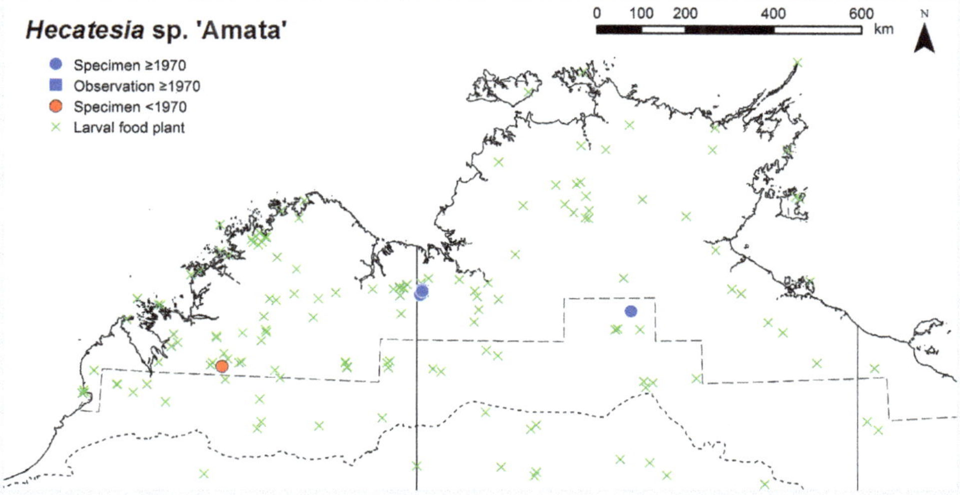

*Hecatesia* sp. 'Amata'

- ● Specimen ≥1970
- ■ Observation ≥1970
- ● Specimen <1970
- × Larval food plant

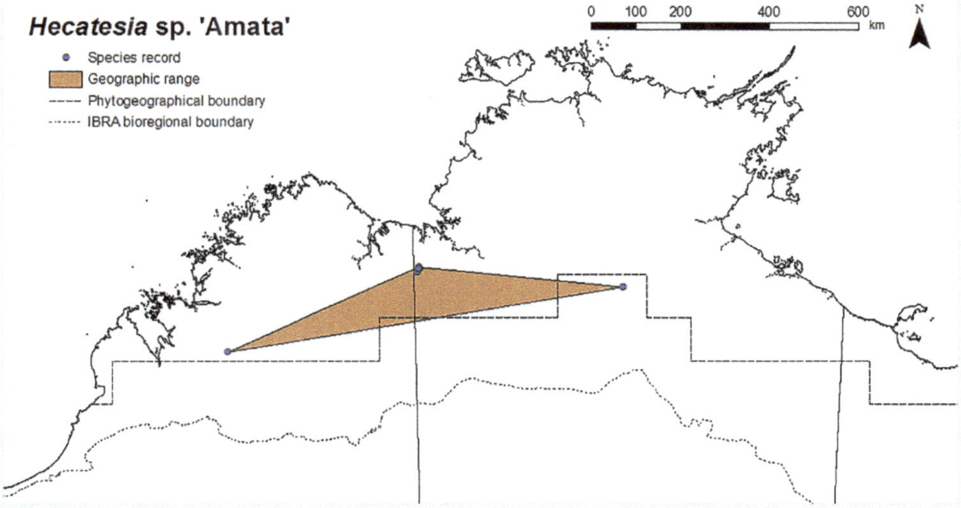

*Hecatesia* sp. 'Amata'

- ● Species record
- ▨ Geographic range
- --- Phytogeographical boundary
- ···· IBRA bioregional boundary

| Month | J | F | M | A | M | J | J | A | S | O | N | D |
|-------|---|---|---|---|---|---|---|---|---|---|---|---|
| Egg   |   |   |   |   |   |   |   |   |   |   |   |   |
| Larva |   |   |   |   |   |   |   |   |   |   |   |   |
| Pupa  |   |   |   |   |   |   |   |   |   |   |   |   |
| Adult |   |   |   |   |   |   |   |   |   |   |   |   |

# Central Spotted Day-moth

## *Mimeusemia centralis* (Rothschild, 1896)

Plate 224 Lee Point, Casuarina Coastal Reserve, NT
Photo: M. F. Braby

## Distribution

This species is known from only a single record in the Top End of the study region, where a specimen was collected at Casuarina Coastal Reserve near Darwin, NT, in December 2007 (Braby 2014a). Outside the study region, *M. centralis* occurs in northern Queensland.

## Habitat

The breeding habitat of *M. centralis* has not been recorded in the study region. Two adult males were observed in coastal semi-deciduous monsoon vine thicket (Braby 2014a) in which the species no doubt breeds.

## Larval food plants

Not recorded in the study region.

## Seasonality

The seasonal abundance and breeding phenology of this species are not understood. The individuals were recorded during the early wet season (December) following the pre-monsoon storms. Presumably, adults are highly seasonal and local in occurrence.

## Breeding status

*Mimeusemia centralis* is assumed to be resident in the study region.

## Conservation status

DD. The species *M. centralis* is currently known from only one location in the study region (AOO is likely to be < 2,000 sq km, with spatial buffering of records providing a first approximation of 700 sq km). Although it has been recorded from a conservation reserve (Casuarina Coastal Reserve), further investigations are required to clarify the ecological requirements, critical habitat and distribution of this species.

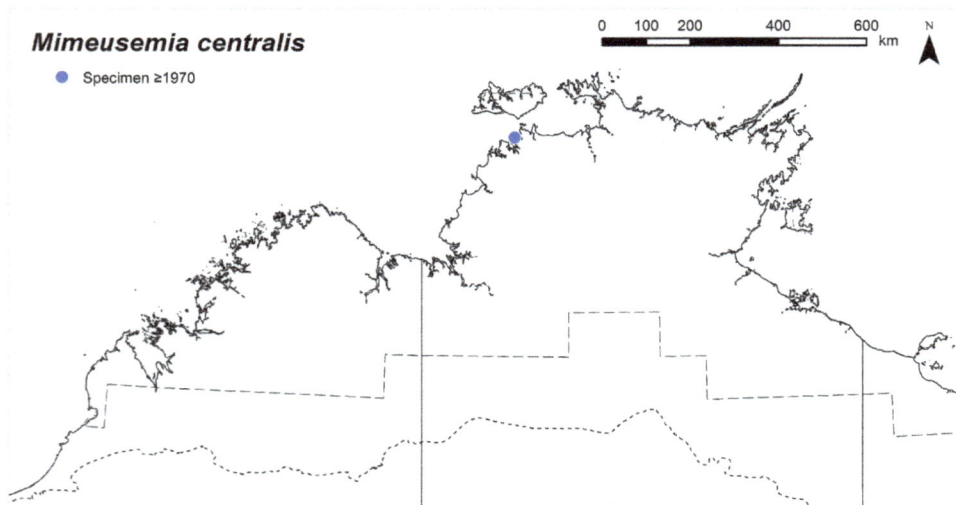

*Mimeusemia centralis*

● Specimen ≥1970

| Month | J | F | M | A | M | J | J | A | S | O | N | D |
|-------|---|---|---|---|---|---|---|---|---|---|---|---|
| Egg   |   |   |   |   |   |   |   |   |   |   |   |   |
| Larva |   |   |   |   |   |   |   |   |   |   |   |   |
| Pupa  |   |   |   |   |   |   |   |   |   |   |   |   |
| Adult |   |   |   |   |   |   |   |   |   |   |   | ■ |

# Rustic Day-moth

## *Mimeusemia econia* Hampson, 1900

Plate 225 Coen, Qld
Photo: M. F. Braby

## Distribution

This species is known from only a small series of specimens (11) collected from two locations (Seaflower Bay and 'Teranadon Bay') in Vansittart Bay near Cape Bougainville, WA, in the northern Kimberley of the study region in 2017 (J. E. and A. Koeyers). Outside the study region, *M. econia* occurs in Christmas Island, the Lesser Sunda Islands, Sulawesi and north-eastern Australia.

## Habitat

The breeding habitat of *M. econia* has not been recorded in the study region. The specimens were collected at night at a light trap adjacent to coastal semi-deciduous monsoon vine thicket in which the species no doubt breeds.

## Larval food plants

Not recorded in the study region.

## Seasonality

The seasonal abundance and breeding phenology of this species are not understood. All adults have been recorded during the early wet season (December and January) immediately following the first pre-monsoon rains. Presumably, adults are highly seasonal and local in occurrence.

## Breeding status

*Mimeusemia econia* is assumed to be resident in the study region.

## Conservation status

DD. The species *M. econia* is currently known from only one location in the study region (AOO is likely to be < 2,000 sq km, with spatial buffering of records providing a first approximation of 1,400 sq km). Further investigations are required to clarify the ecological requirements, critical habitat and distribution of this species.

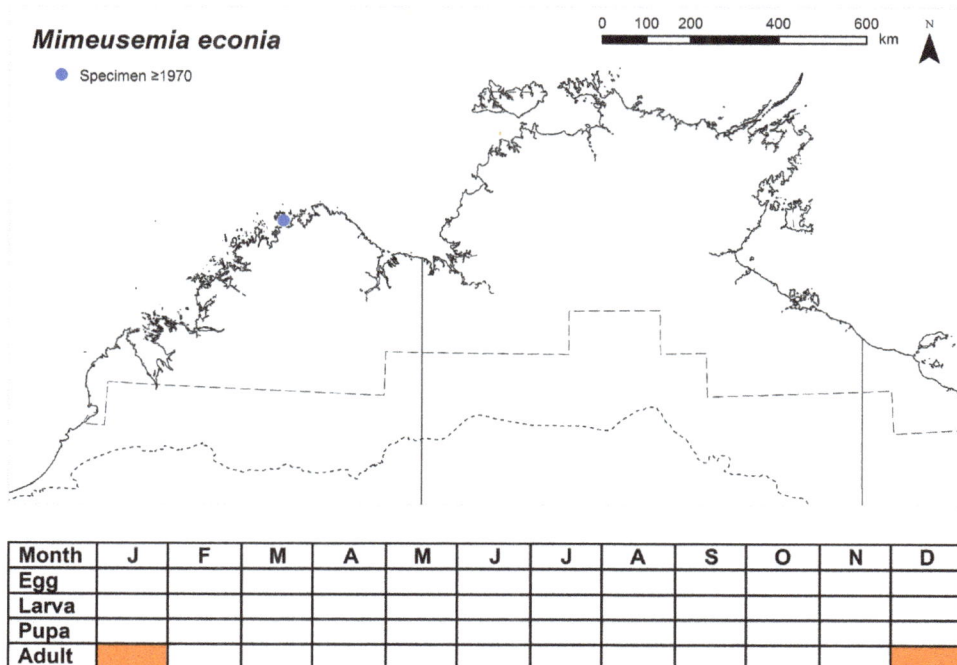

*Mimeusemia econia*

● Specimen ≥1970

| Month | J | F | M | A | M | J | J | A | S | O | N | D |
|-------|---|---|---|---|---|---|---|---|---|---|---|---|
| Egg   |   |   |   |   |   |   |   |   |   |   |   |   |
| Larva |   |   |   |   |   |   |   |   |   |   |   |   |
| Pupa  |   |   |   |   |   |   |   |   |   |   |   |   |
| Adult | ■ |   |   |   |   |   |   |   |   |   |   | ■ |

# Painted Day-moth
*Agarista agricola* (Donovan, 1805)

Plate 226 Kakadu National Park, NT
Photo: Rod Kennett

Plate 227 Mt Burrell, NT
Photo: M. F. Braby

## Distribution

This species is represented by two subspecies: *A. agricola biformis* Butler, 1884, which is endemic to the study region, and *A. agricola agricola* (Donovan, 1805). Both subspecies occur in the north of the Northern Territory, where they are restricted to the higher rainfall areas (> 1,300 mm mean annual rainfall). *Agarista agricola biformis* occurs in the north-western corner of the Top End, reaching its southernmost occurrence at Daly River, NT, whereas *A. agricola agricola* appears to be geographically separated, confined to Gove Peninsula, where it has been recorded at Drimmie Head, Nhulunbuy (L. Wilson) and Macassan Beach (B. Hoffmann), NT. The geographic range of *A. agricola* corresponds well with the spatial distribution of the known and putative larval food plants (*Leea* spp.). The food plants, however, also occur on the Tiwi Islands and Groote Eylandt, NT; thus, further field surveys are required to determine whether the species is present in these areas. Outside the study region, *A. agricola* occurs from Timor, through mainland New Guinea to north-eastern and eastern Australia.

## Habitat

*Agarista agricola* breeds in the ecotone between savannah woodland and monsoon vine thicket or along the very edge of semi-deciduous monsoon vine thicket on rock outcrops where the larval food plant (*Leea rubra*) grows as a deciduous perennial shrub 1–3 m high (Braby 2011a). Adults have also been recorded near evergreen monsoon vine forest where the putative larval food plant (*Leea indica*) grows in the understorey; the larvae readily feed on this plant in captivity, suggesting *A. agricola* probably breeds in this habitat.

## Larval food plants

*Leea rubra* (Vitaceae); probably *Leea indica*.

## Seasonality

Adults are seasonal, occurring only during the 'build-up' and wet season (October–May), but there are too few records (n = 23) to assess temporal changes in abundance. In general, adults tend to be more numerous during the 'build-up' (October and November) following the first substantial pre-monsoon storms and then again during the late wet season (March and April). The breeding phenology is not well understood. The immature stages have been recorded from October to December. The larvae feed on the foliage of the food plant, which is seasonally deciduous, dropping its leaves as the dry season progresses. Presumably, at least two generations are completed annually and the species survives the dry season in pupal diapause.

## Breeding status

This species is resident in the study region.

## Conservation status

*Agarista agricola biformis*: LC. *Agarista agricola agricola*: DD. The subspecies *A. agricola biformis* is a short-range endemic (EOO = 8,860 sq km) and occurs in several conservation reserves, including Fogg Dam Conservation Reserve, Mary River National Park (Mt Bundy) and Kakadu National Park, NT. Despite its restricted occurrence, there are no known threats facing the taxon. However, the subspecies *A. agricola agricola* has a short range in the study region (AOO is likely to be < 2,000 sq km, with spatial buffering of records providing a first approximation of 1,400 sq km) and is known from only two locations on Gove Peninsula, NT. Targeted field surveys to clarify the extent of its distribution and to determine key threatening processes should be a priority for this subspecies.

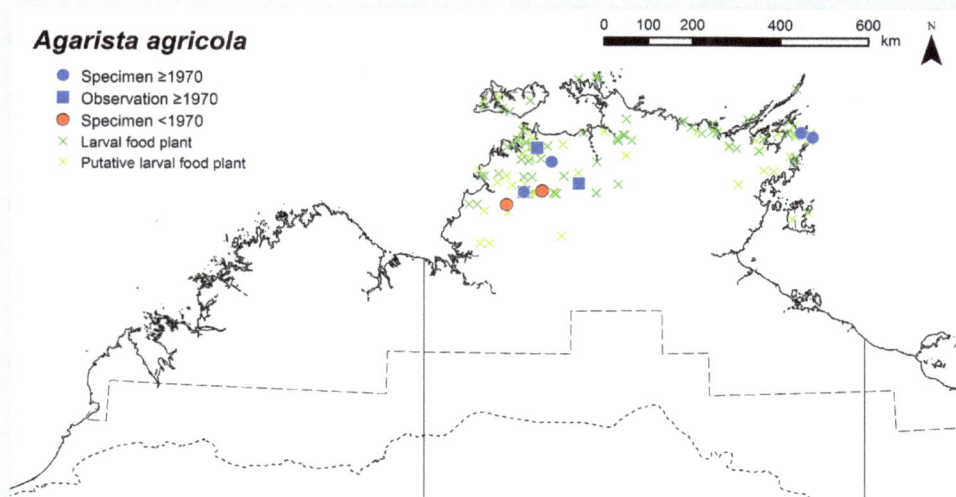

*Agarista agricola*

- ● Specimen ≥1970
- ■ Observation ≥1970
- ● Specimen <1970
- × Larval food plant
- × Putative larval food plant

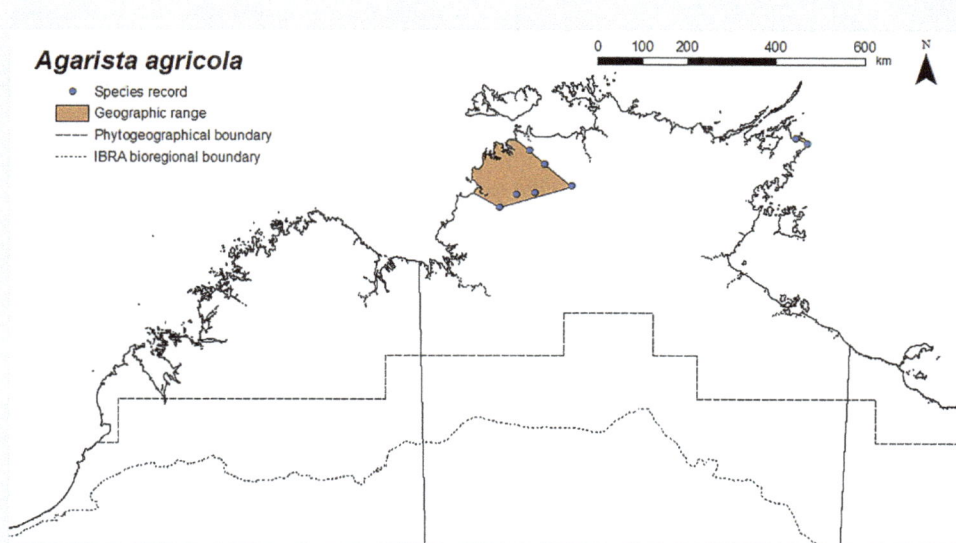

*Agarista agricola*

- ● Species record
- ▨ Geographic range
- --- Phytogeographical boundary
- ···· IBRA bioregional boundary

| Month | J | F | M | A | M | J | J | A | S | O | N | D |
|-------|---|---|---|---|---|---|---|---|---|---|---|---|
| Egg | | | | | | | | | | ▓ | | |
| Larva | | | | | | | | | | ▓ | ▓ | |
| Pupa | | | | | | | | | | | ▓ | ▓ |
| Adult | ▓ | ▓ | ▓ | ▓ | ▓ | | | | | ▓ | ▓ | ▓ |

# Graceful Day-moth

*Cruria donowani* (Boisduval, 1832)

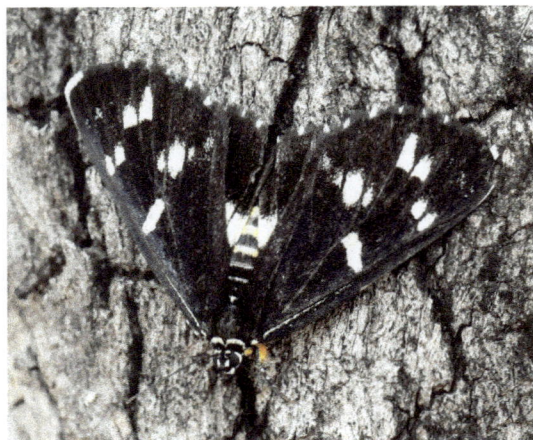

Plate 228 Mt Burrell, NT
Photo: M. F. Braby

## Distribution

This species occurs in the northern and eastern Kimberley, the Top End, Northern Deserts and western Gulf Country of the study region. It extends from moist coastal areas to drier inland areas of the semi-arid zone (c. 700 mm mean annual rainfall), where it has been recorded as far south as 35 km north of Dunmarra and McArthur River, NT. It has also been recorded at Walker Creek 36 km east of Karumba, Qld, just outside the eastern boundary of the study region. The geographic range falls within the spatial distribution of its larval food plant, which occurs more widely in the southern Kimberley, Tiwi Islands, eastern half of the Top End and Groote Eylandt. Further field surveys are thus required to determine whether *C. donowani* occurs in these areas. Outside the study region, *C. donowani* occurs in Timor and throughout eastern Australia.

## Habitat

*Cruria donowani* breeds in savannah woodland where the larval food plant grows as a vine, often at the base of tree trunks (Braby 2015e). Adults are readily attracted to flowers of *Erythrophleum chlorostachys* on which males seek females for mating.

## Larval food plant

*Cayratia trifolia* (Vitaceae).

## Seasonality

Adults are seasonal, occurring only during the 'build-up', wet season and early dry season (October–May), but they are most abundant from November to January, depending on the timing of the pre-monsoon storms, and before the arrival of monsoon rains. The immature stages (eggs and larvae) have been recorded in December. The larvae feed on the new soft leaf growth of the food plant, which regenerates from an underground tuber during the 'build-up' following the pre-monsoon storms and/or an increase in humidity. The food plant is a short-lived seasonal perennial vine and senesces during the dry season. The duration of the pupal stage is variable and there is probably only a single generation annually, followed by a partial second generation in the late wet season. The species survives the dry season in the pupal stage, which may remain in diapause for up to 10 months.

## Breeding status

This species is resident in the study region.

## Conservation status

LC.

## Cruria donowani

- ● Specimen ≥1970
- ■ Observation ≥1970
- ● Specimen <1970
- × Larval food plant

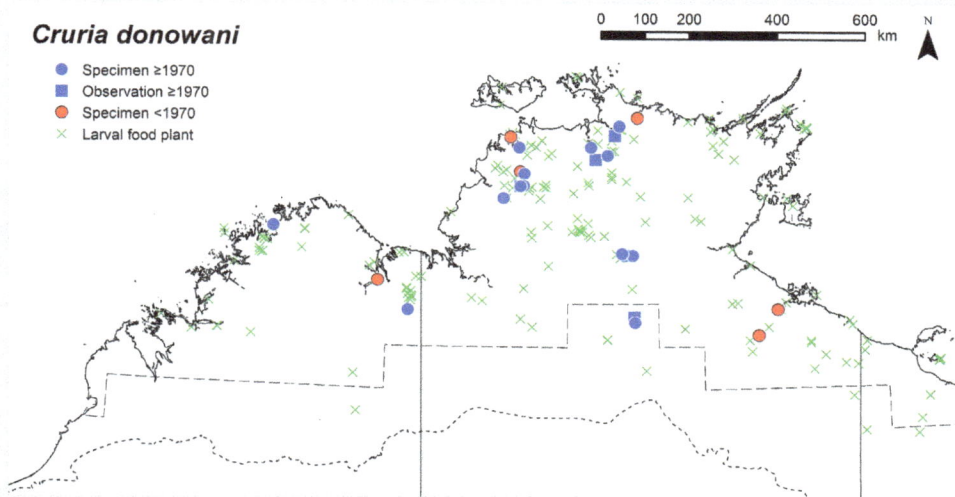

## Cruria donowani

- ● Species record
- ▨ Geographic range
- — — Phytogeographical boundary
- ········ IBRA bioregional boundary

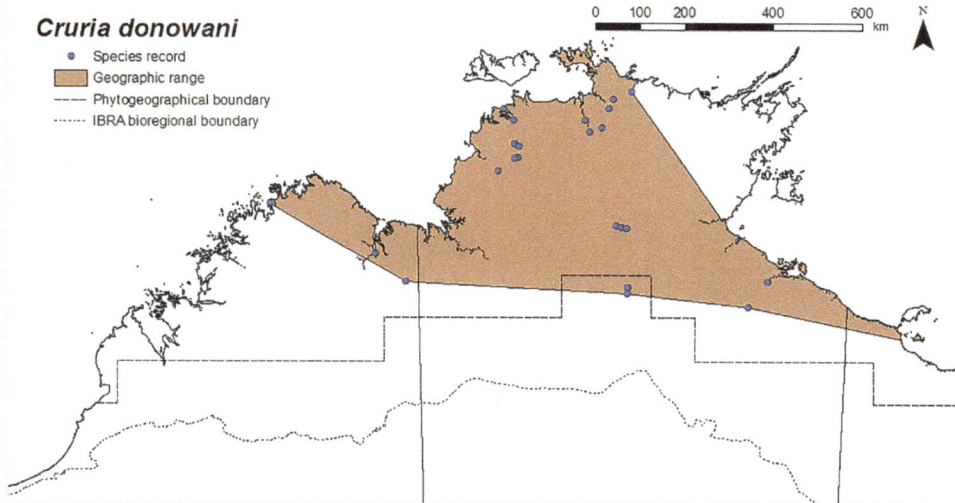

### Cruria donowani (n = 29)

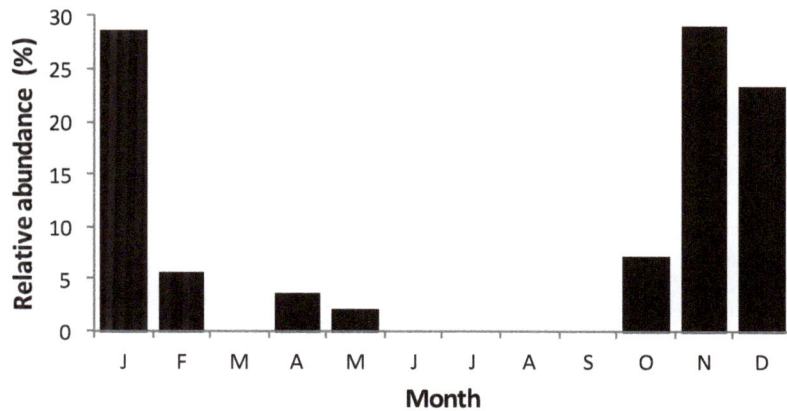

| Month | J | F | M | A | M | J | J | A | S | O | N | D |
|-------|---|---|---|---|---|---|---|---|---|---|---|---|
| Egg   |   |   |   |   |   |   |   |   |   |   |   |   |
| Larva |   |   |   |   |   |   |   |   |   |   |   |   |
| Pupa  |   |   |   |   |   |   |   |   |   |   |   |   |
| Adult |   |   |   |   |   |   |   |   |   |   |   |   |

# Tiger Day-moth
## *Cruria darwiniensis* (Butler, 1884)

Plate 229 Howard Springs Nature Park, NT
Photo: M. F. Braby

Plate 230 Howard Springs Nature Park, NT
Photo: M. F. Braby

## Distribution

This species is endemic to the study region. It occurs in the north of the Northern Territory, where it is restricted to the higher rainfall areas (> 900 mm mean annual rainfall) of the Top End, reaching its southernmost limit at Katherine, NT. The geographic range is wider than the spatial distribution of its known larval food plant (*Typhonium flagelliforme*), indicating that several other (as yet unreported) food plants are used. In particular, *Typhonium cochleare* occurs commonly in the southern and eastern areas of the range of *C. darwiniensis*.

The taxonomic status and relationships of *Cruria darwiniensis* require further investigation. Kiriakoff (1977) treated the taxon as a polytypic species, with two allopatric subspecies: *C. darwiniensis darwiniensis* (Butler, 1884) from the Northern Territory and *C. darwiniensis platyxantha* (Meyrick, 1891) from northern Queensland. However, Edwards (1996) treated *C. darwiniensis platyxantha* as a junior synonym of *C. tropica* (T. P. Lucas, 1891), which was considered a distinct species, whereas *C. darwiniensis* was treated as a synonym of *C. donowani* (Boisduval, 1832). In this work, *C. darwiniensis*, which is clearly a species distinct from *C. donowani*, is tentatively treated as a full species endemic to the Top End pending further investigation of its relationship with *C. tropica*.

## Habitat

*Cruria darwiniensis* breeds in mixed paperbark–pandanus swampland where the known larval food plant grows as a seasonal geophytic herb in low-lying water (Braby 2011a). Adults have also been collected in savannah woodland and mixed semi-deciduous monsoon vine thicket on limestone, where it almost certainly breeds on alternative food plants, especially *Typhonium cochleare*, which typically grows in eucalypt open forest on seasonally waterlogged sandy soils and occasionally in monsoon vine thicket.

## Larval food plants

*Typhonium flagelliforme* (Araceae); probably other *Typhonium* spp., including *T. cochleare*.

## Seasonality

Adults are seasonal, occurring only during the wet season (November–April), but they are most abundant in January and February. The immature stages (eggs or larvae) have been recorded from January to March during and following the monsoon rains. The larvae feed on the foliage of the food plant, which is a short-lived seasonal herb. The duration of the pupal stage is variable and there are probably only two generations annually. The species survives the dry season in the pupal stage, which may remain in diapause for up to nine months.

## Breeding status

This species is resident in the study region.

## Conservation status

LC.

## Cruria darwiniensis

- ● Specimen ≥1970
- ■ Observation ≥1970
- ● Specimen <1970
- ✕ Larval food plant
- ✕ Putative larval food plants

## Cruria darwiniensis

- ● Species record
- ▨ Geographic range
- — Phytogeographical boundary
- ⋯ IBRA bioregional boundary

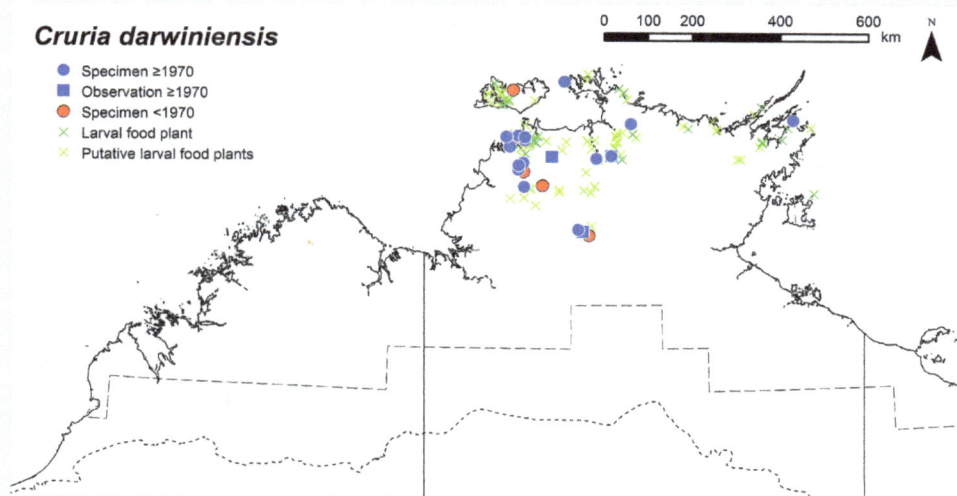

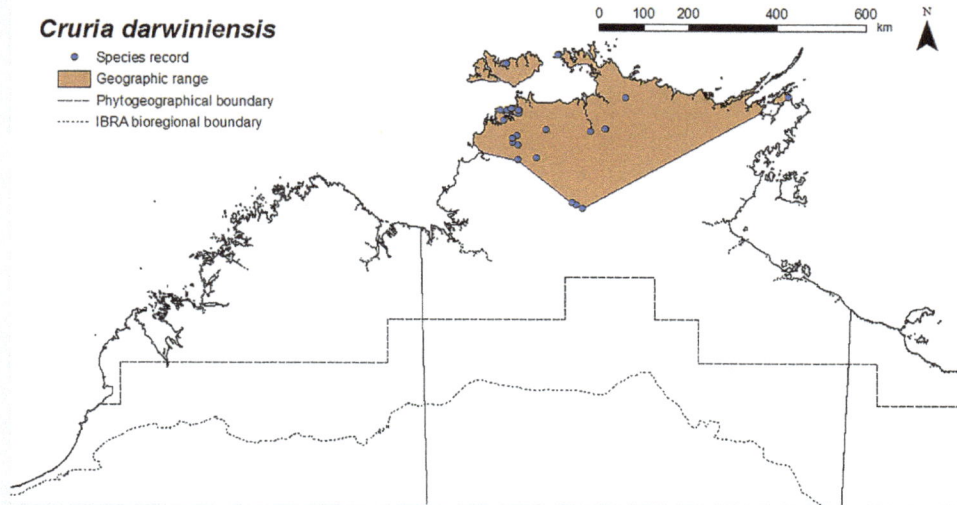

### Cruria darwiniensis (n = 33)

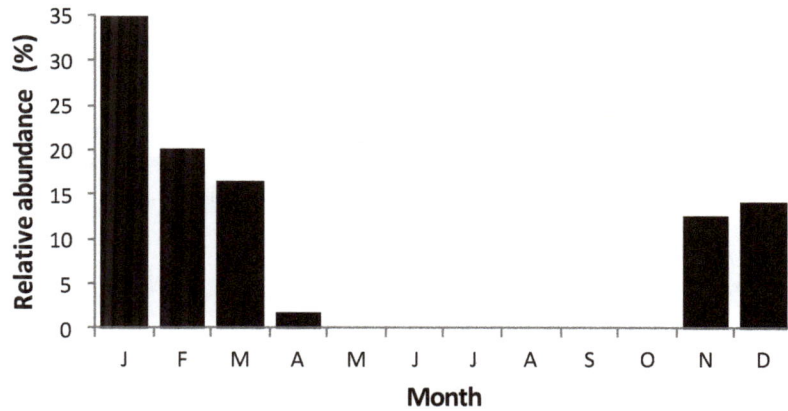

| Month | J | F | M | A | M | J | J | A | S | O | N | D |
|-------|---|---|---|---|---|---|---|---|---|---|---|---|
| Egg   |   |   |   |   |   |   |   |   |   |   |   |   |
| Larva |   |   |   |   |   |   |   |   |   |   |   |   |
| Pupa  |   |   |   |   |   |   |   |   |   |   |   |   |
| Adult |   |   |   |   |   |   |   |   |   |   |   |   |

# Mistletoe Day-moth

## *Comocrus behri* (Angas, 1847)

Plate 231 Cocoparra National Park, NSW
Photo: M. F. Braby

Plate 232 Cocoparra National Park, NSW
Photo: M. F. Braby

## Distribution

This species occurs in the northern Kimberley at Drysdale River Station, WA (S. Craswell), the Top End, Northern Deserts and western Gulf Country of the study region, its presence detected only as recently as 2008 (Braby 2011a). It extends from moist coastal areas to drier inland areas of the semi-arid zone (< 600 mm mean annual rainfall), where it has been recorded as far south as Limbunya Station and Kalkaringi in the Victoria River District and the Favenc Range, NT (Braby 2011a, 2015e). The larval food plants are considerably more widespread than the known geographic range, occurring also in the southern Kimberley, Tiwi Islands, Groote Eylandt and western Queensland in the Gulf Country; thus, further field surveys are required to determine whether *C. behri* is established in these areas. Outside the study region, *C. behri* occurs throughout eastern and southern Australia, as well as in the arid zone of central Australia (Braby 2011a).

## Habitat

*Comocrus behri* breeds in a variety of eucalypt woodland, open woodland and low open woodland habitats where the larval food plants grow as mistletoes (parasitic shrubs) on various trees, including *Eucalyptus*, *Corymbia* and *Acacia* (Braby 2011a, 2015e).

## Larval food plants

*Amyema bifurcata*, *A. sanguinea*, *A. villiflora*, *Diplatia grandibractea* (Loranthaceae).

## Seasonality

Adults have been recorded during most months of the year, but there are too few records (n = 19) to assess any seasonal changes in abundance. In general, adults tend to be more numerous during the second half of the year (June–December), particularly during the mid to late dry season (similar to *Delias argenthona* and *Ogyris* spp., which also use Loranthaceae as their larval food plants); however, adults of *C. behri* are usually rare and are never observed in large numbers or high densities, unlike in the temperate areas of south-eastern Australia. The immature stages (larvae or pupae) have been recorded sporadically from May to November, when adults are more abundant. Presumably, the species breeds continuously and several generations are completed annually.

## Breeding status

This species is resident in the study region.

## Conservation status

LC.

## Comocrus behri

- • Specimen ≥1970
- ■ Observation ≥1970
- × Larval food plants

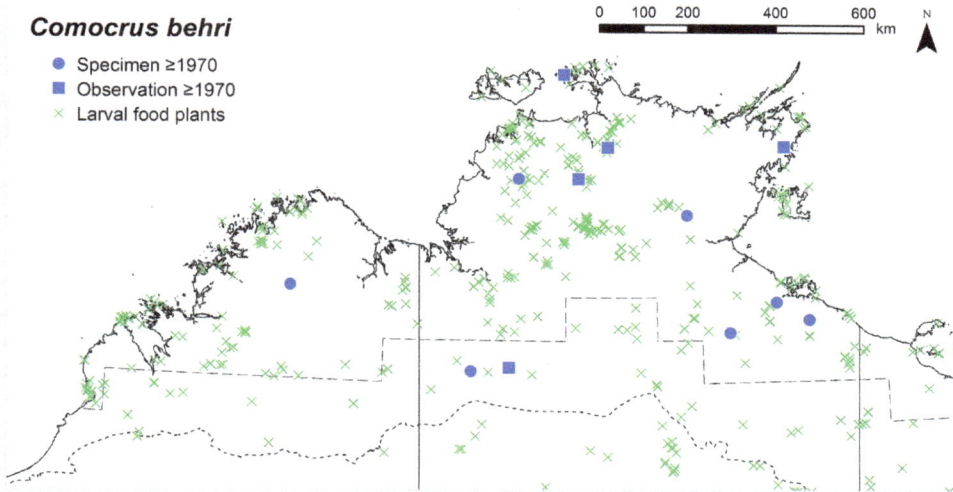

0  100  200      400      600 km

N

## Comocrus behri

- • Species record
- ▨ Geographic range
- --- Phytogeographical boundary
- ⋯⋯ IBRA bioregional boundary

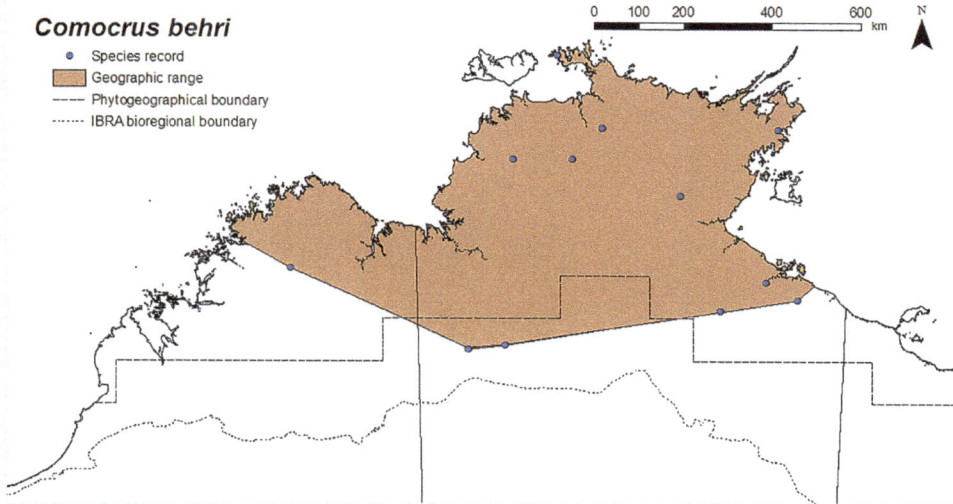

0  100  200      400      600 km

N

| Month | J | F | M | A | M | J | J | A | S | O | N | D |
|-------|---|---|---|---|---|---|---|---|---|---|---|---|
| Egg   |   |   |   |   |   |   |   |   |   |   |   |   |
| Larva |   |   |   |   |   |   |   |   |   |   |   |   |
| Pupa  |   |   |   |   |   |   |   |   |   |   |   |   |
| Adult |   |   |   |   |   |   |   |   |   |   |   |   |

# Addendum

The lycaenid butterfly *Candalides heathi* (Rayed Blue) was recently recorded from the study region. A male in good condition was collected from the Northern Deserts at Camfield Station at the northern edge of the Tanami Desert, NT (17°06'14.1"S, 131°31'27.9"E) by Jared Archibald on 15 September 2018. The species was previously unknown from the study region, having been recorded from arid and semi-arid areas further south and east – the nearest localities being at 270 km ENE of Alice Springs, NT and at 20 km SSW of Normanton, Qld near the Gulf of Carpentaria. This additional species brings the total list for the monsoon tropics of northern Australia to 164 species (133 butterflies and 31 diurnal moths).

# List of larval food plants

## Notes:

1. Asterisk (*) denotes introduced plant.
2. Taxa in blue denote putative larval food plants that require confirmation. These records are based on published Lepidoptera–plant associations adjacent to the study region or are considered very likely based on the spatial distribution of the species or on other evidence from the study region.

## ACANTHACEAE

| | |
|---|---|
| *Alternanthera denticulata* | *Hypolimnas bolina* |
| *Asystasia gangetica* | *Hypolimnas alimena, Hypolimnas bolina, Junonia orithya* |
| *Brunoniella australis* | *Hypolimnas alimena* |
| *Brunoniella linearifolia* | *Hypolimnas alimena* |
| *Dipteracanthus bracteatus* | *Yoma sabina* |
| *Hygrophila angustifolia* | *Junonia hedonia, Junonia villida, Zizula hylax* |
| *Pseuderanthemum variabile* | *Hypolimnas alimena, Junonia orithya* |
| *Thunbergia arnhemica* | *Junonia orithya* |

## AMARANTHACEAE

| | |
|---|---|
| *Tecticornia halocnemoides* | *Theclinesthes sulpitius* |
| *Tecticornia australasica* | *Theclinesthes sulpitius* |
| *Tecticornia indica* | *Theclinesthes sulpitius* |

## ANACARDIACEAE

| | |
|---|---|
| *Buchanania obovata* | *Arhopala eupolis* |
| *Semecarpus australiensis* | *Prosotas dubiosa* |

## ANNONACEAE

| | |
|---|---|
| *Annona muricata* | *Graphium eurypylus* |
| *Hubera nitidissima* | *Graphium eurypylus* |
| *Meiogyne cylindrocarpa* | *Graphium eurypylus* |
| *Melodorum rupestre* | *Graphium eurypylus, Protographium leosthenes* |
| *Miliusa brahei* | *Graphium eurypylus* |
| *Miliusa traceyi* | *Graphium eurypylus* |
| *Monoon australe* | *Graphium eurypylus* |
| *Polyalthia longifolia* | *Graphium eurypylus* |

## APOCYNACEAE

| | |
|---|---|
| *Adenium obesum* | *Euploea corinna* |
| *Brachystelma glabriflorum* | *Danaus petilia* |
| *Calotropis procera* | *Danaus petilia* |
| *Cynanchum carnosum* | *Danaus affinis, Danaus petilia, Tirumala hamata, Danaus chrysippus* |
| *Cynanchum christineae* | *Danaus petilia* |
| *Cynanchum floribundum* | *Danaus petilia* |
| *Cynanchum leibianum* | *Danaus petilia* |
| *Cynanchum pedunculatum* | *Danaus petilia* |
| *Gymnanthera oblonga* | *Euploea alcathoe, Euploea corinna* |
| *Ichnocarpus frutescens* | *Euploea corinna* |
| *Marsdenia australis* | *Danaus petilia* |
| *Marsdenia geminata* | *Euploea corinna, Euploea sylvester* |
| *Marsdenia glandulifera* | *Euploea alcathoe, Tirumala hamata* |
| *Marsdenia velutina* | *Tirumala hamata* |
| *Marsdenia viridiflora* | *Danaus affinis, Euploea corinna* |
| *Oxystelma esculentum* | *Danaus genutia, Danaus petilia* |
| *Parsonsia alboflavescens* | *Euploea alcathoe, Euploea corinna, Euploea sylvester* |
| *Rhyncharrhena linearis* | *Danaus petilia* |
| *Sarcolobus hullsii* | *Danaus affinis, Euploea corinna* |
| *Sarcostemma viminale* | *Euploea corinna* |
| *Secamone elliptica* | *Euploea corinna, Tirumala hamata* |
| *Tylophora flexuosa* | *Danaus petilia* |
| *Rhyncharrhena linearis* | *Danaus petilia* |

## ARACEAE

| | |
|---|---|
| *Typhonium cochleare* | *Cruria darwiniensis* |
| *Typhonium flagelliforme* | *Cruria darwiniensis* |

## ARECACEAE

| | |
|---|---|
| *Carpentaria acuminata* | *Cephrenes augiades* |
| *Livistona benthamii* | *Cephrenes augiades, Cephrenes trichopepla* |
| *Livistona humilis* | *Cephrenes trichopepla* |
| *Livistona inermis* | *Cephrenes trichopepla* |
| *Livistona lorophylla* | *Cephrenes trichopepla* |
| *Livistona nasmophila* | *Cephrenes trichopepla* |
| *Livistona rigida* | *Cephrenes trichopepla* |
| *Livistona victoriae* | *Cephrenes trichopepla* |

## ARISTOLOCHIACEAE

| | |
|---|---|
| *Aristolochia acuminata* | *Cressida cressida* |
| *Aristolochia holtzei* | *Cressida cressida* |
| *Aristolochia indica* | *Cressida cressida* |
| *Aristolochia pubera* | *Cressida cressida* |
| *Aristolochia thozetii* | *Cressida cressida* |

## ASTERACEAE

| | |
|---|---|
| *Synedrella nodiflora* | *Hypolimnas bolina* |
| *Tridax procumbens* | *Hypolimnas bolina* |

## CANNABACEAE

| | |
|---|---|
| *Celtis australiensis* | *Charaxes sempronius, Libythea geoffroyi* |

## CAPPARACEAE

| | |
|---|---|
| *Capparis jacobsii* | *Belenois java, Cepora perimale* |
| *Capparis lasiantha* | *Belenois java, Elodina padusa* |
| *Capparis sepiaria* | *Belenois java, Cepora perimale, Elodina walkeri, Leptosia nina* |
| *Capparis umbonata* | *Belenois java, Cepora perimale* |

## CELASTRACEAE

| | |
|---|---|
| *Salacia chinensis* | *Bindahara phocides* |

## CHRYSOBALANACEAE

| | |
|---|---|
| *Maranthes corymbosa* | *Arhopala eupolis, Hypochrysops ignitus* |

## CLUSIACEAE

*Calophyllum inophyllum*        *Arhopala micale*

## COMBRETACEAE

| | |
|---|---|
| *Lumnitzera racemosa* | *Hypochrysops apelles* |
| *Terminalia carpentariae* | *Arhopala eupolis* |
| *Terminalia catappa* | *Petrelaea tombugensis* |
| *Terminalia ferdinandiana* | *Arhopala eupolis* |
| *Terminalia microcarpa* | *Badamia exclamationis* |
| *Terminalia muelleri* | *Petrelaea tombugensis* |

## CONVOLVULACEAE

*Evolvulus alsinoides*        *Junonia villida*

## CYPERACEAE

| | |
|---|---|
| *Cyperus javanicus* | *Hesperilla sexguttata* |
| *Scleria sphacelata* | *Hesperilla crypsigramma* |

## DILLENIACEAE

| | |
|---|---|
| *Hibbertia brevipedunculata* | *Idalima aethrias* |
| *Hibbertia brownii* | *Idalima leonora* |
| *Hibbertia candicans* | *Idalima leonora, Idalima* sp. 'Arnhem Land' |
| *Hibbertia cistifolia* | *Idalima aethrias* |
| *Hibbertia dilatata* | *Idalima metasticta* |
| *Hibbertia juncea* | *Idalima metasticta* |

## EBENACEAE

*Diospyros maritima*        *Graphium eurypylus*

## EUPHORBIACEAE

| | |
|---|---|
| *Adriana tomentosa* | *Theclinesthes albocinctus* |
| *Excoecaria agallocha* | *Delias aestiva* |
| *Excoecaria ovalis* | *Delias aestiva* |
| *Mallotus nesophilus* | *Catopyrops florinda* |

## FABACEAE

| | |
|---|---|
| *Acacia aneura* | *Jalmenus icilius* |
| *Acacia aulacocarpa* | *Sahulana scintillata* |
| *Acacia auriculiformis* | *Prosotas dubiosa, Theclinesthes miskini* |
| *Acacia difficilis* | *Theclinesthes miskini* |
| *Acacia hemsleyi* | *Charaxes sempronius* |
| *Acacia holosericea* | *Theclinesthes miskini* |
| *Acacia leptocarpa* | *Hypochrysops ignitus* |
| *Acacia platycarpa* | *Theclinesthes miskini* |
| *Acacia plectocarpa* | *Nacaduba biocellata* |
| *Acacia scopulorum* | *Prosotas dubiosa* |
| *Acacia tetragonophylla* | *Jalmenus icilius* |
| *Acacia torulosa* | *Nacaduba biocellata* |
| *Acacia tumida* | *Hypochrysops ignitus, Nacaduba biocellata* |
| *Bossiaea bossiaeoides* | *Jamides phaseli* |
| *Cajanus aromaticus* | *Catochrysops panormus, Jamides phaseli* |
| *Cajanus pubescens* | *Catochrysops panormus* |
| *Canavalia rosea* | *Jamides phaseli* |
| *\*Cassia fistula* | *Catopsilia pomona* |
| *\*Cassia siamea* | *Catopsilia pomona* |
| *Cathormion umbellatum* | *Theclinesthes miskini* |
| *Chamaecrista mimosoides* | *Eurema herla* |
| *Chamaecrista nigricans* | *Eurema herla* |
| *Chamaecrista* sp. | *Eurema laeta* |
| *Crotalaria cunninghamii* | *Lampides boeticus* |
| *Crotalaria novae-hollandiae* | *Lampides boeticus* |
| *Cullen badocanum* | *Papilio demoleus* |
| *Cullen balsamicum* | *Papilio demoleus* |
| *Cullen cinereum* | *Papilio demoleus* |
| *\*Dalbergia sissoo* | *Prosotas dubiosa* |
| *\*Delonix regia* | *Anthene seltuttus* |
| *Derris trifoliata* | *Hasora hurama* |
| *Desmodium heterocarpon* | *Everes lacturnus* |
| *\*Desmodium triflorum* | *Zizina otis* |
| *\*Desmodium* sp. | *Anthene lycaenoides* |
| *Flemingia lineata* | *Catochrysops panormus, Freyeria putli* |
| *Galactia tenuiflora* | *Eurema alitha* |
| *Indigofera linifolia* | *Freyeria putli* |
| *\*Macroptilium atropurpureum* | *Euchrysops cnejus, Famegana alsulus* |
| *\*Macroptilium lathyroides* | *Euchrysops cnejus* |
| *Millettia pinnata* | *Anthene lycaenoides, Anthene seltuttus, Hasora chromus, Jamides phaseli, Prosotas dubiosa* |

*Mimosa pigra* — *Charaxes sempronius*
*Neptunia gracilis* — *Eurema smilax*
*Neptunia monosperma* — *Eurema smilax*
*Peltophorum pterocarpum* — *Charaxes sempronius*
*\*Senna alata* — *Catopsilia pomona*
*Senna artemisioides* — *Jalmenus icilius, Eurema smilax*
*Senna leptoclada* — *Catopsilia scylla*
*Senna magnifolia* — *Catopsilia pomona*
*\*Senna occidentalis* — *Catopsilia pomona, Catopsilia pyranthe*
*Senna oligoclada* — *Catopsilia scylla*
*Senna planitiicola* — *Catopsilia pyranthe*
*Senna surattensis* — *Anthene lycaenoides, Catopsilia scylla*
*Senna venusta* — *Catopsilia pomona, Catopsilia pyranthe*
*Sesbania cannabina* — *Theclinesthes miskini*
*Sesbania simpliciuscula* — *Catochrysops panormus, Jamides phaseli*
*Tephrosia* sp. — *Famegana alsulus*
*Tephrosia spechtii* — *Jamides phaseli*
*Vigna lanceolata* — *Euchrysops cnejus, Famegana alsulus*
*Vigna marina* — *Euchrysops cnejus*
*Vigna radiata* — *Euchrysops cnejus, Famegana alsulus*
*\*Vigna unguiculata* — *Euchrysops cnejus*
*Vigna vexillata* — *Euchrysops cnejus, Famegana alsulus*

## FLAGELLARIACEAE

*Flagellaria indica* — *Telicota augias, Anthene lycaenoides*

## IRIDACEAE

*Patersonia macrantha* — *Mesodina gracillima*

## LAMIACEAE

*Clerodendrum floribundum* — *Hypolycaena phorbas, Hypochrysops ignitus*
*Vitex acuminata* — *Charaxes sempronius*

## LAURACEAE

*Cassytha capillaris* — *Candalides delospila, Candalides erinus, Candalides geminus, Hecatesia* sp. 'Amata'

*Cassytha filiformis* — *Candalides erinus, Candalides geminus, Hecatesia* sp. 'Arnhem Land'

*Cryptocarya cunninghamii* — *Graphium eurypylus*

## LECYTHIDACEAE

| | |
|---|---|
| *Barringtonia acutangula* | *Anthene lycaenoides* |
| *Planchonia careya* | *Chaetocneme denitza, Hypochrysops ignitus, Hypolycaena phorbas* |

## LOGANIACEAE

| | |
|---|---|
| *Strychnos lucida* | *Deudorix smilis* |

## LORANTHACEAE

| | |
|---|---|
| *Amyema benthamii* | *Ogyris amaryllis, Ogyris zosine* |
| *Amyema bifurcata* | *Comocrus behri, Ogyris oroetes, Ogyris zosine* |
| *Amyema miquelii* | *Candalides margarita, Delias argenthona, Ogyris zosine* |
| *Amyema quandang* | *Ogyris amaryllis, Ogyris barnardi* |
| *Amyema sanguinea* | *Birthana cleis, Candalides margarita, Comocrus behri, Delias argenthona, Ogyris amaryllis, Ogyris iphis, Ogyris zosine* |
| *Amyema thalassia* | *Ogyris amaryllis* |
| *Amyema villiflora* | *Candalides margarita, Comocrus behri, Ogyris zosine* |
| *Decaisnina signata* | *Birthana cleis, Candalides margarita, Delias argenthona, Hypolycaena phorbas, Ogyris zosine* |
| *Decaisnina triflora* | *Candalides margarita* |
| *Dendrophthoe glabrescens* | *Birthana cleis, Candalides margarita, Delias argenthona* |
| *Dendrophthoe odontocalyx* | *Birthana cleis, Candalides margarita, Delias argenthona* |
| *Diplatia grandibractea* | *Comocrus behri, Ogyris amaryllis, Ogyris zosine* |

## MALVACEAE

| | |
|---|---|
| *Brachychiton diversifolius* | *Arhopala micale* |
| *Brachychiton paradoxus* | *Hypochrysops ignitus* |
| *Sida acuta* | *Hypolimnas bolina* |
| *Sterculia quadrifida* | *Arhopala micale* |

## MOLLUGINACEAE

| | |
|---|---|
| *Glinus lotoides* | *Zizeeria karsandra* |
| *Glinus oppositifolius* | *Zizeeria karsandra* |

## MORACEAE

| | |
|---|---|
| *Ficus virens* | *Euploea corinna* |
| *Trophis scandens* | *Euploea darchia* |

## MYRTACEAE

| | |
|---|---|
| *Corymbia bella* | *Arhopala eupolis, Theclinesthes miskini* |
| *Corymbia disjuncta* | *Anthene seltuttus, Arhopala eupolis, Theclinesthes miskini* |
| *Corymbia ferruginea* | *Theclinesthes miskini* |
| *Eucalyptus miniata* | *Arhopala eupolis* |
| *\*Syzygium aqueum* | *Anthene lycaenoides, Arhopala eupolis* |

## OROBANCHACEAE

| | |
|---|---|
| *Buchnera asperata* | *Junonia orithya* |
| *Buchnera gracilis* | *Junonia orithya* |
| *Buchnera linearis* | *Junonia orithya* |

## PASSIFLORACEAE

| | |
|---|---|
| *Adenia heterophylla* | *Acraea andromacha, Cethosia penthesilea* |
| *\*Passiflora foetida* | *Acraea andromacha, Acraea terpsicore* |

## PHYLLANTHACEAE

| | |
|---|---|
| *Breynia cernua* | *Eurema hecabe, Hypolycaena phorbas* |
| *Bridelia tomentosa* | *Anthene lycaenoides* |
| *Glochidion apodogynum* | *Hypochrysops ignitus* |
| *Phyllanthus* sp. | *Eurema hecabe* |

## POACEAE

| | |
|---|---|
| *\*Andropogon gayanus* | *Telicota colon* |
| *Aristida macroclada* | *Hypocysta adiante* |
| *Arundinella nepalensis* | *Hypocysta adiante* |
| *\*Axonopus compressus* | *Hypocysta adiante, Ocybadistes walkeri, Ypthima arctous* |
| *\*Cenchrus pedicellatus* | *Borbo impar, Melanitis leda, Pelopidas lyelli, Taractrocera ina* |
| *\*Chloris* sp. | *Hypocysta adiante* |
| *Chrysopogon aciculatus* | *Neohesperilla crocea* |
| *Chrysopogon elongatus* | *Pelopidas lyelli* |
| *Chrysopogon fallax* | *Synemon* sp. 'Roper River', *Synemon* sp. 'Kimberley' |
| *Chrysopogon latifolius* | *Synemon phaeoptila* |
| *Chrysopogon pallidus* | *Synemon* sp. 'Roper River' |
| *Chrysopogon setifolius* | *Synemon wulwulam* |
| *\*Cymbopogon citratus* | *Taractrocera ina* |
| *Cymbopogon procerus* | *Taractrocera ina* |
| *\*Cynodon dactylon* | *Hypocysta adiante, Ocybadistes flavovittatus* |

| | |
|---|---|
| *Cynodon radiatus* | Melanitis leda |
| Dichanthium sericeum | Mycalesis perseus |
| Digitaria gibbosa | Hypocysta adiante |
| Eriachne triodioides | Pelopidas lyelli |
| Eulalia aurea | Taractrocera anisomorpha |
| Heteropogon contortus | Mycalesis perseus |
| Heteropogon triticeus | Mycalesis perseus |
| Hymenachne acutigluma | Borbo impar |
| Imperata cylindrica | Melanitis leda, Suniana lascivia, Telicota colon, Ypthima arctous, Mydosama sirius, Ocybadistes walkeri |
| Ischaemum australe | Hypocysta adiante, Melanitis leda, Suniana lascivia, Telicota colon, Mydosama sirius, Ocybadistes hypomeloma |
| Ischaemum tropicum | Hypocysta adiante |
| *Megathyrsus maximus | Borbo impar, Melanitis leda, Pelopidas lyelli, Suniana lascivia, Suniana sunias, Taractrocera ina |
| *Melinis repens | Ocybadistes walkeri |
| Micraira adamsii | Taractrocera ilia |
| Micraira brevis | Taractrocera psammopetra |
| Micraira compacta | Taractrocera ilia |
| Micraira dentata | Taractrocera ilia |
| Micraira dunlopii | Taractrocera psammopetra |
| Micraira inserta | Taractrocera ilia |
| Micraira lazaridis | Taractrocera psammopetra |
| Micraira multinervia | Taractrocera ilia |
| Micraira pungens | Taractrocera ilia |
| Micraira sp. 'Purnululu' | Taractrocera psammopetra |
| Micraira spiciforma | Taractrocera psammopetra |
| Micraira spinifera | Taractrocera ilia |
| Micraira subspicata | Taractrocera ilia |
| Micraira tenuis | Taractrocera ilia |
| Micraira viscidula | Taractrocera ilia |
| Mnesithea rottboellioides | Pelopidas lyelli, Telicota colon |
| *Oryza sativa | Borbo impar, Melanitis leda, Parnara amalia, Pelopidas lyelli, Telicota colon |
| *Paspalum scrobiculatum | Telicota colon |
| Rottboellia cochinchinensis | Borbo cinnara |
| Schizachyrium pachyarthron | Neohesperilla crocea |
| Schizachyrium perplexum | Neohesperilla xiphiphora |
| *Sorghum bicolor | Taractrocera anisomorpha |
| Sorghum intrans | Neohesperilla xiphiphora |
| Sorghum macrospermum | Taractrocera ina |

| | |
|---|---|
| *Sorghum plumosum* | *Synemon wulwulam* |
| *\*Sporobolus* sp. | *Hypocysta adiante* |
| *Themeda triandra* | *Mycalesis perseus, Neohesperilla senta, Ocybadistes hypomeloma, Ypthima arctous* |
| *Triodia bitextura* | *Proeidosa polysema* |
| *Triodia microstachya* | *Proeidosa polysema* |
| *Triodia pungens* | *Proeidosa polysema* |
| *\*Urochloa mosambicensis* | *Hypocysta adiante* |
| *Whiteochloa airoides* | *Borbo impar* |
| *\*Zea mays* | *Borbo impar* |

## PORTULACACEAE

| | |
|---|---|
| *Portulaca oleracea* | *Hypolimnas misippus* |

## PRIMULACEAE

| | |
|---|---|
| *Embelia curvinervia* | *Nacaduba kurava* |

## PUTRANJIVACEAE

| | |
|---|---|
| *Drypetes deplanchei* | *Appias albina, Appias paulina* |

## RHAMNACEAE

| | |
|---|---|
| *Alphitonia excelsa* | *Hypochrysops ignitus* |

## RHIZOPHORACEAE

| | |
|---|---|
| *Carallia brachiata* | *Dysphania numana* |
| *Ceriops australis* | *Hypochrysops apelles* |

## RUBIACEAE

| | |
|---|---|
| *Gardenia megasperma* | *Hypochrysops ignitus* |
| *\*Oldenlandia corymbosa* | *Periopta diversa* |
| *\*Spermacoce articularis* | *Periopta diversa* |
| *Spermacoce phalloides* | *Periopta diversa* |

## RUTACEAE

| | |
|---|---|
| *\*Citrus* sp. | *Papilio aegeus, Papilio anactus, Papilio fuscus* |
| *Boronia kalumburuensis* | *Nesolycaena caesia* |
| *Boronia lanceolata* | *Nesolycaena urumelia* |
| *Boronia lanuginosa* | *Nesolycaena urumelia* |

| | |
|---|---|
| *Boronia laxa* | *Nesolycaena urumelia* |
| *Boronia wilsonii* | *Nesolycaena caesia, Nesolycaena urumelia* |
| *Glycosmis trifoliata* | *Papilio fuscus* |
| *Micromelum minutum* | *Papilio aegeus, Papilio fuscus* |
| *Zanthoxylum parviflorum* | *Papilio fuscus* |

## SALICACEAE

| | |
|---|---|
| *\*Flacourtia inermis* | *Phalanta phalantha* |
| *\*Flacourtia rukam* | *Phalanta phalantha* |
| *Flacourtia territorialis* | *Phalanta phalantha* |

## SANTALACEAE

| | |
|---|---|
| *Santalum lanceolatum* | *Theclinesthes miskini* |

## SAPINDACEAE

| | |
|---|---|
| *Atalaya hemiglauca* | *Theclinesthes miskini* |
| *Atalaya variifolia* | *Theclinesthes miskini* |
| *Cupaniopsis anacardioides* | *Anthene lycaenoides, Anthene seltuttus, Arhopala eupolis, Arhopala micale, Prosotas dubiosa, Sahulana scintillata* |
| *Dodonaea hispidula* | *Catopyrops florinda* |
| *\*Litchi chinensis* | *Anthene seltuttus* |

## SMILACACEAE

| | |
|---|---|
| *Smilax australis* | *Hypochrysops ignitus, Hypolycaena phorbas* |

## URTICACEAE

| | |
|---|---|
| *Pipturus argenteus* | *Pollanisus* sp. 7 |

## VIOLACEAE

| | |
|---|---|
| *Hybanthus aurantiacus* | *Acraea andromacha* |
| *Hybanthus enneaspermus* | *Acraea andromacha, Acraea terpsicore* |

## VITACEAE

| | |
|---|---|
| *Ampelocissus acetosa* | *Periopta ardescens, Pseudosesia oberthuri, Radinocera* sp. 'Sandstone', *Radinocera vagata* |
| *Ampelocissus frutescens* | *Periopta ardescens, Pseudosesia oberthuri, Radinocera vagata* |
| *Cayratia maritima* | *Radinocera vagata* |
| *Cayratia trifolia* | *Cruria donowani, Radinocera vagata* |

*Leea indica*       *Agarista agricola*
*Leea rubra*       *Agarista agricola*

## ZYGOPHYLLACEAE

*Tribulopis bicolor*       *Zizeeria karsandra*
*Tribulus cistoides*       *Zizeeria karsandra*

# Conservation status evaluation for taxa with small geographic range sizes

Conservation evaluation is based on IUCN Red List categories and criteria, particularly criterion B: geographic range size according to extent of occurrence (EOO) or area of occupancy (AOO), number of locations and evidence of decline due to threatening processes. Justifications and actions needed are listed for threatened (Vulnerable: VU), Near Threatened (NT) and Data Deficient (DD) taxa. Symbol (†) denotes taxa endemic to the study region and therefore the status evaluation is global; for non-endemic taxa, the assessment is at the regional level and applies to the relevant population(s) within the study region.

| Taxon | Geographic range (sq km) | Distribution parameter (and method) | No. of locations | Cons. status | Justification | Actions needed |
|---|---|---|---|---|---|---|
| †*Ogyris iphis doddi* | c. 1,400 | AOO (spatial buffering) | 2 | VU | Criterion B2ab(ii)(iii) | Clarify distribution and habitat, and population structure within critical habitat. |
| †*Euploea alcathoe enastri* | 9,103 | EOO (minimum convex polygon) | 11 | NT | Criterion A3c: projected population decline over 10 years based on decline of habitat quality through inappropriate fire regime and disturbance by feral animals. | Monitor extent and/ or quality of critical habitat and butterfly occupancy, and manage threats from fire and feral animals. |
| †*Hypochrysops apelles* ssp. 'Top End' | 4,325 | EOO (minimum convex polygon) | 3 | NT | Criterion D2: projected loss of locations through decline of habitat from coastal development. | Clarify distribution and monitor butterfly occupancy of critical habitat. |
| †*Idalima* sp. 'Arnhem Land' | c. 1,400 | AOO (spatial buffering) | 2 | NT | Criteria A3c, D2: projected loss of locations and/or population decline over 10 years based on decline of larval food plant through inappropriate fire regime. | Monitor abundance of moth and food plant and occupancy of critical habitat. |

| Taxon | Geographic range (sq km) | Distribution parameter (and method) | No. of locations | Cons. status | Justification | Actions needed |
|---|---|---|---|---|---|---|
| †*Hecatesia* sp. 'Arnhem Land' | c. 1,400 | AOO (spatial buffering) | 2 | NT | Criteria A3c, D2: projected loss of locations and/or population decline over 10 years based on decline of larval food plant through inappropriate fire regime. | Monitor abundance of moth and food plant and occupancy of critical habitat. |
| †*Nesolycaena caesia* | 27,900 | EOO (minimum convex polygon) | 5 | DD | Uncertain threats: possible decline of larval food plant and/or butterfly through inappropriate fire regime. | Monitor abundance of butterfly and food plant and occupancy of critical habitat in relation to fire regimes. |
| †*Hesperilla crypsigramma* ssp. 'Top End' | 20,981 | EOO (minimum convex polygon) | 9 | DD | Uncertain threats: possible decline of larval food plant and/or butterfly through inappropriate fire regime. | Monitor abundance of butterfly and food plant and occupancy of critical habitat in relation to fire regimes. |
| *Ogyris barnardi barnardi* | 9,095 | EOO (minimum convex polygon) | 6 | DD | Uncertain threats: possible decline of habitat through poor land management. | Monitor extent and/or quality of critical habitat and butterfly occupancy, and identify key threatening processes. |
| *Petrelaea tombugensis* | c. 1,400 | AOO (spatial buffering) | 2 | DD | Few records and uncertain distribution; deficient ecological information. | Clarify extent of distribution and determine larval food plant. |
| *Synemon* sp. 'Kimberley' | c. 1,400 | AOO (spatial buffering) | 2 | DD | Few records and uncertain distribution; deficient ecological information. | Clarify extent of distribution, determine critical habitat and larval food plant. |
| *Mimeusemia econia* | c. 1,400 | AOO (spatial buffering) | 2 | DD | Few records and uncertain distribution; deficient ecological information. | Clarify extent of distribution, determine critical habitat and larval food plant. |
| *Agarista agricola agricola* | c. 1,400 | AOO (spatial buffering) | 2 | DD | Few records and uncertain population status and/or distribution; uncertain threats. | Clarify extent of distribution and identify key threatening processes. |
| *Badamia exclamationis* | c. 700 | AOO (spatial buffering) | 1 | DD | Deficient ecological information and uncertain threats. | Clarify extent of breeding distribution, determine critical habitat and identify key threatening processes. |
| †*Suniana lascivia lasus* | c. 700 | AOO (spatial buffering) | 1 | DD | Few records and uncertain population status and/or distribution; insufficient ecological information; uncertain threats. | Clarify extent of distribution, determine critical habitat and identify key threatening processes. |
| *Acrodipsas myrmecophila* | c. 700 | AOO (spatial buffering) | 1 | DD | Few records and uncertain population status and/or distribution. | Clarify extent of distribution. |

| Taxon | Geographic range (sq km) | Distribution parameter (and method) | No. of locations | Cons. status | Justification | Actions needed |
|-------|--------------------------|-------------------------------------|------------------|--------------|---------------|----------------|
| †*Acrodipsas decima* | c. 700 | AOO (spatial buffering) | 1 | DD | Few records and uncertain population status and/or distribution; insufficient ecological information. | Clarify extent of distribution and determine critical habitat. |
| *Bindahara phocides* | c. 700 | AOO (spatial buffering) | 1 | DD | Uncertain population status and/or distribution; insufficient ecological information. | Determine whether population is established and clarify distribution and larval food plant. |
| *Theclinesthes albocinctus* | c. 700 | AOO (spatial buffering) | 1 | DD | Uncertain threats. | Clarify extent of distribution and identify key threatening processes. |
| †*Pollanisus sp. 7* | c. 700 | AOO (spatial buffering) | 1 | DD | Old record and uncertain population status and/or distribution; deficient ecological information. | Determine whether population is established and clarify distribution and larval food plant. |
| *Euchromia creusa* | c. 700 | AOO (spatial buffering) | 1 | DD | Few records and uncertain distribution; deficient ecological information. | Clarify extent of distribution and determine larval food plant. |
| *Mimeusemia centralis* | < 700 | AOO (spatial buffering) | 1 | DD | Few records and uncertain distribution; deficient ecological information. | Clarify extent of distribution and determine critical habitat and larval food plant. |
| †*Candalides geminus gagadju* | 39,942 | EOO (Arnhem Land Plateau) | 7 | LC | | |
| †*Borbo impar lavinia* | 35,356 | EOO (modified α-hull) | 20 | LC | | |
| †*Pseudosesia oberthuri* | 29,990 | EOO (modified α-hull) | 6 | LC | | |
| †*Hasora hurama territorialis* | 28,050 | EOO (modified α-hull) | 7 | LC | | |
| †*Nacaduba kurava felsina* | 25,246 | EOO (minimum convex polygon) | 10 | LC | | |
| *Jalmenus icilius* | 22,528 | EOO (minimum convex polygon) | 3 | LC | | |
| *Appias albina albina* | 22,372 | EOO (modified α-hull) | 6 | LC | | |
| *Yoma sabina* | 14,900 | EOO (modified α-hull) | 13 | LC | | |
| †*Agarista agricola biformis* | 8,860 | EOO (minimum convex polygon) | 6 | LC | | |

| Taxon | Geographic range (sq km) | Distribution parameter (and method) | No. of locations | Cons. status | Justification | Actions needed |
|---|---|---|---|---|---|---|
| †*Leptosia nina* ssp. 'Kimberley' | 5,925 | EOO (modified α-hull) | 9 | LC | | |
| †*Taractrocera ilia* | 3,582 | EOO (minimum convex polygon) | 5 | LC | | |
| †*Protographium leosthenes geimbia* | 1,697 | EOO (minimum convex polygon) | 4 | LC | | |
| †*Radinocera* sp. 'Sandstone' | c. 1,400 | AOO (spatial buffering) | 2 | LC | | |

# Bibliography

Aduse-Poku, K., Brattström, O., Kodandaramaiah, U., Lees, D. C., Brakefield, P. M. & Wahlberg, N. 2015. Systematics and historical biogeography of the old world butterfly subtribe Mycalesina (Lepidoptera: Nymphalidae: Satyrinae). *BMC Evolutionary Biology* 15: 167. doi.org/10.1186/s12862-015-0449-3.

Andersen, A. N., Cook, G. D., Corbett, L. K., Douglas, M., Eager, R. W., Russell-Smith, J., Setterfield, S. A., Williams, R. J. & Woinarski, J. C. Z. 2005. Fire frequency and biodiversity conservation in Australian tropical savannas: Implications from the Kalpaga fire experiment. *Austral Ecology* 30: 155–167. doi.org/10.1111/j.1442-9993.2005.01441.x.

Andersen, A. N., Hertog, T. & Woinarski, J. C. Z. 2006. Long-term fire exclusion and ant community structure in an Australian tropical savanna: Congruence with vegetation succession. *Journal of Biogeography* 33: 823–832. doi.org/10.1111/j.1365-2699.2006.01463.x.

Andersen, A. N., Humphrey, C. & Braby, M. F. 2014. Threatened invertebrates in Kakadu National Park. In: Winderlich, S. and Woinarski, J. C. Z. (eds) *Kakadu National Park Landscape Symposia Series. Symposium 7: Conservation of threatened species, 26–27 March 2013, Bowali Visitor Centre, Kakadu National Park*. Internal Report 623, June. pp. 48–57. Supervising Scientist: Darwin.

Andersen, A. N., Parr, C. L., Lowe, L. M. & Muller, W. J. 2007a. Contrasting fire-related resilience of ecologically dominant ants in tropical savannas of northern Australia. *Diversity and Distributions* 13: 438–446. doi.org/10.1111/j.1472-4642.2007.00353.x.

Andersen, A. N. & Reichel, H. 1994. The ant (Hymenoptera: Formicidae) fauna of Holmes Jungle, a rainforest patch in the seasonal tropics of Australia's Northern Territory. *Journal of the Australian Entomological Society* 33: 153–158. doi.org/10.1111/j.1440-6055.1994.tb00942.x.

Andersen, A. N., van Ingen, L. T. & Campos, R. I. 2007b. Contrasting rainforest and savanna ant faunas in monsoonal northern Australia: A rainforest patch in a tropical landscape. *Australian Journal of Zoology* 55: 363–369. doi.org/10.1071/ZO07066.

Andersen, A. N., Woinarski, J. C. Z. & Parr, C. L. 2012. Savanna burning for biodiversity: Fire management for faunal conservation in Australian tropical savannas. *Austral Ecology* 37: 658–667. doi.org/10.1111/j.1442-9993.2011.02334.x.

Anderson, S. J. & Braby, M. F. 2009. Invertebrate diversity associated with tropical mistletoe in a suburban landscape from northern Australia. *Northern Territory Naturalist* 21: 2–23.

Angel, F. M. 1951. Notes on the Lepidoptera of the Northern Territory of Australia, with description of new species. *Transactions of the Royal Society of South Australia* 74: 6–14.

Anonymous 1909. A local entomological collection. In *Northern Territory Times and Gazette*, 22 May. pp. 82–83. Darwin.

Archibald, J. & Braby, M. F. 2017. *Bradshaw Field Training Area Bush Blitz: Diurnal Lepidoptera (butterflies and moths)*. Report to Australian Biological Resources Study, Canberra. Museum & Art Gallery of the Northern Territory: Darwin.

Asher, J., Warren, M. S., Fox, R., Harding, P., Jeffcoate, G. & Jeffcoate, S. 2001. *The Millenium Atlas of Butterflies in Britain and Ireland*. Oxford University Press: Oxford.

Atkins, A. F. 1991. Observations on the biology of *Taractrocera anisomorpha* (Lower) (Hesperiidae: Hesperiinae). *Australian Entomological Magazine* 18: 121–123.

Atlas of Living Australia 2017. *Atlas of Living Australia*. NCRIS and CSIRO. Available from: www.ala.org.au (accessed 17 May 2017).

Australian Plant Census 2017. *Australian Plant Census*. Australian National Botanic Gardens, Centre for Australian National Biodiversity Research and Australian Biological Resources Study: Canberra. Available from: www.anbg.gov.au/chah/apc/about-APC.html (accessed 17 May 2017).

Australia's Virtual Herbarium 2017. *Australia's Virtual Herbarium*. Atlas of Living Australia. Available from: avh.chah.org.au (accessed 17 May 2017).

Bailey, W. J. & Richards, K. T. 1975. A report on the insect fauna of the Prince Regent River Reserve, north-west Kimberley, Western Australia. In: Miles, J. M. and Burbidge, A. A. (eds) *A Biological Survey of the Prince Regent River Reserve, North-West Kimberley, Western Australia, in August, 1974*. pp. 101–112. Department of Fisheries and Wildlife: Perth.

Bisa, D. 2013. New locations of butterflies from northern Arnhem Land, Northern Territory. *Northern Territory Naturalist* 24: 2–13.

Bland, L. M., Bielby, J., Kearney, S., Orme, C. D. L., Watson, J. E. M. & Collen, B. 2017. Toward reassessing data-deficient species. *Conservation Biology* 31: 531–539. doi.org/10.1111/cobi.12850.

Bowman, D. M. J. S. 1996. Diversity patterns of woody species on a latitudinal transect from the monsoon tropics to desert in the Northern Territory, Australia. *Australian Journal of Botany* 44: 571–580. doi.org/10.1071/BT9960571.

Bowman, D. M. J. S. 2000. *Australian Rainforests: Islands of green in a land of fire*. Cambridge University Press: Cambridge. doi.org/10.1017/CBO9780511583490.

Bowman, D. M. J. S. 2002. The Australian summer monsoon: A biogeographic perspective. *Australian Geographical Studies* 40: 261–277. doi.org/10.1111/1467-8470.00179.

Bowman, D. M. J. S., Brown, G., Braby, M. F., Brown, J., Cook, L., Crisp, M. D., Ford, F., Haberle, S., Hughes, J. M., Isagi, Y., Joseph, L., McBride, J., Nelson, G. & Ladiges, P. Y. 2010. Biogeography of the monsoon tropics. *Journal of Biogeography* 37: 201–216. doi.org/10.1111/j.1365-2699.2009.02210.x.

Braby, M. F. 1995a. Reproductive seasonality in tropical satyrine butterflies: Strategies for the dry season. *Ecological Entomology* 20: 5–17. doi.org/10.1111/j.1365-2311.1995.tb00423.x.

Braby, M. F. 1995b. Seasonal changes in relative abundance and spatial distribution of Australian lowland tropical satyrine butterflies. *Australian Journal of Zoology* 43: 209–229. doi.org/10.1071/ZO9950209.

Braby, M. F. 1996. A new species of *Nesolycaena* Waterhouse and Turner (Lepidoptera: Lycaenidae) from northeastern Australia. *Australian Journal of Entomology* 35: 9–17. doi.org/10.1111/j.1440-6055.1996.tb01356.x.

Braby, M. F. 1997. Occurrence of *Eurema alitha* (C. & R. Felder) (Lepidoptera; Pieridae) in Australia and its distinction from *E. hecabe* (Linnaeus). *Australian Journal of Entomology* 36: 153–157. doi.org/10.1111/j.1440-6055.1997.tb01448.x.

Braby, M. F. 2000. *Butterflies of Australia. Their identification, biology and distribution.* CSIRO Publishing: Melbourne.

Braby, M. F. 2008a. Biogeography of butterflies in the Australian monsoon tropics. *Australian Journal of Zoology* 56: 41–56. doi.org/10.1071/ZO08021.

Braby, M. F. 2008b. Taxonomic review of *Candalides absimilis* (C. Felder, 1862) and *C. margarita* (Semper, 1879) (Lepidoptera: Lycaenidae), with descriptions of two new subspecies. *The Beagle, Records of the Museums and Art Galleries of the Northern Territory* 24: 33–54.

Braby, M. F. 2009. The life history and biology of *Euploea alcathoe enastri* Fenner, 1991 (Lepidoptera: Nymphalidae) from northeastern Arnhem Land, Northern Territory, Australia. *The Australian Entomologist* 36: 51–62.

Braby, M. F. 2010a. Conservation status and management of the Gove Crow, *Euploea alcathoe enastri* Fenner, 1991 (Lepidoptera: Nymphalidae), a threatened tropical butterfly from the indigenous Aboriginal lands of north-eastern Arnhem Land, Australia. *Journal of Insect Conservation* 14: 535–554. doi.org/10.1007/s10841-010-9282-6.

Braby, M. F. 2010b. The merging of taxonomy and conservation biology: A synthesis of Australian butterfly systematics (Lepidoptera: Hesperioidea and Papilionoidea) for the 21st century. *Zootaxa* (2707): 1–76.

Braby, M. F. 2011a. New larval food plant associations for some butterflies and diurnal moths (Lepidoptera) from the Northern Territory and eastern Kimberley, Australia. *The Beagle, Records of the Museums and Art Galleries of the Northern Territory* 27: 85–105.

Braby, M. F. 2011b. Revised checklist of Australian butterflies (Lepidoptera: Hesperioidea and Papilionoidea): Addendum and errata. *Zootaxa* (3128): 67–68.

Braby, M. F. 2012a. *Butterflies and diurnal moths of Wongalara Station.* Final Report to Department of Sustainability, Environment, Water and Communities. Department of Land Resource Management: Darwin.

Braby, M. F. 2012b. The butterflies of El Questro Wilderness Park, with taxonomic remarks on the Kimberley fauna, Australia. *Records of the Western Australian Museum* 27: 161–175. doi.org/10.18195/issn.0312-3162.27(2).2012.161-175.

Braby, M. F. 2012c. The taxonomy and ecology of *Delias aestiva* Butler, 1897 stat. rev. (Lepidoptera: Pieridae), a unique mangrove specialist of Euphorbiaceae. *Biological Journal of the Linnean Society* 107: 697–720. doi.org/10.1111/j.1095-8312.2012.01970.x.

Braby, M. F. 2014a. Remarks on the spatial distribution of some butterflies and diurnal moths (Lepidoptera) in the Top End of the Northern Territory, Australia. *Northern Territory Naturalist* 25: 29–49.

Braby, M. F. 2014b. Taxonomic status of *Delias aestiva smithersi* Daniels, 2012 (Lepidoptera: Pieridae) comb. nov. from the gulf country of northern Australia, with description of the female. *Records of the Australian Museum* 66: 241–246. doi.org/10.3853/j.2201-4349.66.2014.1633.

Braby, M. F. 2015a. A butterfly found again. *Wildlife Australia* 52: 38–41.

Braby, M. F. 2015b. First record of *Petrelaea tombugensis* (Röber) (Lepidoptera: Lycaenidae) from Western Australia. *The Australian Entomologist* 42: 127–128.

Braby, M. F. 2015c. A further record of *Danaus chrysippus cratippus* (C. Felder, 1860) (Lepidoptera: Nymphalidae: Danainae) from the Northern Territory, Australia. *The Australian Entomologist* 42: 253–255.

Braby, M. F. 2015d. New distribution records for some butterflies from the Gulf of Carpentaria, Queensland. *Northern Territory Naturalist* 26: 67–75.

Braby, M. F. 2015e. New larval food plant associations for some butterflies and diurnal moths (Lepidoptera) from the Northern Territory and Kimberley, Australia. Part II. *Records of the Western Australian Museum* 30: 73–97. doi.org/10.18195/issn.0312-3162.30(2).2015.073-097.

Braby, M. F. 2016a. *The Complete Field Guide to Butterflies of Australia.* 2nd edn. CSIRO Publishing: Melbourne.

Braby, M. F. 2016b. Migration records of butterflies (Lepidoptera: Papilionidae, Hesperiidae, Pieridae, Nymphalidae) in the 'Top End' of the Northern Territory. *The Australian Entomologist* 43: 151–160.

Braby, M. F. 2016c. The princess flash, *Deudorix smilis* Hewitson, 1863 (Lepidoptera: Lycaenidae), in northern Australia. *Butterflies* 73: 40–47.

Braby, M. F. 2017. A new subspecies of *Candalides geminus* Edwards & Kerr, 1978 (Lepidoptera: Lycaenidae) from the Northern Territory, Australia. *Records of the Western Australian Museum* 32: 207–216.

Braby, M. F. & Archibald, J. 2016. *Diurnal Lepidoptera of Judbarra/Gregory National Park, western Top End, Northern Territory*. Final Report to Bush Blitz, Australian Biological Resources Study. Museum and Art Galleries of the Northern Territory: Darwin.

Braby, M. F., Bertelsmeier, C., Sanderson, C. & Thistleton, B. 2014a. Spatial distribution and range expansion of the tawny coster butterfly, *Acraea terpsicore* (Linnaeus, 1758) (Lepidoptera: Nymphalidae), in South-East Asia and Australia. *Insect Conservation and Diversity* 7: 132–143. doi.org/10.1111/icad.12038.

Braby, M. F., Farias Quipildor, G. E., Vane-Wright, R. I. & Lohman, D. J. 2015. Morphological and molecular evidence supports recognition of *Danaus petilia* (Stoll, 1790) (Lepidoptera: Nymphalidae) as a species distinct from *D. chrysippus* (Linnaeus, 1758). *Systematics and Biodiversity* 13: 386–402. doi.org/10.1080/14772000.2014.992378.

Braby, M. F. & Kessner, V. 2012. *Butterflies/diurnal moths and land/freshwater snails of Fish River Station*. Final Report to Department of Sustainability, Environment, Water and Communities. Department of Land Resource Management: Darwin.

Braby, M. F., Lane, D. A. & Weir, R. P. 2010a. The occurrence of *Appias albina albina* (Boisduval, 1836) (Lepidoptera: Pieridae: Pierinae) in northern Australia: Phenotypic variation, life history and biology, with remarks on its taxonomic status. *Entomological Science* 13: 258–268. doi.org/10.1111/j.1479-8298.2010.00377.x.

Braby, M. F. & Nielsen, J. 2011. Review of the conservation status of the atlas moth, *Attacus wardi* Rothschild, 1910 (Lepidoptera: Saturniidae) from Australia. *Journal of Insect Conservation* 15: 603–608. doi.org/10.1007/s10841-011-9402-y.

Braby, M. F., Thistleton, B. M. & Neal, M. J. 2014b. Host plants, biology and distribution of *Acraea terpsicore* (Linnaeus, 1758) (Lepidoptera: Nymphalidae): A new butterfly for northern Australia with potential invasive status. *Austral Entomology* 53: 288–297. doi.org/10.1111/aen.12078.

Braby, M. F. & Westaway, J. 2016. Rediscovery of the spinifex sand-skipper (*Proeidosa polysema*) in the Darwin area, Northern Territory. *Northern Territory Naturalist* 27: 121–125.

Braby, M. F. & Williams, M. R. 2016. Biosystematics and conservation biology: Critical scientific disciplines for the management of insect biological diversity. *Austral Entomology* 55: 1–17. doi.org/10.1111/aen.12158.

Braby, M. F., Worsnop, A., Yata, O. & Tupper, A. 2010b. First record of *Appias albina infuscata* Fruhstorfer, 1910 (Lepidoptera: Pieridae) from Australia. *The Australian Entomologist* 37: 157–162.

Braby, M. F. & Zwick, A. 2015. Taxonomic revision of the *Taractrocera ilia* (Waterhouse) complex (Lepidoptera: Hesperiidae) from north-western Australia and mainland New Guinea based on morphological and molecular data. *Invertebrate Systematics* 29: 487–509. doi.org/10.1071/IS15028.

Brock, J. 2001. *Native Plants of Northern Australia*. Reed New Holland: Sydney.

Cameron, R. A. D., Pokryszko, B. M. & Wells, F. E. 2005. Alan Solem's work on the diversity of Australasian land snails: An unfinished project of global significance. *Records of the Western Australian Museum* (Supp. 68): 1–10. doi.org/10.18195/issn.0313-122x.68.2005.001-010.

Campbell, J. O. 1947. Some notes on Lepidoptera collected at Darwin during November, 1945. *Australian Zoologist* 11: 159–160.

Cardoso, P., Borges, P. A. V., Triantis, K. A., Ferrández, M. A. & Martín, J. L. 2011a. Adapting the IUCN Red List criteria for invertebrates. *Biological Conservation* 144: 2432–2440. doi.org/10.1016/j.biocon.2011.06.020.

Cardoso, P., Erwin, T. L., Borges, P. A. V. & New, T. R. 2011b. The seven impediments in invertebrate conservation and how to overcome them. *Biological Conservation* 144: 2647–2655. doi.org/10.1016/j.biocon.2011.07.024.

Catullo, R. A., Lanfear, R., Doughty, P. & Keogh, J. S. 2014. The biogeographical boundaries of northern Australia: Evidence from ecological niche models and a multi-locus phylogeny of *Uperoleia* toadlets (Anura: Myobatrachidae). *Journal of Biogeography* 41: 659–672. doi.org/10.1111/jbi.12230.

Clarkson, C., Jacobs, Z., Marwick, B., Fullagar, R., Wallis, L., Smith, M., Roberts, R. G., Hayes, E., Lowe, K., Carah, X., Florin, S. A., McNeil, J., Cox, D., Arnold, L. J., Hua, Q., Huntley, J., Brand, H. E. A., Manne, T., Fairbairn, A., Shulmeister, J., Lyle, L., Salinas, M., Page, M., Connell, K., Park, G., Norman, K., Murphy, T. & Pardoe, C. 2017. Human occupation of northern Australia by 65,000 years ago. *Nature* 547: 306–310. doi.org/10.1038/nature22968.

Coleman, N. C. & Monteith, G. B. 1981. Life history of the north Queensland day-flying moth, *Alcides zodiaca* Butler (Lepidoptera: Uraniidae). *North Queensland Naturalist* 45: 2–6.

Collaborative Australian Protected Area Database 2014. *Collaborative Australian Protected Area Database.* Australian Government, Department of the Environment and Energy: Canberra. Available from: www.environment.gov.au/land/nrs/science/capad/2014 (accessed 17 May 2017).

Common, I. F. B. 1973. Lepidoptera (moths and butterflies). *Alligator Rivers Region: Environmental fact finding study: Entomology.* Appendix 9. CSIRO Division of Entomology: Canberra.

Common, I. F. B. 1978. The distinction between *Hypolimnas antilope* (Cramer) and *H. anomala* (Wallace) (Lepidoptera: Nymphalidae), and the occurrence of *H. anomala* in Australia. *Australian Entomological Magazine* 5: 41–44.

Common, I. F. B. 1981. Part IV. Insects. In: McKenzie, N. L. (ed.) *Wildlife of the Edgar Ranges area, south-west Kimberley, Western Australia.* Wildlife Research Bulletin of Western Australia. No. 10. pp. 60–67. Department of Fisheries and Wildlife: Perth.

Common, I. F. B. & Upton, M. S. 1977. Part XI. A report on insects collected in the Drysdale River National Park north Kimberley, Western Australia. In: Kabay, E. D. and Burbidge, A. A. (eds) *A biological survey of the Drysdale River National Park north Kimberley, Western Australia in August 1975.* Wildlife Research Bulletin of Western Australia. No. 6. pp. 121–131. Department of Fisheries and Wildlife: Perth.

Common, I. F. B. & Waterhouse, D. F. 1981. *Butterflies of Australia.* Rev. edn. Angus & Robertson: Sydney.

Couchman, L. E. 1951. Notes on a collection of Hesperiidae made by F. M. Angel in the Northern Territory. *Transactions of the Royal Society of South Australia* 74: 15–17.

Couchman, L. E. & Couchman, R. 1977. The butterflies of Tasmania. *Tasmanian Year Book, 1977* 11: 66–96.

Cracraft, J. 1991. Patterns of diversification within continental biotas: Hierarchical congruence among the areas of endemism of Australian vertebrates. *Australian Systematic Botany* 4: 211–227. doi.org/10.1071/SB9910211.

Criscione, F. & Köhler, F. 2013. More on snails and islands: Molecular systematics and taxonomic revision of *Setobaudinia* Iredale (Gastropoda: Camaenidae) from the Kimberley, Western Australia, with description of new taxa. *Invertebrate Systematics* 27: 634–654. doi.org/10.1071/IS13027.

Criscione, F., Law, M. L. & Köhler, F. 2012. Land snail diversity in the monsoon tropics of northern Australia: Revision of the genus *Exiligada* Iredale, 1939 (Mollusca: Pulmonata: Camaenidae), with description of 13 new species. *Zoological Journal of the Linnean Society* 166: 689–722. doi.org/10.1111/j.1096-3642.2012.00863.x.

Crisp, M. D., Laffan, S., Linder, H. P. & Monro, A. 2001. Endemism in the Australian flora. *Journal of Biogeography* 28: 183–198. doi.org/10.1046/j.1365-2699.2001.00524.x.

Crosby, D. F. 1986. *Preliminary Distribution Maps of Butterflies in Victoria.* Entomological Society of Victoria: Melbourne.

Daniels, G. 2005. The butterflies of the eastern and southern Gulf of Carpentaria, Queensland, taken during the Flinders Bicentenary Expedition, October–November 2002. In: Comben, L., Preker, M. and Long, K. (eds) *Gulf of Carpentaria Scientific Study Report.* pp. 133–135. The Royal Geographical Society of Queensland Inc.: Brisbane.

Daniels, G. 2012. A new subspecies of *Delias mysis* (Fabricius) (Lepidoptera: Pieridae) from the Gulf of Carpentaria, Queensland, Australia. *The Australian Entomologist* 39: 273–276.

Daniels, G. & Edwards, E. D. 1998. Butterflies from Lawn Hill National Park and Musselbrook Reserve, Queensland. In: Comben, L., Long, S. and Berg, K. (eds) *Musselbrook Reserve Scientific Study Report.* Geography Monograph Series No. 4. pp. 89–91. The Royal Geographical Society of Queensland Inc.: Brisbane.

d'Apice, J. W. C. & Miller, C. G. 1992. The genus *Nesolycaena* Waterhouse and Turner (Lepidoptera: Lycaenidae) with a description of a new species. *Australian Entomological Magazine* 19: 75–80.

De Baar, M. 2004. Notes on the status of some *Elodina* C. & R. Felder species (Lepidoptera: Pieridae). *The Australian Entomologist* 31: 37–42.

De Baar, M. & Hancock, D. L. 1993. The Australian species of *Elodina* C. & R. Felder (Lepidoptera: Pieridae). *The Australian Entomologist* 20: 25–43.

Dodd, W. D. 1935a. Meanderings of a naturalist. No. 6. *The North Queensland Register*, 9 March. Cairns.

Dodd, W. D. 1935b. Meanderings of a naturalist. No. 14. *The North Queensland Register*, 11 May. Cairns.

Dodd, W. D. 1935c. Meanderings of a naturalist. No. 18. *The North Queensland Register*, 8 June. Cairns.

Dodd, W. D. 1935d. Meanderings of a naturalist. No. 19. *The North Queensland Register*, 22 June. Cairns.

Douglas, M. M. & Setterfield, S. A. 2005. Impacts of exotic tropical grasses: Lessons from gamba grass in the Northern Territory. In: Vogler, W. D. (ed.) *Proceedings of the Eighth Queensland Weed Symposium.* pp. 69–73. The Weed Society of Queensland Inc.: Townsville.

Drosdowsky, W. & Wheeler, M. C. 2014. Predicting the onset of the north Australian wet season with the POAMA dynamical prediction system. *Weather and Forecasting* 29: 150–161. doi.org/10.1175/WAF-D-13-00091.1.

Dunn, K. L. 1980. A Northern Territory–Western Australia safari. *Victorian Entomologist* 10: 4–6.

Dunn, K. L. 1985. Specimens of interest in the J. C. Le Souef collection of Australian butterflies. *The Victorian Naturalist* 102: 94–98.

Dunn, K. L. 2009. Notes on *Cephrenes augiades* (C. Felder) in the Northern Territory, Australia (Lepidoptera: Hesperiidae). *Victorian Entomologist* 39: 53–56.

Dunn, K. L. 2013. Field notes: Major extensions to the known distribution of the bright purple azure, *Ogyris barnardi* (Miskin 1890) in Queensland (Lepidoptera: Lycaenidae). *Metamorphosis Australia. Magazine of the Butterfly and Other Invertebrates Club* 68: 26–32.

Dunn, K. L. 2015a. New distribution records for nymphalid butterflies (Lepidoptera: Nymphalidae) in Queensland. *Metamorphosis Australia. Magazine of the Butterfly and Other Invertebrates Club* 76: 18–34.

Dunn, K. L. 2015b. New distribution records for hesperiine butterflies (Lepidoptera: Hesperiidae: Hesperiinae) in Australia. *Metamorphosis Australia. Magazine of the Butterfly and Other Invertebrates Club* 77: 17–32.

Dunn, K. L. 2016. New distribution records for thecline butterflies (Lepidoptera: Lycaenidae) in Australia. Part I: *Jalmenus, Hypochrysops, Hypolycaena* and *Rapala. Metamorphosis Australia. Magazine of the Butterfly and Other Invertebrates Club* 81: 21–34.

Dunn, K. L. 2017a. New and important distribution records for *Candalides* butterflies in Australia, with observations on their biology and adult food plants. *North Queensland Naturalist* 47: 1–5.

Dunn, K. L. 2017b. New distribution records for Polyommatinae butterflies (Lepidoptera: Lycaenidae) in Australia, including biological notes. Part I: *Zizeeria* and *Famegana. Metamorphosis Australia. Magazine of the Butterfly and Other Invertebrates Club* 84: 12–21.

Dunn, K. L. 2017c. New distribution records for Polyommatinae butterflies (Lepidoptera: Lycaenidae) in Australia, including biological notes. Part II: *Nacaduba* and *Prosotas. Metamorphosis Australia. Magazine of the Butterfly and Other Invertebrates Club* 85: 14–22.

Dunn, K. L. 2017d. New distribution records for Polyommatinae butterflies (Lepidoptera: Lycaenidae) in Australia, including biological notes. Part III: *Anthene, Catopyrops, Catochrysops, Euchrysops* and *Freyeria. Metamorphosis Australia. Magazine of the Butterfly and Other Invertebrates Club* 87: 18–28.

Dunn, K. L. & Dunn, L. E. 1991. *Review of Australian Butterflies: Distribution, life history and taxonomy. Parts 1–4.* Self-published: Melbourne.

Eastwood, R. G., Boyce, S. L. & Farrell, B. D. 2006. The provenance of Old World swallowtail butterflies, *Papilio demoleus* (Lepidoptera: Papilionidae), recently discovered in the New World. *Annals of the Entomological Society of America* 99: 164–168.

Eastwood, R. G., Braby, M. F., Lohman, D. J. & King, A. 2008. New ant–lycaenid associations and biological data for some Australian butterflies (Lepidoptera: Lycaenidae). *The Australian Entomologist* 35: 47–56.

Eastwood, R. G. & Fraser, A. M. 1999. Associations between lycaenid butterflies and ants in Australia. *Australian Journal of Ecology* 24: 503–507.

Ebach, M. C. 2012. A history of biogeographical regionalisation in Australia. *Zootaxa* (3392): 1–34.

Ebach, M. C., Murphy, D. J., González-Orozco, C. E. & Miller, J. T. 2015. A revised area taxonomy of phytogeographical regions within the Australian Bioregionalisation Atlas. *Phytotaxa* 208: 261–277. doi.org/10.11646/phytotaxa.208.4.2.

Edwards, E. D. 1977. *Junonia erigone* (Cramer) (Lepidoptera: Nymphalidae) recorded from Australia. *Australian Entomological Magazine* 4: 41–43.

Edwards, E. D. 1980. The early stages of *Adaluma urumelia* Tindale and *Candalides geminus* Edwards and Kerr (Lepidoptera: Lycaenidae). *Australian Entomological Magazine* 7: 17–20.

Edwards, E. D. 1987. A new species of *Mesodina* Meyrick from the Northern Territory (Lepidoptera: Hesperiidae). *Australian Entomological Magazine* 14: 4–12.

Edwards, E. D. 1996. Noctuidae. In: Nielsen, E. S., Edwards, E. D. and Rangsi, T. V. (eds) *Checklist of the Lepidoptera of Australia.* Monographs on Australian Lepidoptera Vol. 4. pp. 291–336, 370–380. CSIRO Publishing: Collingwood, Melbourne.

Edwards, E. D., Newland, J. & Regan, L. 2001. *Lepidoptera: Hesperioidea, Papilionoidea. Zoological Catalogue of Australia. Vol. 31.6.* CSIRO Publishing: Collingwood, Melbourne.

Edwards, R. D., Crisp, M. D., Cook, D. H. & Cook, L. G. 2017. Congruent biogeographical disjunctions at a continent-wide scale: Quantifying and clarifying the role of biogeographic barriers in the Australian tropics. *PLoS ONE* 12(4): e0174812. doi.org/10.1371/journal.pone.0174812.

Eldridge, M. D. B., Potter, S. & Cooper, S. J. B. 2012. Biogeographic barriers in north-western Australia: An overview and standardisation of nomenclature. *Australian Journal of Zoology* 59: 270–272. doi.org/10.1071/ZO12012.

Executive Steering Committee for Australian Vegetation Information 2003. *Australian Vegetation Attribute Manual: National vegetation information system, Version 6.0.* Department of the Environment and Heritage: Canberra.

Fenner, T. L. 1991. A new subspecies of *Euploea alcathoe* (Godart) (Lepidoptera: Nymphalidae) from the Northern Territory, Australia. *Australian Entomological Magazine* 18: 149–155.

Field, A. R. 2017. Arrival of tawny coster butterflies on the east Australian coast coinciding with the winds of Tropical Cyclone Debbie. *North Queensland Naturalist* 47: 28–31.

Field, R. P. 1990a. New and extended distribution records of butterflies from the Northern Territory. *Victorian Entomologist* 20: 40–44.

Field, R. P. 1990b. Range extensions and the biology of some Western Australian butterflies. *Victorian Entomologist* 20: 76–82.

Field, R. P. 2013. *Butterflies: Identification and life history.* Museum Victoria: Melbourne.

Fisher, C. & Calaby, J. 2009. The top of the Top End: John Gilbert's manuscript notes for John Gould on vertebrates from Port Essington and Cobourg Peninsula (Northern Territory, Australia); with comments on specimens collected during the settlement period 1838 to 1849, and subsequently. *The Beagle, Records of the Museums and Art Galleries of the Northern Territory* Supplement 4: 1–239.

Fisher, R. H. 1992. A new distribution record for the pierid butterfly *Elodina perdita* Miskin. *Victorian Entomologist* 22: 45.

FloraBase 2017. *FloraBase, The Western Australian Flora.* Western Australian Herbarium, Department of Parks and Wildlife: Perth. Available from: florabase.dpaw.wa.gov.au (accessed 17 May 2017).

Forster, P. I. 1991. Host records (Family Asclepiadaceae) and distribution of *Danaus chrsyippus petilia* (Stoll) (Lepidoptera: Nymphalidae) in Australia. *Australian Entomological Magazine* 18: 97–98.

Forster, P. I. & Martin, G. 1990. Host records (Family Asclepidaceae) and distribution of *Tirumala hamata hamata* (W. S. Macleay) (Lepidoptera: Nymphalidae) in Australia. *Australian Entomological Magazine* 17: 131–132.

Franklin, D. C. 2007. Dry season observations of butterflies in the 'Gulf Country' of the Northern Territory and far north-west Queensland. *Northern Territory Naturalist* 19: 9–14.

Franklin, D. C. 2011. Butterfly counts at Casuarina Coastal Reserve in the seasonal tropics of northern Australia. *Northern Territory Naturalist* 23: 18–28.

Franklin, D. C. & Bate, P. J. 2013. *Brachychiton megaphyllus*, the red-flowered kurrajong. *Northern Territory Naturalist* 24: 81–88.

Franklin, D. C., Binns, D. & Mace, M. 2007a. Glistening Line-blue: Fifth record of this butterfly in the Northern Territory. *Nature Territory. Newsletter of the Northern Territory Field Naturalists Club Inc.* (February): 4–5.

Franklin, D. C., Brocklehurst, P. S., Lynch, D. & Bowman, D. M. J. S. 2007b. Niche differentiation and regeneration in the seasonally flooded *Melaleuca* forests of northern Australia. *Journal of Tropical Ecology* 23: 457–467. doi.org/10.1017/S0266467407004130.

Franklin, D. C., Michael, B. & Mace, M. 2005. New location records for some butterflies of the Top End and Kimberley regions. *Northern Territory Naturalist* 18: 1–7.

Franklin, D. C., Morrison, S. C. & Wilson, G. W. 2017. A colourful new Australian reaches Talaroo: The tawny coster butterfly, *Acraea terpsicore*. *North Queensland Naturalist* 47: 10–13.

Froggatt, W. W. 1934. A naturalist in Kimberley in 1887. *The Australian Naturalist* 9: 69–82.

Gaffney, D. O. 1971. *Seasonal rainfall zones in Australia*. Working Paper No. 141. Bureau of Meteorology: Canberra.

Garnett, S. T. & Williamson, G. 2010. Spatial and temporal variation in precipitation at the start of the rainy season in tropical Australia. *The Rangeland Journal* 32: 215–226. doi.org/10.1071/RJ09083.

Garnett, S. T., Woinarski, J. C. Z., Crowley, G. M. & Kutt, A. S. 2010. Biodiversity conservation in Australian tropical rangelands. In: du Toit, J. T., Kock, R. and Deutsch, J. C. (eds) *Wild Rangelands: Conserving wildlife while maintaining livestock in semi-arid ecosystems*. Conservation Science and Practice Series. pp. 191–234. Wiley-Blackwell: Oxford. doi.org/10.1002/9781444317091.ch8.

Gaston, K. J. & Fuller, R. A. 2009. The sizes of species' geographic ranges. *Journal of Applied Ecology* 46: 1–9. doi.org/10.1111/j.1365-2664.2008.01596.x.

Gibb, W. F. 1977. The rediscovery of *Hypolimnas antilope* (Cramer) (Lepidoptera: Nymphalidae) in Australia. *Australian Entomological Magazine* 4: 39.

González-Orozco, C. E., Ebach, M. C., Laffan, S., Thornhill, A. H., Knerr, N. J., Schmidt-Lebuhn, A. N., Cargill, C. C., Clements, M., Nagalingum, N. S., Mishler, B. D. & Miller, J. T. 2014. Quantifying phytogeographical regions of Australia using geospatial turnover in species composition. *PLoS ONE* 9: e9258. doi.org/10.1371/journal.pone.0092558.

Grund, R. 1996. The distribution of *Theclinesthes albocincta* (Waterhouse) and *Theclinesthes hesperia littoralis* Sibitani & Grund, based on herbarium records of eggs (Lepidoptera: Lycaenidae). *The Australian Entomologist* 23: 101–10.

Grund, R. 1998. New foodplant recordings and biological observations for some Western Australian butterflies. *Victorian Entomologist* 28: 65–68.

Grund, R. 2005. Some new butterfly observations for Central Australia. *Victorian Entomologist* 35: 111–114.

Grund, R. & Hunt, L. 2001. Some butterfly observations for the Kimberley and Tanami regions, Western Australia. *Victorian Entomologist* 31: 19–23.

Gullan, P., Crosby, D. F. & Quick, W. N. B. 1996. *Victorian Butterfly Database*. CD-ROM. Viridans Biological Databases: Melbourne.

Hall, M. C. 1976. Native food plant records for some Northern Territory butterflies. *Australian Entomological Magazine* 3: 41.

Hall, M. C. 1981. Notes on the life history of *Cethosia penthesilea paksha* Fruhstorfer (Lepidoptera: Nymphalidae). *Australian Entomological Magazine* 7: 89–90.

Harrison, M. 2010. The zodiac moth: A discovery and study. *Metamorphosis Australia. Magazine of the Butterfly and Other Invertebrates Club* 57: 4–7.

Harvey, M. S. 2002. Short-range endemism among the Australian fauna: Some examples from non-marine environments. *Invertebrate Systematics* 16: 555–570. doi.org/10.1071/IS02009.

Harvey, M. S., Rix, M. G., Framenau, V. W., Hamilton, Z. R., Johnson, M. S., Teale, R. J., Humphreys, G. & Humphreys, W. F. 2011. Protecting the innocent: Studying short-range endemic taxa enhances conservation outcomes. *Invertebrate Systematics* 25: 1–10. doi.org/10.1071/IS11011.

Hopkinson, M. & Bourne, S. A. 2018. Life history notes and observations on *Petrelaea tombugensis* (Röber) (Lepidoptera: Lycaenidae). *The Australian Entomologist* 45: 11–16.

Hutchinson, J. F. 1973. A list of butterflies collected at Daly River crossing, Northern Territory. *Victorian Entomologist* 3: 11–14.

Hutchinson, J. F. 1978. Butterflies of the Daly River area, Northern Territory. *Victorian Entomologist* 8: 15–19.

International Union for Conservation of Nature (IUCN) 2001. *IUCN Red List Categories: Version 3.1.* IUCN Species Survival Commission: Gland, Switzerland.

International Union for Conservation of Nature (IUCN) 2012. *Guidelines for Application of IUCN Red List Criteria at Regional and National Levels: Version 4.0.* IUCN: Gland, Switzerland.

International Union for Conservation of Nature (IUCN) Standards and Petitions Subcommittee 2016. *Guidelines for Using the IUCN Red List Categories and Criteria: Version 12.* Prepared by the Standards and Petitions Subcommittee of the IUCN Species Survival Commission: Gland, Switzerland.

Johnson, S. J. 1993. Butterfly records of interest from northern Australia. *The Australian Entomologist* 20: 75–76.

Johnson, S. J. & Valentine, P. S. 1988. Butterflies collected on Mt White in January 1988. *News Bulletin of the Entomological Society of Queensland* 16: 12–13.

Johnson, S. J. & Valentine, P. S. 1989. The life history of *Libythea geoffroy nicevillei* Olliff (Lepidoptera: Libytheidae). *Australian Entomological Magazine* 16: 59–62.

Johnson, S. J. & Valentine, P. S. 2004. Butterfly field notes from northwestern Australia including new distributions of several species. *Victorian Entomologist* 34: 58–59.

Jones, R. E. 1987. Reproductive strategies for the seasonal tropics. *Insect Science and Application* 8: 515–521. doi.org/10.1017/s1742758400022566.

Jones, R. E. & Rienks, J. 1987. Reproductive seasonality in the tropical genus *Eurema* (Lepidoptera: Pieridae). *Biotropica* 19: 7–16. doi.org/10.2307/2388454.

Kallies, A. 2001. New records and a revised checklist of the Australian clearwing moths (Lepidoptera: Sesiidae). *Australian Journal of Entomology* 40: 342–348. doi.org/10.1046/j.1440-6055.2001.00251.x.

Kemp, D. J. 2001. Reproductive seasonality in the tropical butterfly *Hypolimnas bolina* (Lepidoptera: Nymphalidae) in northern Australia. *Journal of Tropical Ecology* 17: 483–494. doi.org/10.1017/ S0266467401001365.

Kenneally, K. F., Keighery, G. J. & Hyland, B. P. M. 1991. Floristics and phytogeography of Kimberley rainforests, Western Australia. In: McKenzie, N. L., Johnston, R. B. and Kendrick, P. G. (eds) *Kimberley Rainforests of Australia.* pp. 93–131. Surrey Beatty & Sons: Sydney.

Kikkawa, J. & Monteith, G. B. 1980. *Animal ecology of monsoon forests of the Kakadu region, Northern Territory.* Queensland Museum consultancy report to Director, Australian National Parks and Wildlife Service: Canberra.

Kikkawa, J., Monteith, G. B. & Ingram, G. 1981. Cape York Peninsula, major region of faunal interchange. In: Keast, A. (ed.) *Ecological Biogeography of Australia.* pp. 1695–1742. Dr W. Junk bv Publishers: The Hague.

Kiriakoff, S. G. 1977. Lepidoptera Noctuiformes. Agaristidae I (Palaearctic and Oriental genera). *Das Tierreich* 97: i–ix, 1–180.

Koch, L. E. 1957. East Kimberley butterflies. *Western Australian Naturalist* 11: 83–84.

Koch, L. E. 1975. *Graphium* butterflies at Kooland Island. *Western Australian Naturalist* 13: 64.

Koch, L. E. & van Ingen, F. C. 1969. The butterflies of Koolan Island, Western Australia. *Western Australian Naturalist* 11: 98.

Kodandaramaiah, U., Lees, D. C., Müller, C. J., Torres, E., Karanth, K. P. & Wahlberg, N. 2010. Phylogenetics and biogeography of a spectacular Old World radiation of butterflies: The subtribe Mycalesina (Lepidoptera: Nymphalidae: Satyrini). *BMC Evolutionary Biology* 10: 72.

Köhler, F. 2010. Uncovering local endemism in the Kimberley, Western Australia: Description of new species of the genus *Amplirhagada* Iredale, 1933 (Pulmonata, Camaenidae). *Records of the Australian Museum* 62: 217–284. doi.org/10.3853/j.0067-1975.62.2010.1554.

Köhler, F. & Criscione, F. 2013. Small snails in a big place: A radiation in the semi-arid rangelands in northern Australia (Eupulmonata, Camaenidae, Nanotrachia n. gen.). *Zoological Journal of the Linnean Society* 169: 103–123. doi.org/10.1111/zoj.12051.

Kudrna, O., Harpke, A., Lux, K., Pennerstorfer, J., Schweiger, O., Settele, J. & Wimers, M. 2011. *Distribution Atlas of Butterflies in Europe.* Geselleschaft für Schmetterlingsschutz: Halle, Germany.

Lambkin, T. A. 2006. Clinal variation in female *Hypolycaena phorbas phorbas* (Fabricius) (Lepidoptera: Lycaenidae) and revision of the status of *H. p. ingura* Tindale. *The Australian Entomologist* 33: 81–92.

Lambkin, T. A. & Kendall, R. 2016. The status of *Yoma algina* (Boisduval, 1832) and *Y. sabina* (Cramer, 1780) (Lepidoptera: Nymphalidae: Nymphalinae) in Australia. *The Australian Entomologist* 43: 211–234.

Lambkin, T. A., Braby, M. F., Eastwood, R. G. & Zalucki, M. P. 2017. Taxonomic revision of the *Euploea alcathoe* complex (Lepidoptera: Nymphalidae) from Australia and New Guinea. *Austral Entomology.* doi.org/10.1111/aen.12299.

Lazarides, M. 2005. Micraira. In: Mallet, K. (ed.) *Flora of Australia. Volume 44B: Poaceae 3.* pp. 120–131. ABRS/CSIRO Publishing: Melbourne.

Le Souëf, J. C. 1971. Winter insect collecting in the Northern Territory. *The Victorian Naturalist* 88: 350–356.

Le Souëf, J. C. & Tindale, N. B. 1970. A new subspecies of *Virachola smilis* (Hewitson) (Lepidoptera: Lycaenidae) from northern Australia. *Journal of the Australian Entomological Society* 9: 219–222. doi.org/10.1111/j.1440-6055.1970.tb00794.x.

Lewis, O. T. & Senior, M. J. M. 2011. Assessing conservation status and trends for the world's butterflies: The sampled Red List Index approach. *Journal of Insect Conservation* 15: 121–128. doi.org/10.1007/s10841-010-9329-8.

Liddle, D. T., Russell-Smith, J., Brock, J., Leach, G. J. & Connors, G. T. 1994. *Atlas of the Vascular Rainforest Plants of the Northern Territory.* Australian Biological Resources Study: Canberra.

Lokkers, C. 1986. The distribution of the weaver ant, *Oecophylla smaragdina* (Fabricius) (Hymenoptera: Formicidae). *Australian Journal of Zoology* 34: 683–687. doi.org/10.1071/ZO9860683.

Macleay, W. S. 1826. Annulosa. Catalogue of insects, collected by Captain King, R. N. Appendix B. Containing a list and descriptions of the subjects of natural history collected during Captain King's survey of the intertropical and western coasts of Australia. In: King, P. P. (ed.) *Narrative of a Survey of the Intertropical and Western Coasts of Australia, Performed between the Years 1818 and 1822. Volume II.* pp. 438–469. John Murray: London.

McCubbin, C. 1971. *Australian Butterflies.* Nelson: Melbourne.

McGeoch, M. A. 1998. The selection, testing and application of terrestrial insects as bioindicators. *Biological Reviews* 73: 181–201. doi.org/10.1111/j.1469-185X.1997.tb00029.x.

McGeoch, M. A. 2007. Insects and bioindication: Theory and progress. In: Stewart, A. J. A., New, T. R. and Lewis, O. T. (eds) *Insect Conservation Biology. Proceedings of the Royal Entomological Society's 23rd Symposium.* pp. 144–174. CABI Publishing: Wallingford, UK. doi.org/10.1079/9781845932541.0144.

McKay, L. 2017. *A Guide to Wildlife of the Top End.* The Environment Centre Northern Territory: Darwin.

McKenzie, N. L., Belbin, L., Keighery, G. J. & Kenneally, K. F. 1991. Kimberley rainforest communities: Patterns of species composition and Holocene biogeography. In: McKenzie, N. L., Johnston, R. B. and Kendrick, P. G. (eds) *Kimberley Rainforests of Australia.* pp. 423–451. Surrey Beatty & Sons: Sydney.

McKenzie, N. L., Fontanini, L., Lindus, N. V. & Williams, M. R. 1995. Biological inventory of Koolan Island, Western Australia 2. Zoological notes. *Records of the Western Australian Museum* 17: 249–266.

Merckx, T., Huertas, B., Basset, Y. & Thomas, J. A. 2013. A global perspective on conserving butterflies and moths and their habitats. In: Macdonald, D. W. and Willis, K. J. (eds) *Key Topics in Conservation Biology 2.* pp. 239–257. John Wiley & Sons: London. doi.org/10.1002/9781118520178.ch14.

Meyer, C. E. 1995. Notes on the life history of *Danaus genutia alexis* (Waterhouse and Lyell) (Lepidoptera: Nymphalidae: Danainae). *The Australian Entomologist* 22: 137–139.

Meyer, C. E. 1996a. Butterfly larval food plant list for the Northern Territory and the Kununurra district in Western Australia. *Victorian Entomologist* 26: 66–72.

Meyer, C. E. 1996b. Notes on the life history of *Nacaduba kurava felsina* Waterhouse and Lyell (Lepidoptera: Lycaenidae). *The Australian Entomologist* 23: 73–74.

Meyer, C. E. 1996c. A new record of *Nesolycaena caesia* d'Apice & Miller (Lepidoptera: Lycaenidae) from north-eastern Western Australia. *The Australian Entomologist* 23: 79.

Meyer, C. E. 1996d. Notes on the immature stages of *Euploeae darchia darcia* (W. S. Macleay) (Lepidoptera: Nymphalidae). *The Australian Entomologist* 23: 81–82.

Meyer, C. E. 1997a. Notes on the life history and variations in adult forms of *Euploea sylvester pelor* Doubleday (Lepidoptera: Nymphalidae: Danainae). *The Australian Entomologist* 24: 73–77.

Meyer, C. E. 1997b. Notes on the life history of *Borbo impar lavinia* (Waterhouse) (Lepidoptera: Hesperiidae). *The Australian Entomologist* 24: 78–80.

Meyer, C. E., Weir, R. P. & Brown, S. S. 2013. Some new and interesting butterfly (Lepidoptera) distribution and temporal records from Queensland and northern Australia. *The Australian Entomologist* 40: 7–12.

Meyer, C. E., Weir, R. P. & Brown, S. S. 2015. A new subspecies of *Hasora hurama* (Butler, 1870) (Lepidoptera: Hesperiidae: Coeliadinae) from the Northern Territory, Australia. *The Australian Entomologist* 42: 133–47.

Meyer, C. E., Weir, R. P. & Wilson, D. N. 2006. Butterfly (Lepidoptera) records from the Darwin region, Northern Territory. *The Australian Entomologist* 33: 9–22.

Meyer, C. E. & Wilson, D. N. 1995. A new distribution record for *Theclinesthes sulpitius* (Miskin) (Lepidoptera: Lycaenidae) in the Northern Territory and notes on the life history. *The Australian Entomologist* 22: 63.

Miller, C. G. & Lane, D. A. 2004. A new species of *Acrodipsas* Sands (Lepidoptera: Lycaenidae) from the Northern Territory. *The Australian Entomologist* 31: 141–146.

Mollet, B. & Tarmann, G. M. 2010. Notes on the ecology, phenology, and distribution of *Pollanisus eumetopus* Turner (Lepidoptera: Zygaenidae, Procridinae, Artonini). *The Australian Entomologist* 37: 63–67.

Monteith, G. B. 1982. Dry season aggregations of insects in Australian monsoon forests. *Memoirs of the Queensland Museum* 20: 533–543.

Monteith, G. B. 1991. *The Butterfly Man of Kuranda, Frederick Parkhurst Dodd.* Queensland Museum: South Brisbane.

Monteith, G. B. & Wood, G. B. 1987. *Endospermum*, ants and uraniid moths in Australia. *Queensland Naturalist* 28: 35–41.

Morgun, D. V. & Wiemers, M. 2012. First record of the Lime Swallowtail *Papilio demoleus* Linneaus, 1758 (Lepidoptera: Papilionidae) in Europe. *Journal of Research on the Lepidoptera* 45: 85–89.

Moritz, C., Ens, E. J., Potter, S. & Catullo, R. A. 2013. The Australian monsoonal tropics: An opportunity to protect unique biodiversity and secure benefits for Aboriginal communities. *Pacific Conservation Biology* 19: 343–355. doi.org/10.1071/PC130343.

Moss, J. T. 2010. Hostplants of the zodiac moth in Australia. *Metamorphosis Australia. Magazine of the Butterfly and Other Invertebrates Club* 57: 7–9.

Müller, C. J., Olive, J. & Lambkin, T. A. 1998. New records for *Petrelaea tombugensis* (Röber) (Lepidoptera: Lycaenidae) in Queensland. *The Australian Entomologist* 25: 61–63.

Naumann, I. D., Weir, T. A. & Edwards, E. D. 1991. Insects of Kimberley rainforests. In: McKenzie, N. L., Johnston, R. B. and Kendrick, P. G. (eds) *Kimberley Rainforests of Australia.* pp. 299–332. Surrey Beatty & Sons: Sydney.

New, T. R. 1997. Are Lepidoptera an effective 'umbrella group' for biodiversity conservation? *Journal of Insect Conservation* 1: 5–12. doi.org/10.1023/A:1018433406701.

Nielsen, E. S., Edwards, E. D. & Rangsi, T. V. 1996. *Checklist of the Lepidoptera of Australia.* Monographs on Australian Lepidoptera Vol. 4. CSIRO Publishing: Collingwood, Melbourne.

Normand, S. 2009. *Marranbala Country.* Simon Normand: Melbourne.

Oberprieler, S. K., Jennings, D. & Oberprieler, R. G. 2016. Captain King's lost weevil: Alive and well in the Northern Territory? *Northern Territory Naturalist* 27: 106–120.

Oliver, P. M., Laver, R. J., De Mello Martins, F., Pratt, R. C., Hunjan, S. & Moritz, C. 2017. A novel hotspot of vertebrate endemism and an evolutionary refugium in tropical Australia. *Diversity and Distributions* 23: 53–66.

Parr, C. L., Lehmann, C. E. R., Bond, W. J., Hoffmann, W. A. & Andersen, A. N. 2014. Tropical grassy biomes: Misunderstood, neglected, and under threat. *Trends in Ecology and Evolution* 29: 205–213. doi.org/10.1016/j.tree.2014.02.004.

Paton, G. 2013. A new host plant for the satin azure (*Ogyris amaryllis meridionalis*). *Metamorphosis Australia. Magazine of the Butterfly and Other Invertebrates Club* 70: 37–38.

Pepper, M. & Keogh, J. S. 2014. Biogeography of the Kimberley, Western Australia: A review of landscape evolution and biotic response in an ancient refugium. *Journal of Biogeography* 41: 1443–1455. doi.org/10.1111/jbi.12324.

Peters, J. V. 1969. Notes on the distribution of Australian Hesperoidea and Papilionoidea. *Australian Zoologist* 15: 178–184.

Peters, J. V. 2006. New distribution records for Australian butterflies (Lepidoptera). *The Australian Entomologist* 33: 113–114.

Peters, J. V. 2008. New distribution records for Australian butterflies (Lepidoptera: Lycaenidae) from Broome, Western Australia. *The Australian Entomologist* 35: 66.

Pierce, F. 2008. Range extension records for various butterflies throughout Australia. *Victorian Entomologist* 38: 15–16.

Pierce, F. 2010. More range extension records for various butterflies throughout Australia. *Victorian Entomologist* 40: 133–134.

Pierce, F. 2011. 'Corrigenda' to the editors. *Victorian Entomologist* 41: 38.

Pierce, F. 2017. Range extension records for various butterflies throughout Australia. *Victorian Entomologist* 47: 131–133.

Pollard, E. & Yates, T. J. 1993. *Monitoring Butterflies for Ecology and Conservation*. Chapman & Hall: London.

Powell, R. J. 1993. The use of two species of *Parietaria* (Urticaceae) as food plants by the butterfly *Vanessa itea* (Fabricius) in south-western Australia. *The Australian Entomologist* 20: 57–58.

Press, T., Brock, J. & Andersen, A. N. 1995. Fauna. In: Press, T., Lea, D., Webb, A. and Graham, A. (eds) *Kakadu: Natural and cultural heritage and management*. pp. 167–216. Australian Nature Conservation Agency and the North Australian Research Unit, The Australian National University: Darwin.

Prout, L. B. 1921–34. The Indo–Australian Geometridae. Hemitheinae. In: Seitz, A. (ed.) *The Macrolepidoptera of the World. Volume 12*. pp. 44–142. Alfred Kernen Verlag: Stuttgart.

Puccetti, M. 1991. Butterflies of Doomadgee, northwestern Queensland. *Victorian Entomologist* 21: 142–147.

Reichel, H. & Andersen, A. N. 1996. The rainforest ant fauna of Australia's Northern Territory. *Australian Journal of Zoology* 44: 81–95. doi.org/10.1071/ZO9960081.

Rosauer, D. F., Bloma, M. P. K., Bourke, G., Catalanob, S., Donnellanb, S., Gillespie, G., Mulder, E., Oliver, P. M., Potter, S., Pratt, R. C., Rabosky, D. L., Skipwith, P. L. & Moritz, C. 2016. Phylogeography, hotspots and conservation priorities: An example from the Top End of Australia. *Biological Conservation* 204: 83–93. doi.org/10.1016/j.biocon.2016.05.002.

Rosauer, D. F., Laffan, S., Crisp, M. D., Donnellan, S. C. & Cook, L. G. 2009. Phylogenetic endemism: A new approach for identifying geographical concentrations of evolutionary history. *Molecular Ecology* 18: 4061–4072. doi.org/10.1111/j.1365-294X.2009.04311.x.

Rossiter, N. A., Setterfield, A. A., Douglas, M. M. & Hutley, L. B. 2003. Testing the grass–fire cycle: Alien grass invasion in the tropical savannas of northern Australia. *Diversity and Distributions* 9: 169–176. doi.org/10.1046/j.1472-4642.2003.00020.x.

Russell-Smith, J. 1991. Classification, species richness, and environmental relations of monsoon rain forest in northern Australia. *Journal of Vegetation Science* 2: 259–278. doi.org/10.2307/3235959.

Russell-Smith, J. & Bowman, D. M. J. S. 1992. Conservation of monsoonal vine-forest isolates in the Northern Territory, Australia. *Biological Conservation* 59: 51–63. doi.org/10.1016/0006-3207(92)90713-W.

Russell-Smith, J., McKenzie, N. L. & Woinarski, J. C. Z. 1992. Conserving vulnerable habitat in northern and north-western Australia: The rainforest archipelago. In: Moffatt, I. and Webb, A. (eds) *Conservation and Development issues in Northern Australia*. pp. 63–68. North Australia Research Unit, The Australian National University: Darwin.

Russell-Smith, J., Ryan, P. G. & Cheal, D. C. 2002. Fire regimes and the conservation of sandstone heath in monsoonal northern Australia: Frequency, interval, patchiness. *Biological Conservation* 104: 91–106. doi.org/10.1016/S0006-3207(01)00157-4.

Russell-Smith, J., Ryan, P. G., Klessa, D., Waight, G. & Harwood, R. 1998. Fire regimes, fire-sensitive vegetation and fire management of the sandstone Arnhem Plateau, monsoonal northern Australia. *Journal of Applied Ecology* 35: 829–846. doi.org/10.1111/j.1365-2664.1998.tb00002.x.

Russell-Smith, J. & Yates, C. P. 2007. Australian savanna fire regimes: Context, scale, patchiness. *Fire Ecology* 3: 48–63. doi.org/10.4996/fireecology.0301048.

Samson, P. R. 1991. The life history of *Everes lacturnus australis* Couchman (Lepidoptera: Lycaenidae). *Australian Entomological Magazine* 18: 71–74.

Samson, P. R. & Lambkin, T. A. 2003. The immature stages and seasonality of *Petrelaea tombugensis* (Rober) (Lepidoptera: Lycaenidae). *The Australian Entomologist* 30: 181–183.

Samson, P. R. & Wilson, D. N. 1995. The life history of *Candalides gilberti* Waterhouse (Lepidoptera: Lycaenidae). *The Australian Entomologist* 22: 71–73.

Sands, D. P. A. 1986. *A Revision of the Genus* Hypochrysops *C. & R. Felder (Lepidoptera: Lycaenidae).* E. J. Brill/Scandinavian Science Press: Leiden & Copenhagen.

Sands, D. P. A. & New, T. R. 2002. *The Action Plan for Australian Butterflies.* Environment Australia: Canberra.

Scheermeyer, E. 1993. Overwintering of three Australian danaines: *Tirumala hamata hamata, Euploea tulliolus tulliolus,* and *E. core corinna.* In: Malcolm, S. B. and Zalucki, M. P. (eds) *The Biology of Conservation of the Monarch Butterfly.* pp. 345–353. Natural History Museum of Los Angeles County: Los Angeles.

Scheermeyer, E. 1999. The crows, *Euploea* species, with notes on the blue tiger, *Tirumala hamata* (Nymphalidae: Danainae). In: Kitching, R. L., Scheermeyer, E., Jones, R. E. and Pierce, N. E. (eds) *Biology of Australian Butterflies.* Monographs on Australian Lepidoptera Vol. 6. pp. 191–216. CSIRO Publishing: Collingwood, Melbourne.

Setterfield, S. A., Rossiter-Rachor, N. A., Douglas, M. M., Wainger, L., Petty, A. M., Barrow, P., Shepherd, I. J. & Ferdinands, K. B. 2013. Adding fuel to the fire: The impacts of non-native grass invasion on fire management at a regional scale. *PLoS ONE* 8(5): e59144. doi.org/10.1371/journal.pone.0059144.

Setterfield, S. A., Rossiter-Rachor, N. A., Hutley, L. B., Douglas, M. M. & Williams, R. J. 2010. Turning up the heat: The impacts of *Andropogon gayanus* (gamba grass) invasion on fire behaviour in northern Australian savannas. *Diversity and Distributions* 16: 854–861. doi.org/10.4996/fireecology.0301048.

Smithers, C. N. 1978. Migration records in Australia. 2. Hesperiidae and Papilionidae (Lepidoptera). *Australian Entomological Magazine* 5: 11–14.

Smithers, C. N. 1983a. Migration records in Australia. 3. Danainae and Acraeinae (Lepidoptera: Nymphalidae). *Australian Entomological Magazine* 10: 21–27.

Smithers, C. N. 1983b. Migration records in Australia. 4. Pieridae (Lepidoptera) other than *Anaphaeis java teutonia* (F.). *Australian Entomological Magazine* 10: 47–54.

Smithers, C. N. 1985. Migration records in Australia. 5. Lycaenidae and Nymphalidae (Lepidoptera). *Australian Entomological Magazine* 11: 91–97.

Smithers, C. N. & McArtney, I. B. 1970. Record of a migration of the chequered swallowtail *Papilio demoleus sthenelus* Macleay (Lepidoptera: Papilionidae). *North Queensland Naturalist* 37: 8.

Smithers, C. N. & Peters, J. V. 1977. A record of migration and aggregation in *Alcides zodiaca* (Butler) (Lepidoptera: Uraniidae). *Australian Entomological Magazine* 4: 44.

Start, A. N. 2013. Mistletoe flora (Loranthaceae and Santalaceae) of the Kimberley, a tropical region in Western Australia, with particular reference to fire. *Australian Journal of Botany* 61: 309–321. doi.org/ 10.1071/BT13021.

Talbot, G. 1928–37. *A Monograph of the Pierine Genus Delias. Parts I–VI.* British Museum (Natural History): London.

Tarmann, G. M. 2004. *Zygaenid Moths of Australia: A revision of the Australian Zygaenidae (Procridinae: Artonini).* Monographs of Australian Lepidoptera Vol. 9. CSIRO Publishing: Collingwood, Melbourne.

Thackway, R. & Cresswell, I. D. 1995. *An Interim Biogeographic Regionalisation for Australia: A framework for establishing the national system of reserves, Version 4.0.* Australian Nature Conservation Agency: Canberra.

Tindale, N. B. 1922. On a new genus and species of Australian Lycaeninae. *Transactions and Proceedings of the Royal Society of South Australia* 46: 537–538.

Tindale, N. B. 1923. On Australian Rhopalocera. *Transactions and Proceedings of the Royal Society of South Australia* 47: 342–354.

Tindale, N. B. 1927. A new butterfly of the genus *Papilio* from Arnhem Land. *Records of the South Australian Museum* 3: 339–341.

van Swaay, C., Nowicki, P., Settele, J. & van Strien, A. 2008. Butterfly monitoring in Europe: Methods, applications and perspectives. *Biodiversity and Conservation* 17: 3455–3469. doi.org/10.1007/s10531-008-9491-4.

Virtue, J. & McQuillan, P. 1994. *Butterflies of Tasmania.* Tasmanian Field Naturalists Club Inc.: Hobart.

Wade, A. 1978. Notes on the habits of the northern jezabel butterfly. *Northern Territory Naturalist* 1: 13–14.

Walker, F. 1854. List of the specimens of Lepidopterous insects in the collection of the British Museum. *Lepidoptera Heterocera* 1: 1–278.

Warham, J. 1957. West Kimberley butterflies. *Western Australian Naturalist* 5: 229–230.

Waterhouse, G. A. 1932. Australian Hesperiidae. II. Notes and descriptions of new forms. *Proceedings of the Linnean Society of New South Wales* 57: 218–238.

Waterhouse, G. A. 1933. Australian Hesperiidae. IV. Notes and descriptions of new forms. *Proceedings of the Linnean Society of New South Wales* 58: 461–466.

Waterhouse, G. A. 1937a. Australian Hesperiidae. VI. Descriptions of new subspecies. *Proceedings of the Linnean Society of New South Wales* 62: 32–34.

Waterhouse, G. A. 1937b. Presidential address: The biology and taxonomy of the Australian butterflies. *Report of the Australian and New Zealand Association for the Advancement of Science, Auckland Meeting, January 1937* 23: 101–133.

Waterhouse, G. A. 1938. Notes on Australian butterflies in the Australian Museum. *Records of the Australian Museum* 20: 217–222. doi.org/10.3853/j.0067-1975.20.1938.570.

Waterhouse, G. A. & Lyell, G. 1909. New and rare Australian butterflies of the genus *Miletus*. *The Victorian Naturalist* 26: 110–116.

Waterhouse, G. A. & Lyell, G. 1914. *The Butterflies of Australia: A monograph of the Australian Rhopalocera.* Angus & Robertson: Sydney.

Weir, R. P., Meyer, C. E. & Brown, S. S. 2011. Notes on the biology of *Ogyris zosine* (Hewitson, 1853) (Lepidoptera: Lycaenidae: Theclinae), including the first record of the purple female form from the Northern Territory, Australia. *The Australian Entomologist* 38: 97–100.

White, D. L., O'Brien, E. A., Fejo, D. M., Yates, R. W., Goodman, A. A., Harvey, M. & Wightman, G. 2009. *Warray plants and animals: Aboriginal flora and fauna knowledge from the upper Adelaide and upper Finniss rivers, northern Australia.* Northern Territory Botanical Bulletin No. 33. Department of Natural Resources, Environment, the Arts and Sport: Palmerston, NT.

Williams, A. A. E., Hay, R. W. & Bollam, H. H. 1992. New records for six lycaenid butterflies in Western Australia (Lepidoptera: Lycaenidae). *Australian Entomological Magazine* 19: 25–27.

Williams, A. A. E. & Powell, R. J. 1998. The butterflies (Lepidoptera) of East and West Wallabi islands, Western Australia. *The Australian Entomologist* 25: 107–112.

Williams, A. A. E. & Powell, R. J. 2006. The butterflies (Lepidoptera) of Middle, Mondrain, Sandy Hook, Woody and Goose islands in the Recherche Archipelago, Western Australia. *The Australian Entomologist* 33: 39–48.

Williams, A. A. E., Williams, M. R., Edwards, E. D. & Coppen, R. A. M. 2016. The sun-moths (Lepidoptera: Castniidae) of Western Australia: An inventory of distribution, larval food plants, habitat, behaviour, seasonality and conservation status. *Records of the Western Australian Museum* 31: 90–162. doi.org/10.18195/issn.0312-3162.31(2).2016.090-162.

Williams, A. A. E., Williams, M. R. & Swann, G. 2006. Records of butterflies (Lepidoptera) from the Kimberley region of Western Australia. *Victorian Entomologist* 36: 9–16.

Williams, M. R. 2014. *Butterflies of Home Valley, Durack River and Karunjie stations, East Kimberley region, Western Australia*. Final Report to Department of Sustainability, Environment, Water and Communities. Department of Parks and Wildlife: Perth.

Williams, M. R., Lamont, B. B. & Henstridge, J. D. 2009. Species-area functions revisited. *Journal of Biogeography* 36: 1994–2004. doi.org/10.1111/j.1365-2699.2009.02110.x.

Willmott, K. R., Lamas, G. & Huertas, B. 2011. *Priorities for Research and Conservation of Tropical Andean Butterflies*. McGuire Center for Lepidoptera and Biodiversity: Gainsville, Fl.

Wilson, P. R. 2016. Tawny coster arrives in Queensland. *News Bulletin of the Entomological Society of Queensland* 44: 121–122.

Wilson, P. R. & Johnson, I. R. 2017. Five new butterfly life histories (Lepidoptera) from Christmas Island, Australia. *The Australian Entomologist* 44: 181–195.

Woinarski, J. C. Z. 1992. Biogeography and conservation of reptiles, mammals and birds across north-western Australia: An inventory and base for planning an ecological reserve system. *Wildlife Research* 19: 665–705. doi.org/10.1071/WR9920665.

Woinarski, J. C. Z., Fisher, A. & Milne, D. 1999. Distribution patterns of vertebrates in relation to an extensive rainfall gradient and variation in soil texture in the tropical savannas of the Northern Territory, Australia. *Journal of Tropical Ecology* 15: 381–398. doi.org/10.1017/S0266467499000905.

Woinarski, J. C. Z., Hempel, C., Cowie, I., Brennan, K., Kerrigan, R., Leach, G. & Russell-Smith, J. 2006. Distributional pattern of plant species endemic to the Northern Territory, Australia. *Australian Journal of Botany* 54: 627–640. doi.org/10.1071/BT05041.

Woinarski, J. C. Z., Mackey, B. G., Nix, H. A. & Traill, B. 2007a. *The Nature of Northern Australia: Natural values, ecological processes and future prospects*. ANU E Press: Canberra.

Woinarski, J. C. Z., Pavey, C., Kerrigan, R., Cowie, I. & Ward, S. 2007b. *Lost from Our Landscape: Threatened species of the Northern Territory*. Northern Territory Government: Darwin.

Woinarski, J. C. Z., Williams, R. J., Price, O. F. & Rankmore, B. 2005. Landscapes without boundaries: Wildlife and their environments in northern Australia. *Wildlife Research* 32: 377–388. doi.org/10.1071/WR03008.

Yata, O. 1985. Part 1: Pieridae. In: Tsukada, E. (ed.) *Butterflies of the South East Asian Islands. II: Pieridae, Danaidae*. pp. 205–438. Plapac: Tokyo.

Yeates, D. K. 1990. New records of butterflies from the east Kimberley, Western Australia. *Australian Entomological Magazine* 17: 73–74.

Young, D. A. 2005. Spring–summer–autumn 2004–2005; field and other notes. *Victorian Entomologist* 35: 87–97.

# Index

www.ingramcontent.com/pod-product-compliance
Lightning Source LLC
Chambersburg PA
CBHW050241220326
41598CB00048B/7475